# 3rd International Conference on Aeronautical Materials and Aerospace Engineering (AMAE 2019)



IOP Conference Series: Materials Science and Engineering
Volume 608

Shanghai, China
24 – 26 May 2019

ISBN: 978-1-5108-9485-3
ISSN: 1757-8981

**Printed from e-media with permission by:**

Curran Associates, Inc.
57 Morehouse Lane
Red Hook, NY 12571

**Some format issues inherent in the e-media version may also appear in this print version.**



Institute of Physics
Dirac House, Temple Back
Bristol BS1 6BE UK

Phone: 44 1 17 929 7481
Fax: 44 1 17 920 0979

techtracking@iop.org

**Additional copies of this publication are available from:**

Curran Associates, Inc.
57 Morehouse Lane
Red Hook, NY 12571 USA
Phone: 845-758-0400
Fax: 845-758-2633
Email: curran@proceedings.com
Web: www.proceedings.com

# Table of contents

## Volume 608

**3rd International Conference on Aeronautical Materials and Aerospace Engineering (AMAE 2019)**

**24–26 May 2019, Shanghai, China**

**Accepted papers received: 21 June 2019**
**Published online: 27 August 2019**

### Preface



### Papers

#### Mechanical Engineering in Aerospace

**Aeronautical Technology**

IOP Conf. Series: Materials Science and Engineering **608** (2019) 011001 doi:10.1088/1757-899X/608/1/011001

# PREFACE

2019 International Conference on Aeronautical Materials and Aerospace Engineering (AMAE 2019) was held in Shanghai, China from May 24 to 26, 2019. AMAE 2019 was organized Hong Kong Society of Mechanical Engineers (HKSME). The conference provides a useful and wide platform both for display the latest research and for exchange of research results and thoughts in Aeronautics and Astronautics and other topics. The participants of the conference were from almost every part of the world, with background of either academia or industry, even well-known enterprise. The success and prosperity of the conference is reflected high level of the papers received.

The proceedings are a compilation of the accepted papers and represent an interesting outcome of the conference. This book covers 3 chapters: Mechanical Engineering in Aerospace; Aeronautical Technology; Transformers; Aerospace Technology.

We would like to acknowledge all of those who supported AMAE 2019. Each individual and institutional help were very important for the success of this conference. Especially we would like to thank the organizing committee for their valuable advices in the organization and helpful peer review of the papers.

We sincerely hope that AMAE 2019 will be a forum for excellent discussions that will put forward new ideas and promote collaborative researches. We are sure that the proceedings will serve as an important research source of references and the knowledge, which will lead to not only scientific and engineering progress but also other new products and processes.

Prof. Ian McAndrew, Capitol Technology University, USA
Prof. Kretov Anatolii, Nanjing University of Aeronautics and Astronautics, China

IOP Conf. Series: Materials Science and Engineering **608** (2019) 011002 doi:10.1088/1757-899X/608/1/011002

# COMMITTEES

## Conference Committee Co-Chairs

Prof. Ian McAndrew, Capitol Technology University, USA

Prof. Kretov Anatolii, Nanjing University of Aeronautics and Astronautics, China

## Program Co-Chairs

Prof. Liu Qiang, Beihang University, China

Prof. Xiaohui Su, Dalian University of Technology, China

## International Technical Committee

Prof. Nickolay Zosimovych, Shantou University, China

Dr. Jielong Wang, Beijing Aeronautical Science & Technology Research Institute, China

Dr. A Ram (Bella) Kim, Iowa State University, USA

Dr. James Chen, University at Buffalo, USA

Dr. Fidel Khouli, Carleton University, Canada

Dr Kin Hing Lo, University of Surrey, UK

Prof. M. Chandrasekaran, Vels University, India

Prof. Shinn-Liang Chang, National Formosa University, Taiwan

Prof. K. K. Viswanathan, Kuwait College of Science and Technology, Kuwait

Dr. Wael G. Abdelrahman, King Fahd University of Petroleum & Minerals, Saudi Arabia

Dr. Fairuz Izzuddin Romli, Universiti Putra Malaysia, Malaysia

Dr. Santanu Ghosh, Indian Institute of Technology Madras, India

IOP Conf. Series: Materials Science and Engineering **608** (2019) 011003 doi:10.1088/1757-899X/608/1/011003

# Peer review statement

All papers published in this volume of *IOP Conference Series: Materials Science and Engineering* have been peer reviewed through processes administered by the proceedings Editors. Reviews were conducted by expert referees to the professional and scientific standards expected of a proceedings journal published by IOP Publishing.

# Numerical Simulation of the Reliability of In-Situ Eddy Current Testing Considering the Influence of Penetration Depth

**P Q Tian[1, *], S Z Zhan[2], Y W Feng[1], X L Zhang[1] and X F Xue[1]**

[1]Northwestern Polytechnical University, Xi'an, China
[2]Aviation Industry Aircraft Strength Research Institute, Xi'an, China

[*]peiqiangtian@163.com

**Abstract.** In order to study the reliability of in-situ eddy current testing, the factors affecting the in-situ eddy current detection and the signal response values are firstly determined. Then the numerical simulation model of eddy current detection is established by COMSOL, and the different values of the device and the tested structural are substituted into the simulation model to obtain the signal response value. Finally, the probability of detection (POD) curve is obtained by numerical simulation method based on function model. In this paper, based on the numerical simulation method of probability density function, the numerical simulation of eddy current testing reliability is formed. Through the in-situ eddy current testing reliability numerical simulation case, the detection probability of cracks in the structure is quantitatively affected by the penetration depth.

## 1. Introductions

The purpose of non-destructive testing is mainly to determine whether there is a crack in the object to be tested. Properties such as physical parameters of cracks are analyzed when cracks are present in the structure.[1][2] In practical detection applications, eddy current testing determines the state of the test piece being tested by detecting changes in the impedance of the detecting coil. At the same time, eddy current detection is affected by many factors, such as the parameters of the detection equipment, the size of the object to be inspected and the cracks in the structure. In order to describe the detection of limit cracks by inspection equipment, detection probabilities are currently generally used to characterize the uncertainty of detecting small cracks. [3][4]

## 2. The theory of Eddy current testing

Eddy current detection technology is a non-destructive testing technology based on the principle of electromagnetic induction. When the exciting coil is connected with the alternating exciting current, the induced current is generated inside the conductor. This induced current is called eddy current [5]. In the impedance analysis method, when the excitation coil is placed above the metal conductor to be tested, the model formed by the interaction is equivalent to the relationship between the primary coil and the secondary coil. The probe excitation coil is used as the primary coil, and the eddy current loop in the conductor to be tested is used as the secondary coil. The model of the excitation coil is in the form of a series connection of a resistor and an inductor with an alternating excitation. The model of the detected conductor is in the form of a closed resistor in series with the inductor [6]. The complex impedance of the primary coil is expressed as:

$$Z=R+jX \tag{1}$$

Where R, X- the resistance and inductive reactance of the primary coil;

## 3. Research content

In this paper, by analyzing the basic principle of eddy current testing and impedance analysis, the test equipment parameters and the tested structural parameters are determined and substituted into the COMSOL numerical simulation model firstly. Then the quantitative effects of crack size and penetration depth on in-situ eddy current testing were studied respectively, and the corresponding crack detection probability (POD) curve was obtained. The two-parameter Weibull function was used to fit the POD curve. Finally, the crack size and penetration depth were obtained. The factors are combined to obtain the crack detection probability (POD) function under the influence of multiple factors. It can inversely calculate the minimum detectable size under a certain detection probability, which provides ideas for numerical simulation of in-situ eddy current testing reliability.

## 4. Establishment of simulation model for in-situ eddy current testing

Because the stress on the hole edge of the aircraft structure is concentrated and easy to be destroyed, the numerical simulation of the reliability of the in-situ eddy current test is carried out by using the common hole edge crack on the aircraft structure as the research object. The numerical simulation model of eddy current testing is established by using COMSOL simulation software. The parameters of the detection device and the detected structure is set, as shown in Table 1.

**Table 1.** Input parameter table for simulation model

| Parameters | Values |
|---|---|
| Material of the detectded structure | Aluminum Alloy |
| Material of the coil | Copper |
| Inner diameter of the coil | Φ3mm |
| Outer diameter of the coil | Φ6mm |
| Frequency of probe | 200kHz |

By substituting the input parameters into the simulation model, a magnetic induction distribution map of the crack (0.39 mm*0.13 mm) can be obtained, as shown in Figure 1.

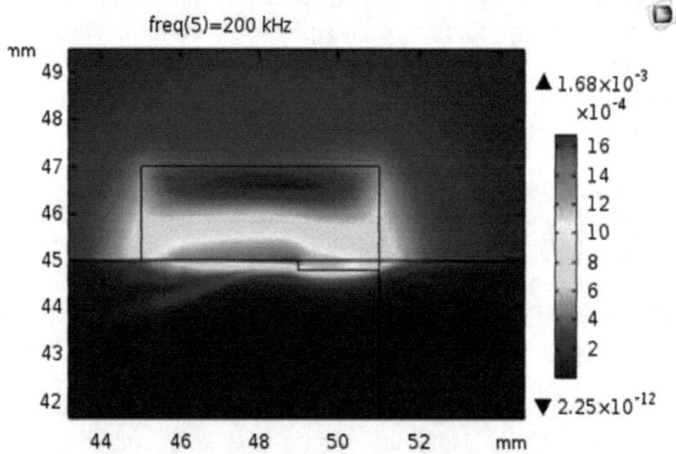


**Figure 1.** The distribution map of magnetic induction.

After measuring the impedance value $R_i$ with the probe of COMSOL, the impedance value R=(1.5680+13.037i) when the penetration depth d=0 mm, and the crack size a=(0.39*0.13) mm is obtained.

## 5. Research on numerical simulation of eddy current testing reliability

### *5.1. Numerical simulation of reliability considering the influence of penetration depth*

As the penetration depth between the detecting device and the detected structure is greater, the electromagnetic field generated by the device coil has a weaker influence on the induced magnetic field formed by detected the structure. It means the greater the penetration depth, the smaller the impedance value, thereby affecting the probability of the crack detection.

Taking the hole edge crack as 0.39mm*0.13mm as an example, when there is no crack at the edge of the hole, the impedance value measured by the simulation is the threshold value $R_0$=0.3336Ω. When in-situ eddy current testing is performed on the edge crack, the penetration depth should be randomized. It is assumed that the penetration depth between the detection device and the structure under inspection obeys the normal distribution N~(10.5,0.1), and the penetration depth data subject to the normal distribution is substituted into the eddy current detection numerical simulation to obtain the probability density of the corresponding impedance value. By integrating the probability density, the detection probability value of the hole edge crack at the penetration depth d=10.5 mm is obtained, as shown in Figure 2.

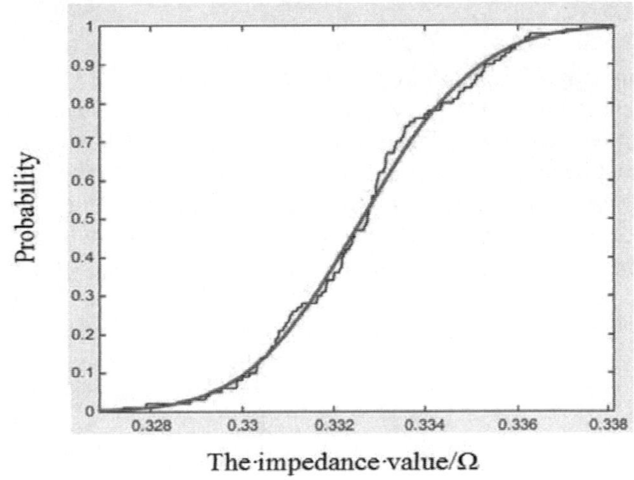


**Figure 2.** The probability curve of Crack detection when d=10.5mm.

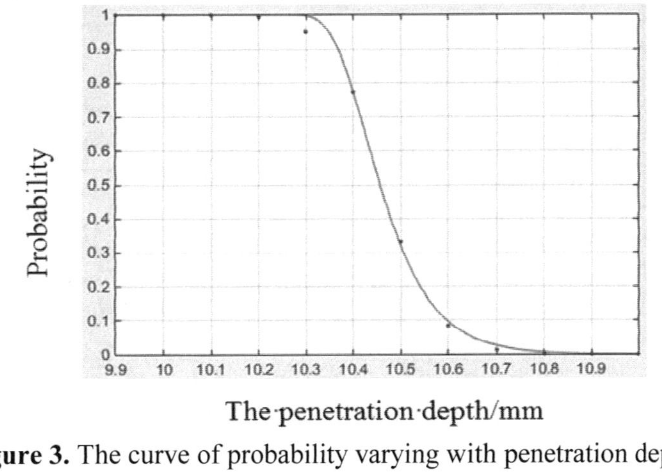


**Figure 3.** The curve of probability varying with penetration depth.

When the penetration depth d=10.5 mm, the crack detection probability $P_h$=0.33. The crack detection probability values at different penetration depths can be obtained by repeating the above process, as shown in Table 2.

**Table 2.** The detection probability of cracks at different penetration depths.

| **Penetration depths/mm** | 9.9 | 10.0 | 10.1 | 10.2 | 10.3 | 10.4 |
|---|---|---|---|---|---|---|
| **The detection probability** | 1 | 0.9999 | 0.9998 | 0.9941 | 0.9504 | 0.7730 |
| **Penetration depths /mm** | 10.5 | 10.6 | 10.7 | 10.8 | 10.9 | 11 |
| **The detection probability** | 0.3322 | 0.0829 | 0.0138 | 0.0005 | 1E-05 | 0 |

Using the two-parameter Weibull function to fit the detection probability values at different penetration depths, the curve of crack detection probability with penetration depth can be obtained, as shown in Figure 3.

Then the probability of crack detection with the change of penetration depth is expressed as:

$$P = 1 - \exp\left[-\left(\frac{d}{10.43}\right)\right]^{-140.8} \quad (2)$$

*5.2. Function of POD curve and determination of minimum crack*

Since the research content is under in-situ conditions, it is affected by many factors when performing in-situ eddy current testing. Therefore, the probability of in-situ eddy current detection under the influence of crack size and penetration depth should be studied next. In the in-situ eddy current testing of the hole edge crack, firstly, the numerical simulation of multiple cracks at the same penetration depth and one crack at multiple penetration depths was carried out by using COMSOL. The impedance values are obtained by the simulation. Then the same process as 5.1 can be used to find the detection probability of different cracks at different penetration depths. Then use the two-parameter Weibull function to fit the POD curve (as shown in Figure 4) and the corresponding POD function (as shown in Table 3).

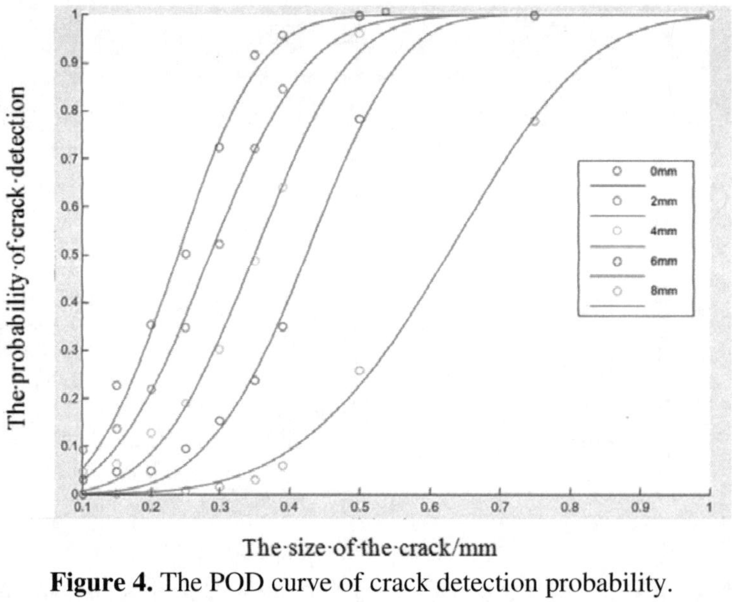


**Figure 4.** The POD curve of crack detection probability.

IOP Conf. Series: Materials Science and Engineering **608** (2019) 012001 doi:10.1088/1757-899X/608/1/012001

**Table 3.** Summary of POD functions and Detectable minimum size.

| **Penetration depth/mm** | | **0** | **2** | **4** | **6** | **8** |
|---|---|---|---|---|---|---|
| **Weibull function:** $P=1-\exp\left[-\left(\frac{a}{\beta}\right)\right]^{-\alpha}$ | $\alpha$ | **2.90** | **2.95** | **3.65** | **4.48** | **4.41** |
| | $\beta$ | **0.27** | **0.32** | **0.39** | **0.46** | **0.68** |
| **Detectable minimum size /mm** | | 0.36 | 0.43 | 0.49 | 0.56 | 0.82 |

Therefore, The detection probability of different cracks at different penetration depths can be quantitatively calculated by the POD function in Table 3. Finally, the POD function obtained by the Weibull function fitting inversely calculates the minimum crack size detectable under a certain detection probability, as shown in Table 3. As can be seen from Table 3, the minimum detectable crack size gradually increases with the penetration depth increases and the minimum crack size can be quantitatively calculated.

## 6. Conclusions and Prospect

In order to quantify the influence of penetration depth on the reliability of in-situ eddy current testing, the crack detection probability (POD) under different conditions was obtained by numerical simulation of in-situ eddy current testing. The conclusions are summarized as follows:

1) By analyzing the basic principle of in-situ eddy current testing, the numerical model of in-situ eddy current detection was established by COMSOL. The impedance values at different crack sizes and penetration depths are obtained, and the probability density function of the impedance values is obtained. When the impedance value is greater than the threshold value, the crack can be detected, otherwise it cannot be detected. The probability of detecting the crack is obtained at the same time. The two-parameter Weibull function is used to fit the crack detection probability to obtain the detection probability (POD) function. The numerical simulation method of in-situ eddy current detection reliability is formed.

2) The probability of detection under different penetration depths is obtained by case calculation. The POD of different cracks can be calculated by the POD function, and the influence of penetration depth on the reliability of eddy current detection is quantitatively characterized.

3) The case shows that when the detection probability of the crack in the structure is 90%, the minimum size of the detectable crack increases from 0.36mm to 0.43mm as the penetration depth changes from 0mm to 2mm. The minimum size of the detected crack increases by 19.4%.

In addition to eddy current testing, there are various non-destructive testing techniques such as ultrasonic testing, radiation detection and magnetic particle testing. Therefore, various non-destructive testing reliability studies can be carried out for different defects in other structures.

**Acknowledgements**
Thanks for the project support from the Aviation Industry Strength Institute.

**References**

[1] Jiawei L and Ji zhen C 2012 *Mechanical Industry Press*. 46
[2] Adams R D, Cawley P 2008 *NDT international*. 21(4): 208-222
[3] Jilin R, Junming L 2011 *Mechanical Industry Press*. 105
[4] Brasche L 2010 *Electromagnetic non-destructive evaluation*. 33: 129
[5] Wan J, Yusa N and Pan H 2013 *Materials Transactions*. 54(1): 90-95.
[6] Peter W and Krause T 2015 *IEEE Sensors Journal*. 35(3):23-26

# Cotton-Bamboo Composite Material and Structure Applications

**JingBo Sun[1], ZePing Wu[1], WeiHua Zhang[*1], DongHui Wang[1] and Shuo Su[2]**

[1]College of Aerospace Science and Engineering, National University of Defense Technology, No.109 DeYa Street, ChangSha, HuNan, P.R. China
[2] School of Mechanical Engineering & Automation, Beihang University, Beijing 100191, P.R. China

[*]zhangweihua@nudt.edu.cn

**Abstract.** Green material and sustainable technologies have been becoming a heated topic recent years. This study tries to invent cotton-bamboo fiber reinforced composite material, a fully biodegrade green material, based on the preview study and explore the mechanical performance and its structure application. Two sets of experiments were carried out in this research, which are single-shear tensile loading test under ASTM standard and bending test of wind turbine blade. Results revealed that this biodegrade green material have good mechanical properties and can satisfy the requirement of some low-intensity industrial sectors. Further research is needed to apply this cotton-bamboo fiber material in broader market.

## 1. Introduction

Material is the foundation and precursor of the progress of human material civilization. The development of material civilization for thousands of years, especially the development of modern industry for more than 200 years, has made great achievements in the development of materials. Now more than 50,000 materials are available for human use, and new materials with high performance, multi-function and intellectualization are being developed one after another. With the deep integration of new materials and modern high technology, the extraction, synthesis, manufacture, modification and application of new materials have reached an unprecedented level, showing a very bright prospect for the future of mankind. On the other hand, all materials are exchanged for resources. There are only two kinds of global resources, one is non-renewable resources, the other is renewable resources. At present, the situation of global resources is that the non-renewable resources are exhausted day by day, and the renewable resources have not been fully exploited and utilized, which is the reason why green materials and materials have sprung up and become popular. Because the composite materials have the characteristics of multi-phase compounding and designable properties, the green development of the composite materials is more congenital advantages, and the prospect is very broad[1]. The so-called green materials and materials refer to the material from product design, raw material selection, processing and manufacturing, packaging and transportation, service use, recycling and reuse of the entire life cycle, the highest utilization rate of resources, the smallest impact on the environment[2].

Angel P M[3] and his partners has studied some properties about bamboo-polylactic acid(PLA) composite material before. In their study, the morphological study, the manufacture process and some mechanical tests have been researched. However, whether this material can be applied in industrial sectors remains a problem to solve.

IOP Conf. Series: Materials Science and Engineering **608** (2019) 012002 doi:10.1088/1757-899X/608/1/012002

This paper introduces a new natural fiber composite material, Cotton-bamboo fiber reinforced composite material, based on the bamboo-PLA material[3] and the structure application of this material.

## 2. Materials

*2.1 Bamboo and matrix*

Angel has found that the Guadua bamboo has better mechanical properties, with an average tensile strength of 202.7MPa (coefficient of variation COV=16.2%) and an average Young's modulus of 25.52GPa (COV=6.4%).After his study, the Guadua species has been selected to manufacture the bamboo composite[3].

Ingeo PLA Biopolymer 4032D PLA as 0.025-mm-thick film sheets were selected by Angel for using in the selected manufacturing process of resin film infusion due to the properties of full-biodegradability, bio-based sourcing, commercial availability, and compatibility with the manufacturing process[3].

The preparation of the bamboo layer need bamboo pieces of the certain thickness width, and then the bamboo pieces are placed on the PLA layer without gaps, and after the laying is completed, another PLA is added on the other side. And using a hot air blower, blowing at 180 ° for 10-15 minutes, so that the PLA slightly sticks the bamboo piece. Back up. As shown in figure 1.

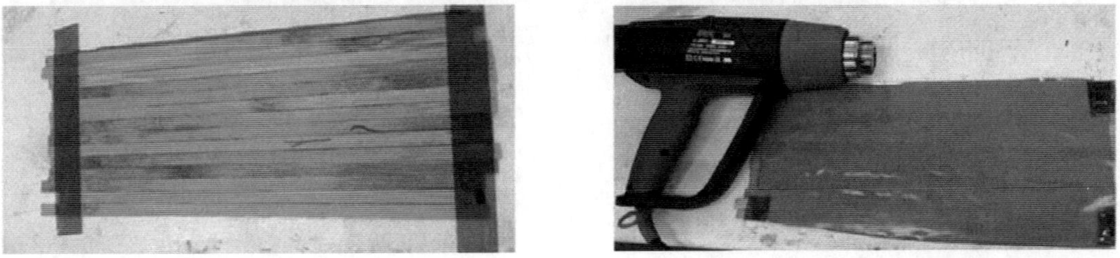

(a) (b)

**Figure 1.** Manufacture process of bamboo layer

*2.2 Flex layer*

The flex layer is made up by the cotton and a kind of natural transparent resin, shown in figure 2. The addition of the flex layer is aim to reduce the weight and increase flexibility and manipulability. The resin is made from potatoes and rice. So the flex layer is also fully biodegradable material.

The preparation of the flex layer is shown in figure 3. In outface of the self-made circular automatic roller, the mold cloth is first laid off firstly, and then a layer of cotton thread is evenly wound under the driving of the motor, and then two layers of resin are laid; 3 to 4 times, put into a 100 °C incubator for 15 to 20 minutes, until the resin layer is evenly melted in the flex layer. Take out and back up.

**Figure 2.** Flex layer

**Figure 3.** Manufacture process of flex layer

## 3. Single-shear tensile test

IOP Conf. Series: Materials Science and Engineering **608** (2019) 012002 doi:10.1088/1757-899X/608/1/012002

*3.1 Geometric design*

The single-shear type bolt connection method was chosen to test the properties. Therefore, the coupon geometry was designed on the basis of the ASTM Standard D5961 Procedure 'C'. The Dimensions for the coupon geometry are in conformance with ASTM D5961 test method for the single shear one-piece specimen Procedure 'C', shown in figure 4. The geometric parameters are shown in table 1.

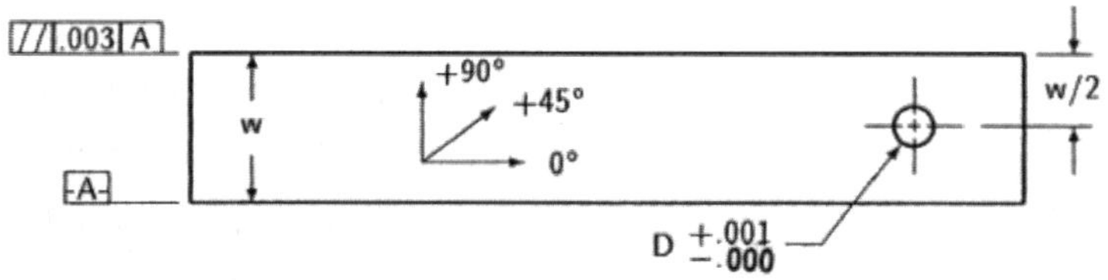


**Figure 4.** Standard coupon geometry for procedure 'C'

**Table 1.** Coupon geometry basic parameter

| Parameter | Value |
|---|---|
| Bolt Diameter (d) | 0.25 inch |
| Coupon Hole diameter (D) | 0.250 inch |
| Coupon Thickness range (h) | 0.08-0.208 inch |
| Coupon Length (L) | 5.5 inch |
| Width of the Coupon (w) | 1.5 inch |
| Distance from Hole Center to Edge (e) | 0.75 inch |

*3.2 Lay up arrangement*

Since natural fiber reinforced composites do not have molded prepregs like other composites, such as carbon fiber composites and glass fiber composites, laminates of resin and adhesive should be added between layers manually. The layup design should base on symmetry principle and the layup sequence is shown in table 2 below.

**Table 2.** Laying sequence of natural fiber composites

| N° | Laying up Sequence |
|---|---|
| 1,25 | Resin*2 |
| 2,24 | +45°flex*1 |
| 3,23 | Resin*3 |
| 4,22 | -45°flex*1 |
| 5,21 | Resin*3 |
| 6,20 | 90°flex*1 |
| 7,19 | Resin*3 |
| 8,18 | 90°flex*1 |
| 9,17 | Resin*3 |
| 10,16 | +45°flex*1 |
| 11,15 | PLA*1 |
| 12,14 | 0°bamboo*1 |
| 13 | PLA*1 |

*3.3 Coupon Preparation*

Panel curing is carried out in a 180 ° incubator at atmospheric pressure for two hours after laying up and making the standard vacuum bag, shown in figure 5.

**Figure 5.** Under one atmosphere, placed in the incubator

Because the half laps of the joint are eccentric to each other, a packing piece (equal to the thickness of a half lap) is bonded to the end of each half lap remote from the joint. This ensures that the joint remains initially straight when it is clamped between the grips of the test machine. As the grips of the test machine move apart the joint is subjected to tension and bending with the latter being equal (initially) to the tensile load multiplied by the thickness of a half lap[4].

Some cheap materials was used to manufacture the packing block of same thicknesses as the composite laminates. Then the adhesive is used for bonding. It should be noted that the surface to which the backing plate and the test piece are bonded should be sanded in advance (as shown in figure 6).

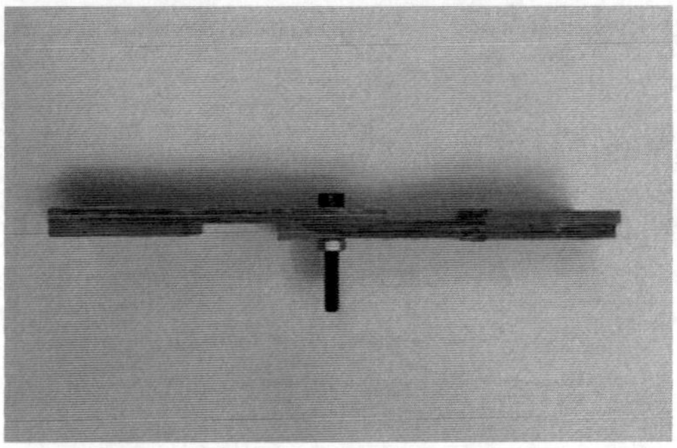

**Figure 6.** Final assembly of the specimen

*3.4 Experimental results and discussion*

The experimental device is a servo hydraulic tensile testing machine, as shown in figure 7.

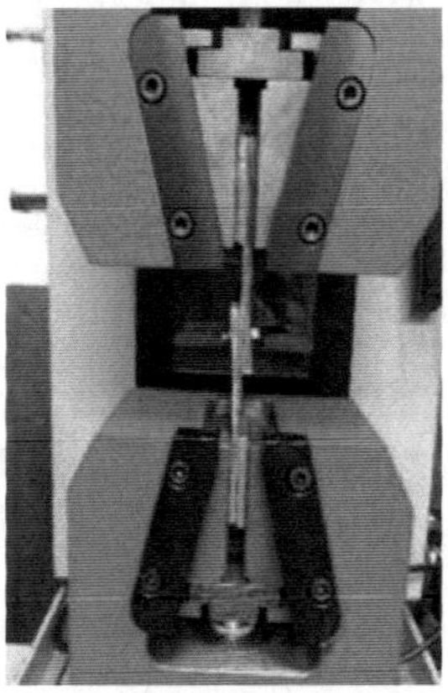

**Figure 7.** Servo hydraulic tensile testing machine

The thickness size of the specimen is 5.35mm. Three tests have been done. The maximize failure stress is 130MPa, and the average failure stress is around 125MPa.

It can be seen from the experimental phenomenon (shown in figure 8) and curves that the inherent damage mechanism is extremely complicated. Since natural fiber reinforced composite materials are still under study, many phenomena cannot be solved at present.

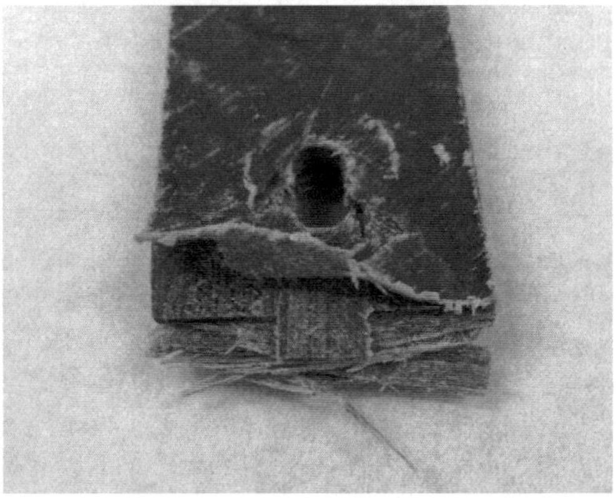

**Figure 8.** Experimental phenomena of specimen

Angel had pointed that many factors could influence the properties, especially the manufacturing process[3]. And the potential application of the developed material for aerospace are very restricted, but the present strength could fulfill the requirements of other industrial sectors, such as the wind turbine blades. In the next section, the application of the wind turbine blades will be introduced.

## 4. The application of the wind turbine blades

It is proved that small wind turbines have the characteristics of low cost, high power generation rate, flexible regional distribution and less land occupation. Small wind turbines can effectively fill the gap in the low wind energy market due to the low wind speed range requirements. And it is meaningful to environment that the green natural material was used in the wind turbine blades instead of the traditional material.

### *4.1 Selection of the airfoil*

In this section, the main parameters concerned are the pressure distribution of airfoil surface, the lift coefficient and the lift-to-drag ratio. Therefore, these parameters were selected as the standard for the best airfoil with low Reynolds number.

Ronit K, Singh and his partners have pointed out that the airfoil AF300 was optimized from existing low Re airfoils through Xfoil code[5],the geometric shape of the AF300 airfoil is shown in figure 9.

**Figure 9.** Geometric shape of the AF300 airfoil

### *4.2 Structure analysis*

The foam structure was selected for its excellent performance that it is more responsive to structural and strength requirements in complex stress conditions and it has a lower water absorption rate that can reduce maintenance costs for its closed cell structure[6][7].

The main role of foam in the blade is to ensure the stability of the blade while reducing the quality of the blade, so that the blade increases the wind-capturing area while satisfying the stiffness. Focusing on cost factors, PU foam has a great advantage in cost.

*4.3 Manufacturing process*

As for the lay up arrangement, [-45° F, +45° F, B, 90° F, B, +45° F, -45° F](F is flex layer, B is bamboo layer)sequence was firsty design. However, the two layers of bamboo had lead to the difficult to carry out the close operation because of the thickness. Therefore, the number of the bamboo layer was reduced. And the final layup arrangement of the wind turbine blade was[-45° F, +45° F, 90° F, B, 90° F, +45° F, -45° F]. In order to get enough resin to stick together and improve the performance, alternated layers of transparent resin and the PLA were applied. The detailed layup arrangement was shown in figure 10. It is important to note that the compaction is necessary during the laying up process.

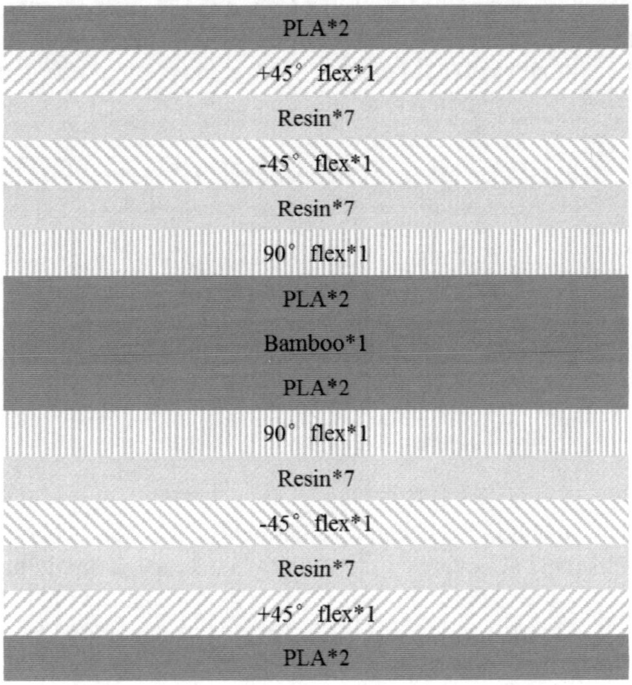


**Figure 10.** Wind turbine blade layup arrangement

The manufacture of the wind turbine blade is not easy for its geometric complexity. Getting the bamboo layer, laying up, making the vacuum bag and final cutting are the main difficult steps. Shown in figure 11-figure 13.

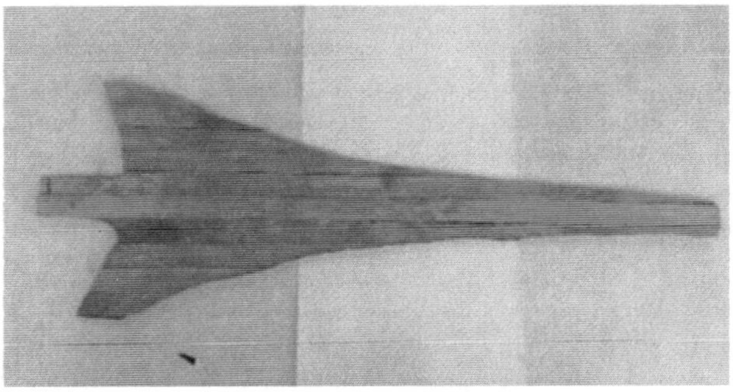

**Figure 11.** Bamboo layer

IOP Conf. Series: Materials Science and Engineering **608** (2019) 012002 doi:10.1088/1757-899X/608/1/012002

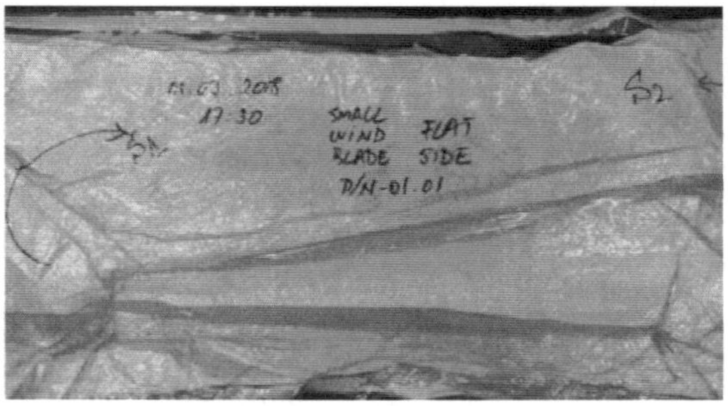

**Figure 12.** Vacuum bag

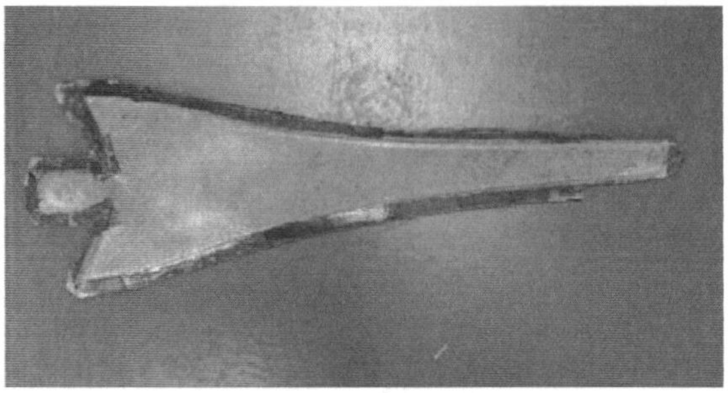

**Figure 13.** Blade

The curing temperature of the adhesive was 120 ° C, and the resin curing temperature was 180 ° C. The adhesive is first cured so that each laminate can be bonded to the mold. When the temperature reaches 180 ° C, the resin will melt into a liquid, but will not move the position of each layer, the detail temperature of curing process is shown in figure 14 .

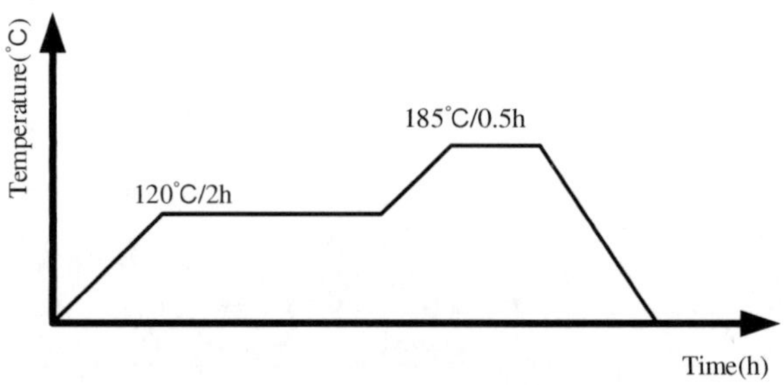


**Figure 14.** Curing temperature

*4.4 Bending test*

Bending test was carried out to validate whether this blade made by the biodegrade material could fulfill the request of the related industrial sectors. Fixe the blade root, and gradually increases the weight in a little until the damage occurs, the experiment process and damage of the blade are shown in figure 15 and figure 16, respectively. To ensure that the blade roots were completely fixed during the test, 20 layers of carbon fiber prepreg were placed on the blade roots and secured to the brackets with pins, as shown. Due to the significant bearing capacity of the mechanical properties of the three materials, the minimum weight of the material used in the test was 1 kg. The blade deformation displacement was recorded each time a 1 kg weight was loaded. The results show that when the weight of the load is 5Kg, the natural fiber reinforcement blade is destroyed, and the damage is located at the root of the blade, and its deformation displacement is 16mm. This strength can withstand the torque load of low-speed wind turbine.

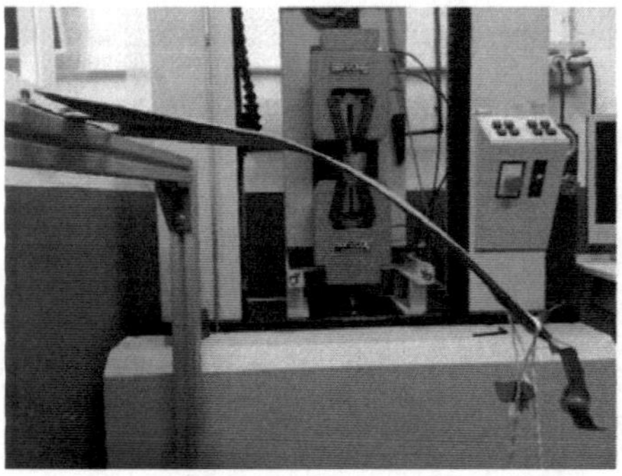

**Figure 15.** Experiment process

**Figure 16.** Blade damage

IOP Conf. Series: Materials Science and Engineering **608** (2019) 012002 doi:10.1088/1757-899X/608/1/012002

## 5. Conclusion

In this investigation, the aim is to explore the properties and structure application of the Cotton-bamboo fiber reinforced composites based on the sustainable technologies. And the present study has shown that this biodegradable composite material has good mechanical performance and could fulfill the request of low-speed wind turbine blade. In the future, we can try to explore more application in low-intensity industrial sectors. As well as it is totally green nature material and very friendly for the environment. Therefore, it is potential to get more development and widely application in the future.

As for the future work, firstly, wind tunnel test can be carried out based on the test described above to get a further performance validation; Secondly, more research and study need to be done to improve the mechanical performance. And because of the complexity of the manufacture process, the environment condition plays an important role and could be paid more attention in the later work; Thirdly, more application in industrial sectors could be explored based on the present study.

## References

[1] Mazumdar S K 2002 Composite manufacturing: materials product and process engineer (USA: CRC Press)

[2] Jianmao T 2015 A brief review of green composites in aerospace applications *(Spacecraft Environment Engineer* Vol 32*)* p 457-463

[3] Angel P M, Alfredo G, Antonio F L, Veronica C, Sonia R L 2017 Bamboo-Polylactic acid(PLA) composite material for structural application *(Material* Vol 10*)* No.10

[4] Sun H T, Chang F K and Qing X 2002 *The response of composite joints with bolt-clamping loads* pp 69-91

[5] Singh R K, Ahmed M R, Zullah M A, Lee Y-H 2012 Design of a low reynolds number airfoil for small horizontal axis wind turbine *(Renewable Energy* Vol 42*)*pp 66-76

[6] Pei H 2010 Design of sandwich structure and selection of foam core material for aircraft *(Aeronautical Manufacture Technology* Vol 17*)* pp 94-96

[7] Micheal C Y Niu 1992 *Composite airframe structures* pp 285-349

# A Method of Remote Support Jamming Based on the Chaff Clouds Scattering

**Shidai Qiu[1], Dengkai Yao[1] and Guhao Zhao[1*]**

[1]Air Traffic Control and Navigation College, Air Force Engineering University, 710051 Xi'an Shanxi, China

[*]Email: zghlupin@163.com

**Abstract.** Electronic Jamming is an important means to cover the penetration formation and improve survival probability of penetration aircraft when Air force performs ground penetrating mission. A compound interference method of jamming surface-to-air missile guidance radar by scattering long –range jamming signals though chaff clouds is proposed, which uses jammer to irradiate chaff clouds scattered on the penetration route, and injects jamming energy into the main lobe of surface-to-air missile guidance radar by scattering chaff clouds to make better interference effect.

## 1. Introduction

Aviation in the process of penetration task execution, often need to use the electronic jamming equipment such as machine of long distance support jamming enemy air defense missile guidance radar should be jammed, can to a certain extent, reduce the risk of penetration formation was found and shot down, effectively ensure the safety of the flight crew and mission success.

For chaff interference problem existing relevant personnel of different levels of study. The research and application of chaff jamming in foreign countries has been started since the Second World War. Foreign researchers have done a lot of research on the separation characteristics of chaff, chaff cloud diffusion model, descent model, RCS rule, etc. [1-3], which has been applied in many wars and achieved good results. Domestic researchers have also done a lot of work. H. Zheng [4], J. Li etc. [5] make the thorough analysis of the chaff cloud polarization and energy characteristics, chaff interference principle and application development; H.Q. Hu etc. [6] of airborne chaff bomb is given under different conditions the optimal timing of using calculation method; W.Y. Cai et al. [7] established chaff motion model in atmospheric environment; Y. Li et al. [8] proposed a new compound jamming method to solve the problem of chaff jamming's decreasing jamming ability and effectiveness to the new radar system; J.H. Zhou et al. [9] studied the influence of chaff as an important way of passive jamming on radar detection.

This paper proposes a method of Compound Jamming using chaff cloud scattering to support jamming signal in long distance. A front projection is established. The multitime spraying model of antitime chaff bomb can quickly and effectively determine the location of each spraying chaff bomb in penetration formation. In order to verify the validity of the model, simulation verification is carried out with an example.

## 2. A method for the implementation of remote jamming signal scattered by chaff clouds

The jammer's strong interference and strong scattering of chaff clouds, which can transfer interference energy from radar antenna's main lobe into the receiver. Reached a good interference effect, at the same time, avoid the radar echo caused by low velocity of chaff have been radar filter with dynamic target display technology, thus greatly expand the application range. In jammer screen penetration formation process, when penetration formation reaches the boundary of enemy air defense fire circle. Spreading Chaff Clouds on the Line of Guidance Radar and Our Penetration Formation.

According to Radar Principle [10], the scattered echo power of penetration formation received by radar receiver is as follows:

$$P_r = \frac{P_t G_z^2 \lambda^2 \sigma}{(4\pi)^3 R^4} \tag{1}$$

The interference signal power of chaff cloud scattering jammer injecting strong interference energy into radar antenna main lobe is:

$$P_{rg} = \frac{P_g G_g \sigma_b A_r}{(4\pi)^2 D^2 C^2} = \frac{P_g G_g G_z \lambda^2 \sigma_b}{(4\pi)^3 D^2 C^2} \tag{2}$$

Among them, $P_t$ is radar transmitting power, $G_z$ is antenna main lobe gain, $\lambda$ is radar operating wavelength, $A_r$ is effective area for current radar antenna, σ is equivalent scattering area of penetration formation, $R$ is the maximum detection distance of radar, $\sigma_b$ is equivalent scattering area of chaff cloud, $D$ is the distance between chaff cloud and radar, $C$ is the distance between jammer and chaff cloud.

Assuming that the radar receiver can detect the target signal successfully when the signal-to-interference ratio is greater than 10, the main lobe gain of the radar antenna is equal to the main lobe gain of the jamming antenna, i.e. $G_z = G_g$, and there is no other noise source except the jamming energy, then the signal-to-interference ratio can be expressed as:

$$\frac{P_r}{P_{rg}} = \frac{10 D^2 C^2 P_t \sigma}{R^4 P_g \sigma_b} \geq 10 \tag{3}$$

Then the maximum detection range of radar can be expressed as:

$$R = \sqrt[4]{\frac{D^2 C^2 P_t \sigma}{P_g \sigma_b}} \tag{4}$$

In the process of penetration of enemy ground-to-air missile guidance radar, the position of jammer in airspace and ground-to-air missile guidance radar is relatively fixed, the position distance between jammer and enemy ground-to-air missile guidance radar is $L$, and the interference angle between jammer and ground-to-air missile guidance radar and penetration route is $\theta$. The maximum detection range of radar can be expressed as:

$$R \leq \sqrt[4]{\frac{D^2 P_t \sigma}{P_g \sigma_b}(D^2 + L^2 - 2DL\cos\theta)} \tag{5}$$

Suppose the jammer jamming power is $P_g$=100 kw, the radar transmitting power $P_t$=100 kw, the equivalent scattering area of chaff cloud formed by a single spraying of n chaff bullets is $\sigma_b$=40n m$^2$, the equivalent scattering area of penetration formation $\sigma$=10 m$^2$, $L$=220 km. When the angle of interference is equal to 45 degrees, the jammer penetrates the ground-to-air missile guidance radar frontally, then the maximum detection range of the radar is the distance from the chaff cloud radar. The relationship between the equivalent scattering area of chaff cloud and chaff cloud is shown in Figure 1.

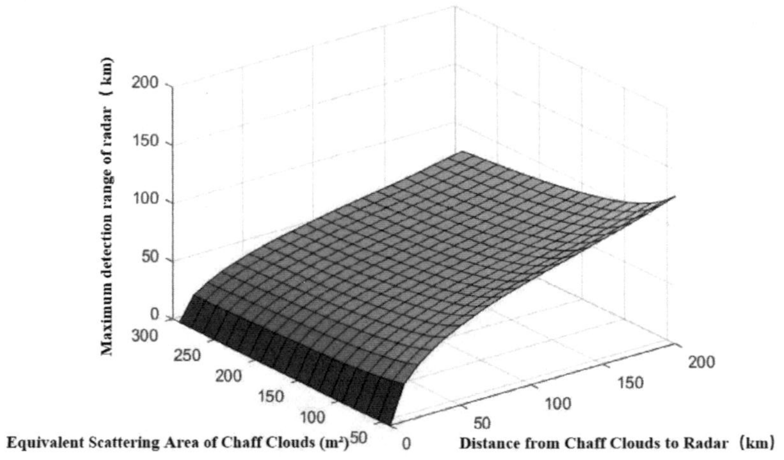


**Figure 1.** The relation between $R$ and $D$ and $\sigma_b$

If a long-range jamming aircraft injects jamming directly into the radar side-lobe from the distance radar antenna $L$, and $G_p$ is the gain of the radar antenna side lobe, then the effective area of the radar antenna becomes $A_f$, and the definition of the remaining variables remains unchanged, the jamming power entering the radar receiver can be written as follows:

$$P_z = \frac{P_g G_g A_f}{4\pi L^2} = \frac{P_g G_g G_p \lambda^2}{(4\pi)^2 L^2} \tag{6}$$

The SIR can be expressed as:

$$\frac{P_r}{P_z} = \frac{P_t G_z^2 \lambda^2 \sigma}{(4\pi)^3 R^4} / \frac{P_g G_g G_p \lambda^2}{(4\pi)^2 L^2} = \frac{L^2 P_t G_z \sigma}{4\pi R^4 P_g G_p} \geq 10 \tag{7}$$

At this time, the maximum detection range of radar is expressed as:

$$R \leq \sqrt[4]{\frac{L^2 P_t G_z \sigma}{40\pi P_g G_p}} \tag{8}$$

Assuming that the average side-lobe level of radar with side-lobe cancellation technology is $G_p/G_z$=-40dB, the distance of long-range support jamming aircraft is $L$=220 km, and the transmitting power of radar is $P_t$=100 kw, in which the scattering area of penetration formation is $\sigma$=10 m$^2$, the side-lobe cancellation technology can provide about 10dB cancellation ratio, thus further reducing the side-lobe level, the relationship between the maximum detection range of radar and the jamming power is shown in Figure 2.

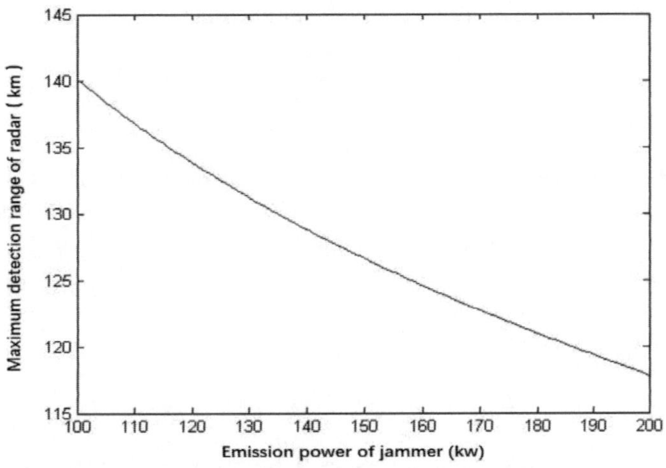


**Figure 2.** Relationship between $P_g$ and $R$ under average side lobe level

Comparing the jamming effect of two different methods of Figure 1 and Figure 2, it can be found that the detection distance of radar to penetration formation can be greatly reduced by using chaff cloud to transmit long-range jamming signal, and the jamming effect is better. The jamming effect of chaff cloud transmitting remote support jamming signal is affected by the distance $D$ from chaff cloud to radar and the equivalent scattering area $\sigma_b$ of chaff cloud. The closer the chaff cloud is to radar, the larger the equivalent scattering area, the better the jamming effect. However, it is more effective to shorten the distance between chaff cloud and radar.

## 3. Multiple spraying model of chaff bomb based on compound interference method

In front penetration, when the penetration formation flies to the detection range boundary of the enemy ground-to-air missile guidance radar, the penetration formation throws some chaff bombs. The rear long-distance support jammer begins to illuminate the chaff cloud formed by the chaff bomb explosion, jamming the enemy ground-to-air missile guidance radar, suppressing the maximum detection range of the enemy ground-to-air missile guidance radar, making it unable to detect the penetration formation. When the penetration formation flew to the edge of the maximum detection range of the enemy surface-to-air missile guidance radar after jamming suppression, a number of chaff bombs were thrown again to shorten the maximum detection range of the enemy surface-to-air missile guidance radar, repeating the above process until the maximum detection range of the enemy surface-to-air missile guidance radar was suppressed to the required range. The spraying process is shown in Figure 3.

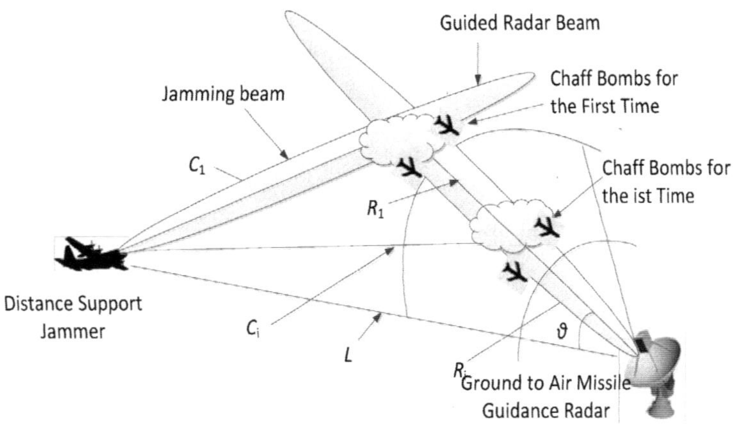


**Figure 3.** Chaff cartridge throwing timing diagram

The maximum detection range of the enemy ground-to-air missile guidance radar after each spraying of chaff bomb is:

$$R_{i+1} \le \sqrt[4]{\frac{R_i^2 C_i^2 P_t \sigma}{P_g \sigma_b}} \tag{9}$$

$R_{i+1}$(i=1~n) is the maximum detection range of a certain surface-to-air missile guidance radar after the ith chaff bomb throwing. $C_i$(i=1~n) is the distance between the jammer and the chaff cloud after the (i-1)th chaff bomb throwing. It is required to suppress the maximum detection range of the enemy surface-to-air missile guidance radar to $K$.

Then the maximum detection range of the enemy ground-to-air missile guidance radar after each spraying of chaff bomb can be expressed as follows:

$$R_{i+1} \le \sqrt[4]{\frac{R_i^2 P_t \sigma}{P_g \sigma_b}(R_i^2 + L^2 - 2R_i L\cos\theta)} \tag{10}$$

The first time to throw chaff bombs is when the penetration formation first flies to the edge of the unjammed detection range of the enemy surface-to-air missile guidance radar.

$$X = R_1 \tag{11}$$

$X$ is the distance between penetration formation and enemy ground-to-air missile guidance radar.

Then the penetration formation flies forward from $S_i$ and sprays chaff bullets for the ($i$+1)th time, that is:

$$S_i = R_i - R_{i+1} \tag{12}$$

## 4. Case simulation analysis

Assuming that the detection range of a certain enemy surface-to-air missile guidance radar is $R_1$=200 km, the launch power $P_t$=100 kw, the equivalent scattering area of a certain fighter penetration formation $\sigma$=10 m$^2$, the equivalent scattering area $\sigma_b$ =40 m$^2$ of each chaff bomb, the jammer is arranged on side of our side to the enemy surface-to-air missile guidance radar $L$=220 km, the jamming angle $\theta$=45 degrees, and the jammer's jamming power $P_g$=100 kw. Assuming the maximum launching distance of air-to-ground weapons is 60 km, it is required that the maximum detection range of enemy-to-air missile guidance radar should be suppressed within 60 km by the number of chaff bombs thrown three times. Due to the restriction of weight and other factors, the interference resources that aircraft can carry are limited. Therefore, the method with the least number of chaff bombs and the best suppression effect should be preferred.

In the whole process of penetration, there are five methods to satisfy the requirement of three chaff bombs thrown by the penetration formation. As shown in Table 1, the total number of chaff bombs used at least is 8. The second method can suppress he smallest range of radar detection. Therefore, the second method is used to suppress the maximum range of radar detection to 56.8 km by using eight chaff bombs, as shown in Table 2.

**Table 1.** Three times chaff cartridge throwing

| Method | Number of Dispersions in Position 1 | Number of Dispersions in Position 2 | Number of Dispersions in Position 3 | Range of Radar Detection | Totals |
|---|---|---|---|---|---|
| First | 1 | 1 | 6 | 58.1km | 8 |
| Second | 1 | 2 | 5 | 56.8km | 8 |
| Third | 1 | 3 | 4 | 57.7km | 8 |
| Fourth | 2 | 2 | 4 | 58.4km | 8 |
| Fifth | 2 | 1 | 5 | 59.9km | 8 |

IOP Conf. Series: Materials Science and Engineering **608** (2019) 012003 doi:10.1088/1757-899X/608/1/012003

**Table 2.** the second method of chaff cartridge throwing

| Data Type | Location 1 Data Size | Location 2 Data Size | Location 3 Data Size |
|---|---|---|---|
| The distance from the spraying position to the radar | 200km | 127.2km | 84.33km |
| Number of chaff bombs thrown | 1 | 2 | 5 |
| Equivalent scattering area of chaff cloud cluster | 40m² | 80m² | 200 m² |
| The maximum detection range of radar | 127.2km | 84.33km | 56.8km |

The interference process can be shown in Figure 4:

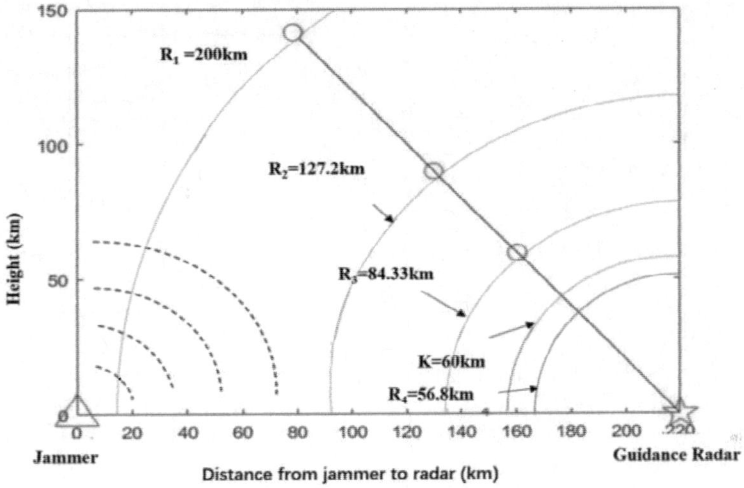


**Figure 4.** Schematic diagram of jam process

When the effective detection range of radar becomes 56.8 km, which is less than the required maximum detection range of radar 60 km. No longer continue to throw chaff bombs, penetration formations can start to implement ground penetration tasks. In actual combat, the number and position of chaff bombs are calculated according to the operational requirements and the number of chaff bombs that penetration formation can actually load, so as to achieve practical and anticipated interference effect.

## 5. Conclusion

(1) In this paper, a method of transmitting long-range jamming signals by chaff clouds is proposed. The method injects jamming energy from the main lobe of radar by scattering chaff clouds. Compared with the method of long-range side-lobe jamming of jammer, the method can achieve better jamming effect.

(2) A chaff bomb spraying model for frontal penetration is established. The results show that the model can suppress the maximum detection range of enemy ground-to-air missile guidance radar continuously and effectively during penetration, and provide effective protection for aircraft formation for penetration mission.

## References

[1] W.S. Dong, J.N. Hyun, J. Oh, IEEE Transactions on Aerospace an-d Electronic Systems, 48(2012)

IOP Conf. Series: Materials Science and Engineering **608** (2019) 012003 doi:10.1088/1757-899X/608/1/012003

[2] R.J. Puskar, Prod of IEEE 1974 National Aerospace and Electronics Conference, 177-183(1974)
[3] J. Brunk, D. Mihora, P. Jaffe, *Chaff Aerodynamics*(1975)
[4] H. Zheng, X.H. Wang, X. Zhang. Firepower and Command Control, **37**(2012)
[5] J. Li., ship electronic countermeasures, **26**(2003)
[6] H.Q. Hu, W. Zhong, Space Control, **26**(2008)
[7] W.Y. Cai, X. Li, S.H. Wan, Systems Engineering and Electronic Technology, **31**(2009)
[8] Y. Li, H. Wu, H.W. Wang, S.Y. Cheng, Journal of Air Force Engineering University (Natural Science Edition), **12** (2011)
[9] J.H. Zhou, M. Gao, J. Dai, Science and Technology Information (2012)
[10] L.F. Ding, F.L. Geng, J.C. Chen, *Radar Principle* (2009)

IOP Conf. Series: Materials Science and Engineering **608** (2019) 012004 doi:10.1088/1757-899X/608/1/012004

# Identification of Key Conflict Nodes Based on Complex Network Theory

**Li Jiawei[1*], Wen Xiangxi[1], Zhang Libiao[1] and Liu Fei[1]**

[1]Air Control and Navigation College, Air Force Engineering University Xi'an, China

[*]Email: Jiaweilee036@gmail.com

**Abstract.** The current methods of detecting flight conflicts are difficult to grasp the overall conflict situation in the air, which is not conducive to the controllers to make accurate judgment on different conflict situations. In this paper, a method to identify key nodes of flight conflict based on complex network theory is proposed. Firstly, a conflict situation network model was built based on the ACAS protection area model. Then node degree centrality, closeness centrality and PageRank were used to evaluate the comprehensive importance of all nodes in the conflict situation network by AHP method. And we can find out the key conflict locations and conflict aircrafts based on the evaluation results of the index. The simulation results show that the model can effectively identify the key conflict aircrafts and positions in the airspace. And It can help controllers fully grasp the air security situation by classifing the air conflict situation levels.

## 1. Introduction

At present, with the rapid development of China's aviation industry, air traffic flow is increasing. Air collision defense is becoming more urgent. How to effectively mediate the air conflict between aircraft and reduce the probability of flight conflict has become an urgent problem to be solved.

In recent years, domestic and foreign scholars have conducted a large number of studies on this problem: HAN Dong et al [1]. used SVM classification learning to solve the problem of conflict detection and relief in air traffic control. Krozel et al[2] used the optimal control theory to propose a method for conflict detection and relief of two aircraft in free flight. LIU Xin et al.[3] used vector method to construct the information conflict detection method based on ads-b technology for the detection of two aircraft conflicts. Liu Yang et al.[4] took low altitude as the research object and constructed the low altitude conflict detection algorithm under the condition of free flight. Most of the above research methods focus on a small number of aircraft. And there is a big gap between them and the actual airspace operation. So it is impossible to comprehensively grasp the flight conflict situation in the complex airspace. In addition, when these methods are used for conflict judgment, all aircrafts in the airspace need to be pairwise judged, and the calculation process is complex.

To improve these problems, this paper converts the conflict relationship between the aircraft to the relationship between network nodes, build a conflict situation network model based on the ACAS flight reserve. And on the basis of the model, we built node comprehensive evaluation index to identify key conflict nodes. It can realize the key conflict nodes in conflict situation network selection. We hoped that this method can quickly find out the aircraft with high conflict level and the conflict location in the airspace, so as to provide information for the air controller to quickly grasp the status of conflict in the airspace.

## 2. Construction of conflict situation network model

In this paper, the ACAS flight protection area is selected as the criterion for conflict determination. And the conflict situation in the air is abstracted into a complex network.

### *2.1. Model construction of ACAS flight reserve*

ACAS is a device that detects and retrieves the surrounding aircraft, issues queries to the aircraft that may be at risk of invasion.[5]. The detection range of the ACAS antenna is usually set at 15-30 nautical miles in front and 3 kilometers in the longitudinal direction. In this case, ellipsoid conservation area model is usually adopted to establish the flight conservation area model of a single aircraft under the ACAS system. Determine whether the two aircraft there is a conflict or potential conflict rule is: when the horizontal distance is less than the minimum interval standard 10 km, and the vertical distance is less than the minimum vertical interval standard 0.3 km. When the horizontal interval is more than 10 km and less than 15 km, said the potential conflicts between the two aircraft. On the basis of this, the established protected area model is shown in Figure 1.

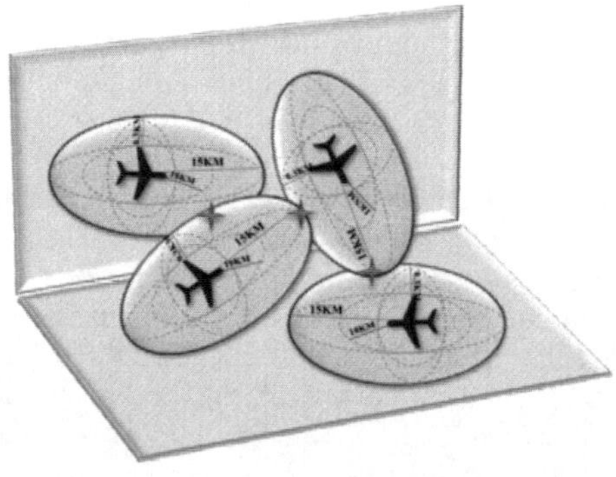

**Figure 1.** ACAS flight protection zone model.

In Figure 1, four aircrafts are active in a certain airspace. Each aircraft uses the flight protection area model based on ACAS. Based on the ellipsoid model, the center point of the aircraft was placed at the post-ellipsoid focus P, giving the aircraft a larger pre-detection distance of 15 km. And the longitudinal intervals of the protection zone were all set at 0.3 km. The conflict domain of this model is:

$$\frac{(x-x_0-m)^2}{a^2}+\frac{(y-y_0-n)^2}{b^2}+\frac{(z-z_0)^2}{c^2}\leq 1 \tag{1}$$

In the conflict domain equation, offset m and n are introduced. The model also divides the detection levels into the protection zone. It takes the 15 km range of the protection zone as the limit to judge the invasion level according to the invasion distance of the two aircraft. The greater the invasion distance is, the higher the corresponding conflict level will be. At this time, ACAS will make corresponding anti-collision warning or countermeasures.

### *2.2. Construction of Flight Conflict Network*

By building the model of a single aircraft flight protection zone, we can abstract the conflict situation in the airspace into a complex network[6]. each aircraft is regarded as a node in the network. If there are aircraft intruding into each other's protection zone, there will be an edge between the two nodes. As shown in Figure 2, there are four conflict nodes in the airspace, which are $v_1$, $v_2$, $v_3$, $v_4$.

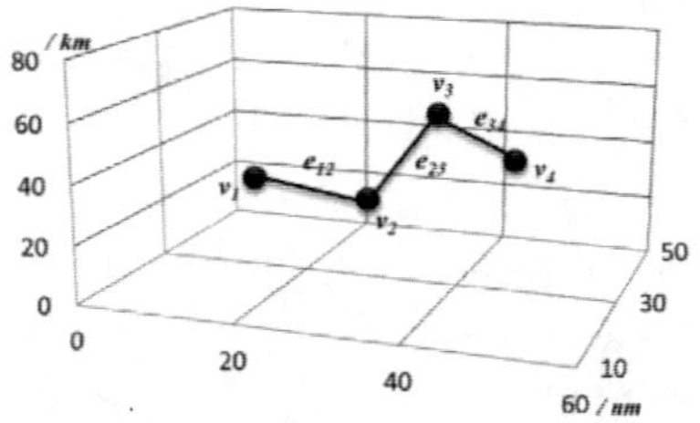

**Figure 2.** Conflict Network Corresponding to Scenarios of Figure 1.

In conflict detection based on ACAS protection zone, the horizontal and vertical spacing between aircraft plays a critical role in aircraft conflict detection. Therefore, we introduce collision edge weights [7] to reflect the influence of two kinds of spacers on the network, and construct the flight conflict weighting network. In Figure 3, when the two aircraft intrude into the protected area, the edge right of the generated conflict is $\omega_{ij}$. According to the fact that the closer the distance between two aircraft is, the greater the probability of conflict between them is, the edge weight designed in Figure 3. Where $d_{La}$ represents the horizontal distance of the two aircraft from which the invasion occurred, and $d_{Lo}$ represents the vertical distance.

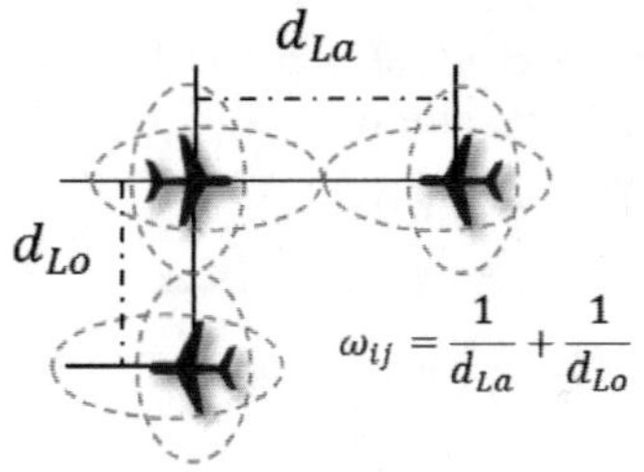

**Figure 3.** Edge weight setting.

The conflict situation in airspace is abstracted as a weighted network. The network can be expressed in the form of the weighting adjacency matrix. For example, the conflict situation weighted network shown in Figure 2 can be expressed as:

$$Q' = \begin{bmatrix} 0 & 0.39 & 0 & 0 \\ 0.39 & 0 & 0.65 & 0 \\ 0 & 0.65 & 0 & 0.51 \\ 0 & 0 & 0.51 & 0 \end{bmatrix} \tag{2}$$

## 3. Key Conflict Nodes Identification Method

In the conflict situation network, a few aircraft play the role of 'key nodes', they can have a key impact on the robustness of the conflict network. For air traffic management activities, the deployment of these key aircraft nodes can quickly alleviate air traffic congestion. How to find out these key conflict nodes accurately is the key problem to be solved in this paper.

### *3.1. Selection of Evaluation Indicators for Node Importance*

In order to effectively evaluate the status of each aircraft node in the entire conflict situation network, we select some indicators to evaluate it. In the selection of indicators, the following factors are mainly considered: 1) the selected indicators can reflect the characteristics of the nodes. 2) the selected indicators can reflect the role of nodes in the network. 3) the selected index can reflect the relationship between nodes and other nodes. Therefore, we selected node degree, near-centrality and PageRank value as evaluation indicators.

**Defination 1** Degree Centrality

The ratio of the number of edges associated with node $i$ to the maximum number of edges that node $i$ may have. The expression for degree centrality is:

$$DC_i = k_i / (N-1) \tag{3}$$

Where $k_i$ represents the number of edges associated with node $i$ in the network. The definition of degree centrality indicates the ability of a node to directly associate with other nodes.[8-9] The larger the value, the greater the role it plays in the network. Corresponding to the flight conflict network, the index focuses on reflecting the extent and influence of conflict spread of the aircraft itself.

**Defination 2** Closeness Centrality

Assuming that $d_{ij}$ represents the number of edges contained in the shortest path with node $i$ as the starting node and $j$ as the ending node, the proximity centrality of node $i$ can be expressed as the reciprocal of the sum of its distances to all other nodes in the network, namely:

$$CC_i = N / \sum_{j=1}^{N} d_{ij} \tag{4}$$

The greater the value of the closeness centrality, the greater the degree of the node in the center of the network, and the more important it is.[10-11] This index can effectively identify aircraft in the core zone of conflict.

**Defination 3** *PageRank*

*PageRank* method is realized through a similar voting method. If a node obtains a higher score (weight), the higher the value of the node, and its weight calculation expression is:

$$\Pr(A) = \Pr(A)/C(B) + \Pr(C)/C(C) + \cdots \tag{5}$$

Where, $\Pr(A)$ represents the updated weight value of node A, $\Pr(B)$, $\Pr(C)$ represents the weight value assigned to A by all other nodes connected to node A, and $C(B)$ and $C(C)$ represent the number of outlinks of node B and C [12]. This index can effectively identify the important conflict nodes by combining the conflict distance edge weight in airspace conflict situation network.

### *3.2. Evaluation of Node Importance Based on AHP*

In this paper, we adopted AHP method to determine the index weight and construct the comprehensive evaluation index for identifying key conflict nodes. This method has low computational complexity and strong systematicness. It can better show the influence of each index on the whole.

Among the selected indexes, the *PageRank* takes into account the connection condition and distance of nodes comprehensively. Here we consider it as the most important, followed by the degree of nodes.

Obtaining a judgment matrix $A$:

$$A = \begin{bmatrix} 1 & 3 & 5 \\ 1/3 & 1 & 3 \\ 1/5 & 1/3 & 1 \end{bmatrix} \tag{6}$$

The corresponding eigenvector $\omega$ of the maximum eigenroot $\lambda_{\max}$ is calculated and normalized to obtain the weight vector:

$$W = [W_1 \quad W_2 \quad W_3] = [0.6370 \quad 0.2583 \quad 0.1047] \tag{7}$$

Carry out consistency check and calculate the maximum eigenvalue:

$$\lambda_{\max} = \frac{1}{n}\sum_{i=1}^{n}\frac{\sum_{j=1}^{n} a_{ij}W_j}{W_i} = 3.0385 \tag{8}$$

The consistency index $CI$ is:

$$CI = \left(\frac{\lambda_{\max} - n}{n-1}\right) = 0.0192 \tag{9}$$

Consistency specific gravity *CR* is:

$$CR = \frac{(\lambda_{\max} - n)/(n-1)}{RI} = 0.0370 < 0.1 \tag{10}$$

Where *RI* is a random consistency index, when $n = 3$, *RI* is 0.58.

This judgment matrix satisfies the consistency test.

In order to eliminate the order of magnitude difference among the three indicators, the maximum and minimum normalization method is adopted:

$$DC_i = \frac{DC(v_i) - \min DC(v)}{\max DC(v) - \min DC(v)} \tag{11}$$

The node importance comprehensive evaluation $P_i$ result is obtained:

$$P_i = 0.6370 \times \Pr_i + 0.2583 \times DC_i + 0.1047 \times CC_i \tag{12}$$

## 4. Key Conflict Node Identification Process

Through the description in the first three sections, the key conflict node identification process of the training airspace conflict situation network established in this paper is presented as shown in Figure 4.

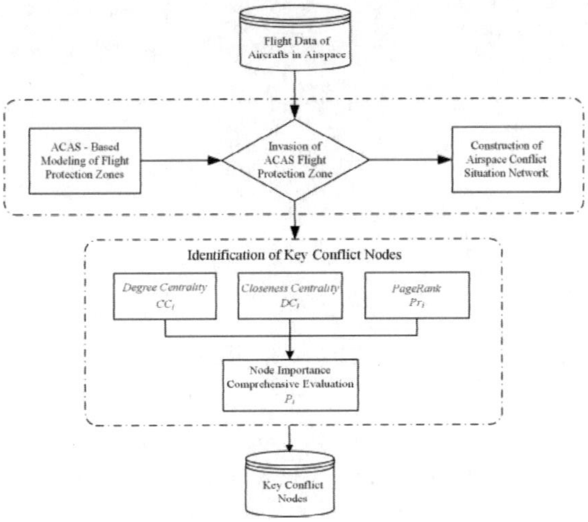


**Figure 4.** Key collision nodes identification process.

The identification of key conflict nodes in conflict situation network mainly consists of three parts: data acquisition, construction of conflict situation network, and evaluation and identification of key conflict nodes.

Part I: Data acquisition, collection of real-time flight parameters of all aircraft in the current airspace, including flight speed, space coordinate position, etc.

Part II: Construction of flight conflict network. On the basis of the acquired aircraft flight data and the ACAS protection area of each aircraft, the connection and edge weight between nodes are determined to construct the conflict network.

Part III: identification of key conflict nodes. The degree centrality, closeness centrality and PageRank were used to evaluate the importance of each node in the conflict situation network and identify the key conflict nodes.

## 5. Experimental analysis

In this paper, Matlab is used to generate an appropriate number of node sets in three-dimensional space to simulate the space position of aircraft in normal flight. It is assumed that the airspace range is a space with a length and width of 400 km and a height of 1500 m. In Figure 5, there are 50 aircrafts in the airspace.

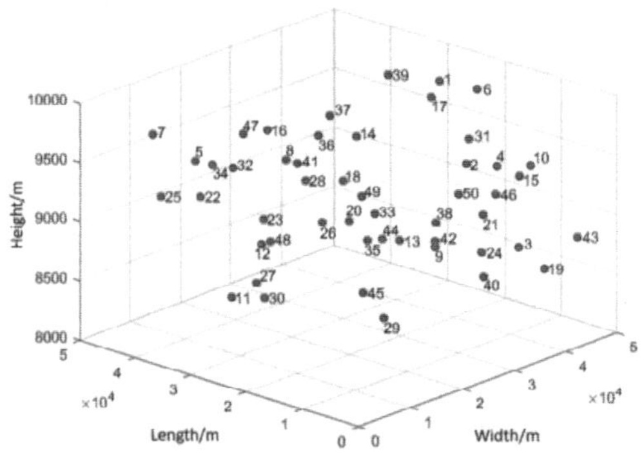

**Figure 5.** Simulation training airspace state.

According to the position relationship of all aircraft in the airspace, the conflict situation network model in the airspace is constructed according to the ACAS flight protection zone model, as shown in Figure 6.

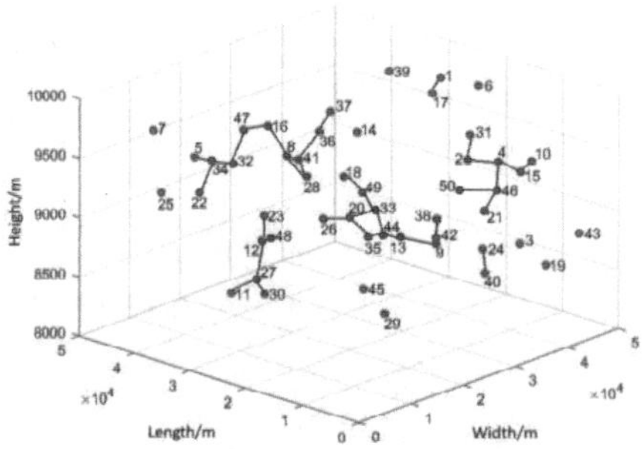

**Figure 6.** Construction of flight conflict network.

On the basis of the airspace conflict situation network, we respectively analyzed the degree, closeness centrality and *PageRank* value of the network. And the comprehensive importance evaluation $P_i$ value through AHP were obtained. The results are shown in Table.1.

**Table.1** Evaluation of importance of collision nodes

| Aircraft number | Degree Centrality | Closeness Centrality | *PageRank* | $P_i$ |
|---|---|---|---|---|
| A1 | - | - | - | - |
| A2 | 0.2857 | 0.4401 | 0.1421 | 0.3105 |
| A3 | - | - | - | - |
| A4 | 0.4286 | 0.9940 | 0.1977 | 0.5505 |
| A5 | 0.1000 | 0.0844 | 0.0561 | 0.0913 |
| A6 | - | - | - | - |
| A7 | - | - | - | - |
| A8 | 0.3000 | 0.9172 | 0.1402 | 0.4426 |
| A9 | 0.2000 | 0.4910 | 0.0950 | 0.2642 |
| A10 | 0.1423 | 0.0001 | 0.0791 | 0.0990 |
| … | … | … | … | … |
| A20 | 0.3000 | 0.5893 | 0.1270 | 0.3566 |
| … | … | … | … | … |

In Table.1, each nodes' the conflict situation was evaluated. There are some nodes lacking importance evaluation index values. Because they are not the nodes in a conflict situation network. Their importance evaluation indexes can not be caculate. We found that one node that has a high value of index may not the important node. We also need to combine the results of the other two indexes to draw a conclusion. Accroding the values of $P_i$, we choise the top 10 conflict nodes are identified as the key conflict nodes, which can be used as the focus of flight conflict adjustment. The results are shown in Figure 7:

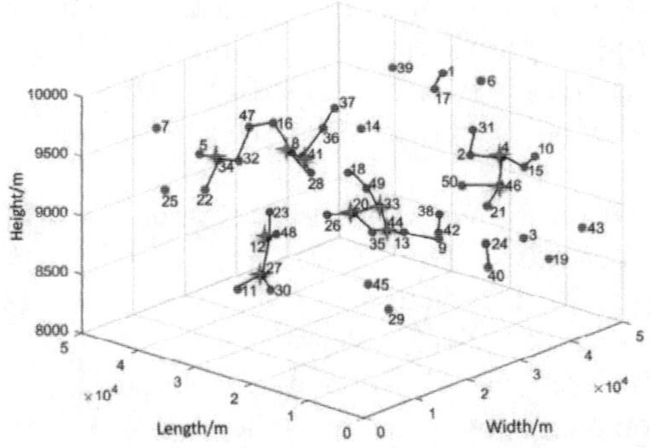


**Figure 7.** Identification results of key collision nodes.

In Figure 7, there are ten key conflict points with high conflict situation level in the conflict situation network. They are node 4, 8, 12, 20, 27, 33, 34, 41, 44 and 46. Nodes of the comprehensive evaluation of node 33 is 0.8925. It is the highest of all nodes' $P_i$. It can be seen that there are three key position around node 33. And the number of aircraft with the adjacent has four. Although No.49 aircraft did not identify the conflict in current, but can be used as a potential conflict object. In addition, No.33 aircraft's local conflicts situation is very complex. The local density is big. We need to focus on. These key conflict nodes often conflict with multiple aircraft and require the attention and priority of controllers. Through the construction of conflict network, this paper effectively avoids the repeated conflict judgment between two aircrafts in the airspace. and reduces the complexity of controller conflict detection in the complex airspace with a large number of aircrafts. In particular, the identification method of key conflict points can provide controllers with high-value conflict situation information. And it assists them to make reasonable deployment plans according to different conflict levels.

## 6. Conclusion

Firstly, this paper netted the aircrafts with conflicts and potential conflicts in the airspace and constructed the airspace conflict situation network model based on the ACAS conservation area model. Then, the AHP method was used to construct the evaluation index of node comprehensive importance by using three key node identification indexes of degree centrality, closeness centrality and PageRank. The key conflict nodes in the simulated airspace situation network were found out. And the conflict situation in the airspace is classified according to the index. The experimental results show that the nodes with higher conflict level are surrounded by multiple additional conflict nodes. The location change of the core conflict aircrafts will have a great influence on the local conflict network. By providing air traffic controllers with conflict situations in the airspace, the method optimizes their attention distribution and provides assistance for the formulation of targeted conflict resolution schemes quickly and efficiently.

## References

[1] Han D, Zhang X, Nie Z, et al, A conflict detection algorithm for low-altitude flights based on SVM, Journal of Beijing University of Aeronautics & Astronautics, 2018.
[2] Krozel, J., & Peters, M., Conflict detection and resolution for free flight, Air Traffic Control Quarterly, 5(1997) 181-212.
[3] Liu X, Yang X P, Tian S J, et al, Flight collision detection algorithm based on ADS-B information spherical model, Transducer & Microsystem Technologies, 6 2017) 1-15.

IOP Conf. Series: Materials Science and Engineering **608** (2019) 012004 doi:10.1088/1757-899X/608/1/012004

[4] Liu Y, Xiang J W, Luo Z P, et al, Short - term Conflict Detection Algorithm for Low - altitude Free Flight, Journal of Beijing University of Aeronautics and Astronautics, 43(2017) 1873-1881.
[5] Jeannin J B, Ghorbal K, Kouskoulas Y, et al, A formally verified hybrid system for safe advisories in the next-generation airborne collision avoidance system, International Journal on Software Tools for Technology Transfer, 230(2015) 1-25.
[6] Newman, M. E., Mathematics of networks. The new Palgrave dictionary of economics, 23 (2016) 1-8.
[7] Muldoon, S. F., Bridgeford, E. W., Bassett, D. S., Small-world propensity and weighted brain networks, Scientific reports, 6 (2016) 22057.
[8] Lü, L., Chen, D., Ren, X. L., Zhang, Q. M., et al, Vital nodes identification in complex networks, Physics Reports, 650(2016) 1-63.
[9] Liu, Y., Wei, B., Du, Y., Xiao, F., & Deng, Y, Identifying influential spreaders by weight degree centrality in complex networks, Chaos, Solitons & Fractals, 86(2016) 1-7.
[10] Du, Y., Gao, C., Chen, X., Hu, Y., et al, A new closeness centrality measure via effective distance in complex networks, Chaos: An Interdisciplinary Journal of Nonlinear Science, 25(2015) 033112.
[11] Park, M., Lee, S. H., Kwon, O. M., & Seuret, A., Closeness-centrality-based synchronization criteria for complex dynamical networks with interval time-varying coupling delays, IEEE transactions on cybernetics, 48(2018) 2192-2202.
[12] Ishii, H., & Suzuki, A. Distributed Randomized Algorithms for PageRank Computation: Recent Advances, In Uncertainty in Complex Networked Systems, Birkhäuser, Cham (2018) 419-447.

# Research on Height and Diameter of Doppler VHF Omnidirectional Beacon in Complex Environment

**Liang Fei [1,*], Hu Yue [1] and Li Yuankai [1]**

[1]The Second Research Institution of CAAC, 610041, Cheng Du, China

[*]liangfei@caacsri.com

**Abstract.** With the development of community around Doppler VHF omnidirectional beacon(DVOR), more and more DVORs are restricted, and even severely disabled. The DVOR's counterpoise can be elevated to improve the coverage. But raising counterpoise will cause some problems such as null and worse bearing error. The paper research the lateral multipath, longitudinal multipath, height and diameter of counterpoise from the null, the first Fresnel zone, signal coverage, bearing error combing surrounding buildings and terrain, providing some technical means for improve or eliminate nulls and bearing error in construction of DVOR, and pre-evaluating whether it can pass the flight inspection after the completion of construction.

## 1 Introduction

VHF omnidirectional beacon(VOR) is a navigational aid that operates in the very high frequency(VHF) band of the radio spectrum and that radiated uniformly in azimuth. Specifically, this facility operates between 108MHz and 118MHz and provides azimuth guidance to the pilot in the form of a visual display. There two general types of VOR systems being using by ICAO, namely, the conventional VOR(CVOR) and the Doppler VOR(DVOR). Because CVOR is more susceptible to multipath interference than DVOR, DVOR is used mainly by CAAC.

With the development of Chinese cities, more and more tall buildings block and reflect DVOR signals, making them limited or even unavailable. Some DVOR around the airport are particularly serious. For example Lianshenwei DVOR of Zhuhai, most angles are restricted including 98°~113°, 330°~010°,135°~158°etc. A similar situation occurred in Guanlan, Pingzhou, Ruian, Xiaoshan DVOR navigation aid. Except reselecting the site of DVOR, the only thing we can do is raise the height of counterpoise while increasing the diameter of counterpoise.

## 2 Multipath effect

Multipath[1] include mainly longitudinal multipath and lateral multipath. Lateral multipath can, in certain of the navigation aids, cause rays emitted from DVOR at different azimuth angles to meet and combine at the airborne equipment. The combination of two different modulation signals yields a resultant generally containing false bearing information.

Longitudinal multipath cannot ordinarily cause errors in the modulating intelligence except, as mentioned, when it causes the serious weakening of the total signal. Similarly, lateral multipath cannot cause nulls in the vertical radiation pattern. Figure 1 illustrates the two different categories of multipath.

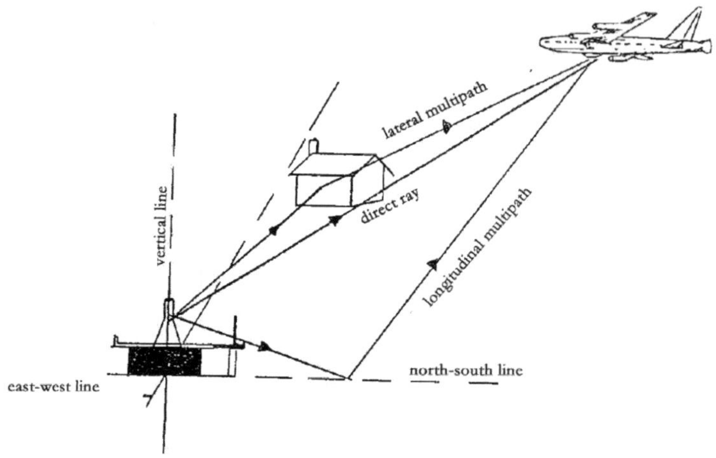


**Figure 1.** longitudinal multipath and lateral multipath

*2.1 Longitudinal multipath*

The longitudinal multipath can result from reflection from the counterpoise and from the earth, causing nulls in the vertical pattern of radiation and this in turn causes loss of signal in the airborne receiver. Hence, it is important in analysis to identify those features that may lead to excessive multipath problems.

Usually, the height of counterpoise is less 10 meter, longitudinal multipath is not obvious and vertical pattern of radiation does not divide severely. However, when tall buildings appear around DVOR, raising the height of counterpoise will result in dramatic longitudinal multipath, and the vertical lobe pattern will have multiple nulls at low elevation angles.

*2.2 Lateral multipath*

Lateral multipath involves the mutual interference between signals radiated at two different azimuth bearings. One is direct ray. The second, the interfering ray, contains different navigational information than does the direct ray. The mix of the two modulations in the airborne receiver results in incorrect information being presented to the pilot. Lateral multipath will causes bearing error and 9960 Hz modulation beyond standard of CAAC.

The requirement for lateral multipath to exist is that there be obstacles in the radiation field that accept radiation at one azimuth angle and reflect it off in a different direction either in a specular or diffuse fashion. Obstacles will exist in the radiation field and may consist of building that cannot be removed, power line, and antenna for other equipment.

The lateral multipath has little relation with the height of counterpoise. Regardless of whether the counterpoise is high or low, as long as there are obstacles around DVOR, the influence of lateral multipath effect should be considered. However, increasing the radius of the counterpoise can improve the lateral multipath caused by the obstacles.

**3 Null and coverage**

There are two kinds of DVOR navigation aids. The first is responsible for approach and departure, which the required coverage distance is no more than 70 nautical miles. The second is the route navigation aids, which required coverage about 150 nautical miles.

Some obstacles exist around DVOR, which is characterized by a short coverage distance and the counterpoise that needs to be elevated. The conflict of coverage and null must be solved combing the flight procedure in actual operation.

*3.1 Required coverage distance*

DVOR coverage[2-3] as specified Annex 10 of International Civil Aviation Convention is "The VOR shall provide signals such as to permit satisfactory operation of a typical aircraft installation at the levels and distances required for operational reasons, and up to an elevation angle of 40 degrees. The field strength or power density in space of VOR signals required to permit satisfactory operation of a typical aircraft installation at the minimum service level at the maximum specified service radius should be 90 microvolts per meter or minus 107dBW/m2."

Every airport has its own flight procedures, each airport has its own DVOR flight procedure, and the DVOR is responsible for guiding distances that may be long or short. DVOR is responsible for the approach and departure flight procedures shown in Figure 2

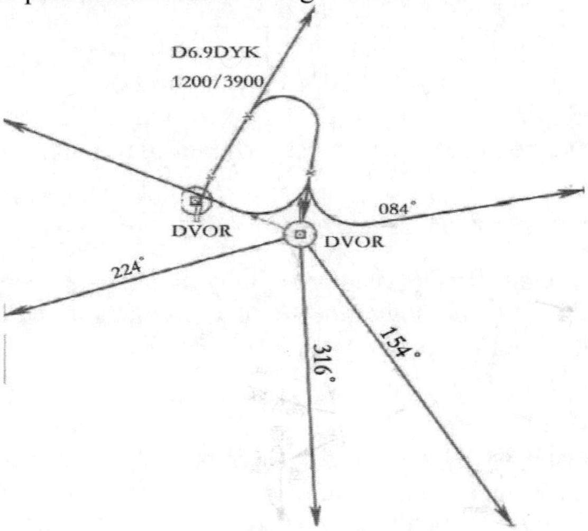


**Figure 2** departure and approach based on DVOR

*3.2 Vertical pattern synthesis*

The vertical lobes of DVOR are composed of the lobes reflected by counterpoise and the lobes reflected by the ground. The lobes reflected by counterpoise are responsible for the coverage of the high elevation angle, while the lobes reflected by the ground are responsible for the low elevation angle, show as Figure 3.

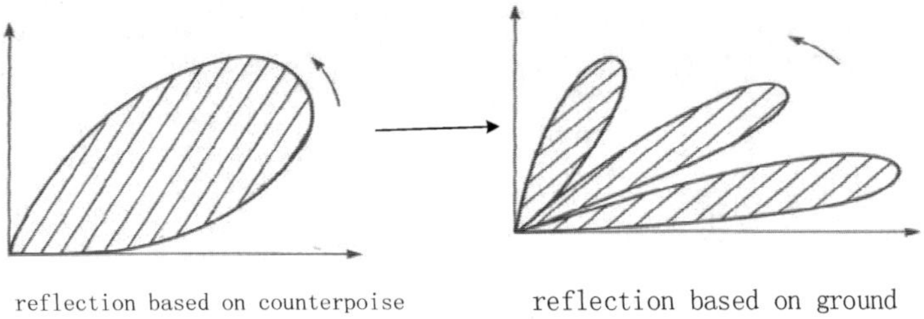


**Figure 3.** vertical lobe throuth counterpoise and ground

When the counterpoise is 10 meters high and counterpoise diameter is 30 meters, the spatial lobes of the DVOR are as shown in Figure 4.

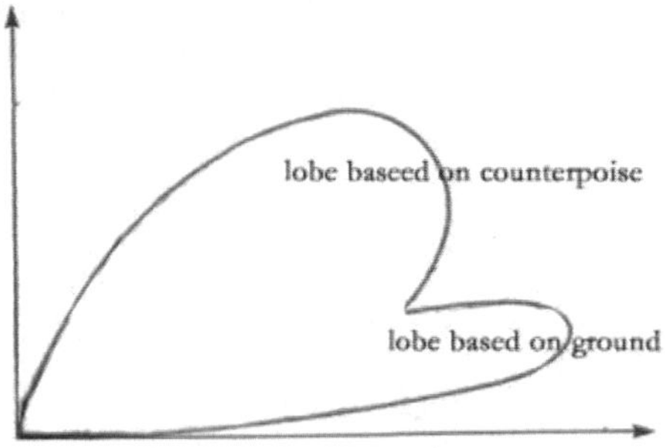


**Figure 4** DVOR synthetic lobe

There is an obvious drop in the local signal of the two lobe. The upper lobes are responsible for approach and departure, and the lower lobes are responsible for routing.

*3.3 Nulls change accompanying counter's height increase*

*3.3.1 Null's position calculating.* DVOR is a horizontally polarized antenna[4], the nulls of DVOR[3] can calculated from equation (1). The higher the counterpoise height, the lower the first null, then the second to the Nth null will appear at a lower elevation angle.

$$\gamma = sin^{-1}[(n\lambda)/(2h)] \quad (1)$$

The DVOR frequency is 108.1MHz, counterpoise is 60 meters, and antenna's height is 1.2 meters from counterpoise, the first, second and third null is 1.32°, 2.64°, 3.96°. if the counterpoise diameter is still the regular 30 meters, the flight altitude is 3000 feet, the nulls will appear in 22nm, 11nm and 7nm, show as Figure 5.

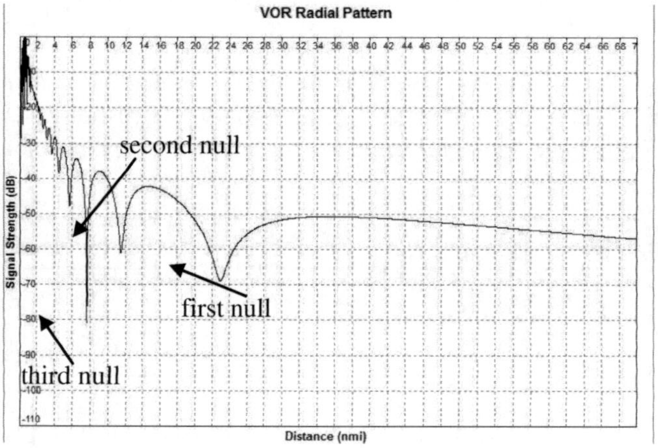


**Figure 5** first three nulls when counterpoise is 60m

The first, second and third null is 1.13°, 2.27°, 3.40°when the counterpoise height is 70m, the nulls will appear in 24nm, 12nm and 8nm, show as Figure 6.

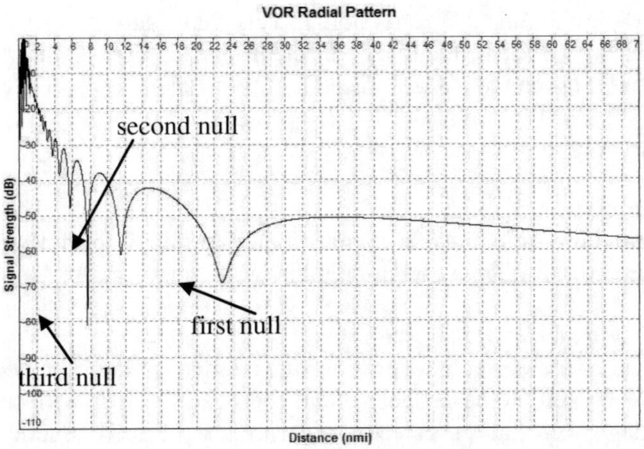


**Figure 6** first three nulls when counterpoise is 70m

*3.3.2 Eliminating nulls.* We can calculate the suitable counterpoise according to the null. When the counterpoise is 60 meters, antenna is 1.2 meters from counterpoise, we can eliminate the third null while the counterpoise may be expanded to 18 meter. The second is 26 meter and the first is 52 meter, show as Figure 7. In practical operation, the diameter of counterpoise is usually 30 meter, so if necessary we mainly eliminate the first few nulls.

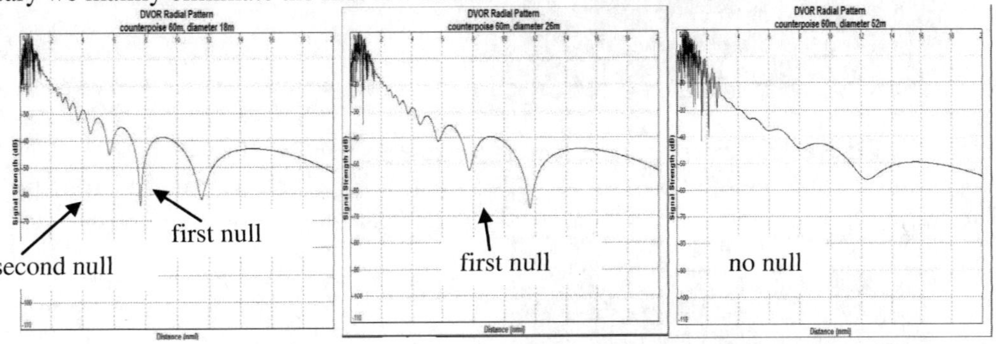


**Figure 7** expanding diameter elimanate nulls

Eliminating all nulls will cause diameter of counterpoise to be too large, resulting in poor economic benefit. In practice, the counterpoise diameter can be determined by the DVOR flight procedure of the actual airport to save economic costs.

## 4 Reflection and bearing error

### *4.1 Bearing error requirement of CAAC*

The duty of DVOR is to guide the bearing of aircraft, so bearing error[4-5] is the most important parameter of DVOR. According to flight inspection centre of CAAC, Bearing error of DVOR include orbit bearing error which average error within ±1°and radial error within ±2°.

Orbit bearing error is checked through a flight inspection method that distance is 20 nautical mile to DVOR centre, and flight orbit is generally performed between 4° and 6° of DVOR elevation angle. Radial bearing error is checked through actual flight procedure of DVOR.

### *4.2 Reflection effect*

Reflection will cause bearing error cannot meet the standard requirements. How to eliminate the effect of reflection of DVOR has always been a topic. After the counterpoise is elevated drastically, this is more and more difficult. Currently there are some measures improving the effect of reflection include that removing obstacle eliminating the reflection is very rude and effective, changing the exterior material of the building to reduce reflection coefficient, and extending the diameter of diameter to screen reflection. The latter two methods is focused by this article.

## 5 Eliminating null and error

### *5.1 Expanding diameter of counterpoise*

The counterpoise can not only reflect useful signals, but also shield obstacles from signal effects. We can improve the bearing error through expanding diameter of counterpoise.

### *5.2 Changing the exterior material of the buildings*

The reflection coefficient of the obstacle determines the strength of the reflected signal, metal obstacle will cause the signal to be completed reflected, and the bearing error will be deteriorated. Figure 8 show effect from a radar tower that the exterior material is reinforced concrete. From this figure, the bearing error exceeds the standard requirements in the azimuth of the radar tower is 75° ~ 120°,

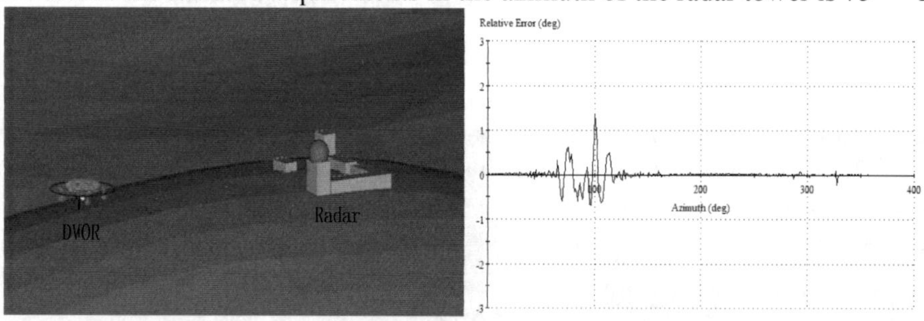


**Figure 8** the exterior material of radar tower is reinforced concrete

For improving the bearing error within 75°~120°, put the brick on the outer wall of the radar tower. The bearing error is improved and meet standard requirement, show as Figure 9.

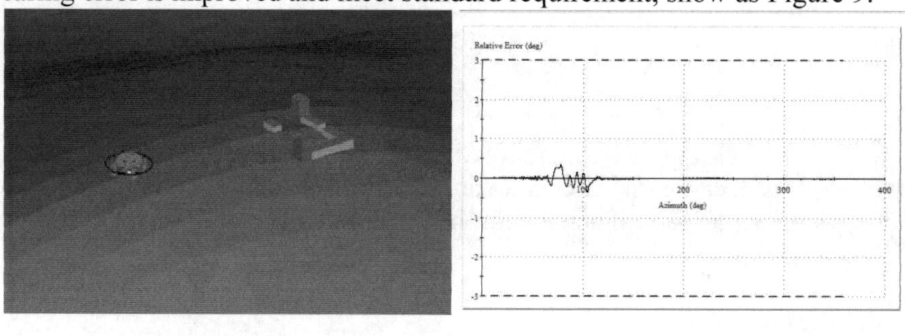

**Figure 9** put the brick on the outer wall of the radar tower

### *5.3 Changing the roughness of external buildings*

Another way to improve the bearing error is to increase the roughness of the wall and change the specular reflection to diffuse reflection. According to Rayleigh roughness criterion, As long as the convexity of the reflective surface is greater than formula 2, the bearing error may be improved.

$$h > \lambda / (16 \sin(\gamma)) \tag{2}$$

According to formula 2, Rayleigh roughness criterion figure can draw through incident angle and wave length, show as Figure 10.

IOP Conf. Series: Materials Science and Engineering **608** (2019) 012005 doi:10.1088/1757-899X/608/1/012005

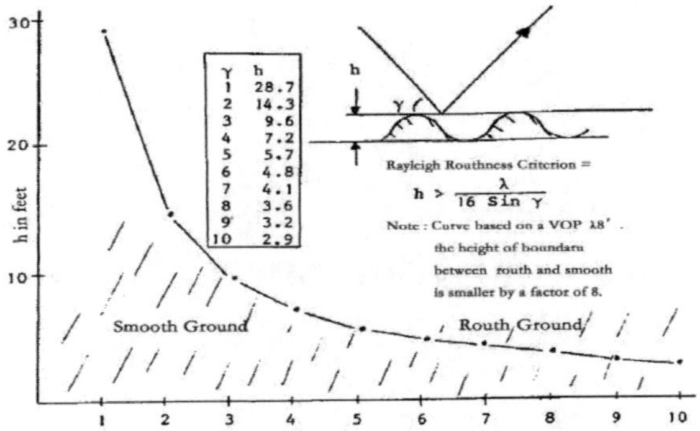


**Figure 10** Rayleigh roughness criterion

For improving the bearing error within 75°~120°in figure 7, change the convexity of the outer wall of the radar tower. The bearing error is not improved and meet standard requirement, the reason may be convexity is not good, show as Figure 11.

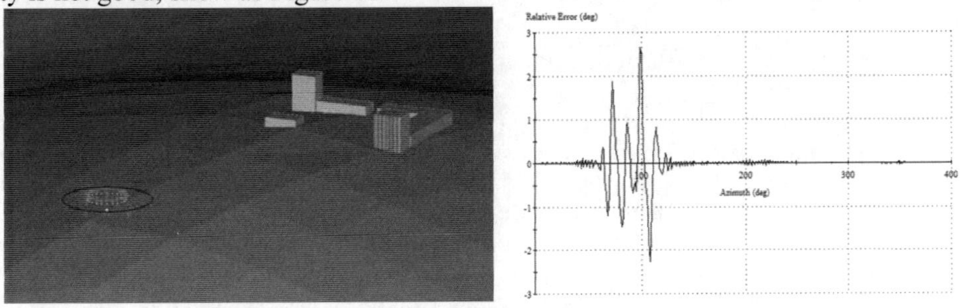

**Figure 11** change the convexity of the outer wall of the radar tower

## 6 Conclusion

As the environment around the DVOR station becomes more and more complex, some navigation stations are surrounded by obstacles. It is impossible to improve the error of some azimuth by expanding the radius of counterpoise and changing the outer wall of the obstacle. In this case, we can only move away from our navigation aids, but this is designed to deal with airspace, land acquisition, route changes and many other issues. Therefore, the best way is to demonstrate the surrounding planning before the construction of the navigation platform, and clearly define the height limit around the DVOR to avoid the deterioration of the signal after the operation.

## Acknowledgments

First of all, I would like to extend my sincere gratitude to my supervisor, Ye jiaquan, for his instructive advice and useful suggestions on my thesis. I am deeply grateful of his help in the completion of this thesis.

I am also deeply indebted to all the other tutor in my DVOR studies for their direct and indirect help to me.

Finally, I am indebted to my wife for their continuous support and encouragement.

## References

[1] Department of transportation federal aviation administration, VOR, VOR/DME and VORTAC SITING CRITERIA(1986)

[2] ICAO Annex 10, Aeronautical telecommunications Volume 1 Radio Navigation Aids(2012).

[3] CAAC, aeronautical radio navigation aids Part 2: Technical requirements for very haigh frequency omnidirectional range(VOR)

[4] Flight inspection centre of CAAC, Civil aviation land-based navigation equipment flight inspection specification(2016)

[5] Ni Yude, Ludan, Wangying, Navigation Principles and Systems(2015)

IOP Conf. Series: Materials Science and Engineering **608** (2019) 012006 doi:10.1088/1757-899X/608/1/012006

# Long Time Integration for High Speed Target Detection Based on Radonkt-LVT

**Jie Chen[1], Qi Wu[1], Yue Zhang[1] and Zengping Chen[1]**

[1]National Key Laboratory of Science and Technology on ATR, College of Computer Science and Technology, National University of Defence Technology, 410073 Changsha, China

E-mail : chenjie13@nudt.edu.cn

**Abstract.** In this paper, a computationally efficient long-time integration and parameters estimation method via Keystone transform(KT), Radon Transform(RT) and Lv's Transform(LVT) is proposed. Firstly, the KT-Radon is used to eliminate the range cell migration(RCM) of the echo after pulse compression, then using LVT to obtain the long-time integration and parameters estimation of high-speed target with acceleration motion. Compared with the GRFT algorithm, the proposed method only needs one dimensional search for range. Compared with the KT-FRFT algorithm, there is no need to search Doppler fold factor in this proposed method. So the proposed method doesn't suffer from huge computation cost and obtain a good tradeoff between the computational cost and detection performance in comparison with the above algorithms. Finally, several experiments are provided to demonstrate the effectiveness of RadonKT -LVT.

## 1. Introduction

With the development of technology, more and more high-speed targets with low radar cross section(RCS) appear in the radar searching area. And with the widely equipped of phase array antennas and ubiquitous technology, the long time integration processing is widely believed to be an effective way to increase the detection/estimation performance. However, the range cell migration(RCM) and Doppler frequency migration(DFM) would occur during the long time integration because of the complex motions of targets (i.e., high speed and acceleration, etc.), which may degrade the performance of traditional moving target detection (MTD).

Usually, the parametric-searching methods such as keystone transform(KT) [1] and Radon-Fourier transform(RFT) [2] are supposed to eliminate the RCM. In particular, KT can correct RCM by rescaling the slow time axis but it needs to do decouple and fold factor searching in the time-frequency domain. However, RFT still needs to perform a jointly searching along range and velocity directions. In order to detect high speed targets, these two methods will suffer from a lot of computation cost in searching parameters.

To remove the DFM induced by acceleration, the generalized Radon Fourier transform (GRFT) [3] is proposed by adding acceleration searching dimension. Meanwhile, a method combining KT and fractional Fourier transform(KT-FRFT) [4] is introduced to do a long time integration. A method combining GRFT and Lv's transform (LVT) [5] is proposed to accumulate in the centroid frequency and chirp rate (CFCR) domain. The computational cost of these two methods are still very huge.

Motivated by the previous works, we propose a computationally efficient method to correct the RCM and estimate motion parameters of the moving target. Firstly, we use RT to indicate the fold factor of KT, then we can do the keystone transform by none factor searching. After that, the coherent accumulation of target energy and parameter estimation about velocity and acceleration could be obtained by the LVT. Compared with the GRFT and KT-FRFT, RadonKT-LVT is computationally efficient and exhibits a better detection ability. Compared with MTD and RFT, RadonKT-LVT can have a better performance in lower SNR because of the correction of RCM and DFM.

The rest is arranged as follows. Section 2 constructs the signal model and Section 3 proposes the RadonKT-LVT algorithm. Section 4 shows the result of experiments. Finally, a conclusion is given in Section 5.

## 2. Signal model

Suppose that the radar transmits a linear frequency modulated (LFM) signal as follows

$$s_t\left(\hat{t},t_m\right)=rect\left(\frac{\hat{t}}{T_p}\right)\exp\left(j\pi K\hat{t}^2\right)\exp\left[j2\pi f_c\left(\hat{t}+t_m\right)\right] \tag{1}$$

where $\hat{t}$ is the fast time, $t_m=mT_r$, $T_r$ denotes the pulse repetition time, $T_p$ is the pulse duration, K is the frequency modulated rate, $f_c$ is the carrier frequency, $m=-\frac{M}{2},\cdots,\frac{M}{2}$ denotes the transmitted pulse number index, M is the number of coherent integrated pulse which is usually used to be the integer power of 2.

The received baseband signal of a moving target can be stated as

$$s_r\left(\hat{t},t_m\right)=A_0 rect\left[\frac{\hat{t}-2R\left(t_m\right)/c}{T_p}\right]\exp\left[j\pi K\left(\hat{t}-2R\left(t_m\right)/c\right)^2\right]\exp\left[-j\frac{4\pi}{\lambda}R\left(t_m\right)\right] \tag{2}$$

where $A_0$ is the amplitude of the received signal, $R\left(t_m\right)$ is the instantaneous range between radar and target and $\lambda$ denotes the wavelength of the transmitted signal.

After pulse compression, the compressed signal in the range time domain can be presented as [6]

$$s_{rm}\left(\hat{t},t_m\right)=A_1\sin c\left[B\left(\tau-\frac{2R\left(t_m\right)}{c}\right)\right]\exp\left[-j\frac{4\pi}{\lambda}R\left(t_m\right)\right] \tag{3}$$

where $A_1$ denotes the signal amplitude and B denotes the bandwidth of the transmitted signal.

Considering a maneuvering target, the $R\left(t_m\right)$ can be expressed as $R\left(t_m\right)=R_0+vt_m+\frac{1}{2}at_m^2$, where $R_0$ is the initial range from radar to the target, and $v$、$a$ denote the target's real radial velocity and acceration. Usually, the pulse repetition frequency(PRF) of the radar system is low and the velocity of the maneuvering target is high lead to the doppler ambiguity. The radial velocity [7] can be expressed as $v=v_0+Kv_{amb}$, where $v_{amb}=\lambda/2T_r$ is blind velocity, $v_0=\mathrm{mod}\left(v,v_{amb}\right)$ is the unambiguous velocity and K is folded factor.

From (3), it can be seen from this that RCM is caused by v and a. During the increasing integration time, RCM will reduce the gain of accumulate. However, the quadratic phase caused by acceleration (DFM) can also diffuse the energy of target.

After performing Fourier transform to the fast time $\hat{t}$, the compressed echo in the range frequency domain can be transfered as

$$s_{rm}(f,t_m)=A_2 rect\left(\frac{f}{B}\right)\exp\left[-j\frac{4\pi}{c}(f+f_c)\left(R_0+v_0t_m+\frac{1}{2}at_m^2\right)\right]$$
$$\times\exp\left(-j\frac{4\pi}{c}fKv_{amb}t_m\right) \tag{4}$$

## 3. Proposed method Based on RadonKT and LVT

### 3.1. Range cell migration correction

From (4), there exists a linear and quadratic coupling between $f$ and $t_m$. Usually the linear one occupies the main position of RCM due to the high speed of target. Therefore, the keystone transform (KT) is used to eliminate the linear coupling and its scaling formula is as follows

$$t_m = f_c/(f+f_c)t_n \tag{5}$$

where $t_n$ denotes the scaled slow time after KT. Applying the (5) to (4), we can obtain

$$s_{KT}(f,t_n)=A_2 rect\left(\frac{f}{B}\right)\exp\left(-j\frac{4\pi R_0}{\lambda}\left(1+\frac{f}{f_c}\right)\right)\exp\left(-j\frac{4\pi v_0 t_n}{\lambda}\right)$$
$$\times\exp\left(-j\frac{4\pi a t_n^2}{\lambda}\frac{f_c}{f+f_c}\right)\exp\left(-j\frac{4\pi K v_{amb}t_n}{\lambda}\frac{f}{f+f_c}\right) \tag{6}$$

In the narrowband of radar system $f \ll f_c$ ,we can have $\frac{f_c}{f+f_c}\approx 1, \frac{f}{f+f_c}\approx\frac{f}{f_c}$. So (6) can be approximately expressed as

$$s_{KT}(f,t_n)\approx A_2 rect\left(\frac{f}{B}\right)\exp\left(-j\frac{4\pi R_0}{\lambda}\left(1+\frac{f}{f_c}\right)\right)\exp\left(-j\frac{4\pi v_0 t_n}{\lambda}\right)$$
$$\times\exp\left(-j\frac{4\pi a t_n^2}{\lambda}\frac{f_c}{f+f_c}\right)\exp\left(-j\frac{4\pi K v_{amb}t_n}{\lambda}\frac{f}{f_c}\right) \tag{7}$$

It can be seen from (7) that the coupling between $v_0$ and $f$ has been eliminated but the migration caused by folded factor K still exists. Usually, we compensate the folded factor by searching the possible $K'$. That is to construct a compensation function as follows

$$H(f,t_n;K') \tag{8}$$

After multiplying (8) by (7) and performing the inverse FFT along the $f$ dimension, we have

$$s_{KT}(\hat{t},t_n;K')=A_3\sin c\left[B\left(\hat{t}-\frac{2R_0}{c}-\frac{2(K-K')v_{amb}t_n}{c}\right)\right]\exp\left[-j\frac{4\pi}{\lambda}\left(R_0+v_0t_n+\frac{1}{2}at_n^2\right)\right] \tag{9}$$

If and only if the $K'=K$ ,the RCM is totally eliminated (assuming that $\left|a(NT_r)^2/2\right|<\Delta r$ , $\Delta r$ is half of one range cell). And the efficient of target's energy integration can reach the max peak. However, due to the higher and higher speed of maneuvering target the searching times will be more and more big. Finally, the computation load will be very heavy and can't meet the instant processing requirements.

So we use the radon transform (RT) to get the real folded factor instead of the searching method. The definition of RT is as follows:

Creating a two-dimensional plane $f(x,y)$ and doing the integration along a line $l$ of $f(x,y)$. The integration in the $\rho-\theta$ domain can be expressed as [8]

$$R(\rho,\theta)=\iint f(x,y)\delta(\rho-x\cos\theta-y\sin\theta)dxdy \tag{10}$$

where $\rho$ is the minimum distance between the line $l$ and the origin of the coordinates in $f(x,y)$, $\theta$ is the angle between the line $l$ and the X axis and $\delta(x)$ is the Dirac's function.

It can be obtained from the (9) that the peak position between adjacent pulse in $\hat{t}$ domain is:

$$\Delta_p=\frac{2Kv_{amb}T_r}{c}=\frac{2KT_r}{c}\bullet\frac{1}{2}\lambda f_r=\frac{K}{f_c}s \tag{11}$$

So in the whole integration time, these peak positions construct a line with a slope $k$ can be represented as $k=Kf_s/f_c$. The slope $k$ which has the maximum value in the RT domain can be transformed to the folded factor $\hat{K}$. The estimation formula can be expressed as:

$$K'=\tan\left\{\arg\max_{\theta}\left|Radon\left(abs\left[s_{KT}\left(\hat{t},t_n\right)\right],\theta\right)\right|\right\}f_c/f_s \tag{12}$$

Then we can construct the compensation function $H(f,t_n;K')$ to make the target echo signal within the same original range cell.

*3.2. DFM correction and parameters estimation*

In the last section, we use the RadonKT method to eliminate the RCM and there only exists the DFM of the target echo signal. It can be obtained from (9) that the signal after KT along the slow time axis at $\hat{t}=t_0$ is a LFM signal. We can use the fast time-frequency analysis algorithm for LFM signal—LVT to correct the DFM and estimate the maneuvering parameters of target. As we can obtain the centroid frequency and chirp rate of LFM signal, we would get the velocity and acceleration of the maneuvering target.

Firstly, we defined the parametric symmetric instantaneous autocorrelation function(PSIAF) as follows [9]

$$\begin{aligned}R_0^C(t_n,\tau_n)&=s_{KT}\left(t_n+\frac{\tau_n+b}{2}\right)s_{KT}^*\left(t_n-\frac{\tau_n+b}{2}\right)\\&=A_3^2\exp\left[j2\pi f_0(\tau_n+b)+j2\pi k_0(\tau_n+b)t_n\right]\end{aligned} \tag{13}$$

where $\tau_n$ denotes a lag variable and b is a constant time delay related to a scaling operator.

And where [10]

$$f_0=f_d=\frac{v_0}{2\lambda}\quad k_0=\frac{2a}{\lambda} \tag{14}$$

It can be seen from the (13) that there exists the coupling between $t_n$ and $\tau_n$. The scaled transform is constructed to remove the coupling as follows

$$t_n=\frac{t_k}{h(\tau_n+b)} \tag{15}$$

where $t_k$ denotes the scaled time and $h$ is the scaled factor. Then we can have

$$R_0^C(t_k,\tau_n)=A_3^2\exp\left[j2\pi f_0(\tau_k+b)+j2\pi\frac{k_0}{h}t_k\right] \tag{16}$$

It can be seen from the above that the coupling has been eliminated. Usually, we set $b=1$ and $h=1$, then perform 2-D FFT on $R_0^C(t_k,\tau_n)$. We can obtain

$$\begin{aligned}L_0(f,\gamma)&=\underset{t_k}{FFT}\left\{\underset{\tau_n}{FFT}\left[R(t_k,\tau_n)\right]\right\}\\&=A_4\exp(j2\pi f)\sin c(f-f_0)\sin c(\gamma-k_0)\end{aligned} \tag{17}$$

where $A_4$ the amplitude after performing 2-D FFT.

It can be seen from the above that the energy of LFM signal is accumulated as a peak in the CFCR domain by using LVT method. And the estimated value of $v_0$ and $a$ can be obtained as

$$\hat{v}_0=2\lambda\hat{f}_0 \quad \hat{a}=\frac{1}{2}\lambda\hat{k}_0 \tag{18}$$

## 4. Computational complexity analysis

The computational complexity of the proposed method is analyzed in the follows in the lights of the number of complex multiplications.

Set that $N_r,N_p,N_v,N_a,N_k$ represent the number of range cells, pulse numbers, searching velocity numbers, searching acceleration numbers and searching folded factor numbers in KT compensation. For Radon-KT, the computation complexity [11] is $O(N_rN_p\log_2 N_p)$. For LVT, the computation complexity [12] is $O(N_p^2\log_2 N_p+N_rN_p)$. Thus, the total computation complexity of RadonKT-LVT is in the order of $O(N_rN_p^2\log_2 N_p)$. For GRFT proposed in [3], the computation complexity is $O(N_rN_vN_aN_p\log_2 N_p)$. As for the KT-FRFT, the computation complexity of FRFT in signal after KT without radon transform is $O(N_kN_vN_p\log_2 N_p)$. Hence, its total computation cost is $O(N_rN_kN_vN_p\log_2 N_p)$.

Under the assumption that $N_r=N_p=N_v=N_a=N$ and $N_k\log_2 N_p=N$, the computation cost of GRFT method is $O(N^4\log_2 N)$ and the KT-FRFT method has computation cost of $O(N^4)$. However, the computation cost of proposed method is $O(N^3\log_2 N)$. It can be seen from above that GRFT has more computation cost than proposed method due to the three-dimension searching process while the proposed method can remove RCM and DFM and the estimate the velocity and acceleration with only one-dimension searching in range dimension.

## 5. Experiments and Discussions

The radar parameters in the following simulation experiments are showed in Table 1

**Table 1** Radar Parameters

| **Simulation parameters** | Carrier frequency | Bandwidth | Pulse width | Sample frequency | Pulse repetition frequency | Pulse number |
|---|---|---|---|---|---|---|
| **values** | $1GHz$ | $2MHz$ | $400us$ | $2MHz$ | $800Hz$ | 1024 |

Suppose there is a point target with $1000m/s$ initial radial velocity and $30m/s^2$ acceleration. After adding Gauss white noise to the echo of target, the SNR before pulse compression (PC) is -20dB.

IOP Conf. Series: Materials Science and Engineering **608** (2019) 012006 doi:10.1088/1757-899X/608/1/012006

First of all, the long time integration ability of RadonKT-LVT for the target is evaluated in Figure 1. Figure 1(a) shows the target echo before PC which is completely buried in noise. The PC result is shown in Figure 1(b), which shows that RCM happens. Figure 1(c) shows the integration result along different polar angles. It can be seen from the figure that there is an obvious peak in the figure. Once we obtain the position of the peak, we can immediately estimate the folded factor according to (13). Then we can perform KT to correct the RCM as is shown in Figure 1(d). After eliminating the RCM, there still exists DFM. So LVT is performed to focus the energy and the result in CFCR domain is shown in Figure 1(e). We estimate the moving parameters of target by confirming the position of the peak in LVT result. The estimated velocity and acceleration are $1000.22m/s$ and $29.67m/s^2$. The result shows that the proposed method has high estimation precision of maneuvering parameters. As for comparison, the classic MTD method is used under the same environment shown in the Figure 1(f).

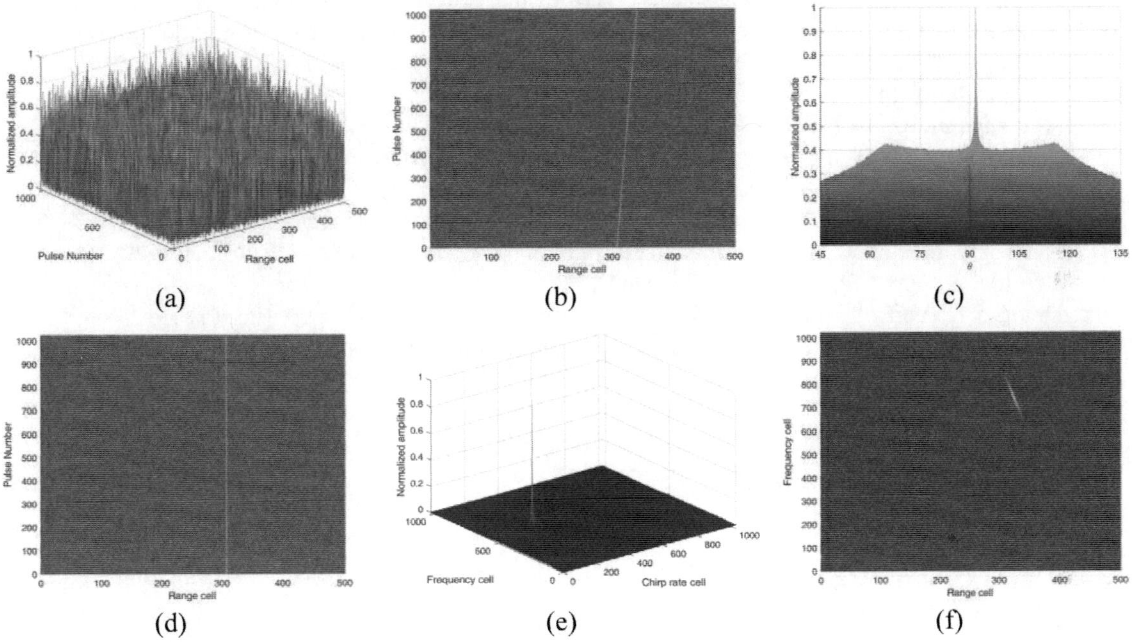


**Figure 1.** Simulation results based on RadonKT-LVT. (a) target echo before PC. (b) PC result. (c)Radon result. (d) KT result. (e) LVT result. (f) MTD result.

However, when the SNR before PC is dropped to -30dB, the integration result with MTD method can't detect the target in Figure 2(a) and the integration with RFT method still has poor detection result in Figure 2(b) with serious blind sidelobe phenomenon. As can be seen from Figure 2(c) that the proposed method can still have obviously better result.

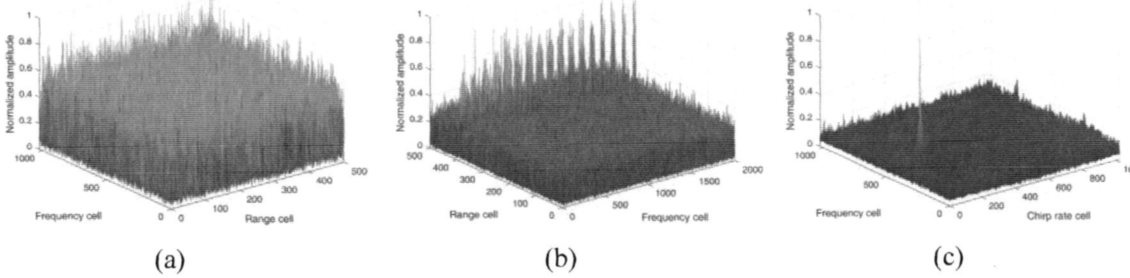


**Figure 2.** Simulation results of different methods. (a) MTD result. (b) RFT result. (c) RadonKT-LVT result

## 6. Conclusions

A long time integration approach combining the RadonKT and LVT (i.e., RadonKT-LVT) is presented to keep a balance between detection ability and computational cost of target with high speed and acceleration motion. The RCM of target echo can be effectively eliminated by using the improved KT based on RT. Then estimating the target's velocity and acceleration. Results of simulation experiments have verified the effectiveness of the RadonKT-LVT algorithm.

## References

[1] Perry R P, R C Dipietro and R L Fante 1999 SAR imaging of moving targets *Ieee.* T. *Aero. Elec.Sys.* **35(1)** 188-200
[2] Xu J, Yu J, Peng Y N and Xia X G 2011 Radon-fourier transform for radar target detection, i:generalized doppler filter bank *Ieee.* T. *Aero. Elec. Sys.* **47(2)** 1186-1202
[3] Xu J, Xia X G, Peng S B, Yu J and Qian L C 2012 Radar maneuvering target motion estimation based on generalized radon-fourier transform *Ieee.* T. *Signal. Proces.* **60(12)** 6190-6201
[4] Tian J, Cui W and Wu S 2013 A novel method for parameter estimation of space moving targets *Ieee. Geosci. Remote.* S. **11(2)** 389-393
[5] Li X, Cui G, Yi W and Kong L J 2015 Coherent integration for maneuvering target detection based on radon-lv's distribution *Ieee. Signal. Proc. Let.* **22(9)** 1-1
[6] Barton D K 2005 Radar system analysis and modelling *Ieee. Aero. El. Sys. Mag.* **20(4)** 23-25
[7] Bai X, Zhou F, Xing M and Bao Z 2012 A novel method for imaging of group targets moving in a formation *Ieee. Geosci. Remote. S.* **50(1)** 221-231
[8] Deans S R 1983 The radon transform and some of its application *Wiley.* **223(1)** 3-4
[9] Lv X, Bi G, Wan C and Xing M 2011 Lv's distribution: principle, implementation, properties, and, performance *Ieee.* T. *Signal. Proces.* **59(8)** 3576-91
[10] Li X, Sun Z, Yi W, Cui G and Kong L 2018 Radar detection and parameter estimation of high speed target based on mart-lvt *Ieee. Sens.* J. **PP(99)** 1-1
[11] Lanari R 1995 A new method for the compensation of the sar range cell migration based on the chirp z-transform *Ieee.* T. *Geosci. Remote.* **33(5)** 1296-99
[12] Li X, Cui G, Yi W and Kong L 2015 Coherent integration for maneuvering target detection `based on radon-lv's distribution *Ieee. Signal. Proc. Let.* **22(9)** 1-1

# A Fast and Collision Avoidance Distributed TDMA Schedule Based on the Multi-Arms Bandit

**ChaoYi Zheng[1,*], ShengChun Huang[1] and TaiLi Li[1]**

[1]College of Electronic Science, National University of Defense Technology, Changsha 410073, China

[*]Corresponding author : zcy754420194@163.com

**Abstract.** In this paper, we propose a novel distributed TDMA protocol based on the Multi-Arms Bandit model for the dynamic UAVs environment. Because of the frequent topology changes of UAVs, we consider a distributed communication protocol, which do not require the maintenance of accurate network topology information. Through the adaptive interaction between nodes, perceiving the surrounding topology environment and learning from historical experience, each node in the network can transmit information in a fast collision-free way. Also, the MAB model is utilized in our protocol, so that the time slot can obtain sufficient multiplexing rate through multiple rounds of node selection. Experiments show that the protocol can achieve better throughput and fast enough convergence speed, even in the case of high network density, and scales well with the size of the network.

## 1. Introduction

In recent years, Unmanned Air Vehicles(UAVs) have developed rapidly, and their applications have been utilized from civilian to military. Because of the complex environment, sending and receiving data requirements, and the long fly distance, the topology of the Unmanned Air Vehicles(UAVs) swarm changes so quickly that we have a high demand for resource scheduling, which needs fast convergence and stabilized. To coordinate the access of the shared wireless resources, a MAC protocol is employed.

MAC design aims at optimizing the performance of communication by formulating the strategies sensor nodes use to access the common channel [1]. At present, the distributed TDMA protocol applies universally for UAVs swarm, which can be broadly divided into two categories : topology-transparent scheduling schemes and topology-based scheduling schemes. The nodes with topology-based scheduling schemes need network topology information to schedule schemes [2][3], and it will cause plenty of computation and communication overhead, which likely consume UAVs excessively. Therefore, topology-transparent scheduling schemes attracted considerable research interest.

Rhee et al. [4] proposed the DRAND scheme which is a randomized distributed time slot scheduling algorithm. It allocated the time slot sequence by coordinating requests among network node. However, the completely randomness decreased the efficiency, and also resulted in the high collision rate of the message in the distribution process. Based on the DRAND, LI et al. [5] made improvement and proposed the E-T DRAND which allocates time slot through the remainder energy for each node, the results show that there is a certain improvement in the scheduling rate and an appropriately decrease in the overhead.

And other researchers discuss mathematics models to schedule scheme and attempt to get a collision-free allocation through probability analysis. In [6][7], Farago et al proposed a topology-transparent algorithm based on Polynomial function. The algorithm's performance depends on the number of nodes in the network and the maximum degree, i.e. the number of neighbors that each a node can have. In [8], Ju and Li improved the algorithm in order to maximize the minimum throughput. They used the Hamming weight and the Hamming distance in order to describe the relationship between the transmission slot assignments of two nodes. Liu et al. [9] proposed a topology-transparent algorithm, i.e. m-MPR-l-code algorithm. The results showed that the algorithm had a certain improvement on the performance of resource scheduling. The polynomial-based approach can achieve considerable improvements in efficiency and robustness in mobile environments [10] by taking advantage of the multi-packet reception capability. However, these approaches introduce complicated polynomial-based algorithms and an important problem is that it exits redundant slots not to be utilized efficiently.

As machine learning developing rapidly, plenty of researchers combined the allocation with Machine Learning and Reinforcement learning models, which is very popular recently. In [11], Liu et al proposed a reinforcement learning model named the Multi-Armed Bandit with Multiple players. And the RL-Mac protocol [10] used the Markov decision to infer the state of other nodes, the results show that the protocol can achieve a high throughput, but the computational complexity and power consumption are some high. Qiao et al. [12] combined the polynomial function with reinforcement learning and gradient descent model, which chooses the slot and reduce redundant time slots. The theoretical performance of the algorithm is better than [13], and there is a certain improvement in throughput and convergence speed.

In the paper, we propose a novel time slot algorithm based on Multi-Armed Bandit model so that each node can be allocated with a collision-free time slot. Considering the surrounding environment and interaction of each node, we enable each node to learn from history knowledge to better performance on convergence and scalability. Our main contributions consist in :

1. We propose a mac protocol based on the MAB model, which is a distributed, modelless, online learning algorithm. In this algorithm, each node only feeds back its own local information, which can reduce the convergence time of the algorithm and acheive a better throughput.
2. A large scale dynamic simulation network is implemented on the matlab, and we compare the algorithm with other algorithms from the aspects of convergence speed and throughput to evaluate the overall scheme performance.

## 2. System model

### *2.1 Network Model*

We consider a wireless communication network that contains M nodes, and the initial density of the network is medium. each node does not need to support multi-packet reception and just knows its own address and the sum of time slots. The network can be represented by an directed graph G(V,E), where V and E are the sets of nodes and edges respectively to indicate some nodes are connected. The degree of a node u, D(u), is defined as the number of its interference neighbors. The maximum node degree $D_{max}$ is defined as $D_{max}$ = max D(u). We assume that $D_{max}$ is smaller than the number of nodes N and remains constant while the network topology changes. In our paper, time is divided into frames of fixed duration, each one consisting of N time slots. At the beginning of each frame each node randomly chooses one of the N available slots to transmit its packet and start the algorithm we propose. Also, We propose a definition referred to as Priority Factor(Pr): $Pr = F(\alpha Dv - \beta Nv)$ , where Nv is the number of time slots occupied by nodes v, $\alpha$ and $\beta$ are coefficients($\alpha, \beta \in [0,1]$). According to the topology, each node will have their own Pr to have positive influence on network 's business and throughput, it will be mentioned in the follow section. We assume that the transmission channel is error-free and the only reason for unsuccessful transmission between nodes is due to

IOP Conf. Series: Materials Science and Engineering **608** (2019) 012007 doi:10.1088/1757-899X/608/1/012007

collisions, for a transmission from node v to its neighbor in slot t, a collision occurs when another node which is within two-hop neighbors range of node k simultaneously intends to transmit in slot t.

*2.2 The MULTI-ARMs BANDIT Model*

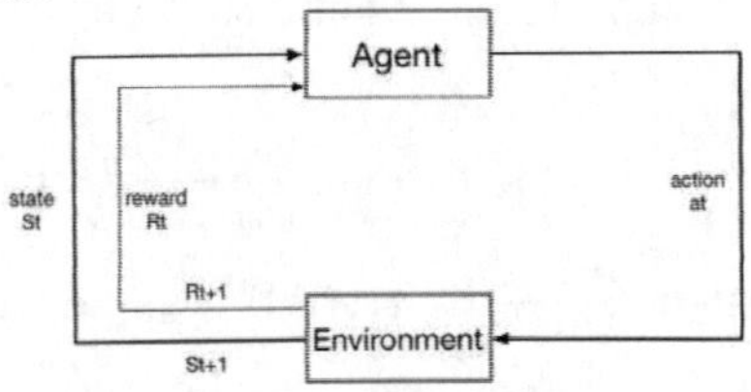


**Figure 1**. the elements of reinforcement learn

Multi-Arms Bandit(MAB) model is essentially a kind of reinforcement learning model(Fig.1). In the classic multi-armed bandit (MAB) problem, there are N independent arms and a single player. Playing arm i (i = 1, · · ·, N) yields i.i.d random rewards with a distribution parameterized by an unknown $\theta_i$. At each time, the player chooses one arm to play, aiming to maximize the total expected reward in the long run. Under an unknown reward model, the essence of the problem lies in the well-known tradeoff between exploitation and exploration.

Besides the single player MAB, a multi-armed bandit with multiple players is put forward. Support that there are N multi-arms bandits(MAB) and players will get reward with an unknown probability while they pull one arm to take N kinds of action. At each time, a player chooses one arm to play based on its local observation and decision history. Players do not exchange information on their decisions and observations. Collisions occur when multiple players choose the same arm, which cause that all players can get positive or negative rewards with different probabilities.

In order to find the optimal way, exploration and exploitation are considered inevitably in the MAB model. Exploration means that the player tries all the arms as much as possible, which makes the stability of the reward poor, and exploitation means that once the player finds the right arm, he will not continue to try, which makes the overall income poor. In the MAB model, for each player, initial probability of arms is set by softmax function :

$$x(a) = \frac{e\hat{}H(a)}{\sum_{b=1}^{k} e\hat{}H(a)} \tag{1}$$

where H(a) is the priority factor of every arm for player a and k is the maximum number of time slots that nodes can choose. We also set a variable e=1/t where t is running round. Hence, player can take some action, namely choose an arm with the probability $p_{max}$ and $p_i$

$$p_{max} = \frac{1-e}{n} \tag{2}$$

$$p_i = \frac{e}{(N-n)} \frac{xi}{1 - xmax * n} \tag{3}$$

where n is the number of maximum probability arm.

The expectation that the player choose the arm b at t time can be shown as:

$$E[Rt] = \sum_b \pi t(b) q_*(b) \tag{4}$$

where $\pi t(b)$ indicates the probability of choosing arm b at t time. Hence the expectation of the reward is :

$$\frac{\partial E[Rt]}{\partial Ht(a)} = \frac{\partial}{\partial Ht(a)} \left[ \sum_b \pi t(b)\, q_*(b) \right]$$

$$= \sum_b q_*(b)\frac{\partial \pi t(b)}{\partial Ht(a)} = \sum_b \big(q_*(b) - Xt\big)\frac{\partial \pi t(b)}{\partial Ht(a)} \tag{5}$$

Let the random variable be At, and change Xt and $q_*(b)$ to $\overline{Rt}$, which is the average reward, and Rt. Therefore, the expectation becomes :

$$= E[(Rt - \overline{Rt})\frac{\partial \pi t(At)}{\partial Ht(a)}/\pi t(At)] \tag{6}$$

according to the softmax function, $\frac{\partial \pi t(At)}{\partial Ht(a)} = \pi t(At)[1_{a=At} - \pi t(a)]$, and we get:

$$= E\{\frac{(Rt - \overline{Rt})\pi t(At)[1_{a=At} - \pi t(a)]}{\pi t(At)}\} \tag{7}$$

Hence, the reward function can be shown as:

$$H_{t+1}(a) = H_t(a) + \alpha E\{(Rt - \overline{Rt})[1_{a=At} - \pi t(a)]\} \tag{8}$$

where $\overline{Rt}$ is the average of the sum of previous R, and $\alpha$ is a variable that can influence the convergence. In this paper, we set $\alpha$=0.5 and it seems to be great in the experience. After repeated cycles, player keeps learning the environment information and update his reward, eventually an optimal way will be found.

## 3.Alogrithm

### *3.1 Alogrithm Description*

In the distributed TDMA system, the node select different time slots with its two-hop neighbors to make the transmission get rid of collision. Before utilizing the algorithm, each node is in the unknown environment and only knows the sum of time slots, i.e. arms, and stores their own Pr, current occupied time slot, and the number of neighbors, where the number of neighbors is collected and updated by receiving ack from other nodes each round. According to the softmax function, every node will get a H factor to calculate the probability. Initially, the probabilities of all arms for every slot are the same.

Each node which can be seen as an independent player follows the MAB model rule every round, and the time slot can be seen as the arm. But what we should notice is that the reward received by taking action is related to other node's action. For example, node u selects slot k in the s state and its neighbors may also select this slot, so collision will be caused which influences the selection of node u in the s+1 state. Hence, the process is like a game that finding the balance among all nodes without a negotiate.

*3.1.1 the First Round.* The reward calculation algorithm of the first round is somewhat different from the other rounds, so we introduce it firstly.

All nodes store the maximum number of ack packets receiving from other nodes which we define as History, and the History factor is set to 1 initially. In order to avoid falling into a local optimum trap, we definite an exploration factor e as 1/t (t is current cycle). The factor is beneficial to the exploration of the entire arms, which can make the node explore more node in the first few cycles. We stipulate that the reward equals 1 while receiving ack packets of node u is more than the number in the History, and updating the History value ; the reward equals -3 while receiving ack packets is less than the number in the History, and there must be some collisions for node u in this situation. If receiving ack packets is the same as the number in the History, reward equals 0. When node u receives some hello packets from its neighbors, it collects these neighbors Pr and compares with its own Pr. If node u's Pr is larger than some of these neighbors, the reward still equals 1 while node u does not receive hello packets from the neighbors whose Pr is less than u.

We stipulate that node u will store the probabilities while the number of receiving ack packets equals to the History value during k cycles (k is related to the density of the network), and improve reward to 5. If the node receives ack packets which is more than History at some later moment, changing the probabilities to the previous one and update the History. Follow the above steps, every node iterates in the order of state, action and reward. We set thresholds T1 and T2.

1. If a probability is more than T1, the node converges to the time slot
2. If a probability is less than T2, the probability turns to 0 and averages its probability to other time slots.

Figure 2 shows the relationship between convergence speed and average throughput with different R.

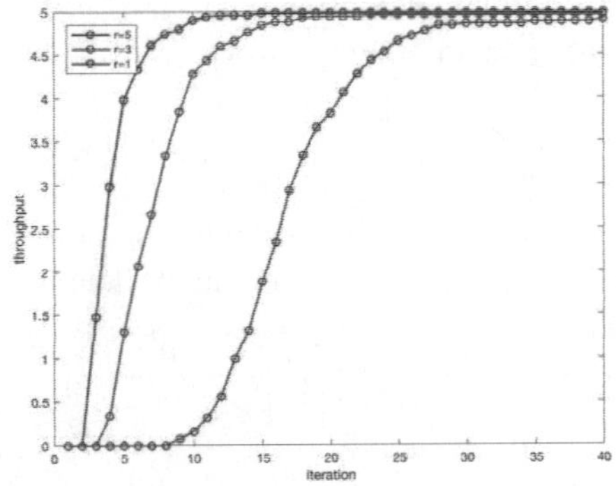


**Figure 2**. the comparison of M-TDMA with different reward

We can see from Fig. 2 that when the positive reward is set to 5, convergence is fastest, and as the setting reward falling, the convergence speed become dramatically slow. But we should notice that if the reward is set too high, there may more than two nodes' probabilities reach the threshold at the same cycle and make mistakes.

*3.1.2 The Following Round.* For each node, when they converge, this means that each node occupies one-time slot without collision, and knows the time slots occupied by its one-hop and two-hop neighbors and the Pr of its one-hop neighbors. According to these information, each node knows its unoccupied time slots. We can think that each node has built a table for the next round distribution.

When the node u has occupied a time slot after the first round, it will send hello packet or data packet in the time slot and listen to other nodes' hello packets so that it can ensure if its neighbors converge. If after k cycle neighbors' occupied slots do not change, node u can cognize its unoccupied time slots and starts the next round. In the next round, for each node we consider to make some changes about the reward algorithm. Since node u has known the number and Pr of its neighbors, taking action as mentioned above : sending hello packet to its neighbors, and comparing the number of ack packets with History.

1. If ack=History, which means there is no collision during the time slot, R=3 ;
2. If ack<History, which means there happened some collisions, R=-3 ;

Furthermore, node u will compare its own Pru with Prn receiving from neighbors in the same slot.

1. If Pru<Prn, the node will not choose this time slot in the next cycle.
2. If Pru>Prn, there will be no effect for the node.

Based on the rule, when there is no enough time slot for all nodes, the nodes with less Pr will be not allocated. The less Pr node has, the less slot node occupies. We also set a value to Pr so that if a node occupied enough slot, it will quit playing, which is conducive to the overall throughput and convergence speed. The thresholds are the same and if there are still some time slots remaining unoccupied, the algorithm will continue until the throughput meet our demand.

*3.2 Throughput Analysis*

If we consider the time slot allocation as a probabilistic problem, it is necessary to estimate the the normalized average node throughput by the proposed time slot allocation algorithm.

We reckon that node u only transmits information packet to its neighbor node v in a time slot in the distributed TDMA. Suppose that there are s time slots in a frame. Therefore, the minimum guaranteed node throughput can be expressed as :

$$T_{min} = \frac{s - kD}{s} \tag{9}$$

where k is the maximum number of collisions between two nodes in the communication rage. s-kD is the minimum number of successful transmissions for each node during one frame. kD is the number of packets which two nodes in the communication rage fail to transmit.

In a round, the probability that node u collides with at least one neighbor is :

$$P_{uc}(l) = \prod_{k=1}^{l} \frac{s^k - k}{s^{k+1} - k}, l \leq D \tag{10}$$

where l is the number of neighbors selecting the same slot with node u.

The probability that other neighbors do not collide with node u is:

$$P_{uc}f(D-l) = \prod_{k=1}^{D-l} \frac{p^{k+1} - p^k - k + l}{p^{k+1} - k + l}, l \leq D \tag{11}$$

Therefore, by jointly considering (10) and (11), the probability that node u collide in a time slot is :

$$P_u = \sum_{l=1}^{D} \binom{D}{l} (P_{uc}(l)) (P_{uc}f(D-l)) \tag{12}$$

Based on the above functions, the normalized average node throughput is:

$$\mathrm{T} = \frac{1}{N} \sum_{i=1}^{N} (1 - P_{ui}) \tag{13}$$

*3.3 Theoretical Analysis*

*Theorem 1*: The M-TDMA algorithm allocates time slots without collision.

*Proof*: In the protocol, we rule that a node will occupies a time slot while its probability of the time slot is bigger than threshold T. Based on the algorithm, if its two-hop neighbors send hello in this time slot, they will be received negative reward and choose other time slots with larger probabilities. Therefore, it can ensure that any pair of two neighbor nodes does not select the same time slot.

*Theorem 2* : The M-TDMA algorithm can converge.

*Proof :* Initially, each node is not sure which time slot is occupied, so selection is random. For a node i, once it has been in the broadcast process without collision in a certain period, it will confirm its maximum ack number. Therefore, node i can get accurate and large enough reward to take the next action. If there are k two-hop neighbors and n time slots in the network, the probability that node i broadcasts without collision is: $(\frac{n-1}{n})^k$, and the probability of collision-free broadcast at least once in t cycles is:

$$p_{no\_collion} = 1 - [\frac{n^k - (n-1)^k}{n^k}]^t \tag{14}$$

We can see that if n and k is large enough, $p_{no\_collion}$ approaches 1. And the probability of node i selecting the same time slot with a two-hop neighbor in a certain period is :

$$\frac{1}{n} \sum_{i=1}^{k} {}_d^i (pi)^i (p'i)^{k-i} < \frac{1}{n} \sum_{i=1}^{k} {}_d^i (\frac{1}{n})^i (\frac{1}{n-1})^{k-i} \tag{15}$$

which is similar to $\frac{1}{n^3}$. If without convergence, the probability is $\frac{1}{n^{3k}}$ in the k period.

Therefor, as the density of the network increasing, the unconverged probability of node i is similar to 1. In the meanwhile, the convergence is like to flood. It starts from the node with least neighbors generally, because the k is less and get its acknumber faster so that it is more likely to converge. And after this, other nodes can occupy time slots with less cycles. There is no interaction process, so the sooner as many nodes converge, the faster the overall convergence can be achieved.

*Theorem 3* : Priority factor(Pr) can coordinate the network time slot allocation.

*Proof* : If there are too many edge nodes, the convergence speed will be slowed down. It is necessary to balance the throughput and convergence speed, and the Pr makes a compromise to coordinate the throughput and convergence speed. The Pr is related to the number of two-hop neighbors and occupied time slot by itself. When the Pr of edge nodes reaches a certain value, they quit to play. Also, in the case where the number of time slots is less enough for the number of nodes, the node with a larger Pr can occupies the slot preferentially, and the node with least Pr will back off. The outcome is obvious that more time slots will be occupied by the center nodes averagely, which has more neighbors, i.e. more Pr, this can benefit the whole network business.

## 4.Evaluation

In the paper, we evaluate the performance of the proposed distributed tdma time slot allocation algorithm based on MAB model, and get the simulation result and model performance. As is shown in Fig.3, the topology of the network consists of 200 nodes which are randomly distributed in a 300 × 300m plane, the line represents there is a link between the two nodes in the presented network. Firstly, we compare the M-TDMA algorithm with DRAND and E-T DRAND. In order to reach the standard, the number of time slots is equal, and is approximated to the minimum number of slots, so that comparing the convergence speed.

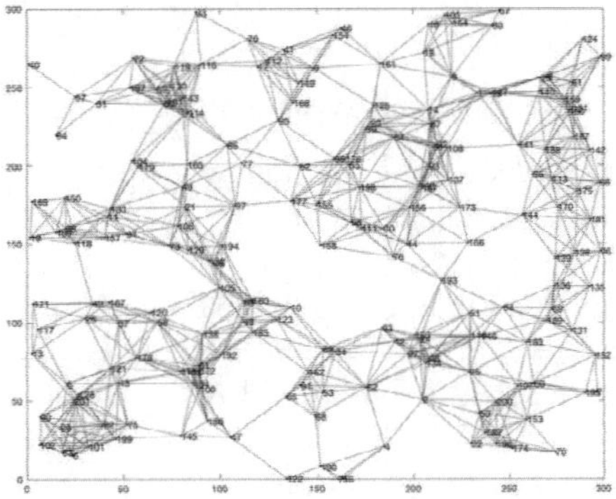

**Figure 3**. the topology structure of simulation

Fig. 4 shows the comparison of the average number of cycles where three algorithms success to allocate the slot based on different maximum number of neighbors. We can observe that when the number of neighbors increases, the collisions during allocation will increase, and their trends are similar. As D increases, the number of cycles required for successful time slot allocation increases. The number of cycles required by the DRAND algorithm is the most, and E-T DRAND is an improved algorithm, which the number of cycles required is less. The cycles they need is observably decrease when D is about 30, and then continue to increase. Observably, compared with the other two protocols, the M-TDMA needs less cycles to converge, and increase in a linearly similar way. Although when D is 30, the M-TDMA is a bit larger than E-T DRAND, its performance is better than E-T DRAND and DRAND on the whole.

IOP Conf. Series: Materials Science and Engineering **608** (2019) 012007 doi:10.1088/1757-899X/608/1/012007

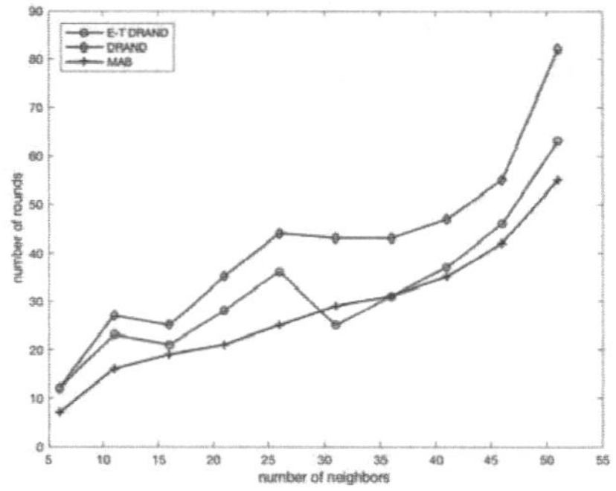


**Figure 4**. the average number of rounds for successful allocating time slot in different size of neighbor nodes

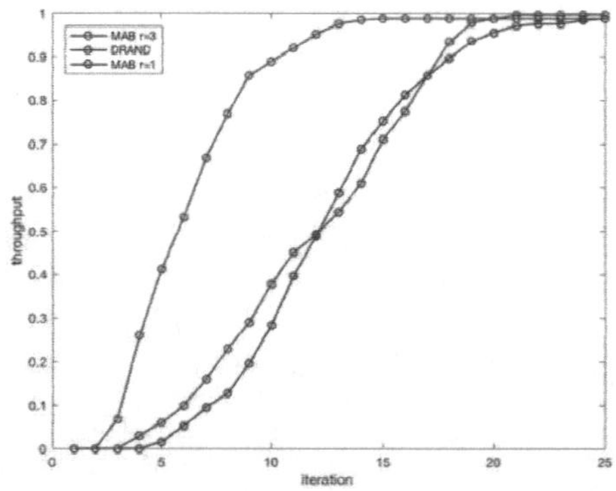


**Figure 5**. the achieved normalized average node throughput in 25 iterations

Fig. 5 performs a statistical analysis on the the achieved normalized average node throughput of protocols within 25 iterations, which the grey line represents the M-TDMA protocol with r=1 initially and the green line represents the MAB protocol with r=3 initially. Through the three curves in the picture, we can observe that during the 15 iterations, the M-TDMA protocol with r=3 raises fastest and its throughput can reach almost 1, i.e. the optimal throughput. The throughput of M-TDMA protocol with r=1 can only reach 0.75, and the DRAND protocol is the worst, which only reaches 0.7. After th 15th iteration, M-TDMA with r=1 and DRAND lead to rise alternately, and in the 20th iteration, the throughput of both of them approaches the maximum. We can observe that when setting a suitable reward, the M-TDMA with a great r can achieve the approximate optimal throughput at a fast rate., and even if the value r is not suitable initially, the throughput of M-TDMA is similar with DRAND. Hence, in the case of high-speed dynamic topology, M-TDMA is better than DRAND.

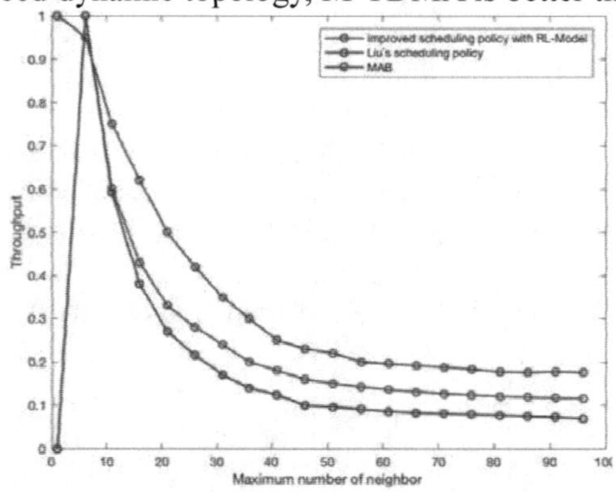


**Figure 6**. The achieved normalized average node throughput versus the maximum node degree.

Fig. 6 shows that the throughput comparison of three protocols which utilize reinforcement learning to allocate the time slot as maximum number of neighbor increasing. When there are more neighbors, the network topology will be more complicated, and the probability of collision raise concomitantly. Hence the throughput of three protocols all drops and stabilizes eventually. However, the performance of MAB protocol is better than the other two apparently. This finding can be explained as follows : the multiple rounds of iterative mechanism of MAB, redundant time slots can be utilized better than the other two protocols, so that the M-TDMA can avoid collision and achieves a larger throughput.

## 5.Conclusion

For swarm UAVs, dynamic topology represents a large amount of challenges to the TDMA schedule scheme, and the conventional algorithm can not solve perfectly. In this paper, we propose the M-TDMA protocol based on the MAB model, which learns from environment by local adaptive between nodes, avoids collision and makes full use of slots through multiple rounds of allocation. Also, we utilize the Pr to accelerate the converge. The experimental results show that the algorithm provides a better throughput and significantly improves the convergence speed.

## References

[1] Nisioti E, Thomos N. Fast reinforcement learning for decentralized MAC optimization[J]. *arXiv preprint arXiv*:1805.06912, (2018).

[2] Omar H A, Zhuang W, Li L. VeMAC: A TDMA-based MAC protocol for reliable broadcast in VANETs[**J**]. *IEEE transactions on mobile computing*, (2013).

[3] Alinaghian M, Ghazanfari M, Norouzi N, et al. A novel model for the time dependent competitive vehicle routing problem: Modified random topology particle swarm optimization[J]. *Networks and Spatial Economics*, (2017).

[4] Rhee I, Warrier A, Min J, et al. DRAND: Distributed randomized TDMA scheduling for wireless ad hoc networks[J]. *IEEE Transactions on Mobile Computing*, (2009).

[5] Li Y, Zhang X, Zeng J, et al. A distributed TDMA scheduling algorithm based on energy-topology factor in Internet of Things[J]. *IEEE Access*, (2017).

[6] Chlamtac and A. Farago, ''Making transmission schedules immune to topology changes in multi-hop packet radio networks,'' *IEEE/ACM Trans. Netw.*, **vol. 2**, no. 1, pp. 23–29, Feb. (1994).

[7] J.N.TsitsiklisandB.VanRoy,''Ananalysisoftemporal-differencelearn- ing with function approximation,'' *IEEE Trans. Autom. Control*, **vol. 42**, no. 5, pp. 674–690, May (1997).

[8] Ju J, Kim D H, Bi L, et al. Four-color DNA sequencing by synthesis using cleavable fluorescent nucleotide reversible terminators[J]. *Proceedings of the National Academy of Sciences*, (2006).

[9] Liu Y, Li V O K, Leung K C, et al. Topology-transparent scheduling in mobile ad hoc networks with multiple packet reception capability[J]. *IEEE Transactions on Wireless Communications*, (2014).

[10] Y. Liu, V. O. K. Li, K.-C. Leung, and L. Zhang, ''Performance improve- ment of topology-transparent broadcast scheduling in mobile ad hoc networks,'' *IEEE Trans. Veh. Technol.*, **vol. 63**, no. 9, pp. 4594–4605, Nov. (2014).

[11] Liu K, Zhao Q. Distributed learning in multi-armed bandit with multiple players[J]. *IEEE Transactions on Signal Processing*, (2010).

[12] Liu Z, Elhanany I. RL-MAC: A QoS-aware reinforcement learning based MAC protocol for wireless sensor networks[C]//*Networking, Sensing and Control, 2006. ICNSC'06. Proceedings of the 2006 IEEE International Conference on. IEEE*, (2006).

[13] Y. Liu, V. O. K. Li, K.-C. Leung, and L. Zhang, ''Topology-transparent scheduling in mobile ad hoc networks with multiple packet recep- tion capability,'' *IEEE Trans. Wireless Commun.*, **vol. 13**, no. 11, pp. 5940–5953, Nov. (2014).

# A Modified KNN Indoor WiFi Localization Method With K-median Cluster

**Wei Lan**[1] **and Hongxin Li**[1]

[1]School of Information Science and Engineering, Lanzhou University, Lanzhou, 730000,China

E-mail :396820924@qq.com

**Abstract:** Because of the serious attenuation and multi-path effect of GPS signal, outdoor-positioning technology can not be applied in complex indoor environment. Through the study of K-Nearest Neighbor applied in WiFi positioning, according to the problem that the time complexity of KNN algorithm increases linearly with the quantity of samples, this paper combined clustering algorithm with KNN optimized the similarity measure in fingerprint feature space and proposed a efficient indoor target location algorithm . Experimental results showed that the algorithm improved the positioning accuracy, had strong robustness to noise and more importantly, the positioning time was effectively shortened and it can meet the requirements of practical applications.

## 1. Introduction

Positioning technology in indoor environment is widely used in logistics management, smart home, modern industry, etc. GPS target location and tracking technology in outdoor environments is developing rapidly and with high positioning accuracy. However, GPS is not well suited for use in closed environments due to limited signal propagation. As a result, a large number of new positioning methods have emerged in this field. The infrared positioning and tracking technology in the literature [1] is susceptible to external interference, and the infrared radiation distance is short, which is not suitable for indoor environments. Although the ultrasonic positioning tracking proposed by the literature [2] has high tracking accuracy and wide positioning range, it is not put into use a lot due to high cost, and it has certain harm to human health for a long time. Document [3] RFID positioning technology proposed by expensive hardware also failed to be accepted by the public. WiFi location tracking technology is a technology that uses WiFi hotpots for target location and tracking. It can locate existing infinite LANs in indoor environments. The existing smart devices are generally equipped with WiFi functions, which do not require additional hardware costs, and the accuracy can meet the requirements, which has been widely concerned by researchers at home and abroad.

WiFi positioning can be divided into three types: (1) Location-based radio wave arrival time; (2) Location-based radio wave arrival angle; (3) Intensity localization based on wireless wave arrival signal[4]. Considering the difficulty of angle and time measurement during radio wave propagation, this paper only discusses the location tracking algorithm based on received signal strength(RSS). Depending on the algorithm used, positioning techniques can be divided into two categories: Positioning based on a signal propagation model and the position location fingerprinting algorithm. The former is modeled through the channel of the spread of WiFi, but the complexity and diversity of the indoor environment is often difficult to achieve. The latter is by constructing the signal strength and location information database match, the best match and the predicted position by the evaluation function, the feasibility of the former is far superior.

The existing KNN method has two problems: large computational complexity and sensitivity to RSS signal noise. Aiming at these two problems, this study builds a fingerprint database, selects the k-median clustering algorithm to cluster the samples, reduces the algorithm overhead, optimizes the similarity measure, and proposes a WiFi localization algorithm based on k-median-WKNN. Experiments were conducted in the laboratory to further compare and analyze the positioning results.

## 2. WiFi fingerprinting algorithm based on location fingerprint

### *2.1. RSS*

RSS is defined as: received signal strength [4-7]. RSS is generally negative, and the stronger the signal, the larger the value. Ideally, the measured RSS values determined under the same general location of the device is substantially unchanged. Because of the complexity of the radio propagation, the same position measured RSS values there is a large deviation of the actual application, and the change with time of the RSS value is also evident, but there may be missing signals in the plurality of sets of data. The arrow in Figure 1 shows a large error point.

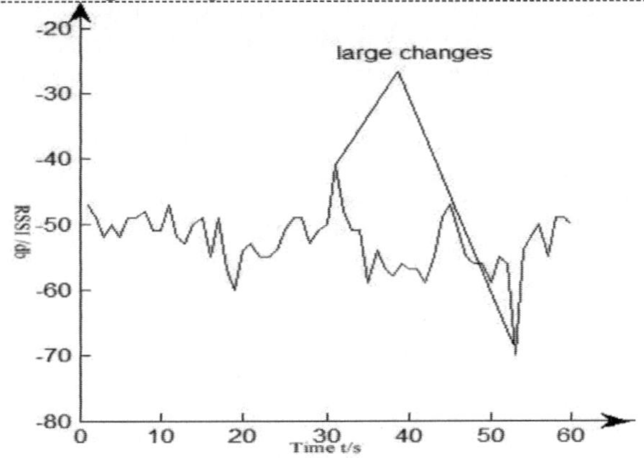


**Figure 1**. RSS distribution over time

### *2.2. WiFi localization algorithm based on K-nearest neighbor method*

K-nearest neighbor method is widely used in the field of data analysis. The algorithm to find the minimum distance from the target sample with the *k* reference points in the feature space. The properties of the target sample are determined by discriminating the properties of the K reference points. Assume that the experimental site selects $\theta$ WiFi hotpots. A total of $\alpha$ samples are collected at each location to form a location fingerprint database, and the data training set is:

$$K = \{(r_1, p_1), (r_2, p_2), \cdots, (r_\alpha, p_\alpha)\} \tag{1}$$

Vector $r_i = (r_{i1}, r_{i2}, \cdots, r_{i\theta}) \in R^\theta$ in the formula means that RSS signal measured by $\theta$ access points. $r_{i1}$ represents the data collected from the first AP. $r_{i2}$ represents the data collected from the second AP. From this analogy, it can be concluded that $r_{i\theta}$ represents the data collected from the $\theta$ Vector $p_i = (x_i, y_i) \in R^2$ represents the corresponding position vector, where $i = 1, 2, \cdots, \alpha$.

During the online positioning phase, a set of RSS signals can be measured at the measurement location. That is, the *r* vector is known, and the position P vector is estimated. First, through Equation 2, *k* reference points closest to the distance of the *r* vector in the location fingerprint library feature space are calculated. The distance between the *r* vector and the training sample $r_i$ is defined as:

$$d = [\sum_{\beta=1}^{\theta} \left| r_{t\beta} - r_{i\beta} \right|^{q}]^{\frac{1}{q}} \tag{2}$$

Where $q$ is a positive integer and Manhattan distance when $q=1$. Euclidean distance when $q=2$. Text uses Euclidean distance to select neighbors. $(x_{k1}, y_{k1}),(x_{k2}, y_{k2}),\cdots,(x_{kk}, y_{kk})$ are its coordinates. Taking the average of these coordinates to get the positioning result is $(x, y)$ .As can be seen from the above, the KNN algorithm needs to calculate the Euclidean distance for each training sample, which does not need to be considered for the small sample set, but when the sample size of the fingerprint database reaches a certain level, the complexity of the algorithm increases, and the practical application is therefore limited.

## 3. Improvement of WiFi Location Algorithm Based on K-median-WKNN

### *3.1. Algorithmic process*

The improved algorithm in Figure 2 corresponds to two processes: the offline learning phase and the online positioning phase. In the offline phase, by collecting the RSS signal and position coordinates, after the data is cleaned, the location fingerprint database is constructed, and then the samples are K-median clustered, and each sample is grouped into one cluster. In the online reasoning stage, the class to which the verification sample belongs is first defined, and the sample points are weighted and averaged to obtain the observation position.

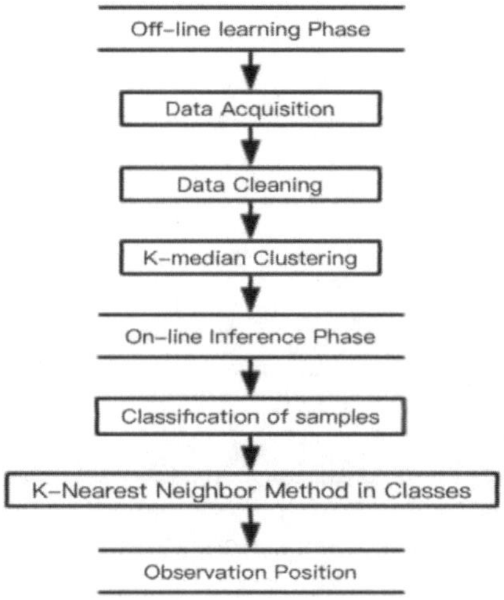


**Figure 2**. Algorithm processing flow

### *3.2. Selection of clustering algorithm*

The K-means clustering algorithm proposed in [5] can quickly cluster the location fingerprint database of the localization area. However, in the WiFi positioning process, the method of averaging the RP point RSS information to update the class center vector often has a large error. For example, the RSS value collected by the same AP at a certain location is(-41,-42,-42,-43,-89). Due to the presence of noise points (outliers), and the mean value of -51 with the actual RSS there is a large variation, due to the complexity of the indoor environment, such noise often unavoidable point.

The K-medoids clustering proposed in [6] updates the class center vector by selecting the point with the smallest extinction error of each sample point distance in the cluster, which avoids the distortion of the class center vector caused by the noise point, and the corresponding cost is It is an increase in time complexity. Considering the particularity and comprehensive requirements of indoor location fingerprint location, this study selects K-median clustering algorithm. The difference between this algorithm and K-means is that when updating the class center vector, the mean value is used instead of the mean value to avoid the noise point. Distortion, while inheriting the fast and simple features of K-means.

The approach to clustering is to measure the similarity between the assignments assigned to each cluster to minimize a specified cost function. The cost function in this experiment is:

$$J=\sum_{n=1}^{\alpha}\sum_{k=1}^{\delta}\left\|r_n-E_k\right\|^2 \tag{3}$$

Where $r_n$ is the RSS vector for each location point, $E_k$ was $r_n$ center vectors of their class. $k$ is the number of clusters. $n$ represents the number of RSS vector samples. The essence of K-median lies in the problem of minimizing the total cluster variance $J$ by continuous classification. The selection of the cluster number K is particularly important.

### *3.3. Positioning algorithm improvement*

*3.3.1 Offline learning phase.* In the offline stage, first of all training data K-median clustering. Assume that the number of samples in the fingerprint database is $\alpha$, each sample point receives the RSS value from $\theta$ APs, divides $\alpha$ reference points into $\delta$ clusters, and defines reference point $c$ to receive the RSS value from the $\theta$ AP as $r_{c\beta}(c=1,\cdots,\alpha;\beta=1,\cdots,\theta)$. The degree of dissimilarity between sample point $a$ and sample point $b$ is defined as:

$$d=[\sum_{\beta=1}^{\theta}\left|r_{a\beta}-r_{b\beta}\right|^2]^{\frac{1}{2}} \tag{4}$$

K-median clustering of samples in the fingerprint database, the clustering process consists of three steps:

Step 1: Select the initial class center reference point.

(1) Randomly select a sample point $A$ as the first class center reference point, calculate the distance $d_A$ from the remaining points to the sample point $A$. The introduction of the weighting factor $1-1/d_A$ makes the probability that the farther point to $A$ is selected is greater.

(2) On the basis of the above, select the second class center reference point $B$, and calculate the distance $\{d_A,d_B\}$ between the remaining points to the sample points $A$ and $B$. the greater the probability that the point farther away from the distances $A$ and $B$ is selected.

(3) Repeat the above two steps until the $\delta$ class center reference points are selected.

Step 2: Update the class center reference point.

The new class center reference point is re-determined in each class by re-determining the new class center vector by taking the median value for each dimension of the sample RSS vector $r_i=(r_{i1},r_{i2},\cdots,r_{i\theta})$ in the class in each class.

Step 3: The reference point is reassigned to the class center reference point.

(1) Calculate the distance from each sample point to all class center vectors and assign them to the nearest class.

(2) Repeat the second step until the class center vector changes within the threshold and the algorithm stops.

*3.3.2 Online positioning stage.* The purpose of this algorithm is to estimate its approximate position vector $(x,y)$ in the fingerprint database by measuring a target sample $J=(r_1,r_2,\cdots,r_\theta)$ representing the signal strength from $\theta$ APs.

Step 1: Delineate the class to which the point to be located belongs.

Calculate the similarity between the $J$ vector and various center vectors $\{E_1,E_2,\cdots,E_\delta\}$, and delineate the cluster to which the $J$ vector belongs. The cosine similarity function proposed in [7] only pays attention to the directional difference of vectors, ignoring the difference in vector amplitude. Therefore, this study introduces Tonimoto coefficients to describe the differences between vectors. The similarity function of the $J$ vector and the central vector $Ei$ of the $i$ class is defined as:

$$TonSim_i = \frac{E_i \cdot J}{\|E_i\|^2 + \|J\|^2 - E_i \cdot J} \tag{5}$$

Step 2: Determine the spatial coordinates in the class to which the point to be located belongs. In this class, the similarity between the $J$ vector and the $S$ training sample RSS vectors in the class is calculated. The similarity function is shown in Equation 5. After the ascending order, select the similarity of the $K$ training samples $\{r_{m1}, r_{m2}, \cdots, r_{mk}\}$ with similar similarities:

$$\{TonSim_{m1}, TonSim_{m2}, \cdots, TonSim_{mk}\} \tag{6}$$

The corresponding coordinates are:

$$\{(x_{m1}, y_{m1}), (x_{m2}, y_{m2}), \cdots, (x_{mk}, y_{mk})\} \tag{7}$$

Considering that the similarity large vector has a great influence on the sample vector prediction result, the introduction similarity normalization weight $\{w_{m1}, w_{m2}, \cdots, w_{mk}\}$ is defined as

$$w_{mi} = \frac{TonSim_{mi}}{\sum_{i=1}^{k} TonSim_{mi}} \tag{8}$$

The coordinate position is then weighted according to Equation 9 to obtain its physical coordinates.

$$\begin{cases} x = w_{m1}x_{k1} + w_{m2}x_{m2} + \cdots + w_{mk}x_{mk} \\ y = w_{m1}y_{k1} + w_{m2}y_{m2} + \cdots + w_{mk}y_{mk} \end{cases} \tag{9}$$

## 4. Experimental result

### *4.1. Data collection and processing*

The experimental site is a closed indoor environment of 12m × 10m. As shown in table 2, five AP points are arranged at the indoor edge position, and 120 position points are selected, and each position point collects 70 RSS data separately at different angles in the upper and lower afternoons. Figure 3 shows that the indoor environment has a large flow of people and there are many obstacles. The signal propagation has multipath effect, so signal noise is unavoidable. The RSS signal error in actual WiFi positioning is the key for researchers to consider. The data cleaning in this study mainly deals with two aspects:1. Missing data processing, assuming that $r_{ti}$ is missing data, the data at time $t_i$ is equal to the data at the previous sampling time. 2. Abnormal data processing, when the RSS value of two adjacent moments is greater than the constant 20db, the data is considered abnormal. Let $r_{ti} = r_{t(i-1)}$ .After the data is cleaned, 6021 more stable data are finally obtained. According to the idea of cross-validation, these data are divided into 8,78 estimated subsets and 1,204 verified subsets according to 8:2.

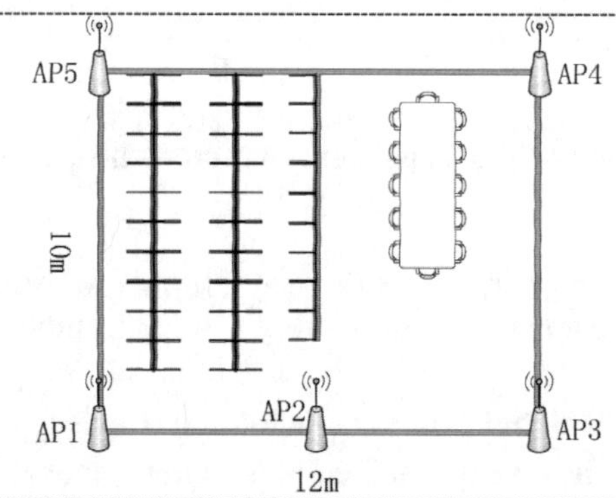


**Figure 3**. Schematic diagram of positioning experimental environment

*4.2. Determination of the number of clusters and the number of neighbors*

The value of the number $k_n$ of the K-median cluster has a great relationship with the training samples. In this experiment, $k_n = 4$ is used after repeated measurements. The training samples are divided into four clusters and there are four class center RSS vectors. Then the target sample is delineated into the most similar cluster, and the WKNN algorithm is used in the cluster. The number $k_n$ of the nearest neighbors in the algorithm needs to be artificially given. In this experiment, the error results of the two algorithms are obtained by measuring different $k_n$ selections. As shown in Figure 4:

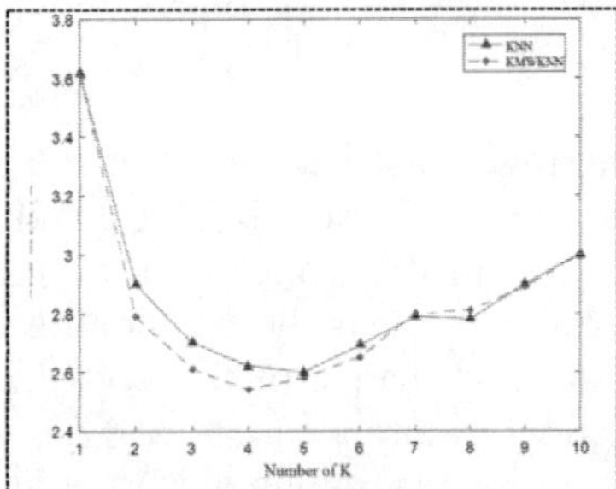

**Figure 4**.Relation between error and Kn

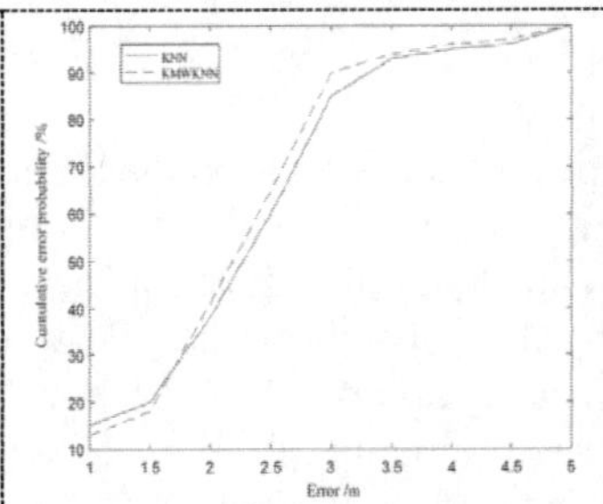

**Figure 5**. Comparison of error accumulation probability

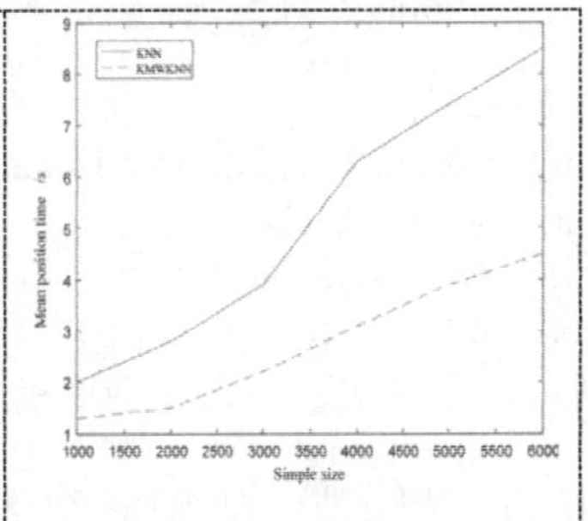

**Figure 6**. Comparison of error accumulation probability

It is not difficult to find from the figure that the value of $k_n$ is too small, the sample is susceptible to noise interference, and the error of observation is larger. When the value of $k_n$ becomes larger, the effect of noise becomes smaller and smaller, and the error also educed. However, when the value of $k_n$ is too large, samples with lower similarity also participate in the weighting of the sample, which in turn leads to an increase in the error of the sample. It is not difficult to see that $k_n = 4$ is the best in this experiment. As can be seen from the above figure, compared with the traditional KNN algorithm, the improved algorithm uses K-median clustering to reduce the sample size, but the difference between the two is less than 0.3m. In this experiment, $k_n = 4$ was taken and the experimental results were further observe.

*4.3. Comparison before and after algorithm improvement*

Figure 5 shows the cumulative probability of different positioning errors of the two algorithms. The former can be seen that the positioning error of the improved algorithm starts to converge within 3m, the probability of error within 2.5m is close to 80%, and the accuracy is slightly better than the KNN algorithm. The accuracy of domestic and foreign research results is in line with practical applications.

Although the accuracy improvement is not obvious, it can be clearly seen in Figure 6. As the sample size increases, the improvement of the positioning time of the improved algorithm becomes more and more significant. The larger the sample size, the more time the traditional KNN algorithm takes. Meet the requirements of indoor positioning real-time.

As shown in Table 1, this experiment will use the traditional KNN method, Deep Belief Network (DBN) algorithm. K-means-KNN algorithm and K-median-WKNN improved algorithm run on the same configuration server, and comprehensively evaluate the difference in performance of the three algorithms through five indicators. It can be seen that the DBN algorithm is significantly better than others in accuracy. The improved K-median-WKNN algorithm is slightly better than the KNN method, but it is not obvious. Therefore, it is often limited in practical applications. At the same time, the clustering algorithm improves the efficiency of the algorithm to some extent. From the perspective of error variance, K-means clustering is susceptible to interference from noise points, resulting in large fluctuations in positioning results. On the whole, the improved algorithm has certain robustness while ensuring the positioning accuracy, and shortens the positioning time, which is in line with practical applications.

IOP Conf. Series: Materials Science and Engineering **608** (2019) 012008 doi:10.1088/1757-899X/608/1/012008

**Table 1.** Comparison of algorithm performance

| Index | KMWKNN | K-means-KNN | KNN | DBN |
|---|---|---|---|---|
| X axis error/m | 1.41 | 1.45 | 1.56 | 1.17 |
| Y axis error/m | 2.02 | 2.01 | 2.05 | 1.53 |
| Error mean/m | 2.46 | 2.56 | 2.62 | 1.81 |
| Error variance | 1.31 | 1.54 | 1.61 | 1.15 |
| Mean time/S | 4.2 | 4.17 | 8.54 | 35.63 |

## 5. Conclusion

By constructing the location fingerprint database, k-median clustering is used to improve the time complexity of the algorithm, and the Tonimoto coefficient is applied to characterize the similarity between RSS vectors. A new WiFi localization algorithm based on k-median-WKNN is proposed. Theoretical derivation and experimental results show that the algorithm shortens the positioning time under the condition of ensuring accuracy, effectively suppresses the disturbance of the noise to the positioning result, and can meet the requirements of indoor positioning. Future research will focus on further optimization of the position fingerprint database and improve the positioning accuracy in this regard.

## References

[1] LU Qi,SHU Guohua. An indoor orientation System based on single chips[J]. Microprocessors, 2006,2,66-68.

[2] YANG Yang,XIAO Jinhong,LIU Zhi.Indoor Three-Dimensional Positioning System Based On Ultrasonic[J].Jonrnal of Jilin University,2012,3(20):267-271.

[3] LI JunHuai, ZHANG GuoMou,YU Lei,ZHANG Jing, Analysis and Simulation of RFID positioning method for indoor environment[J]. Computer Engineering,2012(14): 276-279.

[4] JIA YuQing, WiFi localization algorithm based on deep learning[D]. Nanjing,Nanjing University,2014.

[5] Bai S,Wu T. Analysis of K-Means algorithm on fingerprint based indoor localization system [C] Microwave、Antenna,Propagation and EMC Technologies for Wireless Communications (MAPE),2013 IEEE 5th International Symposium on. IEEE 2013:44-48.

[6] TAO Zheng. Research on Improved WLAN Iocation Fingerprint location Algorithm Based on Chi-square Distance[D].DALIAN: Dalian University of Technology,2016.

[7] LIU Xingchuan,LIN Xiaokang,Fast Wi-Fi Positioning Algorithm Based on Clustering[J].Computer Engineering,2011,37(4):285-287.

IOP Conf. Series: Materials Science and Engineering **608** (2019) 012009 doi:10.1088/1757-899X/608/1/012009

# Electromagnetic Thermal Coupling Simulation Analysis of Silicon Carbide Shock Absorber

**Yin Xiaofang, Liu Xiaoning, Qin Jiayong and Du Chunlin**

Beijing Institute of Spacecraft Environment Engineering, Beijing 100094, China

E-mail: 190064602@qq.com

**Abstract.** This paper introduces a set of absorbing equipment used in 3.2GHz band vacuum high and low temperature environment. Through the finite element method modeling, COMSOL Multiphysics multi-physics simulation software analyzes the electromagnetic thermal coupling capability. The simulation results show that the absorbing properties of the silicon carbide pyramidal absorbing material are good. If the loss caused by the electrical conductivity of silicon carbide is considered, the absorbing performance of the silicon carbide pyramid will be greatly improved to -40dB. The simulation calculation model combines the electromagnetic field with the thermal, and gives the temperature curve and the isothermal contour dynamic diagram of the pyramid. It provides theoretical guidance for practical engineering applications.

## 1. Introduction

With the rapid development of communication satellite technology, transceiver sharing has become a common design method for communication satellite payloads. In this case, the problem of passive intermodulation is particularly prominent. If these weak interference signals enter the receiver with high sensitivity. It is very likely to exceed the thermal noise floor of the receiver, affecting the normal operation of the satellite [1-2]. In order to complete the PIM index test of the whole star and transponder subsystem in a hot vacuum environment, a low PIM absorbing heat sink was developed using silicon carbide absorbing materials, and the absorbing heat sink was used to establish an index of less than 150 dBm. Low PIM test environment with both heat flow simulation and microwave power withstand capability. The electromagnetic thermal coupling capability needs to be simulated and analyzed before the test [3-4].

In this paper, a set of absorbing equipment for vacuum high temperature and low temperature environment in 3.2GHz band is modeled by finite element method, and COMSOL Multiphysics multi-physics simulation software is used to simulate and analyze the electromagnetic thermal coupling capability[5]. The structure of the box is shown in Figure 1. The box body is made of stainless steel frame structure, and the absorbing module is installed inside. During the test, the antenna is facing the mouth of the absorbing device, radiating energy to the absorbing device, and the absorbing device ensures that the energy is absorbed to ensure the normal operation of the antenna.

IOP Conf. Series: Materials Science and Engineering **608** (2019) 012009 doi:10.1088/1757-899X/608/1/012009

**Figure 1** Schematic diagram of the overall structure of the box

## 2. Introduction to the absorbing box

The overall technical requirements for absorbing equipment:

1) Operating temperature range: -190°C ~+100°C
2) Withstand power: greater than 1000 W/m2
3) Absorbing performance: 3.2G band absorbing performance is better than -30Db
4) Surface requirements: Coating hemisphere emissivity $\varepsilon h>0.85$
5) Vacuum suitability: TML (total mass loss) ≤ 1%; CVCM (condensable volatiles) ≤ 0.1%
6) Dimensions: Internal dimensions of absorbing equipment 1600mm *700mm *700mm

In order to improve the reliability, the microwave darkroom absorbing material installation idea is adopted, and the physical installation and the adhesive bonding method are adopted. The absorbing wave tip is stuck on the alloy aluminum plate to form the absorbing module by using the card slot structure (as shown in Fig. 2), and the card slot structure can be realized by extrusion molding.

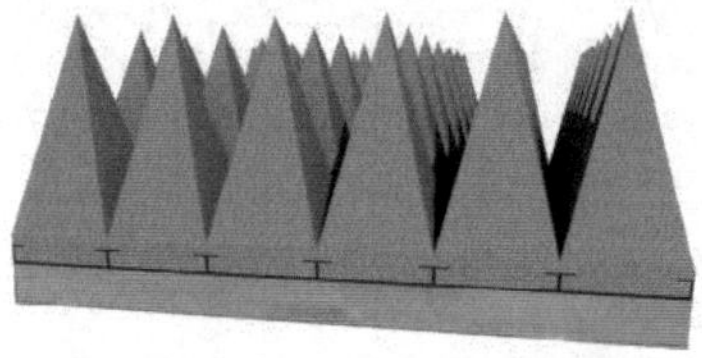

**Figure 2** Absorbing module schematic

The silicon carbide absorbing tip is shown in Figure 3. The absorbing tip is connected by a card slot, and the absorbing tip is neatly arranged without gaps. In order to ensure that the heat in the absorbing body is effectively conducted to the wall surface, the heat-dissipating adhesive (usually silica gel) is used for reinforcement during assembly. The design of the absorbing tip cone is 40 mm (W) × 40 mm (L) × 115 mm (H), which is a hollow structure, and the inner hole is conical.

## 3.The principle of the method

COMSOL Multiphysics multi-physics simulation analysis software is a numerical calculation software based on finite element method, which can be applied to the simulation analysis of electromagnetics, thermal, mechanics, optics and other disciplines. Here is a brief introduction to the relevant basic principles of electromagnetic and thermal coupling.

**Figure 3** Single absorber shape

The essence of solving the electromagnetic problem is to solve the Maxwell equations. In order to facilitate the computer's solution operation, the Maxwell equations are simplified by a series of reasoning, as shown in the following equation.

$$\nabla \times \frac{1}{\mu_r}(\nabla \times \boldsymbol{E}) - k_0^2(\varepsilon_r - \frac{j\sigma}{\omega\varepsilon_0})\boldsymbol{E} = 0 \tag{1}$$

The wave number k0=w/c, ε r , μ r is the relative dielectric constant and the relative magnetic permeability, respectively, σ indicating the conductivity.

For solid heat transfer, the temperature field is solved using the following formula,

$$\rho C_p \mathbf{u} \cdot \nabla T + \nabla \cdot \mathbf{q} = Q + Q_{\mathrm{ted}} \tag{2}$$

$$\mathbf{q} = -k\nabla T \tag{3}$$

Among them ρ is the density of the object, Cp is the heat capacity of the object, u is the velocity field, q is the heat flux density, and k is the thermal conductivity. $Q_{ted}$ is the heat source and is the thermoelastic damping.

In addition to solving electromagnetic problems and heat transfer problems separately, there is also a need to solve the electromagnetic-thermal coupling problem. Considering that the energy of the heat source Q in heat transfer can be composed of temperature boundary and electromagnetic heat loss, the electromagnetic-thermal multiphysics coupling problem can be solved.

$$Q_e = Q_{\mathrm{rh}} + Q_{\mathrm{ml}} \tag{4}$$

$$Q_{\mathrm{rh}} = \frac{1}{2}\mathrm{Re}(\boldsymbol{J} \cdot \boldsymbol{E}^*) \tag{5}$$

$$Q_{\mathrm{ml}} = \frac{1}{2}\mathrm{Re}(j\omega \boldsymbol{B} \cdot \boldsymbol{H}^*) \tag{6}$$

## 4.Simulation calculation modeling

The finite element method is a numerical calculation method that divides the model into a large number of grid subunits and then calculates the field functions in these small unit structures to obtain an approximate solution. The full size of the absorbing box is much larger than the minimum wavelength. Building a full-scale model will cause the number of grid cells to grow exponentially, consume a lot of computing resources, and the calculation efficiency is low. At the same time, according to the electromagnetic field and heat transfer theory, it is not difficult to find that the field function of each small quadrangular pyramid unit structure inside such a large-sized absorbing box structure is approximately periodic, so that only one of the pyramid structures can be analyzed.

The silicon carbide square pyramid has a size of 120 mm (W) × 40 mm (L) × 115 mm (H), and the inner hollow cone has a size of 12 mm (R) × 69 mm (H). The front, back, left, and right sides of the model are set to the period boundary (electromagnetic + temperature). The upper bottom surface is the incident wave port P1, the incident power P=1.8W (withstand power greater than 1000 W/m$^2$), and the lower bottom surface is a perfect electrical conductor. In this model, the temperature boundary conditions are set to the upper and lower bottom surfaces as constant temperature boundary conditions, T=-38. In addition, the material parameters of the silicon carbide used in the calculation of the model are shown in Table 1 below. The constant pressure heat capacity C (Fig. 4) and the conductivity sigma (Fig. 5) are the default values in the COMSOL material database. The thermal conductivity, density, relative permittivity, and relative magnetic permeability are the basic parameters of the test. The real and imaginary parts of the electrical constant are shown in Figures 6 and 7, respectively. In the solid heat transfer module, in addition to the temperature boundary conditions such as the above-mentioned upper and lower bottom surface temperature T and the periodic boundary, it is also necessary to provide a silicon carbide pyramid as a total power dissipation density heat source, that is, electromagnetic loss heat generation.

**Table 1.** Silicon carbide material parameters

| Name | Value | *Unit* |
|---|---|---|
| **Thermal conducity** | 5.5 | W/(m·K) |
| **Heat capacity at constant preasure** | C(T[1/K]) | J/(kg·K) |
| **Electrical conductivity** | 0 or sigma(T[1/K]) | S/m |
| **Density** | 1700 | kg/m³ |
| **Relative permeability** | 1 | 1 |
| **Relative permittivity** | epsilonReal(freq)-i*epsilonImg(freq) | 1 |

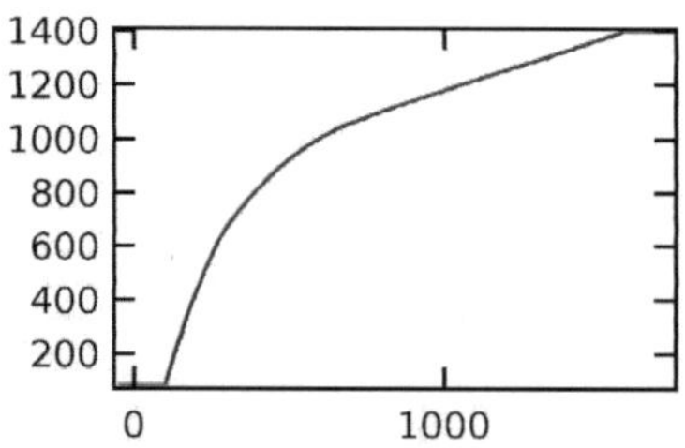


**Figure 4** Silicon carbide constant pressure heat capacity (sigma-temperature)

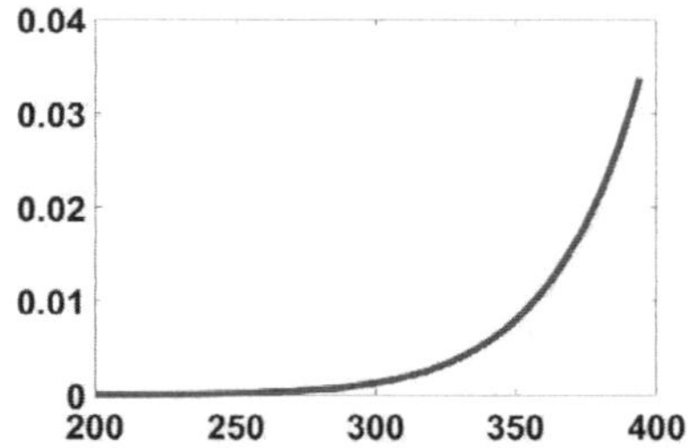


**Figure 5** Silicon carbide conductivity (C-temperature)

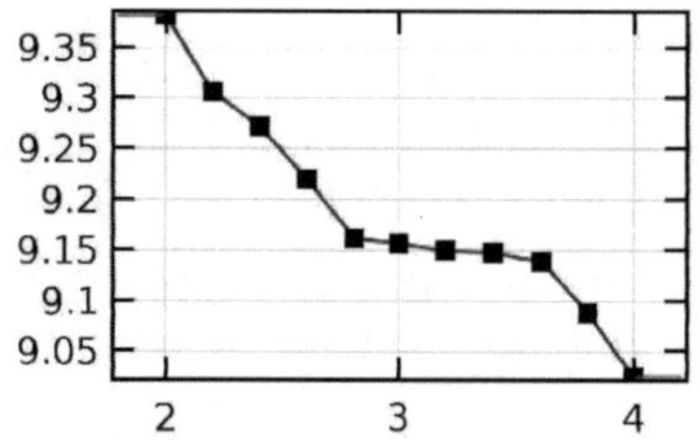


**Figure 6** SiC dielectric constant real part

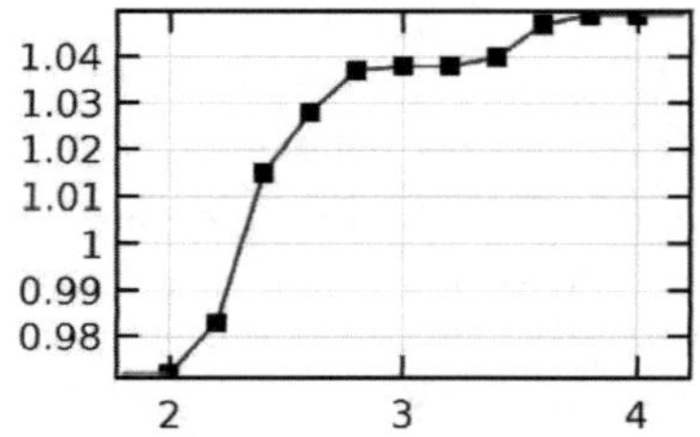


**Figure 7** SiC dielectric constant imaginary part

## 5. Simulation results and analysis

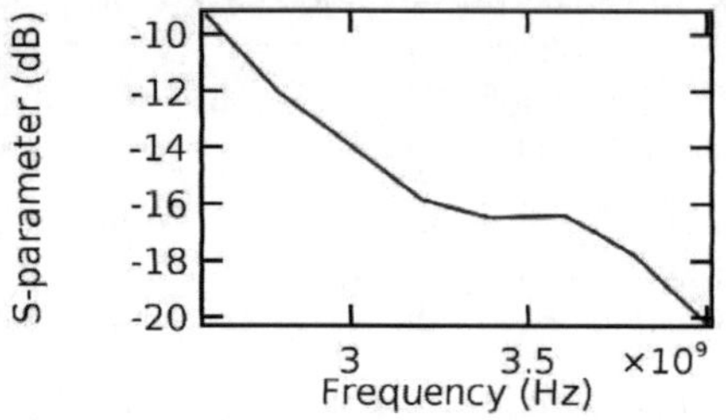


**Figure 8.**S11 reflection coefficient curve, sigma=0

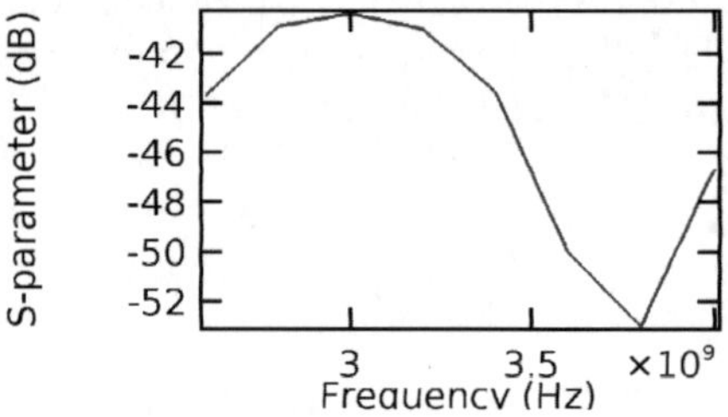


**Figure 9.**S11 reflection coefficient curve, sigma=default

After the simulation model is established as described above, the reflection coefficient is first calculated using the steady-state frequency domain solver. Since the silicon carbide type absorbing material is a dielectric loss absorbing material, the dielectric constant and conductivity sigma are the key parameters of the material, but the sigma parameter value of silicon carbide is not given at present, here sigma= 0 and sigma=default (material library) are calculated. The silicon carbide in the material library in COMSOL is alpha polycrystalline silicon carbide, sigma≠0. The silicon carbide in practical application is a ceramic material, and the electrical conductivity of the material is not zero compared with the silicon carbide in the material library, and tends to be doped to increase the loss and enhance the absorbing effect. Figure 8 shows the S11 reflection coefficient curve for sigma = 0. S11 is less than -10 dB in the 2.65-4 GHz band and decreases as the frequency increases. This is consistent with the results of Figures 6 and 7, because as the frequency increases, the loss tangent angle increases correspondingly, resulting in increased losses. However, if the influence of the material sigma is considered, the S11 reflection coefficient curve will undergo a large change. As shown in Fig. 9, the absorbing effect of the material is significantly enhanced, and the enhancement is 30 dB. It can be seen that the sigma value has a great influence on the absorbing performance. Therefore, in order to obtain more accurate simulation data, further testing of the material and obtaining the accurate value of sigma is an indispensable link.

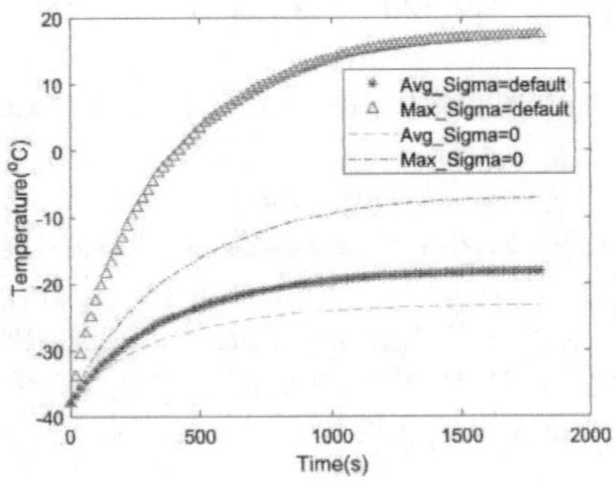


**Figure 10.** Silicon carbide pyramid temperature change

The thermodynamic analysis of the model is performed using a frequency domain transient solver. Fig. 10 is a graph showing the average body temperature Avg of the hollow silicon carbide pyramid and the maximum temperature Max in the body as a function of the electromagnetic wave irradiation time. Regardless of whether the parameter value of the material sigma is 0, the temperature of the pyramid increases rapidly under the irradiation of strong electromagnetic waves, but then the increase of the pyramid temperature tends to be stable due to the existence of the heat sink, and finally reaches the thermal equilibrium state. When the conductivity of the silicon carbide is set to the default value of the material library, ie sigma=default, the average temperature and the maximum temperature of the pyramid are significantly larger than the corresponding sigma=0. This phenomenon is easier to understand. When sigma=default, the loss of electromagnetic waves in the silicon carbide pyramid increases, and the corresponding coupling generates higher thermal energy, so the corresponding average temperature and maximum temperature are greater than the temperature when sigma is zero. In addition, it is not difficult to find that the average temperature difference in the two cases $\Delta$ Avg$<$ 5℃, but the maximum is close to 25℃, which indicates that the transient warming effect caused by sigma is very strong, further illustrating the big influence of sigma value on absorbing performance and temperature change.

## 6. Summary

The basic principle of electromagnetic-thermal coupling simulation calculation and the modeling and simulation process are introduced above. The simulation results show that the absorbing properties of the silicon carbide pyramidal absorbing material are good, and the reflection coefficient of the absorbing body is less than -10 dB even without considering the loss effect caused by sigma. But this still has not reached the technical requirement of -30dB. If we consider the loss caused by the conductivity of silicon carbide, the absorbing performance of the silicon carbide pyramid will be greatly improved to -40dB. Therefore, it is very effective to dope in the silicon carbide material to enhance the loss effect caused by the conductivity, thereby improving the absorbing properties of the silicon carbide material, which is very consistent with the methods reported so far. In addition to the absorbing properties of the absorbing box, the temperature field changes are also the data of interest to the model. The simulation calculation model combines the electromagnetic field with the thermal, and gives the temperature curve and the isothermal contour dynamic diagram of the pyramid. Although the radiation time and temperature conditions of the port (antenna) in the model are not actively controlled, that is, the electromagnetic boundary conditions and temperature boundary conditions are not accurate experimental temperature conditions, but also provide theoretical guidance for practical engineering applications.

## References

[1] CHEN Z C, CHENG L, WANG T X, et al. *Development and status of space thermal environment simulation researches[J]*. Cryogenics, 2016, 44(4): 6-13

[2] Eccosorb SF used in thermal vacuum chambers for ground based satellite testing[EB/OL]. [2017-02-10]. http://www.eccosorb.eu/cases/eccosorb-sf-used-thermalvacuum-chambers-ground-based-satellite-testing, 2014- 06-23

[3] TAO Y, LI B Y, DENG A M, et al. Progress of the high-temperature absorbers and modified SiC absorbing performance[J]. Surface Technology, 2011, 40(3): 88-104

[4] GE K Y, WANG Q, ZHANG X N, et al. Study on property improvement of SiC wave absorbers[J]. Journal of Functional Materials and Devices, 2002, 8(3): 263-26

[5] INCROPERA F P, DEWITT D P, BERGMAN T L, et al. Fundamentals of heat and mass transfer[M]. New York: John Wiley, 2007: 4-17

# Study on Sequent Test Effects of Atomic Oxygen/thermal Cycling

**R Q Zhai, D B Tian, L X Jiang, Y XIA, X Y Li and Y Li**

Beijing Institute of Space Environment Engineering, Beijing 100094, China

lingid_83@163.com

**Abstract.** Two kinds of unidirectional laminates made from carbon/epoxy composite named M40/DFA-1 & M40/DFA-N3(nano $TiO_2$ contained) are selected to study the different effects between different test sequences of atomic oxygen/thermal cycling. Based on the ground simulating test of atomic oxygen and thermal cycling, several kinds of tests and analysis are conducted, such as surface morphology observation, mass loss test, inter-laminar shear strength test, and X-ray photoelectron analysis of surface components. After the exposure of atomic oxygen and thermal cycling, surface erosions is observed on both of the materials, in addition, mechanical properties are both changed. It must be highlight that, under the condition of $1.5\times10^{20}$ atoms/cm$^2$ AO and 200 thermal cycles within the range of -150℃~+150℃, different sequences lead to different results, either the extend or the tendency. In which, samples under AO+TC sequence exhibited relatively higher mass loss and erosion rate as well as more severe mechanical performance degradation.

## 1. Introduction

In the Low Earth Orbit(LEO) environment where some of the spacecraft would be located, surfaces will be exposed to potentially damaging environment such as atomic oxygen, ultraviolet radiation, charged particles radiation, thermal cycling, debris and so on[1,2]. Synergistic effects may exist between different space environment factors[3]. Therefore, under the circumstances of ground-based facilities, basing on the acknowledgement of single factor effects, it is of great importance to discover the principles of performance degradation, and to figure out what's going on.

Atomic oxygen and thermal cycling are referred as the most principal factors among LEO environment[4]. In which, atomic oxygen constitutes the primary components of neutral atmosphere, furthermore, it will harshly impact on the surface of spacecraft at the relative speed of 8km/s, and leads to severe erosion of the surface material[5-8]. Meanwhile, thermal cycling between extreme temperatures is another inevitable environment when spacecraft orbits through the shadow area in and out over and over again.

On the basis of applications in the engineering[9-12], two kinds of unidirectional laminates made from carbon/epoxy composite named M40/DFA-1 & M40/DFA-N3(nano $TiO_2$ contained) are selected in this work, to do the research on the different effects result from different test sequence, focusing on the environment of atomic oxygen and thermal cycling.

## 2. Experimental Procedure

### *2.1. Test Conditions and Materials*

IOP Conf. Series: Materials Science and Engineering **608** (2019) 012010 doi:10.1088/1757-899X/608/1/012010

Atomic oxygen impact fluence is set to be $1.5\times10^{20}$ atoms/cm$^2$.To avoid the effect of micro-cracks on the mechanical properties of testing materials during thermal cycling, temperature range is chosen as -150℃~+150℃ according to other references, with 200 cycles in all.

Epoxy resin composites enhanced by carbon fibers M40/DFA-1, and 3%nano $TiO_2$ contained epoxy ester composites enhanced by carbon fibers M40/DFA-N3 are exposed to the test environments as mentioned respectively.

### *2.2. Instruments and Methods*

The atomic oxygen tests and thermal cycling tests are conducted respectively on the facilities at Beijing Institute of Space Environment Engineering.

The sample's mass loss is measured by German Miro Sartorius precision electronics balance, with the accuracy of $10^{-5}$. Mass data before and after environment test have been recorded respectively, and the mass loss rates are calculated as following:

$$M_{\text{loss}} = [(M_0 - M_t)/M_0]\times 100\% \quad (1)$$

Where, $M_0$ is the original mass of specimen, g; $M_t$ is the mass of specimen after environment test, g.

The inter-laminar shear strength is tested based on the GB/T 30969-2014 standard by applying the three-point short beam method, at the loading rate of 1.3 mm/min, calculated as following:

$$\text{ILSS} = \frac{3P_{\text{max}}}{4bd} \quad (2)$$

Where, $P_{\text{max}}$ is the maximum of the breaking load, N; $b$ is the width of the specimen, mm; $d$ is the thickness of the specimen, mm.

The optical surface morphology is observed by Olympus GX-51 optical microscope; the micro morphology is observed by Scanning Election Microscope（SEM）of the type MX2600FE;the surface element analysis is conducted on the basis of XPS(X-ray Photoelectron Spectroscopy) by VG ESCALAB MKⅡ.

## 3. Results and Analysis

### *3.1. Optical Surface Morphology*

Generally speaking, most of the polymer-matrixed composites are relatively vulnaerable to the atomic oxygen(AO) and thermal cycling(TC). Specimen of both materials are oberserved before and after AO&TC sequent tests, as shown in Fig.1 and Fig.2 In the view of optical microscope, both surfaces have turned dark, but phenomenon can't be distinguished obviously between different test sequence.

(a)original (b)AO+TC (c) TC+AO

**Figure 1.** Surface of the M40/DFA-1 specimen before/after AO&TC sequent tests

(a)original (b)AO+TC (c)TC+AO

**Figure 2.** Surface of the M40/DFA-N3 specimen before/after AO&TC sequent tests

### *3.2. Mass Loss Rate*

Masses of the specimens made from the both materials have been weighed before and after the bidirectional sequent tests, in addition, mass loss has been calculated, as well as the mass loss rate. The results are shown in the Table 1.

**Table 1.** Mass loss rate of the specimen under AO/TC sequence test

| Types of the Materials | Mass Loss Rate | |
|---|---|---|
| | AO+TC | TC+AO |
| M40/DFA-1 | 0.425% | 0.193% |
| M40/DFA-N3 | 0.456% | 0.199% |

As the data reveals, for both materials, the mass loss rates after AO+TC test are obviously higher than that after TC+AO test. To discuss the reason beneath the phenomenon, it may be possible that atomic oxygen brought the roughness to the materials surfaces, which increased the specific surface area, thus, the composites contact area interacted with C,N,H in the air has raised, and released more volatile gases like CO, $CO_2$ and $H_2O$,therefore,raised the mass loss rate.

On the other hand, mass loss after TC+AO sequent test are obviously lower, the primary reason may be the post curing process of the polymers undergoing thermal cycling exposure. The procedure of post curing could enhance the crosslinking density of resin matrix, which tightens the combination between atoms, as a result, strengthens the materials' resistance to atomic oxygen.

The data indicates that, as far as atomic oxygen and thermal cycling is concerned, different test sequences lead to damage with distinction. Given the conditions in this work, the sequence of AO+TC induced higher mass loss than that from TC+AO sequence.

### *3.3. Inter-laminar Shear Strength*

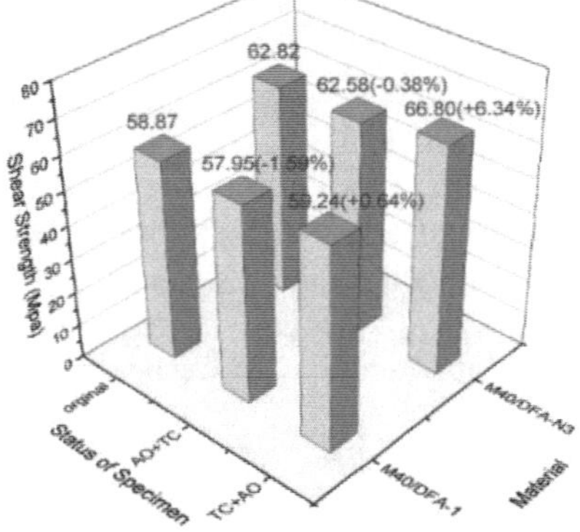


**Figure 3.** ILSS of specimens before/after AO&TC sequent test

As illustrated in Fig.3, is the inter-laminar shear strengths of specimens before and after the sequent tests. The data shows that the inter-laminar strengths of both materials are declined after AO+TC sequent test, of which M40/DFA-1 dropped from 58.87MPa to 57.95MPa, about 1.59% reduction, while M40/DFA-N3 dropped from 62.82MPa to 62.58MPa, about 0.38% reduction. On the other hand, inter-laminar strengths of both materials are enhanced after TC+AO sequent test, of which M40/DFA-1 raised to 59.24MPa at the rate of 0.64%, while M40/DFA-N3 raised to 66.80MPa, at the rate of 6.34%.

Thus, given the conditions of atomic oxygen and thermal cycling test, different test sequence leads to different effects in mechanical property.

Basically, according to the related researches, atomic oxygen leads to the declination of the mechanical performance, whereas, thermal cycling usually strengthens it (in the circumstances without micro-cracks occur). From that point of view, we deduce that, in the AO+TC sequent test, atomic oxygen did more severe damage to the mechanical property than that of thermal cycling, which resulted in mechanical reduction; while in the TC+AO sequent test, exposure in thermal cycling brought performance promotion to the mechanical performance directly in the way of raising the crosslinking density, besides, this procedure itself could also reduce the erosion from atomic oxygen.

### *3.4. Surface Micro Morphology*

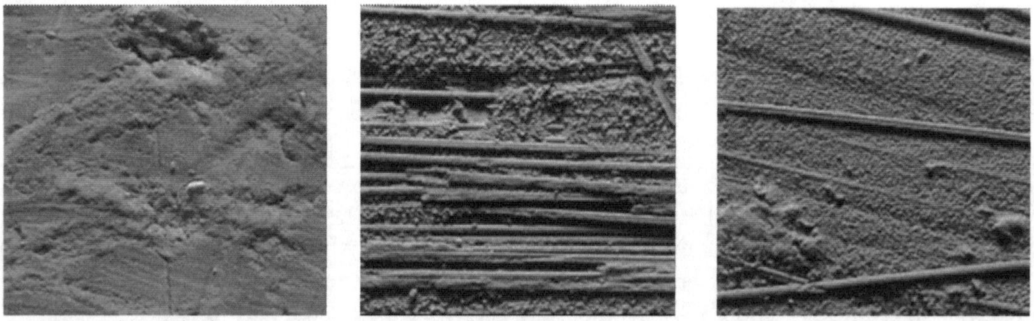

(a) original (b) AO+TC (c) TC+AO

**Figure 4.** Surface morphology of M40/DFA-1 before/after AO&TC sequent tests (SEM)

(a) original (b) AO+TC (c) TC+AO

**Figure 5.** Surface morphology of M40/DFA-N3 before/after AO&TC sequent tests (SEM)

As illustrated in Fig.4 and Fig.5, surface micro morphology was observed before and after bidirectional sequent tests by the Scanning Election Microscope (SEM).

It is obvious that, the resin was denudated seriously on the surface of M40/DFA-1, leaving plenty of near-face fibers exposed outwardly, with several carbon fibers ruptured; otherwise, under the sequent test of TC+AO, resin on the surface is also eroded, leads to "carpet" morphology, but, relatively unconspicuous.

Similar phenomenon appeared on the surface of M40/ DFA-N3.It is visible that single fibers exposed on the surface after AO+TC sequent test, while the erosion is less severe after TC+AO sequence, with no fibers on the surface exposed.

### *3.5. Analysis of Surface Components*

Detailed C1s peak spectrums XPS analysis of M40/DFA-1 before and after sequent tests are illustrated in Fig.6 as follows. The peaks of element of C in organic molecules are usually located at 285 eV(C-C bond), 286.5 eV(C-O-C bond), 287.6 eV(CHO bond), and 288.6 eV(COOH bond)[13]. As is shown in the graphs, C1s spectrums of both materials changed after sequent tests, with distinguished appearances: under the sequence of AO+TC, the specific area of the peak at 285 eV which refers to the C-C bond, shrank sharply, while that of TC+AO seemed little; which refers to the C-O-C bond expanded after AO+TC test, whereas that of TC+AO test diminished; besides, the specific area of the peak at 289 eV referring to the C=O bond heightened slightly after both sequences. Data as mentioned are laid in Table 2.

The result indicates that thermal cycling retained more C-C from de-bonding by atomic oxygen, weakening the effects of atomic oxygen. Basically, if an oxygen atomic breaks in the C-C bond to form a C-O-C bond, it would then reduce the crosslinking density, therefore, in the contrast between AO+TC and TC+AO, the latter maintained a relatively higher crosslinking density, furtherly, the mechanical property. That is matching with the inter-laminar strength measurement.

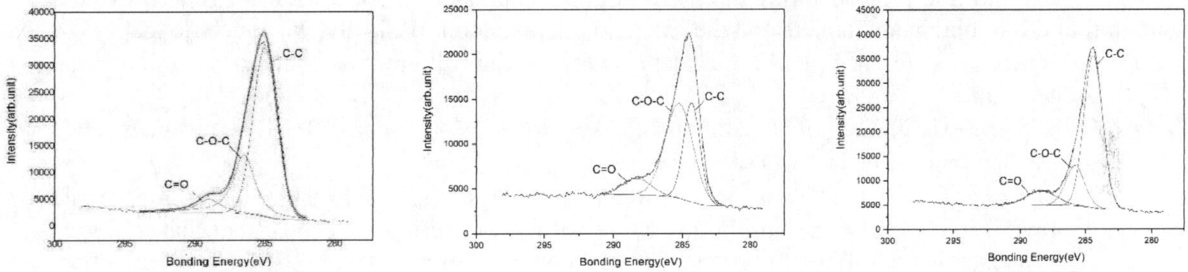


(a) original (b) AO+TC (c) TC+AO

**Figure 6.** C1s spectrum of M40/DFA-1 specimen surface before/after AO&TC sequent tests

**Table 2.** XPS analysis of M40/DFA-1 before/after AO&TC sequent tests

| Functional Groups | Bonding Energy/eV | Relative Percentage of Specific Area/ % original/(AO+TC)/(TC+AO) |
|---|---|---|
| C-C | 285.000 | 70.20/34.83/69.8 |
| C-O-C | 286.500 | 27.82/55.80/19.58 |
| C=O | 289.000 | 6.66/9.37/10.62 |

Detailed C1s peak spectrums XPS analysis of M40/DFA-N3 before and after sequent tests are illustrated in Fig.7 as follows. The result is basically consistence with that of M40/DFA-1: more C-C bonds are retained in TC+AO sequent test than in AO+TC（specific data are laid in Table 3）, which means a higher crosslinking density. So the mechanical property performed higher, which is consistent with measurement.

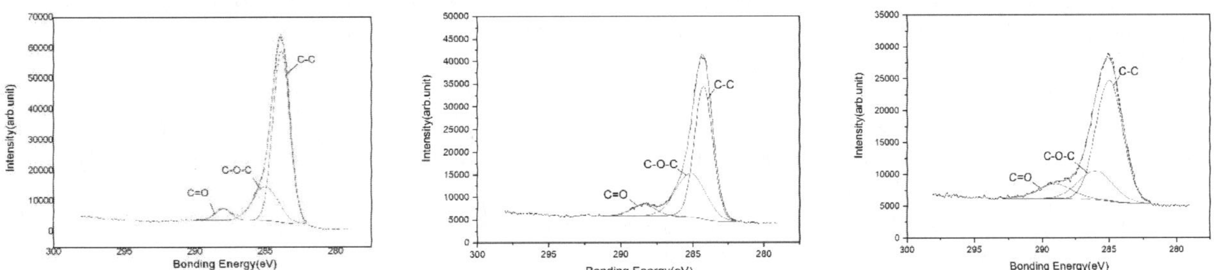


(a) original (b) AO+TC (c) TC+AO

**Figure 7.** C1s spectrum of M40/DFA-N3 specimen surface before/after AO&TC sequent tests

**Table 3.** XPS analysis of M40/DFA-N3 before/after AO&TC sequent tests

| Functional Groups | Bonding Energy/eV | Relative Percentage of Specific Area/ % original/(AO+TC)/(TC+AO) |
|---|---|---|
| C-C | 285.000 | 73.02/61.66/67.32 |
| C-O-C | 286.500 | 22.12/31.96/20.71 |
| C=O | 289.000 | 4.85/6.37/11.96 |

Based on the results demonstrated above, from the angle of surface component analysis, different test sequence generally gives rise to different effects.

## 4. Conclusions

In this work, composites marked as M40/DFA-1 and M40/DFA-N3 were selected to be the test material to discover the effects of sequent tests focusing on atomic oxygen($1.5\times10^{20}$ atoms/cm$^2$) and thermal cycling(200 cycles within -150℃~+150℃), especially the difference in between. Methods such as optical and micro morphology observation, inter-laminar strength test, SEM as well as XPS were introduced to obtain and understand the data and phenomenon. Basically, we conclude that:

- No obvious difference could be identified between different sequences in the optical observation;
- In the SEM examination, it is distinguishing that the erosion under AO+TC sequence is much severe than that under TC+AO sequence；
- Atomic oxygen could weaken the inter-laminar shear strength, while thermal cycling could improve it. Given the condition in this work, for both materials it is consistent that, different test sequence led to different changes in inter-laminar strengths even to the different direction. In which, AO+TC reduced it whereas TC+AO raised it, which is out of our expectation. To discuss the reason, it might be the extend TC brought to the improvement in strength is larger than the extend AO reduced in it. Furtherly, and most critical in the point, post-cuing procedure which fastened the crosslinking combination, thus pre-improved the resistance to atomic oxygen in the sequence of TC+AO.
- According to the result from XPS analysis, for both materials, more C-C bonds were preserved in the TC+AO test. Given the recognition that when oxygen atom breaks in a C-C bond to become a C-O-C bond, it would then lower the resin's crosslinking density, so the result from XPS demonstrates more C-C bonds retained in TC+AO test than that in AO+TC could result in the enhancement in mechanical property. That is matching with the inter-laminar strength measurements.

In conclusion, limited by the ground-based facilities, space environment tests incorporating atomic oxygen and thermal cycling, couldn't be conducted simultaneously so far, bidirectional sequent tests are employed instead. However, results from the sequent tests indicate that, different sequences lead to different results. Thus, we can deduce that, synergistic effects of space environment do not equals to the rough summation of single factor effects, which might be between the effects from the different sequent tests. To discover mechanisms beneath the appearance even to a much broader scope, a large amount of work needs to be conquered in the future.

## References

[1] HUANG B C,TONG J Y. Space Environment Engineering[M]. Beijing: China Science & Technology Press, 2010:3-8
[2] YANG X N, YANG Y. Spacecraft Space Environment Engineering[M].Beijing: Beijing Institute of Technology Press,2018:24-243
[3] ZHAI R Q, REN G H, TIAN D B, et al. Advances in research on low-orbit ultraviolet, charged particle, thermal cycling and atomic oxygen synergistic effects[J]. Vacuum,2019,56(1):72-76

IOP Conf. Series: Materials Science and Engineering **608** (2019) 012010 doi:10.1088/1757-899X/608/1/012010

[4] Anon. Property changes induced by the space environment in composite materials on LDEF Solar array materials passive LDEF_experiment A0171: NAS 1.26192572, NASA-CR-192572 94N11484[R]

[5] MINTON T K, WRIGHT M E, TOMCZAK S J, et al. Atomic oxygen effects on POSS polyimides in low earth orbit: ADA556178[R]. Air Force Research Lab Edwards AFB CA Propulsion Directorate Report, 2012

[6] ZHAI R Q, JIANG H F, TIAN D B, et al. Simulation of atomic oxygen interaction with space station[J]. Equipment Environmental Engineering, 2014,11(3): 35-39

[7] ZHANG W, LI M, SHEN Z, et al. Graphene-reinforced epoxy resin with enhanced atomic oxygen erosion resistance[J]. Mater Sci, 2013(38): 2416-2423

[8] DOOLING D. Material selection guide to limit atomic oxygen effects on spacecraft surfaces[R]: NASA/TP-1999-209260.

[9] JIANG L X, HE S Y, YANG S Q, et al. Application of carbon(graphite) fiber/epoxy composite in the spacecraft and research on their behaviors[J]. Journal of Materials Engineering, 2001(9): 39-46

[10] NG C B, SCHADLER L S, SIEGEL R W. Synthesis and mechanical properties of $TiO_2$-epoxy nanocomposites[J]. Nano Structured Materials, 1999(12): 507-510

[11] ZHANG Z, CHU Q X, LI H Y, et al. One-pot solvothermal synthesis of graphene supported $TiO_2$(B)nano sheets with enhanced lithium storage properties[J]. Journal of Colloid and Interface Science, 2013, 409: 38-42

[12] JIANG L X, HE S Y, YANG D Z. Resistance of nano-$TiO_2$ modified M40/EP648 composite to vacuum ultraviolet irradiation[J]. Chinese Journal of Materials Research, 2003, 17(4): 27-431

[13] LIU Y M, LI M, LIU X P, et al. Effect of atomic oxygen on electric properties of graphene films[J]. Journal of Materials Engineering, 2017, 45(8): 9-13

# Design Considerations for a Composite Chassis of Shielding and Structural Integration

**Bai Yifeng, Tong Wenqing and Jiang Haidong**

The 38th Institute of CETC, Hefei 230031, China

Corresponding author's e-mail address:baiyifenghefei@163.com

**Abstract.** This study describes the designing and shaping technology of a composite chassis. Materials with a capable shielding capacity were chosen to perform shielding designing according to the electromagnetic shielding criteria. To fulfil the high mobility with the shielding material manufacturing requirement, we developed a suitable design to reduce the weight of all structure components to comb the shielding ability with its own structure. The integral shaping technique was used for chassis manufacture. We also further tested the weight, electromagnetic shielding, and environmental adaptation conditions of the chassis. The verification results indicate that our manufacturing protocol met the design requirements of the chassis.

## 1. Introduction

With the increasingly demanding requirements for the miniaturization and light weight of modern military equipment, the use of composite materials has become much more extensive. To complete assignments, a radar chassis needs to frequently change its orientation, such that superior mobility is required[1]. To fulfil this requirement, one must focusona weight reduction design. However, there is limited room for weight reduction in a chassis made of metal. Composite materialsarecharacterized by light weight, highstrength, and remarkable designability. Thus, we decided to design a composite chassis using carbon-fiber materials with electromagnetic shielding (hereinafter referred to as ES) capability after comparing several proposals to realize weight reduction and superior mobility. A large number of programmable logic controllercomponents that are highly sensitive to external electromagnetic interference are mounted in the chassis; hence, the function of protecting internal components from the influence of external electromagnetic fields is a necessity for the chassis to operate normally[2–3]. In addition, the leakage of the electromagnetic field generated inside the chassis must be prevented. Therefore, the mechanism for achieving the ES effect that meets the design requirements is the key to the chassis design.

This work presents the design criteria, structural characteristics, design, and molding process of the chassis and verifies the molded chassis according to the design criteria.

## 2. Chassis overview

### *2.1. Chassis construction*

The overall dimension, including the height of the upper round flange, of the chassis body is 680mm × 540mm × 461mm. The external surface area is approximately 1.2 $m^2$. Each of the two narrow surfaces has an end cover to open or close the chassis, which is connected to the body through a hinge.

Avertical and a horizontal plate inside the body divide the internal space into separateblocks for installation of each system module. One of the body surfaces is configured with an air vent for ventilation and heat dissipation. Figure 1 shows the modelofthe chassis body structure.

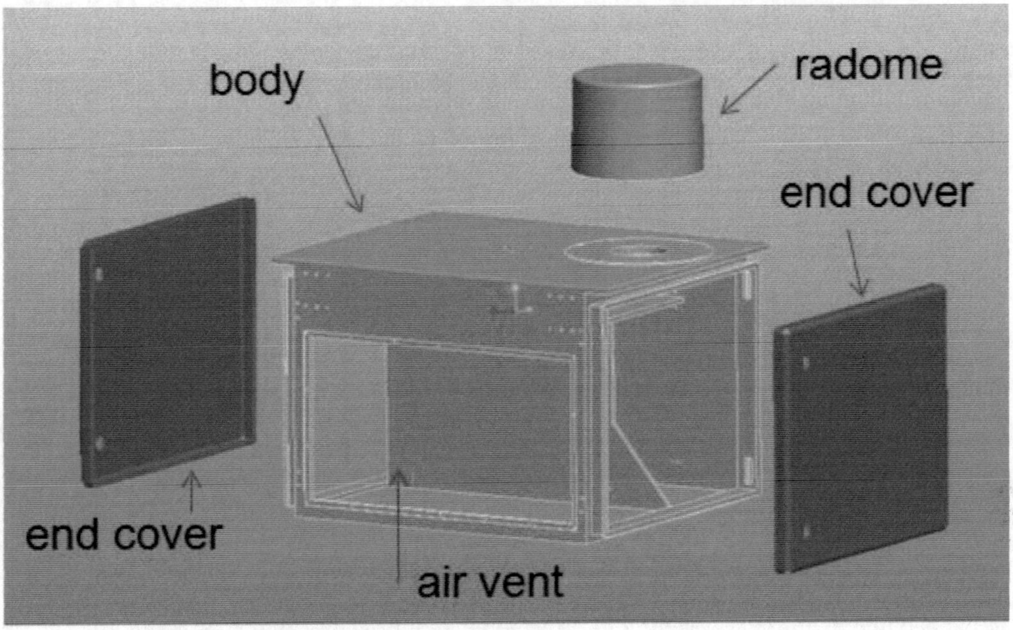


**Figure 1.**Diagram of the body structure model.

*2.2 Chassis technical requirements*

Primary technical criteria:

Chassis weight: ≤10.5 kg (body≤ 8.5 kg, each of end cover ≤ 1 kg)

Service temperature: −40 °C ~ +50 °C

Storage temperature: −45 °C ~ +70 °C

Rainproof requirements: The chassis should be free from water leakage under the conditions of 1 mm/min rainfall intensity, 16 m/s wind speed, and 2 h rain exposure time.

In obtaining the design weight, the chassis should be designed not only to minimize weight, but also to meet the requirements of body strength and internal loading strength of the chassis.Insatisfying thechassisES criteria, ES must be considered when designing the material composition, construction, and edges of adjoining faces. Only by doing so can the product requirementsbe met.

## 3. Overall chassis design

According to the product design criteria, the design purpose of the chassis is conformity to the desired weight, ES, and environmental adaptability,whichis greatly dependent on the material category selected. This section focuses on the design for ES and weight reduction.

### *3.1. Electromagnetic shielding design*

Based on the functions of the electronic equipment inside a chassis, slotted gaps or holes are usually used for ventilation and heat dissipation or serve as the port for removal, installation, and connection with peripheral devices. Therefore, the decisive factors for the chassis ES effectiveness include not only chassis wallthickness and material, but also the holes and gaps.

ES is a practice of blocking both electric and magnetic fields. It generally refers to the shielding of alternating electromagnetic fields above 10 kHz. One of the most common theories explaining the ES mechanism is the transmission line theory. According to this theory,a shield is considered as a transmission line, and the ES mechanism includes three parts, where the first part is the reflection loss of the shield surface; the second part is the absorption loss of the shield material; and the third part is the loss caused by multiple reflections inside the shield (Figure 2).

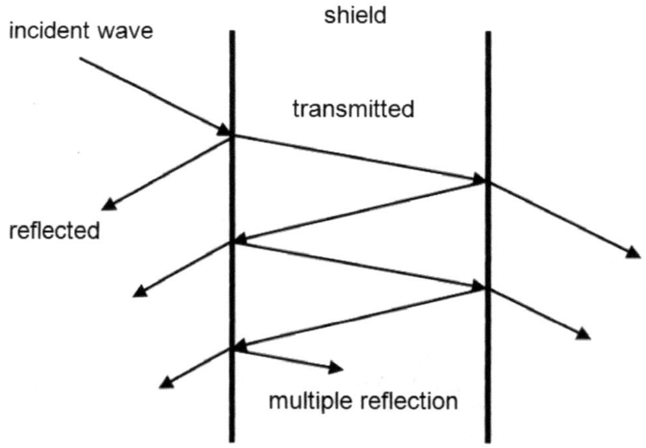


**Figure 2.**Diagram of the ES mechanism.

In transmission line theory, a uniform shielding material is deemed as a transmission line model, and the Schelkunoff formula used for the model solution is the most suitable calculation formula,which is only applicable to uniform and flat shielding materials. This calculation methodapplies in our study because the four walls of the chassis product are made of uniform and even materials. The total shielding effectiveness is calculated by:

$$SE=SE_A+SE_R+SE_M,$$

where $SE_A$ (unit: dB) is the absorption loss of the shielding material; $SE_R$ (in dB) is the single reflection loss of the shield surface; and the $SE_M$ (unit: dB) is the multiple reflection loss inside the shield.

$SE_A$, $SE_R$, and $SE_M$ in the above formula are calculated as follows:

$$SE_A= 131.43t\sqrt{f\mu_r\sigma_r} \quad (1)$$

$$SE_R= \begin{cases} 168.2 + 10lg(\frac{\sigma_r}{f\mu_r})(\text{plane wave}) & (2) \\ 20lg\left(5.35r\sqrt{\frac{f\sigma_r}{\mu_r}} + 0.354 + \frac{1.17\times10^{-2}}{r}\sqrt{\frac{\mu_r}{f\sigma_r}}\right)(\text{magnetic field}) & (3) \\ 3.217+10lg\left(\frac{\sigma_r}{f^3r^3\mu_r}\right)(\text{electric field}) & (4) \end{cases}$$

$$SE_M=10lg[1 - 2 \times 10^{-0.1SE_A} \cos 0.23SE_A + 10^{-0.2SE_A}], \quad (5)$$

where f (unit: Hz) is the electromagnetic wave frequency; t (unit: m) is the shielding material thickness; r (unit: m) is the distance from the field source to the shielding material; $\mu_r$ is the relative magnetic permeability of the shielding material; and s$\sigma_r$ is the relative conductivity of the shielding material.

As can be seen in the formulas, great thickness, outstanding magnetic permeability, and electrical conductivity of the chassis material enable the chassis to obtain a favorable ES effectiveness. Among the composite materials, carbon fiber has a resistivity of 1.3–1.9 × $10^{-5}$ Ω/cm, which shows good electrical conductivity, and the advantages of high specific strength, high specific stiffness, low thermal expansion coefficient, and remarkable damping performance. Therefore, it can be used as the cover material of the chassis. Considering the lightweight requirements, the chassis wall is designed to have an A-sandwich structure configured with a honeycomb-structured interlayer. Aluminum honeycomb is selected as the sandwich material given the cost, shielding effectiveness, and superior conductivity of aluminum, in which resistivity can reach $4 \times 10^{-4}$ Ω/cm. However, the A-sandwich construction consisting of carbon fiber and honeycomb-structured interlayer is not sufficient to produce the desired shielding effect. Among the existing shielding materials, metal conductive materials with certain magnetic properties represented by nickel can combine reflection and absorption loss, thereby surpassing other materials in electromagnetic wave absorption[4]. Based on the nickel ES capability, a layer of nickel mesh was added inside the chassis to enhance the chassis shielding effectiveness. The placement position of the mesh in the sandwich structure was determined through experiments. Finally, the chassis was coated interiorly with a layer of static conductive paint. All of these shielding components made up the shielding system of the chassis. Figure 3 shows the section view of the chassis wall.

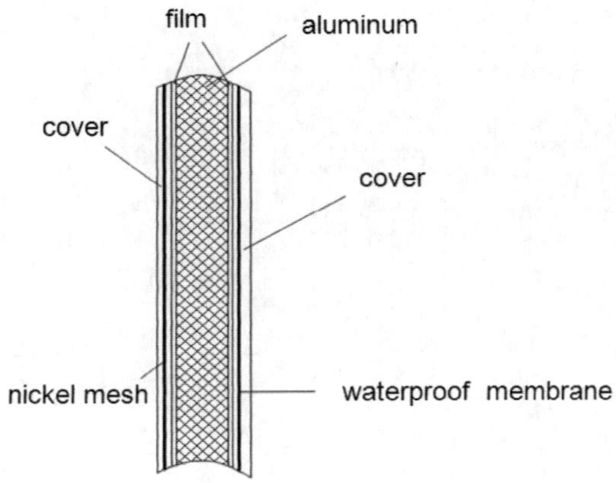


**Figure 3.**Section view of the chassis wall.

Meanwhile, a conductive rubber was added to each joint and connecting hole to avoid the risk of electromagnetic leakage in such a structure. For the large air vent for ventilation and heat dissipation on the chassis, a cut-off waveguide vent was used for ES protection when the conductive rubber was sealed on the air vent for transition. The ES application of the cut-off waveguide vent was well-established. Figure 4 shows the appearance of the vent-equipped chassis.

**Figure 4.**Cut-off waveguide vent used in the chassis air vent.

### *3.2 Weight reduction design*

The total chassis weight should not exceed 10.5 kg (a body ≤ 8.5 Kg and an end cover ≤ 1 kg). By this criteria, we worked on the weight reduction design in combination with the chassis structure and material selections. An A-sandwich structurewasobserved in all chassis faces as well as internal vertical and horizontal partitions, which consisted of a carbon-fiber cover, an aluminum honeycomb panel, a film, a waterproof membrane, and a nickel mesh. The total weight of the single waterproof membrane and the individual nickel meshcannot be altered considering the roles they play; hence, the weight loss design focused on the carbon-fiber cover and aluminum honeycomb panel. According to the target shielding frequency, analuminumhoneycombpanel consisting of 2 mm cells was selected given the relationship between the size of the aluminumhoneycomb cell and the target shielding frequency as well as the honeycomb strength andpurpose of weight reduction. Moreover, in view of the material strengthandthe weight of each module to be installed inside the chassis, the minimum thicknesses of the cover and the honeycomb panel in the A-sandwich structure were confirmed by a simulation analysis to reduce weight (Figure 5).

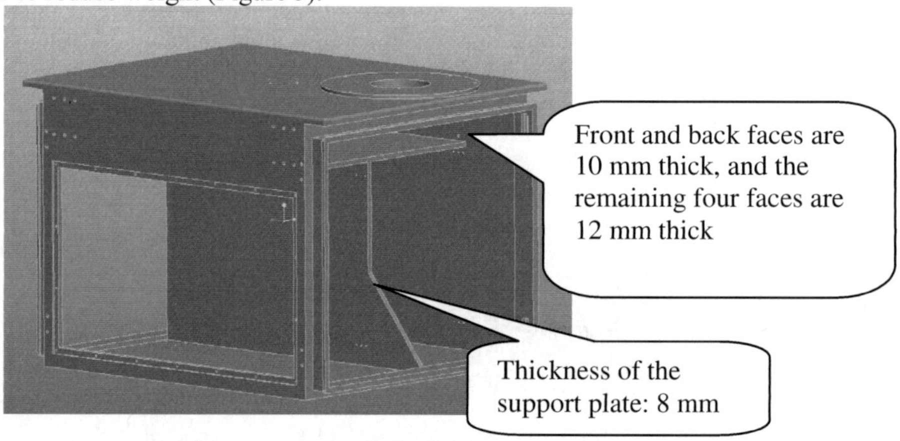


**Figure 5.** Chassis weight reduction design.

Many mounting interfaces can be found on the chassis, which depend on embedding aluminum blocks with threaded holes in the composite structure. A total of 130 embedded parts are included; hence,the aluminum-embedded parts also take up a considerable percentage of weight. Based on the premise that the strength of each embedded part met the design requirements, the size of the embedded parts was optimized to minimize weight, thereby decreasing the total chassis weight.

## 4. Chassis molding process

The various materials to be used in the chassis were determined according to the ES design discussed in the previous section. Table 1 presents the list of materials.

**Table 1.**List of composite materials for ES chassis manufacturing.

| # | Name and model | Specification | Purpose | Manufacturer |
|---|---|---|---|---|
| 1 | Carbon-fiberprepreg | MTM28/CF0300-42% | Cover | Shanghai Kangzhan Composites |
| 2 | Aluminium honeycomb panel | 95-2-0.05 (LF2Y) | Interlayer material | AVIC Composite |
| 3 | Film | J-272 | Adhesion of honeycomb | Heilongjiang Petrochemical |

| | | | | |
|---|---|---|---|---|
| 4 | Shielding mesh | Nickel mesh | Shielding | Heilongjiang Petrochemical |
| 5 | TEDLAR film | TDS15BL3 | Surface film | DuPont USA |
| 6 | Static conductive paint | Polyurethane static conductive film | ES | Jiangsu Jinling Special Paint Co., Ltd. |

The composite chassis molding is mainly based on the overall strength of the chassis and the ES effectiveness.If a gap exists between two adjoining faces, electromagnetic leakage will occur because the chassis has many individual faces. On the contrary, if molded as a whole, the chassis will have excellent mechanical strength. Therefore, we decided to fabricate it using the integral molding technique. In addition, a process path was eventually established given the difficulty of demolding: integrally mold five faces of the chassis; separately mold the top plate; and finally assemble both of them. An autoclave molding process with a higher molding pressure was also adopted considering the bonding strength between the prepreg layers and between the cover and the honeycomb[5]. Figure 6 shows the process flowchart.

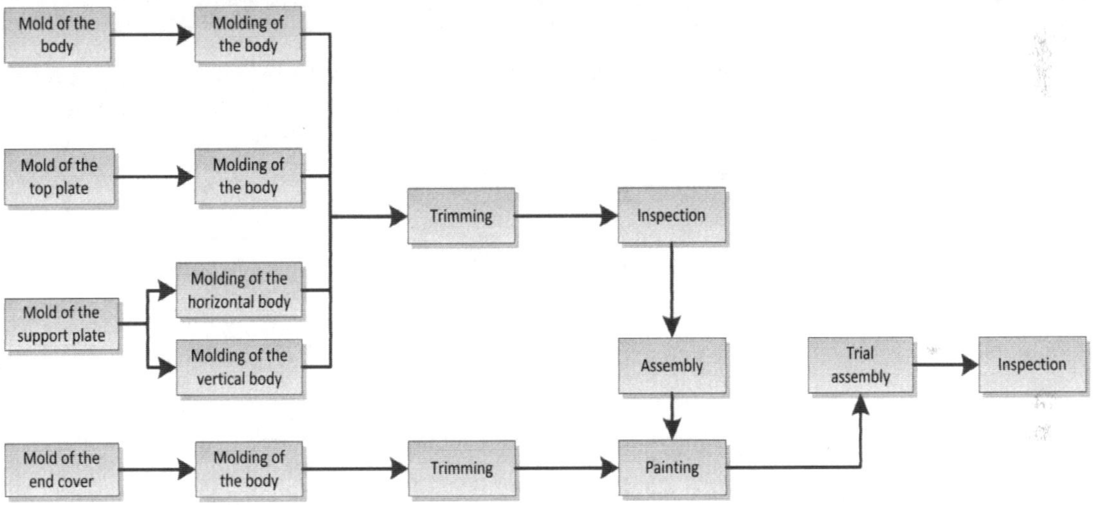


**Figure 6**.Flowchart of the chassis molding process.

The mold was designed to be a five-face male mold, on which the integral molding of the chassis would be performed. The layup design mainly draws on the design principles of composite materials, including “zero coefficient of linear expansion” and “isotropy,” to stabilize the molded chassis. The layup procedure was performed as follows according to the thickness of different areas: lay up four or two layers in a “rotating” way; ensure that the front and back cover layers are symmetrical; and lay the wovencarbon-fiberprepregin the order of 0°/30°/60°.Pay attention to staggering the overlapping positions and reinforcingthecorner overlapping positions. It is a molding process involving a male mold. Therefore, the flange frame used to connect the end cover on the outer surface was separately fabricated and assembled after the inner cover was laid in the chassis. The outer cover was finally laid to shapethe chassis as a whole. The top plate was assembled with the molded five-face body through the blind holes of the angle supports. A conductive sealant was coated on the joint surface at the same time. After the aluminum honeycombpanelwas laid, all the embedded parts were buried in the corresponding positions and fixed using stop pins and holes on the mold for positioning. According to theresinproperties of the selected prepreg, the curing temperature curve was made based on the material parameters and experiment results (Figure 7).

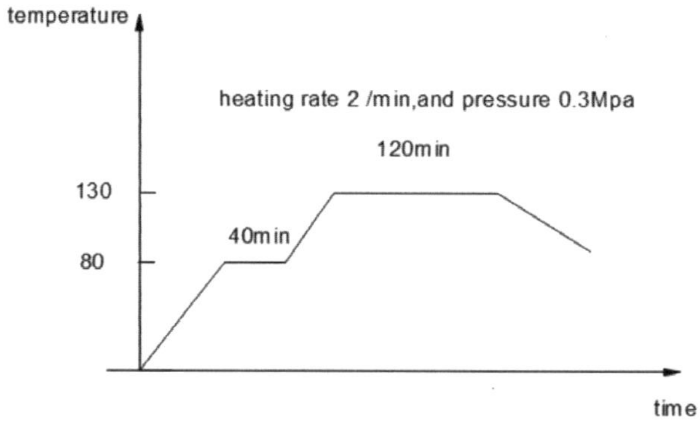


**Figure 7.** Curing temperature curve of the chassis molding

## 5. Verification of the chassis technical criteria

### *5.1 Weight verification*

During the molding process, the body and the end covers were weighed and re-weighed after the chassis was assembled and painted. The results indicate that the chassis total weight and the weight of each component met the design criteria. Table 2 presents the weight and the corresponding criteria.

**Table 2.** Chassis weight verification.

| Name | Actual weight | Criteria | Conclusion |
|---|---|---|---|
| Body | 8.43 kg | ≤8.5 | Accepted |
| End cover 1 | 0.96 | ≤1 | Accepted |
| End cover 2 | 0.98 | ≤1 | Accepted |
| Total weight | 10.37 | ≤10.5 | Accepted |

### *5.2 Verificationof the electromagnetic shielding effectiveness*

The shielding effectiveness of the chassis was tested in conformity with the methods specified in GJB 151B-2013 (*Electromagnetic emission and susceptibility requirements and measurements for military equipment and subsystems*) (Figure 8).

**Figure 8.**Chassis ES effectiveness test.

Four frequencies (i.e., 1 GHz, 2 GHz, 3 GHz, and 4 GHz) were selected as the experiment frequencies according to the target shielding frequency. With each frequency, six faces of the chassis were subject to tests. Table 3 shows the results.

**Table 3.** ES Effectiveness test results of the composite chassis

| | Top plate | Bottom plate | End cover 1 | End cover 2 | Air vent | Chassis wall |
|---|---|---|---|---|---|---|
| 1 GHz | 45 | 43 | 42 | 41 | 41 | 46 |
| 2 GHz | 45 | 43 | 41 | 41 | 41 | 45 |
| 3 GHz | 44 | 42 | 41 | 40 | 41 | 45 |
| 4 GHz | 42 | 41 | 40 | 39 | 40 | 43 |

The test results indicate that the chassis can basically meet the product shielding requirementswith the target frequency. The chassis shielding effectiveness decreased as the frequency increased because the magnetic permeability of the nickel mesh used as a shielding material gradually declined with the frequency increase. This is aninherentproperty of the nickel material. Thus, itsshieldingeffectiveness tapered off.

*5.3 Environmental adaptability verification*

The environmental adaptability was verified according to the chassis environmental criteria through the damp heat test, high and low temperature test, vibration test, and rain test in conformity with GJB 150A. Table 4 presents the results below.

**Table 4.**Chassis environment tests.

| No. | Test name | Test details | Test results |
|---|---|---|---|
| 1 | Damp heat test | Temperature: 35 °C; humidity: 95–98%; and service duration: 48 h | Accepted |
| 2 | High and low temperature test | Stored at high temperature (+70 °C) for 12 h and worked for 2 h<br>Stored at low temperature (−45 °C) for 12 h and worked for 2 h | Accepted |
| 3 | Rain test | Raindrop diameter: 3 mm; wind speed: 16 m/s; and duration: 5 h | Accepted |

As indicated in the environment test results, the chassis made of the ES composite is accepted and meets the product environmental adaptability criteria.

## 6. Conclusion

The electromagnetic environment of the modern radar equipment has become increasingly complicated. Meanwhile, the requirements for the radar electronic chassis are more stringent because of the increasingly demanding requirements for radar use. The chassis requires not only the electromagnetic shielding capability meeting certain criteria, but also a light weight. As a result, composite materials have been used for chassis fabrication. The carbon-fiber composite material with a satisfactory electrical conductivity was chosen as the cover material after selection. Additionally, the weight was reduced using an A-sandwich structure with an aluminum honeycomb-structured interlayer and the optimization of the prepreg thickness. The chassiselectromagnetic shielding

IOP Conf. Series: Materials Science and Engineering **608** (2019) 012011 doi:10.1088/1757-899X/608/1/012011

requirements were satisfied by laying a single nickel mesh inside the chassis. Moreover, an integral molding method was employed to eliminate the electromagnetic leakage risk of the joints and holes in the chassis: the five-face body of the chassis was first molded as a whole, then the top plate was assembled afterwards, followed by a conductive sealant used on the edges of the end-cover mounting face, vent mounting area, and through holes. Under this procedure, the air vent was sealed using the electromagnetic shielding technique of the cut-off waveguide vent. The conformity of the chassis to the design criteria was substantiated through the verification of weight, electromagnetic shielding, and environmental adaptability of the molded chassis. The successful development of the chassis provides an effective alternative for the design of a similar electronic chassis.

**Acknowledgments**
Authors wishing to acknowledge assistance or encouragement from colleagues, special work by technical staff or financial support from organizations should do so in an unnumbered Acknowledgments section immediately following the last numbered section of the paper.

**References**
[1] Tong WQ, J HD .*Manufacturing Process Technology for Radar Composites Cabinet[J]. Aeronautical Manufacturing Technology,2012,18:96-98*
[2] Lu DH, Cheng XY, Guo L. *Design Considerations for a Radar Composite Material Cabinet* [J]. Electromechanical Engineering, 2009, 25 (5): 8–11
[3] Lu DH, Fu J, Cheng XY. *Design of Shielding and Cooling Cabinet for Ground Based Radar* [J]. Electromechanical Engineering, 2010, 26(2): 9–13
[4] Cheng HJ. *Research and application of new preparation technology of nickel-coated copper powder for electromagnetic shielding* [D]. Kunming: Kunming University of Science and Technology 2006.10
[5] Wang FQ, Guo ZM, Liu P, Liang SW, Huan F. *Development of Electromagnetic Shielding Composite Case Based on Autoclave Process[J].Electromechanical Engineering, 2018,34(6):47-50*

IOP Conf. Series: Materials Science and Engineering **608** (2019) 012012 doi:10.1088/1757-899X/608/1/012012

# Initialization Method of Nozzle Flow Field Based on One-Dimensional Isentropic Theory and the Study of Its Application

**HaiLong Zhao[1], Ke Peng[1*, a], WeiHua Zhang[1*, b], YuZhu Fan[2] and JinLiang Jie[1]**

[1]College of Aerospace Science and Engineering, National University of Defense Technology, No.109 DeYa Street, ChangSha, HuNan, P.R. China
[2]Unit 63629, Jiuquan Satellite Launch Center

[*]Corresponding author: [a]pengke.nudt@nudt.edu.cn; [b]zhangweihua@nudt.edu.cn

**Abstract.** The two-dimensional axisymmetric nozzle is initialized by two methods which are the hybrid initialization method of the commercial computational fluid dynamics software Fluent and the initialization program based on the one-dimensional isentropic theory. The first-order upwind format is combined with the SA model to solve the two-dimensional axisymmetric nozzle under the same boundary conditions. The convergence process is obtained in the process of numerical calculation, and the influence of two different initialization methods on the convergence process is analysed. The results show that the initialization method based on the one-dimensional isentropic theory effectively improves the convergence of the nozzle numerical model calculation process, accelerates the convergence speed and improves the calculation efficiency.

## 1. Introduction

The nozzle is an important part of the aircraft for energy conversion. It can transform the internal energy of high temperature and high pressure gas into kinetic energy, generate reaction force, and provide continuous and stable power for the aircraft. Therefore, it is of great significance to study the flow field of the nozzle. At present, numerical simulation, theoretical analysis, experimental research and other methods are often used in the study of flow field problems. These methods are studied from different perspectives and complement each other. For the nozzle flow field research, compared with the experimental research method, the numerical simulation method is more economical and convenient. However, when we use numerical simulation method to study the nozzle flow field problem, it is possible that the process of convergence speed of the nozzle numerical simulation is too slow and sometimes it can be diverging because of the complexity of the nozzle flow filed and the defects of the numerical calculation method.

Therefore, it is of great significance to solve the problems of slow convergence and easy dispersion in numerical simulation of nozzle flow field. As the most simplified flow model, one-dimensional isentropic flow model can give analytical solution or semi-analytical and semi-numerical results, which is helpful to understand the actual three-dimensional flow law. The initialization program of numerical calculation of nozzle is written by using the one-dimensional isentropic theory, and the numerical simulation calculation of nozzle flow field is initialized by using this initialization problem. The results of one-dimensional isentropic flow are taken as the initial conditions of numerical calculation of nozzle, which is helpful to improve the convergence in numerical calculation of nozzle.

## 2. Initialize Method

*2.1 One-dimensional Isentropic theory of Nozzle*

For the one-dimensional flow in the nozzle, the sectional area of the nozzle is set as A, which is function of the axial distance X(the inlet center as the origin point) of the nozzle. The parameters inside the nozzle only change with the change of X. The continuity equation, momentum equation and energy equation of one-dimensional isentropic flow are given below, as shown in (1)-(3).

continuity equation:
$$\rho_1 V_1 A_1 = \rho_2 V_2 A_2 \tag{1}$$

momentum equation:
$$p_1 A_1 + \rho_1 V_1^2 A_1 + \int_{A_1}^{A_2} p\,dA = p_2 A_2 + \rho_2 V_2^2 A_2 \tag{2}$$

energy equation:
$$h_1 + \frac{V_1^2}{2} = h_2 + \frac{V_2^2}{2} \tag{3}$$

In order to give an analytical solution, the complete gas equation (4) and the complete gas enthalpy equation (5) need to be supplemented[1].

complete gas equation:
$$p = \rho RT \tag{4}$$

complete gas enthalpy equation:
$$\mathrm{h} = c_p T \tag{5}$$

The analytical solution can be obtained from the equation (1)-(3).Firstly, the Mach number in the nozzle is only related to the area ratio, and the relational expression is (6).

$$\left(\frac{A}{A^*}\right)^2 = \frac{1}{Ma^2}\left[\frac{2}{\gamma+2}\left(1+\frac{\gamma-1}{2}Ma^2\right)\right]^{\frac{\gamma+1}{\gamma-1}} \tag{6}$$

For a given nozzle, its cross-sectional area A is a function of X, so the relation between the Mach number and X can be obtained from equation (6). After the Mach number is solved, the changes of pressure, density, temperature and others are functions of Mach number, and the equation (7)-(9) can be obtained[2].

$$\frac{P}{P_0} = \left(1+\frac{\gamma-1}{2}Ma^2\right)]^{-\frac{\gamma}{\gamma-1}} \tag{7}$$

$$\frac{\rho}{\rho_0} = (1+\frac{\gamma-1}{2}\mathrm{Ma}^2)^{-\frac{1}{\gamma-1}} \tag{8}$$

$$\frac{T}{T_0} = \left(1+\frac{\gamma-1}{2}Ma^2\right)]^{-1} \tag{9}$$

Among them, $P_0$; $T_0$; $\rho_0$; $\gamma$; $A^*$ are the total pressure at the inlet of the nozzle, the total temperature, the density, the specific heat ratio of the gas, and the area of the throat of nozzle. According to the above conditions, the physical parameters in the nozzle can be determined once the nozzle shape, fluid parameters, and inlet conditions are determined.

*2.2 Initialization Program Based on One-dimensional Isentropic theory*

According to the theory of 1.1, when the gas parameter conditions of the nozzle inlet and the shape of the nozzle are determined, the flow parameters inside the entire nozzle can be obtained. The physical parameter values solved in the one-dimensional isentropic theory are used as the initial values of the nozzle simulation calculation, and the physical parameters obtained by the one-dimensional steady flow theory are analyzed as initial conditions to improve the numerical calculation of the nozzle flow field.

First, after meshing the nozzle, the commercial computational fluid dynamics software Fluent is called to generate the initial IP file, and the coordinate information of all the grid points of the nozzle can be obtained; then a file containing the nozzle generatrix coordinate points and the inlet the gas parameters in the program. The calling program calculates the area of each grid node to obtain the area ratio of the position, and then combines the inlet gas parameters into the formula of equation (1)-(9) to obtain the physical parameters of the mesh node. After looping through all the grid points, an initial nozzle flow field can be obtained, and a new IP file is output, which is imported into Fluent as the initial flow field for numerical simulation calculation, The block diagram is shown in figure 1.

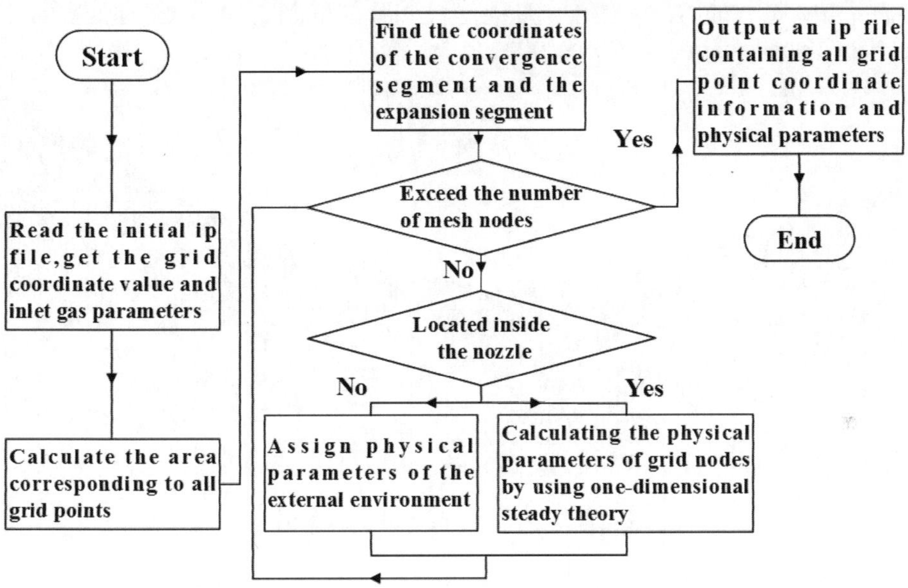


**Figure 1.** Initialization block diagram based on one-dimensional isentropic theory

## 3. Research Object and Method

*3.1 Computational model and meshing*

In this paper, a two-dimensional axisymmetric nozzle is taken as the research object, and the convergence process under the same boundary conditions but different initial conditions are mainly analyzed. The geometry is shown in figure 2 and the geometric parameters are shown in Table 1.

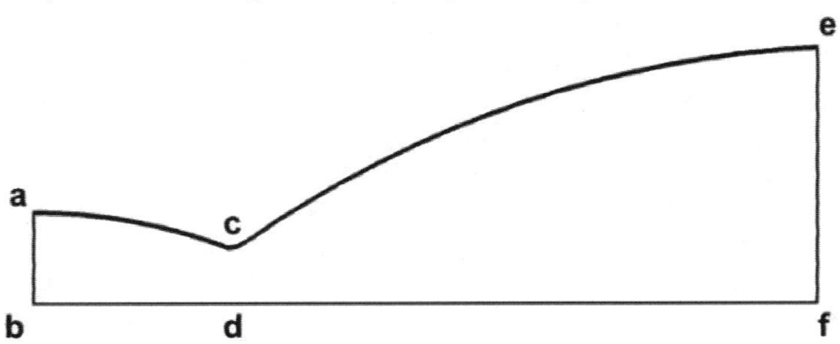

**Figure 2.** Geometric model of the nozzle

**Table 1.** Nozzle geometry basic parameter

| Parameter | Value |
|---|---|
| Nozzle inlet radius(ab) | 0.054(m) |
| Laryngeal radius(cd) | 0.0335(m) |
| Nozzle outlet radius(ef) | 0.15(m) |
| Nozzle length(bf) | 0.465(m) |
| Expansion ratio | 20 |

The mesh division of some areas of the two-dimensional axisymmetric nozzle is shown in figure 3.In order to reduce the influence of the grid on the numerical simulation calculation, a structured grid is used to encrypt the wall, throat, entrance and exit to ensure that the $y^+$ of the first layer is closer to 1[3][4], the total number of grids is 1000×200.

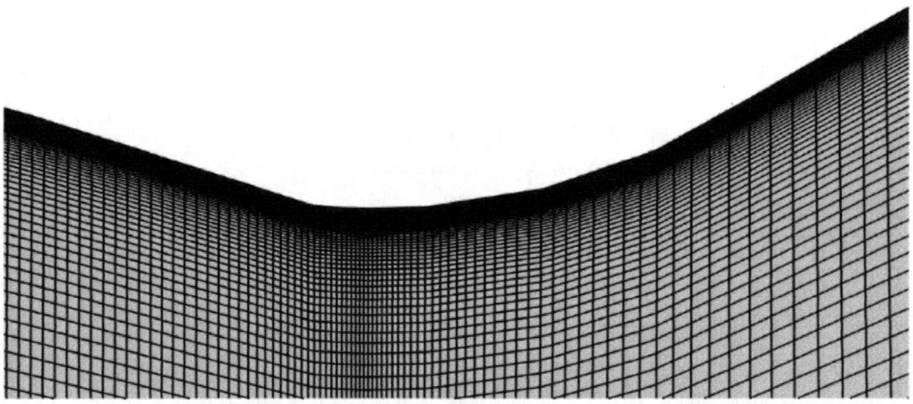

**Figure 3.** Nozzle partial area grid

*3.2 Numerical calculation parameter setting*

In this paper, the SA turbulence model is selected to solve the problem. It is a RANS model with only a small amount of computation for solving a modified eddy viscosity transport equation. In the modified form, the eddy viscosity in the near-wall region is easier to solve and is often used to simulate the inclusion of wall jets[5]. The problem is that the simulation results are better in the boundary layer flow of the inverse pressure gradient. For the discrete format of various parameters, the first-order upwind format is selected. The first-order upwind format has the characteristics of high stability and fast calculation speed, which is beneficial to improve the calculation efficiency. For the selection of the initialization method, one method is to use the hybrid initialization method provided by the commercial computational fluid dynamics software Fluent to initialize, and the other method is to initialize using the initialization program which is written based on the one-dimensional isentropic theory, and compare the convergence process of the two methods. In the numerical calculation, the Coulomb number also has a great influence on the stability and convergence of the calculation, so the Coulomb number in all numerical calculation is set to 0.5.

For this physical model, as shown in figure 2.The nozzle inlet ab is set as a pressure inlet condition, the nozzle outlet ef is set as a pressure outlet condition, the wall surface ace is set as a wall condition, and bdf is set as a symmetry axis condition. In this paper, three pressure inlet conditions are calculated, which are 8MPa,10MPa and 12MPa,and the total temperature is set to 3568.923k.The pressure and temperature of the pressure outlet condition are set to 101325Pa and 300k.

The density of the gas is calculated using the ideal gas model. The constant pressure heat ratio of the gas is set to 3724.088k, and the thermal conductivity is solved by the kinetic-theory. The gas viscosity coefficient is solved by the Sutherland's law of three coefficients, as shown in equation (10).

$$\mu = \mu_0 \left(\frac{T}{T_0}\right)^{3/2} \frac{T_0 + T_S}{T + T_S} \tag{10}$$

In equation (10),the reference temperature $T_0$ is taken as 273.15k; $T_S$ is the Sutherland constant,it is related to the nature of the gas. $\mu_0$ is the reference viscosity coefficient when the temperature is $T_0$.

## 4. Calculation Results and Analysis

First, the flow field pressure and velocity distribution after initialization using two initialization methods are given, when the inlet pressure is 8MPa,as shown in figure 4-figure 5.The physical parameters obtained by the one-dimensional isentropic theory are only related to the area ratio and are a function of the change with X; and the hybrid initialization method is based on the given boundary conditions to pre-iterate the Laplace equation, as shown in equation (11).The initial distribution of velocity and pressure in the calculation region is solved, and other physical quantities( temperature, turbulence ) are automatically obtained from the average interpolation of the flow field.

$$\begin{aligned} &\nabla^2 \varphi = 0 \\ &V = \nabla \varphi \\ &\nabla^2 p = 0 \end{aligned} \tag{11}$$

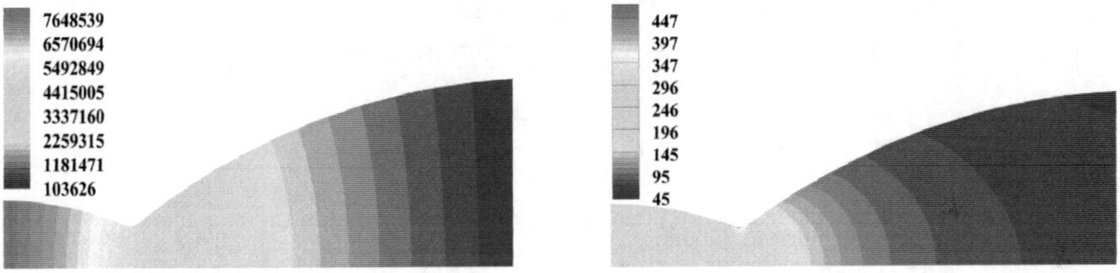


(a) (b)

**Figure 4.** Flow field pressure and velocity distribution after hybrid initialization

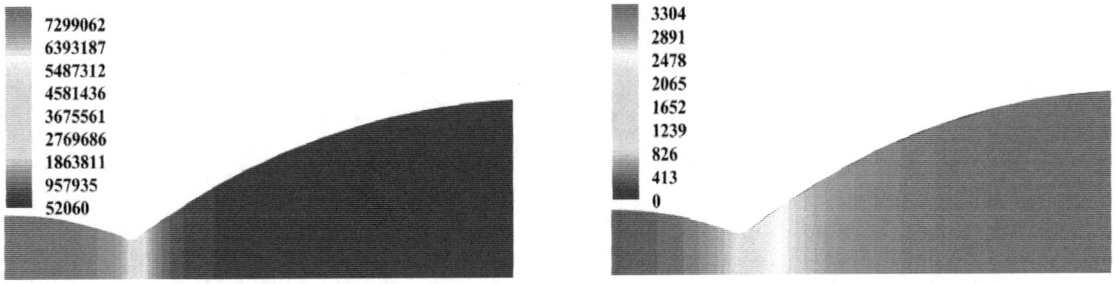


(a) (b)

**Figure 5.** Flow field pressure and velocity distribution after initialization based on isentropic theory

It can be seen from figure 4 and figure 5 that the pressure fields obtained by the two initial methods are similar, but there is a big difference in the velocity field. It can be seen that the velocity field of the hybrid initialization is a certain area as the "source point". Diffusion to the surrounding, the speed gradually decreases; and the velocity field using the one-dimensional isentropic theory is still only a function related to the area ration. In the contraction section of the nozzle, the gas velocity is subsonic, in the expansion section of the nozzle, the gas velocity is supersonic. The continuous residual variable process of two different initialization methods under three inlet pressure conditions is analyzed. As shown in figure 6-figure 8.

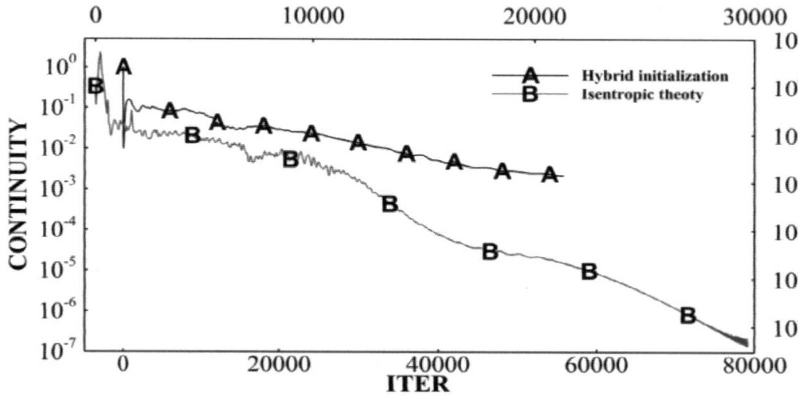


**Figure 6.** 8MPa continuous residual

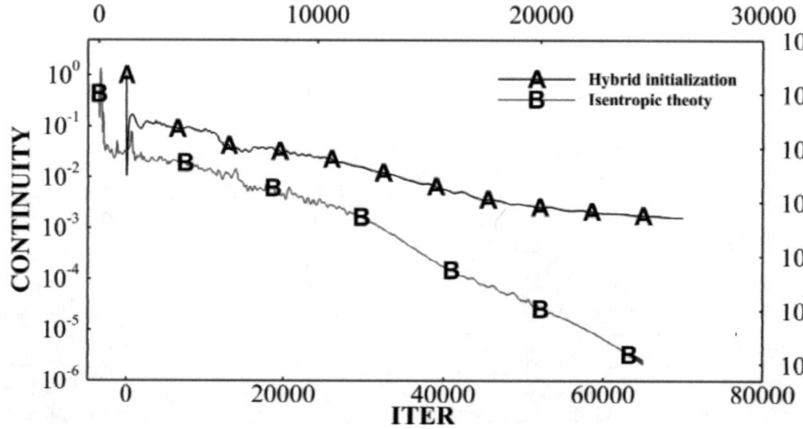


**Figure 7.** 10MPa continuous residual

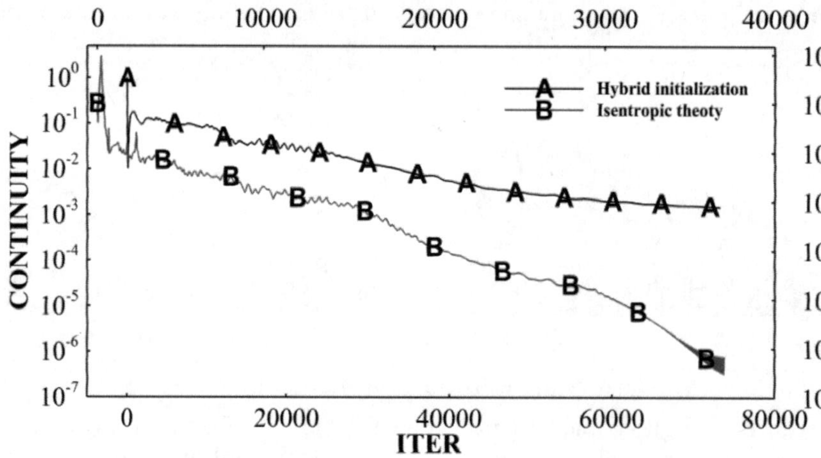


**Figure 8.** 12MPa continuous residual

From figure 6- figure 8, it can be seen that the numerical calculation process initialized using the one-dimensional isentropic theory is significantly accelerated, and the value that can finally converge is reduced by nearly three orders of magnitude compared to the use of hybrid initialization. As the inlet pressure continues to increase, the number of convergence steps of the calculation process using the hybrid initialization method increases from 60000 steps to nearly 80000 steps, while the calculation process using the one-dimensional isentropic theory method has a convergence step of about 30000 steps. There is no significant improvement in the number of steps. It can be seen from figure 4- figure 5 that the initialization method based on the one-dimensional isentropic theory method actually accelerates the convergence speed of the nozzle numerical simulation calculation process, effectively reduces the calculation time and improve the efficiency of the nozzle numerical calculation.

The final calculation results obtained by the two methods are compared below. The inlet pressure condition is 8 MPa, and the calculated pressure, velocity field and streamline diagram are shown in figure 9-figure 10.

(a) (b) (c)

**Figure 9.** After hybrid initialization, pressure, velocity, streamline results

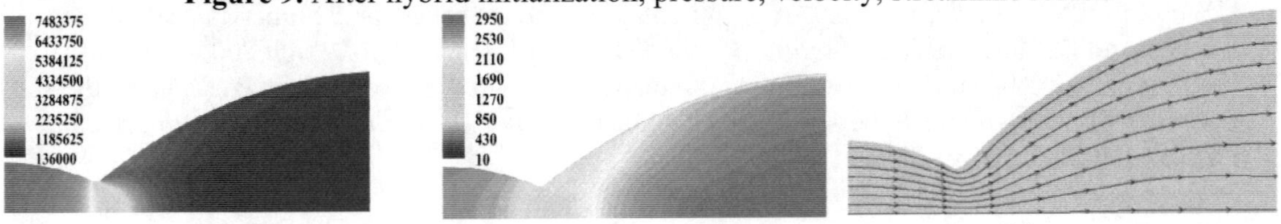


(a) (b) (c)

**Figure 10.** After one-dimensional steady initialization, pressure, velocity, streamline results

Comparing the final results obtained by the two initialization methods, we can see that there are some differences in the numerical results of the two final results, but the difference is not large, and the pressure, velocity, and streamline distribution are basically the same. The results obtained are reasonable. Since the iterative calculation is more sensitive to the initial value. Therefore, the initial conditions are well specified and have a great influence on the convergence of the calculation process. Before the iterative calculation, giving an initial value close to the calculated results is beneficial to improve the convergence of the iterative calculation. Conversely, if the initial value is far from the final value, it may cause convergence difficulties. Comparing figure 10 with 5, we can see that the initial velocity field based on the one-dimensional isentropic theory is closer to the final calculated velocity field, that is, the initial field obtained by the one-dimensional isentropic theory is better than the hybrid initialization method. Therefore, the initialization method based on the one-dimensional isentropic theory has a good effect on improving the convergence of the nozzle numerical simulation calculation process.

## 5. Conclusion

Comparing the two initialization methods, we can see that the one dimensional isentropic theory is used to initialize the Laval nozzle. Under the same conditions, it is better than the Fluent's own hybrid initialization method, which effectively improves the convergence and stability in the calculation process. It also improves the computational efficiency and has a more stable calculation process. Therefore, the numerical simulation of the problem of nozzles, if combined with the one-dimensional isentropic theory, can greatly improve the calculation efficiency and improve the convergence and stability of the calculation process, which is of great significance for the numerical simulation of the nozzle.

The numerical calculation of the nozzle is a complicated process, and its convergence is often not only affected by the initial conditions, but also closely related to various factors such as model simplification, meshing quality, discrete method, solution method and boundary conditions[6][7]. This paper mainly studies the influence of initial conditions on the convergence, so some simplifications are made, and there are still some shortcomings. The factors affecting the convergence of nozzle numerical calculations will be studied in the later stage.

## References

[1] HaiFeng H,FuTing B,Qiang C,Yang L. Flow separation and aeroelastic coupling analysis in overexpanded rocket nozzles.*Journal of Solid Rocket Technology* 2011(6)

[2] QingWei W,Bo L,RuGen W.Numerical Simulation of Two-dimensional Nozzle Using Fluid Injection for Throat Area Control.*Aeronautica Et Astronautica Sinica*.2009(2)

[3] Ping W,ChangPing L,BaiSong C.Laval Nozzle Flow Analysis Based on CFD Computation. *Aeronautical Computing Technique*.2009(2)

[4] WenXiang Z,JinQuan H,RenZhi Z.Improvement of Laval nozzle calculation model and simulative verification in aero-engine performance calculation.*Journal of Aerospace Power* 2009(11)

[5] PengFei L,MinYi X,FeiFei W.*Proficient in CFD Engineering Simulation and Case Combat* (Beijing: Posts and Telecom Press) p123

[6] WenZhi L,NaiRen Z,ChunLin-Z,YongZhong Z.Contour design and numerical calculation of certain solid propellant rocket motor nozzle.*Journal of Engineering Design* 2006(02)

[7] A.A.Verevkin,Yu.M.Tsirkunov.Flow of a dispersed phase in the laval nozzle and in the test section of a two-phase hypersonic shock tunnel *Journal of applied mechanics and technical physics*,2008,5(5):789-798

IOP Conf. Series: Materials Science and Engineering **608** (2019) 012013 doi:10.1088/1757-899X/608/1/012013

# Climatic Environmental Worthiness Model of the Civil Aircraft Based on the Grey Theory

**Zhang Hui[1] and Wu Jingtao[1]**

[1]Aircraft Strength Research Institute, Xi'an, Shan Xi, China

Email: zhanghui07282006@163.com

**Abstract.** The civil aircraft must have the ability to the global operation. The aircraft may experience the high temperature, the low temperature, the high humidity, ice and other extreme climatic environment. Environmental worthiness plays an important role in the integrated capability of the civil aircraft. Environmental worthiness evaluation is a major part of the civil aircraft engineering. According to the environmental worthiness to diversified environmental factors, we can establish the criteria for environment classification. The environmental worthiness model of the civil aircraft was set up using the grey theory. The rationality and feasibility of the model was validated by the example. The results indicate that the environmental adaptability of the aircraft was ranked "poor" to "better". The result can supply the reference data for the environment adaptability requirement for thedesign of the civil aircraft.

## 1. Introduction

The aircraft during the service may experience the extreme climatic environment. The ability of the aircraft adapting to the environment profile during the life cycle can be defined as the environmental worthiness. The environmental worthiness is one of the general features of the aircraft. According to the NSTB database tracing aviation accident, the flight accidents due to the extreme weather account for 20.1%. The US military clearly pointed out the extreme climatic may bring out the equipment in the aircraft losing efficacy. The extreme climatic will influence the function and the performance of the aircraft. Therefore, how to appropriately evaluate the environmental worthiness has become a pivotal case.

Analysis Hierarchy method(AHP)[1-3] was proposed by professor T.L. Safety in 1970s,which is a decision method with both quantitative and quantitative analysis. AHP was used for estimating weights of each criterion. The worthiness model is constructed by the grey theory with the AHP. A sophisticated method is created for the evaluation of the environmental worthiness [4-6].

## 2. The failure of the aircraft in the extreme weather

The extreme climatic will influence the function and the performance of the aircraft. The high temperature, the low temperature, humidity, the freezing, the sleet and the solar radiation can be achieved in the facility, which we can use to evaluate the worthiness of the aircraft to the defined climatic environment.

The high temperature and the solar radiation may deteriorate the properties of the materials, even cause the equipment, the components and the parts failure. The high temperature also causes the fuel losing the lubrication characteristics which may bring the vapor lock. The properties of the electronic equipment may decrease because of the high temperature.

The low temperature may increase the viscosity of the grease and it even may be frozen. The APU, engineer and the electronic equipment can't start in the low temperature.

The humidity may condense nearby the electronical box, which may bring the electronic short circuit.

The flaps and flats may be hard to control in freezing weather.

The sleet drops may permeate through the equipment box and cause the short circuit.

## 3. Building evaluation model

There are so many climatic factors at different levels to impact the climatic environment worthiness. The classification has fuzziness and the different abilities.It is difficult to evaluate the credit of the climatic worthiness[7-9].

A mathematical model of a fuzzy grey [10] comprehensive evaluation can be expressed as follows:

### *3.1. Establish the evaluation Factors*

To evaluate the credit of the environment worthiness, one must first establish a criteria system. This enables an understanding of the relationships between influencing factors so the system can make the desired evaluation.

The scheme of the evaluation is denoted by $U_n = \{U_1, U_2, U_3 \cdots\}$, the second layer is the test climatic factors, which is denoted by:

$$U_1 = \{u_{11}, u_{12}, u_{13}, u_{14}, \cdots, u_{1m}\} \quad (1)$$

$$U_2 = \{u_{21}, u_{22}, u_{23}, u_{24}, \cdots, u_{2m}\} \quad (2)$$

$$U_n = \{u_{n1}, u_{n2}, u_{n3}, u_{n4}, \cdots, u_{nm}\} \quad (3)$$

Determining the comparative judgment matrix:

$$U = \{u_{ij}\}_{n\times m} \quad (4)$$

In formula(4),$n$ is the number of the evaluation number, $m$ is the number of the climatic influence factors,$u_{ij}$ represents the quantized value of the scheme $i$ to the factors $j$.

### *3.2. Normalize the index values*

Normally, the dimension and the range of the evaluation data in the formula (4)are different. Therefore, the datas should be normalized. The normalization equation is defined as follows

$$f_{ij}=\frac{a_{ij}-\min_{1\le i\le n}\{a_{ij}\}}{\max_{1\le i\le n}\{a_{ij}\}-\min_{1\le i\le n}\{a_{ij}\}} \quad (5)$$

Thereinto, i=1,2,…,n, j=1,2,…,m

We get the evaluation normalizationmatrix:

$$F=\begin{bmatrix} F_1 \\ F_2 \\ \cdots \\ F_n \end{bmatrix} = \begin{bmatrix} f_{11} & f_{12} & \cdots & f_{1n} \\ f_{21} & f_{22} & \cdots & f_{2n} \\ & & \cdots & \\ f_{n1} & f_{n2} & \cdots & f_{nn} \end{bmatrix} = [f_{ij}]_{n\times m} \quad (6)$$

### *3.3. Calculating the weight of each indicator according to the judgment matrix*

The weight vector can be confirmed by the method of all evaluative experts on the evaluation indexes or by the entropy method. Considering the indices have little relevance or even no correlation. We use the entropy method.

$$p_{ij} = {f_{ij}}/{\sum_{i=1}^{n} f_{ij}} \quad (7)$$

The information entropy of the index can be obtained using this formula:

$$e_j = \begin{cases} -k\sum_{i=1}^{n} p_{ij}\ln(p_{ij}), p_{ij}\neq 0 \\ e_j=0, p_{ij}=0 \end{cases} \quad i = 1,\cdots,n; j = 1,\cdots,m \; ; \quad (8)$$

In which, $k$ is a coefficient, $k = 1/\ln(n)$.

The redundancy of the information entropy is calculated by:

$$d_j = 1 - e_j \quad (9)$$

The weights of each indicator are calculated as follows:

$$\omega_j = \frac{d_j}{\sum_{j=1}^{m} d_j} \quad (10)$$

*3.4. Calculate the related degree coefficient*

For the evaluation index, the higher the value, the better the ability of the adaptability to the extreme climatic environment. According to the gray system theory, the optimum reference array is $f_0=[f_{01},f_{02},\ldots,f_{0m}]$.

In formula (5), $f_{0j}=\max\{f_{ij}|i=1,2,\ldots,n\}$.

The comparison array is $f_i=[x_i, x_{i2},\ldots,x_{im}]$,then we can get the improved grey related coefficient degree as follows:

$$\xi_{ij} = \frac{\min_i\min_j\{|f_{ij}-f_{0j}|\}+\rho\max_i\max_j\{|f_{ij}-f_{0j}|\}}{|f_{ij}-f_{0j}|+\rho\max_i\max_j\{|f_{ij}-f_{0j}|\}} \quad (11)$$

Where $\rho$ is distinguishing parameter, $0 \le \rho \le 1$,it is usually set as 0.5.

Thenwecan get related degree matrix E:

$$\mathbf{E} = (\xi_{ij})_{n\times m} \quad (12)$$

*3.5. Calculate the comprehensive evaluation matrix*

Using the the grey relation analysis, the comprehensive evaluation result matrix is as follows:

$$R = \omega \times E^T \quad (13)$$

## 4. Example and Analysis

The aircraft during the test will be assessed after experiencing the extreme climatic. Considering the actual operation environment, the extreme climatic include the high temperature, the low temperature, humidity, the freezing, the sleet , the solar radiation. According to the checking result after the test, the original scores can be seen in following table 1-table3, which is the environmental worthiness value to the single environment for different sorties. Each column of the matrix is the aircraft's evaluation value which is given by eight experts. The comprehensive result can be obtained by the the averaging method.

**Table 1.**Original data of the A1 aircraft

| | high temperature | low temperature | solar radiation | humidity | sleet | freezing |
|---|---|---|---|---|---|---|
| 1 | 0.652 | 0.518 | 0.623 | 0.600 | 0.718 | 0.415 |
| 2 | 0.452 | 0.534 | 0.632 | 0.888 | 0.718 | 0.400 |
| 3 | 0.434 | 0.623 | 0.888 | 0.899 | 0.698 | 0.467 |
| 4 | 0.678 | 0.455 | 0.544 | 0.721 | 0.723 | 0.487 |

| | | | | | | |
|---|---|---|---|---|---|---|
| 5 | 0.692 | 0.744 | 0.689 | 0.731 | 0.723 | 0.498 |
| 6 | 0.777 | 0.789 | 0.775 | 0.754 | 0.722 | 0.512 |
| 7 | 0.687 | 0.765 | 0.581 | 0.700 | 0.698 | 0.500 |
| 8 | 0.522 | 0.799 | 0.672 | 0.712 | 0.798 | 0.467 |
| result | 0.612 | 0.653 | 0.676 | 0.751 | 0.725 | 0.468 |

**Table 2.** Original data of the A2 aircraft

| | high temperature | low temperature | solar radiation | humidity | sleet | freezing |
|---|---|---|---|---|---|---|
| 1 | 0.566 | 0.567 | 0.453 | 0.700 | 0.814 | 0.511 |
| 2 | 0.666 | 0.587 | 0.453 | 0.700 | 0.799 | 0.509 |
| 3 | 0.766 | 0.500 | 0.432 | 0.712 | 0.800 | 0.533 |
| 4 | 0.689 | 0.500 | 0.463 | 0.738 | 0.812 | 0.543 |
| 5 | 0.712 | 0.612 | 0.450 | 0.747 | 0.823 | 0.484 |
| 6 | 0.766 | 0.632 | 0.402 | 0.751 | 0.854 | 0.498 |
| 7 | 0.777 | 0.654 | 0.459 | 0.752 | 0.864 | 0.499 |
| 8 | 0.789 | 0.612 | 0.398 | 0.766 | 0.877 | 0.422 |
| result | 0.716 | 0.583 | 0.439 | 0.733 | 0.830 | 0.500 |

**Table 3.** Original data of the A2 aircraft

| | high temperature | low temperature | solar radiation | humidity | sleet | freezing |
|---|---|---|---|---|---|---|
| 1 | 0.594 | 0.700 | 0.623 | 0.722 | 0.758 | 0.324 |
| 2 | 0.584 | 0.720 | 0.613 | 0.732 | 0.760 | 0.344 |
| 3 | 0.590 | 0.750 | 0.623 | 0.762 | 0.761 | 0.360 |
| 4 | 0.576 | 0.770 | 0.653 | 0.722 | 0.768 | 0.373 |
| 5 | 0.568 | 0.730 | 0.663 | 0.782 | 0.762 | 0.354 |
| 6 | 0.578 | 0.750 | 0.673 | 0.722 | 0.700 | 0.359 |
| 7 | 0.590 | 0.720 | 0.663 | 0.722 | 0.794 | 0.350 |
| 8 | 0.594 | 0.730 | 0.653 | 0.712 | 0.757 | 0.353 |
| result | 0.584 | 0.734 | 0.646 | 0.735 | 0.758 | 0.352 |

The judgment matrix of paired-comparisons can be established, which can be seen in table 4:

**Table 4.** The judgment matrix

| Number | high temperature | low temperature | solar radiation | humidity | sleet | freezing |
|---|---|---|---|---|---|---|
| A1 Aircraft | 0.612 | 0.653 | 0.676 | 0.751 | 0.725 | 0.468 |
| A2 Aircraft | 0.716 | 0.583 | 0.439 | 0.733 | 0.830 | 0.500 |
| A3 Aircraft | 0.584 | 0.734 | 0.646 | 0.735 | 0.758 | 0.352 |

According to the formula(5), we can get the normalization matrix:

$$X=\begin{vmatrix} 0.2121 & 0.4636 & 1 & 1 & 0 & 0.7838 \\ 1 & 0 & 0 & 0 & 1 & 1 \\ 0 & 1 & 0.873 & 0.1111 & 0.3143 & 0 \end{vmatrix} \tag{14}$$

The evaluation weight matrixof grey type for respective indexes based on step 3 are as follows:

$$\omega = [0.1952, 0.1458, 0.1254, 0.2379, 0.1687, 0.1270] \quad (15)$$

Computing the related degree Coefficient:

$$\gamma = \begin{vmatrix} 0.3882 & 0.4824 & 1 & 1.000 & 0.3333 & 0.6981 \\ 1 & 0.3333 & 0.333 & 0.3333 & 1.000 & 1 \\ 0.3333 & 1 & 0.7980 & 0.3600 & 0.4217 & 0.333 \end{vmatrix} \quad (16)$$

The information of the entropy based on step are as follows:

$$R = [0.6543, 0.6606, 0.5101] \quad (17)$$

From the results, it can be concluded that the environmental adaptability of the second sortie aircraft (A2) is the best, the first sortie aircraft (A1) is the second, and that of the third sortie aircraft (A3) is the worst.

## 5. Conclusions

The study is mainly focused on the comprehensive evaluation of the worthiness to the extreme climatic which can be achieved in the climatic environment test facility. The paper constructs evaluation model by combining grey evaluation method with AHP. The model provides a new comprehensive method to evaluate worthiness ability of the aircraft. It plays an important role in optimizing Aircraft Design and improving aircraft's worthiness ability.

## Acknowledgments

Authors wishing to acknowledge assistance or encouragement from colleagues, special work by technical staff or financial support from organizations should do so in an unnumbered Acknowledgments section immediately following the last numbered section of the paper.

## References

[1] Xuanzi, Hu .Application of grey clustering theory combined with AHP in railway station evaluation, Proceedings of IEEE ICIA 2006,pp 1289-1294, 2006.

[2] WANG Wenke and Liu Peide. The evaluation of urban public traffic line networkbased on the Grey-AHP method, International Conference on Transportation Engineering 2007,pp.1991-1996,2007.

[3] Minghui, Zhang.A Comprehensive College Coaches Evaluation Model Based on AHP and Grey Correlation Theory, Proceedings 2015, pp.330-334 , 2015.

[4] Guozhong Huang and Siheng SunSafety. Evaluation of Construction Based on the Improved AHP-Grey Model, Wireless Personal Communications, vol 103,pp 209–219,2018.

[5] Shulin Li. Evaluation model of environmental worthiness for military aircraft, Aeronautic Sinica, vol 4,pp.1055-1057,2009.

[6] Zong-Feng Q.I., GuoSheng Wang. Effectiveness evaluation of electronic warfare command and control system based on grey AHP method, Journal of Chemical and Pharmaceutical Research, vol6,pp. 535-542, 2014.

[7] Fei, Lv.Aircraft financial leasing project risk evaluation using AHP and grey comprehensive algorithm, International Journal of Digital Content Technology and its Applications, vol 6,pp 259-267,2012.

[8] Xia, Chunyan.Method of weight assignment for multi-criterion based on grey interval AHP, Advanced Materials Research,vol 243-249,pp 5285-5288,2011.

[9] Sun, Yang, Research on an intelligent behavior evaluation system for unmanned ground vehicles, Energies,vol 11,pp.1-22,2018.

[10] Xiaoxi, Guo.A grey relational analysis model of scientific research ability on music based on AHP and its realization, Journal of Signal Processing, Image Processing and Pattern Recognition,vol8, n 5,pp. 211-222,2015.

IOP Conf. Series: Materials Science and Engineering **608** (2019) 012014 doi:10.1088/1757-899X/608/1/012014

# Research on Safety Risk Evolution of Military Aviation System Based on ANP-SD

**Qiang Wang, Qiuhan Liu*, Ruicong Xia and Yingjie Huang**

No.1, Changle East Road, Baqiao District, Shanxi, China

*E-mail: 774030434@qq.com

**Abstract.** With the development of equipment, the complexity of China's military aviation system continues to increase. The system risk is transformed from the initial independent risk factor to the cross-influence risk factor, which makes the system risk control difficult. In order to maintain the safety level of military aviation system, the factors affecting the safety level of the system are analysed from five aspects: personnel, machine, environment, management and logistics. The analytic hierarchy process (ANP) is used to construct the model of the system risk factors and the system dynamics method is used to establish the risk evolution simulation model, and quantitatively analyse the sensitivity of risk factors to explore the basic rule of risk evolution.

## 1. Introduction

The update of equipment contributes to the complexity of the military aviation system and brings a series of problems. Simple risk factor analysis has not been suitable for nowadays complex military aviation systems, and the interconnection and impact of risks have gradually become the main factors affecting the overall safety level of the system. Therefore, a method that can be used to analyse the risk evolution for the accidents prevention is put forward, and the general mechanism is obtained through system simulation by processing the existing data.

At present, research objects on risk evolution at home and abroad are mostly concentrating on large and complex systems. Jiang Xin et al. constructed a system dynamics model of risk evolution of cross-operation using system dynamics, and simulated the self-organizing formation path of risk evolution [1]; Chen Shuai conducted system dynamics modeling and simulation of risks in civil aviation maintenance system and the relationship between risk factors is studied [2]; Zhao Yiqing et al. constructed a risk evolution model associated with tailings pond accident hidden danger using system dynamics, and characterize the rule of risk evolution of tailings ponds by combining with the complex network methods [3]; Liu Qing et al. constructed the risk evolution model of navigation of ships in the Three Gorges Dam area with the method of system dynamics, and simulated the risk evolution process of navigation of ships under the influence of various factors with VENSIM software [4].

At present, there are few researches on system risk evolution in the field of military aviation, and no mature mechanism has formed. Considering that the military aviation system is of large-scale and complex with nonlinear and multivariable characteristics, this paper combines the analytic hierarchy process (ANP) and the system dynamics method (SD) to construct the safety risk evolution model based on ANP weight calculation, using the SD simulation software VENSIM to analyse the sensitivity of risk factors. Forming the regulation of the risk evolution, and providing a theoretical basis aim at improving the safety management level of military aviation systems.

IOP Conf. Series: Materials Science and Engineering **608** (2019) 012014 doi:10.1088/1757-899X/608/1/012014

## 2. Construction of safety risk evolution indicator system for military aviation systems

### *2.1. The establishment of military aviation system safety risk evolution indicator system*

Military aviation system is a typical complex man-machine system built with complicated structure and various elements. Referring to the theories and models of safety analysis, such as REASON model, SHEL model and FMECA model, etc. and from the perspective of system safety, the influencing factors mainly contain personnel, equipment, environment, management and logistics support. After repeated discussions with 5 experts in the field of aviation safety and combined with the analysis of existing data, 18 risk factors in 5 risk links was finally determined, as shown in Figure 1.

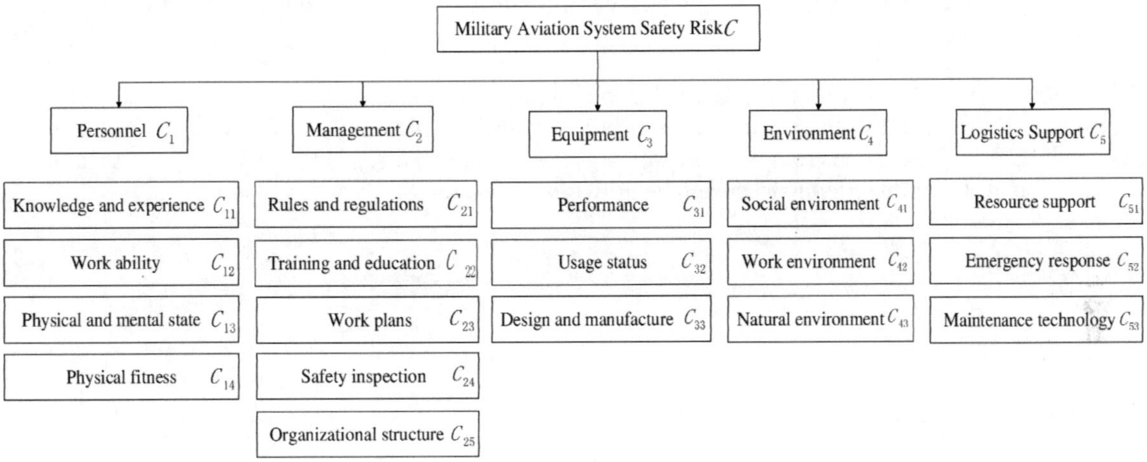


**Figure 1.** Military aviation safety risk evolution indicator system.

### *2.2. The construction of ANP network structure based on indicator system*

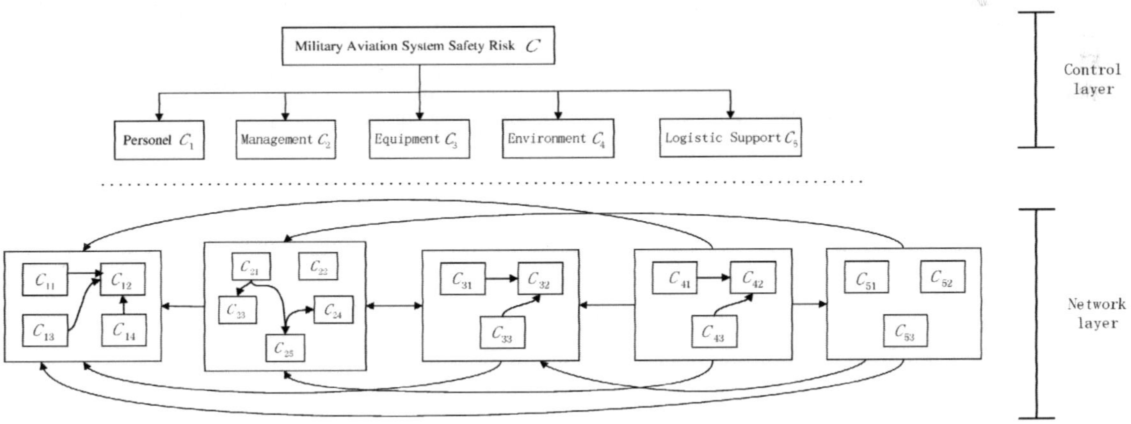


**Figure 2.** ANP structural model of safety risk evolution system.

The ANP is a decision-making method to adapt to the non-independent hierarchical structure and proposed by Professor T.L.Saaty of the University of Pittsburgh in the 1990s. It is developed and extended on the basis of the analytic hierarchy process (AHP). It shows the relationship between each element of a system by a network-like structure, instead of the simple hierarchical structure of AHP [5].

The typical network structure description of ANP usually consists of two parts. The first part is the control layer, including the problem objectives and decision criteria. The second is the network layer, which is composed of all the elements of the control layer and the internal is a network structure where elements interact with each other [6].

According to the safety risk evolution system established in Figure 1, the network structure model of ANP as shown in Figure 2 is established, where the first level indicators are $C_i(i = 1,2,3,4,5)$, the second level indicators are $C_{ij}(i = 1,2,3,4,5; j = 1,2,3,4,5)$.

## 3. Construction and analysis of risk evolution model of military aviation system

System Dynamics is a kind of system analysis technique based on the feedback control theory proposed by professor J.W. forrester of Massachusetts institute of technology (MIT) in 1956, which quantitatively studies nonlinear, multiple feedback and complex time-varying systems by means of computer simulation [7].

### *3.1 Establishment of risk evolution model of military aviation system*

According to Figure 1, it can be divided into five subsystems, and the relationship among every factors in the subsystem is presented through the model. As shown in Figure 3, SD flow diagram of the safety risk evolution of the military aviation system is drawn.

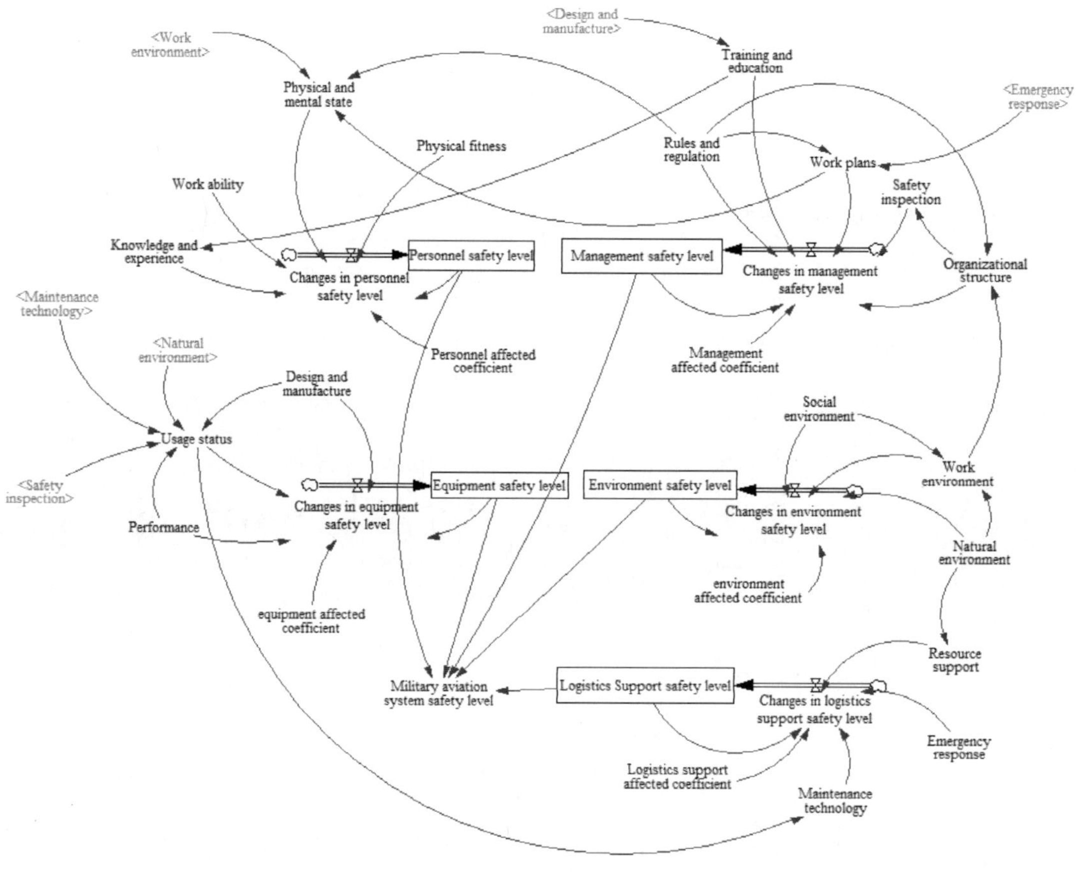


**Figure 3.** Military aviation system safety risk evolution SD flow diagram.

### *3.2 Simulation analysis of safety risk evolution of military aviation systems*

Based on the results of the expert ratings and the weight values, we present the initial values of the horizontal variables in the model, as well as the SD equations for the main variables. The model simulation period is 48 weeks and the step size is 1 week. VENSIM software was used for the sensitivity analysis. The results are shown in Figure 4, Figure 5, Figure 6 and Figure 7.

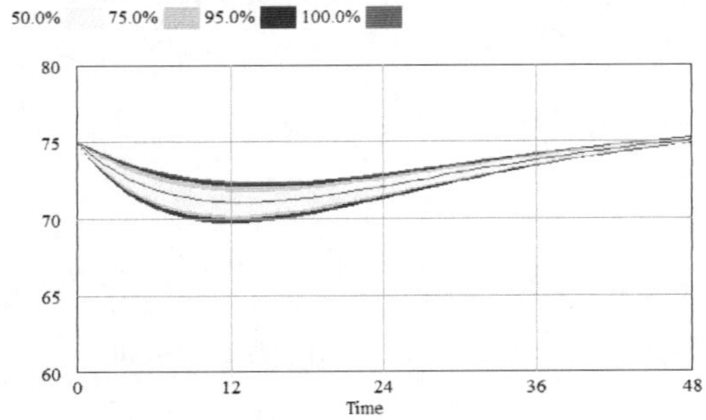


(1) Management impact on personnel

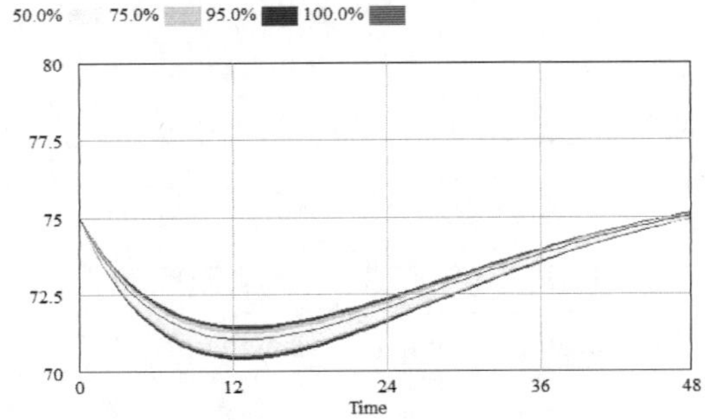


(2) Equipment impact on personnel

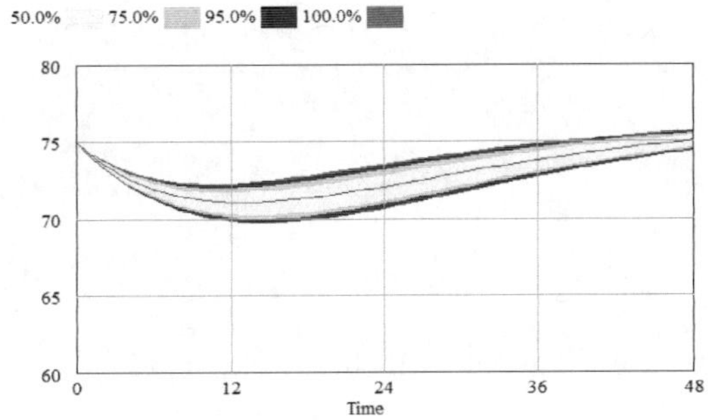


(3) Environment impact on personnel

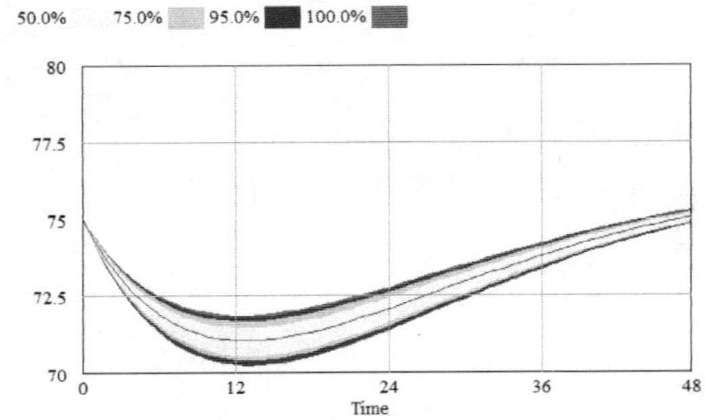


(4) Logistics support impact on personnel

**Figure 4.** Simulation results of various factors affecting personnel safety level

Analysing the simulation result in Figure 4. and the conclusion was drawn as follows.

1) First of all, as can be seen from the four figures, when the safety level of each factor reduced 10%, the safety level of personnel is lower than 70 due to management factors and environmental factors, and their influence is greater than that of equipment and logistics factors.

2) The impact of management and environment factors on personnel safety level were analysed together and it was found that when both of them reduced 10%, the impact of management was appreciable before week 12th, and the safety level of personnel had dropped below 70 at about week 11th. The influence of environment was more noticeable after week 12th, and lasts longer than that of management. From the picture above, combined with equipment and logistics support factors the former influence on personnel safety was less than the latter. Comparing the images of the foregoing with management and environmental factors, it can be drawn that under the 10% rise of four factors, the promotion of equipment and logistics safety level didn't bring significant rise to the personnel safety level while the management and environment factors are just the opposite which can be inferred that they possess greater influence on safety level.

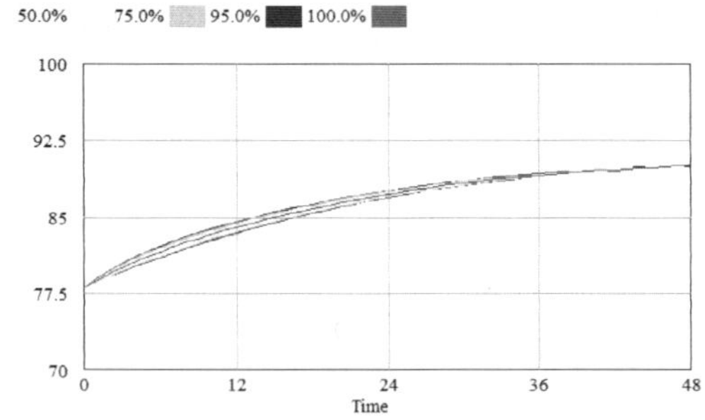


(1) Equipment impact on management

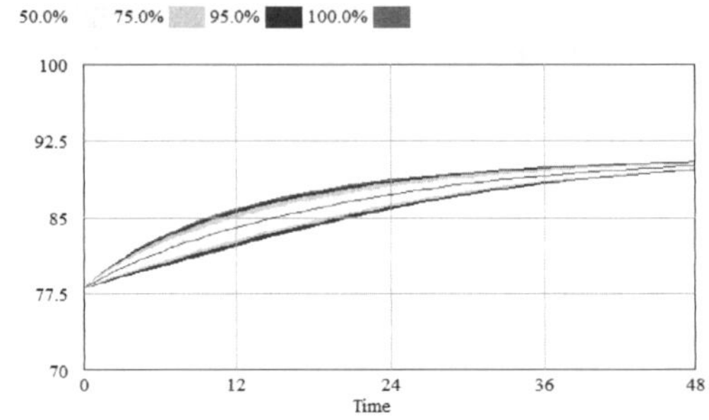


(2) Environment impact on management

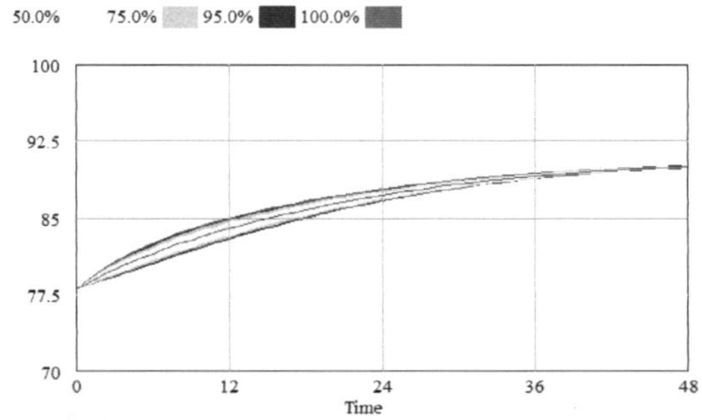


(3) Logistics support impact on management

**Figure 5.** Simulation results of various factors affecting management safety level

In Figure 5, it can be seen as listed below.

3) First of all, the environment has a greater impact on management, followed by logistics support and equipment factors. The main reason for that is the significant impact of the working environment on the management safety level. The working environment includes the basic facilities construction of the workplace, interpersonal relationships, and work enthusiasm. It is necessary to consider the improvement of management level from these aspects.

4) The reason that logistics support factors and equipment factors have little effect on the safety level of management may be that equipment is mainly affected by management factors, and its impact on management factors is small. Logistic support factors are independent of management factors and therefore have less impact.

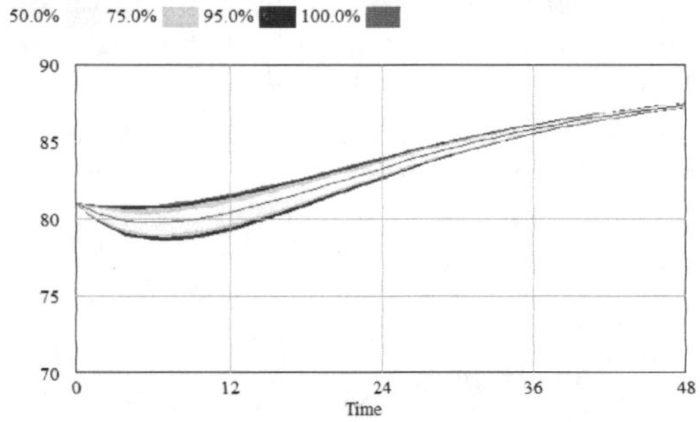


(1) Management impact on equipment

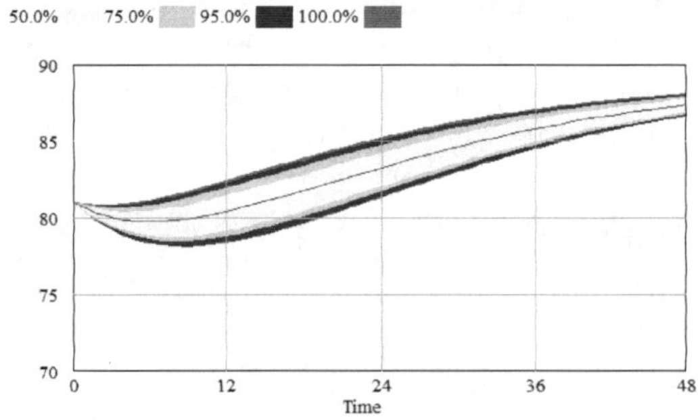


(2) Environment impact on equipment

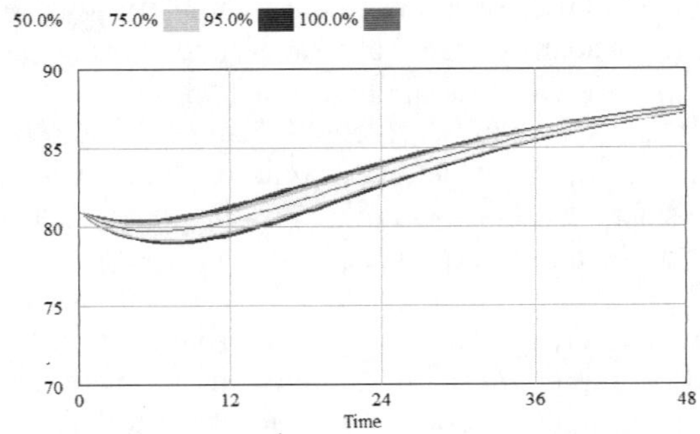


(3) Logistics support impact on equipment

**Figure 6.** Simulation results of various factors affecting equipment safety level

From Figure 6,

1) It can be seen that environmental factors have the greatest impact on equipment safety levels, followed by management factors and logistics support factors. The reason for the greater impact of environmental factors is that the natural environment has a greater impact on the use of equipment.

2) The impact of management and logistics support factors on equipment is slightly different. The impact of management on the safety level of equipment is mainly concentrated around week 8th, and the impact is greater than the logistics support factor. The impact of logistics support factors on equipment safety level is mainly concentrated around week 12th and lasts longer than that of management.

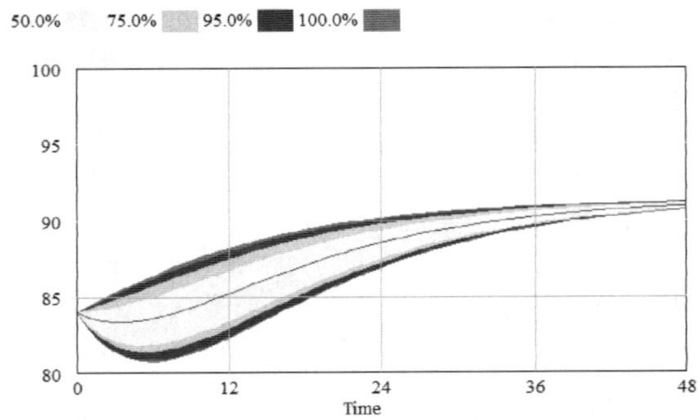


**Figure 7.** Simulation results of environmental factors affecting logistics support safety level

Analysing the simulation result in Figure 7. It can be seen that when the environmental safety level is reduced by 10%, the level of logistical safety is rapidly reduced. However, when the environmental safety level is increased by 10%, the level of logistical safety is growing slowly. The reason may be that natural environment has a great impact on logistical support safety level.

## 4. Conclusion

Based on the analysis of these key aspects of military aviation system safety risk, the indicator system of it was formed. ANP method was used to construct a military aviation system safety risk evolution ANP structure model, and Super Decision software was used to determine the weight of indicators. The risk evolution model is established by using system dynamics method, and the basic rule of risk evolution is quantitatively analysed. From the results of various factors sensitivity simulations, it can be concluded that:

1) The safety level of personnel in military aviation systems is mainly affected by management and environmental factors. The influence of management factors is relatively large and rapid. The influence of environment factors is slow and continuous. The equipment and logistics support factors

have little influence on personnel safety level. Improving the safety level of personnel begins with management. Improving the level of management safety can quickly raise personnel safety level, and it is necessary to continuously improve the level of environmental safety.

2) The level of management safety is mainly affected by environmental factors and is less affected by logistics support factors and equipment factors. Promoting the level of management safety is mainly based on improving the safety level of the work environment, such as optimizing infrastructure construction, maintaining the harmony of interpersonal relationships, and inspiring the enthusiasm of personnel.

3) The level of equipment safety is mainly affected by environmental factors, and management and logistics support factors have less impact on it. The natural environment cannot be changed, but the improvement of equipment can be made according to the characteristics of the environment. For example, in coastal with high humidity, equipment can be regularly rust-proofed. At the same time, the impact of management and logistical support factors cannot be ignored.

4) Logistic support factors are mainly affected by environmental factors. Different equipment support plans can be formulated for different weather conditions to ensure that there are emergency plans in case of special circumstances to ensure the safety of logistics support.

These conclusions provide a reference for managers to correctly analyse and reduce the risk of military aviation.

**Reference**

[1] Jiang Xin and Wu Yuanli 2015 Research on evolution mechanisms of risks in hydropower project group cross operation *China Safety Science Journal* **25** 157

[2] Chen Shuai 2016 Analysis for Civil Aviation Safety Based on Risk Transmission *Nanjing University of Aeronautics and Astronautics*

[3] ZHAO Yiqing, Qin Xuan, LI Zhongxue, TANG Liangyong 2014 Dynamic modeling and simulation hazard and risk evolution for mine tailing ponds *Journal of University of Science and Technology Beijing* **36** 1158

[4] LIU Qing, HAN Dan-dan, CHEN Yanqing, ZHANG Wei 2016 Research on navigation risk evolution in Three Gorges dam area based on system dynamics *China Safety Science Journal* **26** 19

[5] HE Chunchun and WANG Yingming 2014 A Review of the Analytic Network Process *Science and Technology Management Research* **34** 204

[6] Wang Lianfen 2001 The Theory and Algorithm of Analytic Network Process *Systems Engineering-theory & Practice* **3** 44

[7] Xie Hong and Yang Yunqi 2018 SD Flow Diagram for safety behavior influence mechanism *Journal of North China Institute of Science and Technology* **15** 82

IOP Conf. Series: Materials Science and Engineering **608** (2019) 012015 doi:10.1088/1757-899X/608/1/012015

# Application of Fault Diagnosis Expert System for Unmanned Vehicle Safety

**ChunPing Chen[1,2], Lin Zhang[1], Jie Wang[1] and Jie Zhang[1,*]**

[1]Graduate School, Air Force Engineering University, Xi'an 710054, China
[2] Shanghai Institute of Precision Metrology and Testing, Shanghai 200000, China

[*]afeu_zhangjie@163.com

**Abstract.** In order to ensure the safety of the underwater unmanned vehicle (UUV) and reduce the risk of damage or loss during operation, the safety system will be equipped on the UUV. It can accurately locate the fault when the submersible encounters dangerous situations, such as equipment failure and cabin flooding, assess the dangerous situation and make emergency measures, so as to help the underwater unmanned submersible realize self-rescue. Fault diagnosis expert system is a knowledge-based fault diagnosis method, which is suitable for complex systems with incomplete knowledge. However, because the working environment of the underwater unmanned underwater vehicle is very complex and the fault types are diverse and change in real time, the traditional fault diagnosis expert system may not meet the response time requirements of the safety system due to the long response time required. Therefore, the finite state machine technology is applied to the diagnostic expert system to improve the response speed of the diagnostic expert system, and at the same time make the system more flexible and convenient to expand its functions. In addition, the model-based design method was applied to establish the model of the security system in stateflow, and the model verification was carried out. With the help of PLC Coder tool, the code of the target controller was generated and the algorithm was deployed.
**Keywords:** Safety system, Finite state machine, Remote, repair, troubleshooting, Model-based design

## 1. Introduction

Because unmanned submersibles face complex and uncertain working conditions when working underwater, sometimes complex underwater tasks are required, so the risk of underwater submarine unloaded or lost during underwater operations is extremely high. . Historically, the loss of underwater unmanned submersibles has occurred many times, such as the US "Nereus" HROV, the US "ABE" AUV and the British "Autosub-2" AUV [1]. In order to improve the safety of underwater unmanned submersibles and reduce the chance of loss or damage, an efficient and reliable safety system is essential for underwater unmanned submersibles.

A complete security system consists of two parts: troubleshooting and fault decision making. Fault diagnosis and decision-making methods can be divided into signal processing based methods, model analysis based methods and knowledge based methods [2]. The signal processing based method is a quantitative analysis method that analyses the sensor signals to extract eigenvalues such as amplitude, frequency, and variance to determine the type of fault. Model-based analysis methods also require the use of measurement signals from actual sensors to train fault analysis models. However, the actual object is usually a nonlinear system that is difficult to model, and the actual data of the sensor cannot be obtained before the actual test, and this is the input condition necessary to design the diagnostic system based on the signal processing method or the model analysis based method. . In contrast, knowledge-based methods, such as expert systems, fuzzy inference, and pattern recognition, are less likely to be implemented and more suitable because they do not require an accurate object model [3]. This approach can be used to design fault diagnostic systems for complex systems with incomplete knowledge. The expert system is a computer program system that simulates the human reasoning process of the problem, and then completes the representation of human knowledge, and uses heuristic knowledge instead of exact matching to solve the problem, more accurately reflecting most human knowledge. Typically, diagnostic expert systems are designed in three types: rule-based diagnostic expert systems, model-based diagnostic expert systems, and online diagnostic expert systems. In the rule-based diagnostic expert system, the inference strategy includes forward reasoning (that is, the reasoning process that follows the rules from the original data to the conclusion), backward reasoning (that is, the process of inferring from the target to the data through hypothesis verification), and two-way reasoning.

A new expert system emerged, called the model-based diagnostic expert system. The premise of this method is that the measurement sensor is highly accurate, which is also a key issue for model-based diagnostic expert systems. The online diagnostic expert system is a combination of the knowledge base of the traditional expert system and the interaction process with the target object. Because the online expert system is parallel with the dynamic process, response time becomes a key issue in such expert systems [4]. Aiming at this problem, this paper proposes a method to apply the finite state machine algorithm to the online expert system inference engine to improve the inference performance of the online expert system and apply it to the security system of the unmanned submersible. The finite state machine FSM contains a finite number of system states that can be inferred from the state transition table and complete the state transition after the input event is obtained. It has the ability to describe extremely complex logical reasoning relationships in a clear and concise form, and is ideally suited for the design of complex safety systems for underwater unmanned submersibles.

This paper analyses the working modes of underwater unmanned submersibles, sorts out and lists various fault events that may be encountered by underwater unmanned submersibles, and then completes the classification of fault events and the definition of fault levels. Design the working principle of the safety system, determine the state set of the finite state machine and the input event set according to the fault event classification and fault level definition, and complete the state transition diagram of the finite state machine. Finally, the model of the security system is built in the stateflow model tool to verify the model, and the automatic generation and deployment of the controller code is completed.

## 2. Safety system design based on finite state machine

The unmanned submersibles in this article can be divided into two modes: remote mode and autonomous mode. In the remote control mode, the water surface operator remotely operates the unmanned submersible to complete the test task; in the autonomous mode, the unmanned submersible self-completes the binding test task. In either of the above modes, the unmanned submersible may work under normal conditions and emergency conditions. Emergency conditions refer to sudden failures or dangers when unmanned submersibles are working underwater. It is necessary to complete decision-making and self-rescue according to unexpected situations, reducing the risk of damage or loss. At this point, the safety system in the underwater unmanned submersible should play the necessary role.

The safety system in this paper consists of the Emergency Controller Unit (EMU), the actuator and the emergency power supply. Among them, the actuator includes a steering system, a submerged floating system, a high pressure blowing system and a throwing system. The finite state machine based fault diagnosis online expert system proposed in this paper runs in the emergency controller (programmable logic controller PLC).

In practical applications, some fault diagnosis expert systems complete diagnosis and decision making according to the order of fault events, which takes a lot of time [5]. An expert system based on binary decision graph and fault tree classification is an effective fault diagnosis expert system that can guide fault analysis, locate faults and determine emergency strategies [6]. However, the underwater unmanned submersibles in this article may have many concurrent fault events during operation, and even the fault level may change with time. For such dynamic systems, binary decision graphs and fault tree classification methods show their deficiencies in time-varying systems. In order to solve this problem, this paper proposes an improved online fault diagnosis expert system (ODES), which can be used in such complex systems with multiple fault concurrency or fault levels changing with time, and has high flexibility and response speed. Fast, convenient function expansion and so on.

As shown from that, the ODES in this paper has two types of inputs: real-time fault or alarm signals generated by the unmanned submersible master system; and emergency response monitoring signals generated by the emergency controller[7]. They are all stored in the fault database as input to the expert system. The meta-knowledge base and the expert decision rule base constitute the knowledge base of the expert system in the security system.

In the knowledge base of the expert system, the fault event that causes the underwater unmanned submersible to lose working capacity is defined as the top fault category T; the middle fault category A is subdivided into: power and propulsion system fault (A1), hydraulic steering system Fault (A2), submerged and high pressure air system fault (A3), equalization and drain system fault (A4), communication system fault (A5), submersible attitude abnormality (A6), shafting fault (A7). Finally, we can analyse nearly 240 basic fault events X, which correspond to various fault events that may occur in the actual system. Based on past experience and knowledge, the above basic fault events and their combinations can be divided into four levels, including: preventive layer fault events, Class A fault events, Class B fault events, and Class C fault events. The faults corresponding to the four levels are shown in Table 1. When a preventive layer failure occurs, the safety system does not need to be operated, and the master control system on the submersible has the ability to solve the problem by itself. When a Class A, Class B or Class C fault occurs, the emergency controller will take different emergency measures according to the current situation to disengage the submersible.

The inference engine of the online diagnostic expert system in this paper is based on NDFSM design. According to the working mode of the underwater unmanned submersible in this paper, the state collection and input event collection are as follows: 1) State collection: Normal: Normal state; Test: Self In-process; Class A fault status: There is a Class A alarm for steering up and down; Class B fault status: There is a Class B alarm for short circuit blowout; Class C fault status: There is a Class C alarm for dumping; After troubleshooting, it is in standby mode. 2) Input event set: Class A fault alarm: fault1; Class B fault alarm: fault2; Class C fault alarm: fault3; Self-rescue completion; Reset operation; Emergency simulation test starts; Alarm or emergency simulation test ends. Thus, we get the schematic diagram of NDFSM in the security system inference engine.

## 3. Security system model and code deployment

The model-based design approach is an advanced system development method that frees developers from programming. The use of a graphical design approach makes the goal of development clear, clear and unique, as well as easy to communicate and maintain [8]. The core of the model-based design approach is to model and validate the target object and algorithm, followed by data management and parameter configuration to ensure efficient, high-quality, high-availability code generation, and finally to verify the equivalence. Stateflow is a control logic model tool that models the target system through flowcharts in state machines and simulation models [9]. It is a graphical model tool that has unparalleled advantages in model the logic of complex systems and provides great model convenience for the online diagnostic expert system based on finite state machines. In addition, it provides a simulation environment for model validation and finally generates code for the target controller without requiring developers to manually write code to implement the algorithm functionality. The model of the security system inference engine is built in stateflow and packaged. the left side of the model is the input signal, and the right side is the output signal, which is the decision result.

PLC Coder is a PLC-oriented code generation tool in simulink [10]. By setting the IDE to Siemens Step 7, you can automatically generate SCL source files. If the program source file is imported into the portal software, the corresponding PLC program block can be generated [11]. The input and output of this block is consistent with the input and output of the module packaged in stateflow.

## 4. Security system function verification

The safety system is a double closed loop system. Closed loop 1 is the response loop of the safety system to the real-time fault or alarm signal generated by the master control system; Closed loop 2 is the loop of the response of the safety system to the emergency response [12] after taking emergency measures. If the emergency measures obtained through the closed-loop 1 decision are valid, the closed loop 2 will not determine the emergency measures. Otherwise, the closed loop 2 will upgrade the emergency fault level and take further emergency measures to ensure the safety of the unmanned submersible.

In order to verify the function of the safety system developed by applying the model-based design method, the following emergency process is simulated and verified, and the function is judged according to the result of the security system decision.

### *4.1. Emergency process of single failure*

The emergency process of single failure refers to the single failure of the unmanned submersible. The safety system determines the fault level according to the fault event and takes corresponding emergency measures. Taking the case of a type I fault as an example, the expected response process of the safety system is as follows: fault1 → class a fault event → class a fault state → steering up and down → fault release input the same fault event into the designed safety system, system response. The situation. When t = 1 526 s, the unmanned submersible has a type I fault, and the state of the FSM changes from "normal" to "class A fault state". At the same time, the safety system sends the corresponding emergency rudder angle to the underlying motion controller, which indicates that the safety system has taken corrective measures for the input fault.

### *4.2. The evolution of the fault level*

The evolution of the fault level refers to the emergency process in which the unmanned submersible has a fault event and takes corresponding measures. The unmanned submersible encounters another more dangerous fault event. Taking the type II fault after the occurrence of a type I fault as an example, the expected response process of the safety system is as follows: fault1 → class A fault event → class A fault state → steering up and down → fault2 → class B fault event → class B fault state → short circuit blowing out the same fault event into the designed safety system, the system response is shown. The response of the safety system indicates that the system can correctly switch between different fault levels and take corresponding emergency measures.

### *4.3. Emergency failure process*

The emergency failure process refers to the safety system. After the emergency measures are taken, the unmanned submersibles are not out of danger or show no tendency to escape from danger. At this point, the safety system will take further safety measures, such as replacing the rudder with ballast. The expected response process of the safety system is as follows:

Fault1 → Class A fault event → Class A fault state → Steering Upward → Upward failing → Class B fault event → Class B fault condition → Short circuit blowout Enter the same fault event into the safety system. The response of the safety system indicates that the system can correctly identify emergency failure conditions and take further emergency measures.

The simulation results of the emergency failure process show that the security system developed by the model-based design method accurately combines the expert knowledge [13]. Under the various

IOP Conf. Series: Materials Science and Engineering **608** (2019) 012015 doi:10.1088/1757-899X/608/1/012015

decision situations that the security system may encounter, it shows the correctness of the decision and proves that it has the ability to provide effective protection for safe operation of unmanned submersibles.

## 5. Conclusion

In this paper, the finite state machine technology is used to improve the traditional online diagnostic expert system. The model-based design method is applied to facilitate the efficient generation and deployment of controller code, which is faster, clearer and more accurate. A more reliable design method completes the design of the underwater unmanned submersible safety system. The safety system has been verified by a variety of typical decision-making situations, indicating that it has the ability to accurately locate faults and make quick fault decisions. It can provide real-time, effective and reliable basic safety for underwater unmanned submersible operations. At present, the security system is biased towards decision-making and logic control. However, underwater unmanned submersibles may also face more complicated situations. For example, in the process of self-rescue, it is also necessary to consider the avoidance of obstacles and the control of safety posture. In this case, the safety system needs to have the motion control function when the actuator fails, and these have not been considered in the design of the safety system. Therefore, integrated decision-making, logic control, sensor information acquisition and processing, and closed-loop motion control, to develop a more comprehensive and intelligent security system for underwater unmanned submersibles will be the main research content in the future.

## Acknowledgments

This work was supported by the Young Scientists Fund of the National Natural Science Foundation of China under Grant No. 61703412 and the China Postdoctoral Science Foundation under Grant No. 2016M602996.

## References

[1] Liu X, Sheng Z, Yin C, et al. Performance analysis of Routing Protocol for Low power and Lossy Networks( RPL) in large scale networks. IEEE Internet of Things Journal, vol. 4, no. 6, pp. 2172-2185, 2017.
[2] Ancillotti E, Bruno R, Conti M. The role of the RPL routing protocol for smart grid communications. IEEE Communications Magazine, vol. 51, no.1, pp. 75-83, 2013.
[3] Gungor V C, Hancke G P. Industrial wireless sensor networks: Challenges, design principles, and technical approaches. IEEE Transactions on industrial electronics, vol. 56, no. 10, pp. 4258-4265, 2009.
[4] Ghaleb B, Ai-Dubai A, Romdha I, et al. Drizzle: Adaptive and fair route maintenance algorithm for Low-power and Lossy Networks in IoT. Proceedings of the International Conference on Communications ( ICC) , pp. 1-6, 2017.
[5] Winter T, Thubert P, Brandt A, et al. RPL: IPv6 routing protocol for low-power and lossy networks: RFC 6550. Internet Engineering Task Force( IETF) , pp. 1-157, 2012.
[6] Awad A M A, Ab Rahim R, Hashim A H A. Queue backlog as a node metric for RPL protocol. Proceedings of the International Conference on Computer and Communication Engineering( ICCCE) , pp. 246-250, 2016.
[7] Kim H S, Kim H, Paek J, et al. Load balancing under heavy traffic in RPL routing protocol for low power and lossy networks. IEEE Transactions on Mobile Computing, vol. 16, no. 4, pp. 964-979, 2017.
[8] Levis P, Clausen T H. The Trickle algorithm: RFC 6206. Internet Engineering Task Force( IETF), pp.1-13, 2011.
[9] Yao Y K, Liu J B, Ren Z, et al. High-efficient centralized network congestion control for RPL routing protocol. Systems Engineering and Electronics, vol. 39, no. 12, pp. 2810-2816, 2017.
[10] Deveza T, Martins J F. PLC control and Matlab/Simulink simulations: a translation approach.

IEEE Conference on Emerging Technologies & Factory Automation, 2009.
[11] Ha S H. PLC Intelligent Transportation Signal Control Based on Algorithm of Function Block Conversion to Instruction List. International Conference on Intelligent Transportation, 2018.
[12] Iova O, Theoleyre F, Nol T. Exploiting multiple parents in RPL to improve both the network lifetime and its stability. Proceedings of the International Conference on Communications (ICC) , pp. 610-616, 2015.
[13] Miclbee M, Duquennoy S, Quoitin B, et al. Load-balanced data collection through opportunistic routing. 2015 International Conference on Distributed Computing in Sensor Systems IEEE, 2015.

IOP Conf. Series: Materials Science and Engineering **608** (2019) 012016 doi:10.1088/1757-899X/608/1/012016

# Structural Design and Difficulties of Solar UAV

**W W Zhang[1,a], L G Zhang[2,b], Z W Yan[1] and L Wang[1]**
[1] School of Aerospace Engineering, Shenyang Aerospace University, Shenyang, China
[2] General aviation research institute, Shenyang Aerospace University, Shenyang, China

E-mail: [a]zww13579@163.com; [b]1183590317@qq.com

**Abstract:** The design of high-altitude and long-endurance solar aircraft is obviously different from that of conventional aircraft. Its unique energy system and operating conditions make it necessary to carefully weigh the key technical details in design. The structure has a great influence on the endurance performance of solar aircraft. The structure weight can be roughly estimated by the proportion of the airborne energy system to the total weight of the aircraft, so as to optimize the structure. Through the analysis of several typical solar aircraft, it can be concluded that the aerodynamic layout and structure design have a great impact on the flight time and performance of solar aircraft. Finally, the challenges faced by solar aircraft are presented, which can provide reference for the design of solar aircraft in the future and help.

## 1. Introduction

With the continuous consumption of renewable resources and the increasing global warming, most countries begin to focus on non-renewable Green Resources solar energy. Since 1981, when the United States developed the first solar unmanned aerial vehicle, Solar Challenger, many countries have begun to invest a lot of energy in solar unmanned aerial vehicles. Compared with traditional UAVs, solar UAVs have the advantages of environmental protection, high flight altitude, strong stability, wide coverage area and outstanding load capacity. While solar-powered UAV has many advantages, it also faces great difficulties in design, especially in the two aspects of too large force on wing root and too large deformation of wing root. Excessive force on wing root will cause material failure of wing root beam, while too large deformation of wing will cause aerodynamic performance degradation of wing and damage battery, which will affect the endurance of aircraft [1].

In addition to the conversion efficiency of solar cells, the aerodynamic layout of solar UAV also has a great impact on it. There are many aerodynamic layout forms suitable for solar UAV, such as conventional layout, flying wing layout, double strut layout, tandem wing layout, V-tail layout, etc. Each layout has its own characteristics. In order to solve the problems of large aspect ratio wing deformation, flight control difficulties and aeroelasticity, most of the solar UAVs in recent years adopt conventional layout [2].

## 2. Proportion of airborne energy systems

In the design of solar UAV, according to its mission requirements, the total weight, flight speed, lift-drag ratio of UAV can be determined first, and then the weight of storage battery and solar cell can be calculated. From the design principle of solar aircraft, the weight ratio of each part can be calculated.

The weight ratio of energy storage battery is:

IOP Conf. Series: Materials Science and Engineering **608** (2019) 012016 doi:10.1088/1757-899X/608/1/012016

$$\frac{Q_b}{Q_j} = \frac{\left(\frac{Q_j vg}{k\eta_m\eta_B q}\right)t(1+k_b)}{Q_J} = \frac{vg(1+k_b)}{k\eta_m\eta_B q}t \tag{1}$$

Weight ratio of solar cells:

$$\frac{Q_s}{Q_j} = \frac{S_s q_s}{Q_j} = \frac{W_s\frac{\pi q_s}{p_s\eta_s(24-t)}}{Q_j} = \frac{p\left(\frac{t}{\eta_c}+24-t\right)\pi q_s}{p_s\eta_s(24-t)Q_j}(1+k_b) = \frac{vg\left(\frac{t}{\eta_c}+24-t\right)\pi q_s}{k\eta_m\eta_B p_s\eta_s(24-t)}(1+k_b) \tag{2}$$

In the formula, $Q_j$ is the total weight of UAV; $Q_b$ is the weight of storage battery; $v$ is the flight speed; $k$ is the lift-drag ratio; $\eta_m$ is the efficiency of motor; $\eta_B$ is the efficiency of propeller; $\eta_s$ is the efficiency of solar cell; $\eta_c$ is the efficiency of charging and discharging of storage battery; $q$ is the density of storage energy and $p_s$ is the density of solar energy.

When the total weight of the solar UAV, the weight of the storage battery and the weight of the solar panels are all determined, if the weight of the aircraft needs to be increased, the structure of the aircraft can only be improved. The best structure of the aircraft can meet the requirements of strength and stiffness and achieve the minimum quality. For the solar UAV, in order to meet its long-term performance, it usually has a large area of solar cells in the design to meet the energy storage of the solar UAV during day and night flight. At the same time, it uses the form of large aspect ratio to reduce the induced drag of the aircraft and improve its lift-drag ratio. The wing of the aircraft is usually designed relatively large, and the weight increases accordingly. Therefore, the design of the structure is particularly important.

## 3. Structural Analysis of Several Solar Unmanned Aerial Vehicles

### *3.1 "Helios" Series Solar UAV*

"Helios" series solar UAV is a solar UAV developed by the United States. It has gone through five stages, namely "Pathfinder", "Pathfinder+", "centurion", "HP01" and "HP03"[3]. HP03 solar UAV has a main pod, four sub-pods, two hydrogen storage and 14 motors. Its wingspan is 75.3m, chord length is 2.44m, wing area is 183.6$m^2$, and the outer wings on both sides have a 10 degree up angle. "HP03" solar UAV adopts flying-wing configuration, the main reasons are as follows: 1. Flying-wing aircraft are not prone to nose-down dive; 2. Wings will not have the tendency of backward bending; 3. Flying-wing configuration has better balance and operational stability; 4. Its resistance is small, wing load is low, the available space in the wing is large, and the load is large, which is beneficial to large aircraft.

**Figure 1.** "HP03" Solar UAV.

"Helios" series solar UAV adopts the structural form of strong beam and weak wall. Its wing and fuselage are mostly made of composite materials, such as Karaff, Omez, carbon fibers, etc. and high-quality film diaphragm is used as the wing skin. In order to reduce the weight of the wing, holes are arranged on the wing ribs according to the design requirements. The extensive use of composite materials makes the wing flexible and can prevent the wing from breaking during flight. The elastic deformation of the wing is very large. In normal flight, the wing bends upward under the action of aerodynamic force, and the tip of the wing warps up to 4.5m. There are 72 lifts on the trailing edge of the wing, which increases the maneuverability of the aircraft. When turning, it is mainly realized by changing the pull force of four motors on the left and right sides. There are several streamlined flat pods under the wing, which have three main functions: one is to install wheels as landing gear; the other is to load task modules inside; and the third is to play the role of directional stability, as shown in Figure 1.

The HP03 solar UAV trembled for a long time at low altitude, which eventually led to the crash of the unmanned aerial vehicle. The reasons are as follows: first, after modification, 14 motors are reduced to 10 motors. The propellers driven by these 10 motors are scattered over the whole wingspan. When the wing is bent upward by a large margin, the pulling force of the propeller in the outer wing will produce a large downward pitching moment. When the wing bending decreases, the downward pitching moment of the wing decreases sharply, the oscillation of the wing bending and the change of the pitching moment of the wing. The second is that the torsion center of the wing is behind the aerodynamic center of the wing, and when the wing oscillates up and down, the torsion of the wing strengthens and the material fails. Despite the crash of the HP03 solar UAV during its second flight, it is still one of the successful examples of structural design.

### *3.2 "Solar Impulse 2" Solar Aircraft*

The "Solar Impulse 2" aircraft developed by Swiss Solar Impulse Company adopts the conventional layout, the outer wing section is upside down, the fuselage is in the form of short bunker and tail beam, the front fuselage is in the cockpit, and the back half fuselage is triangular fuselage, which makes the fuselage more stable and has better torsion resistance as shown in Figure 2. The wing is 72m long and 2300kg in weight. There are 17 248 high efficiency solar panels installed on the wing surface. In wing design, there is one rib per 50cm, 144 ribs are arranged on the whole wing, and the effective battery installation area of the wing is 270 $m^2$. In order to reduce weight, the battery is directly designed as part of the wing, instead of installing the battery on the wing. Four pods are installed on the lower surface of the wing to install motors, propellers and lithium batteries.

**Figure 2.** "Solar Impulse 2" Solar Aircraft.

The "Solar Impulse 2" aircraft uses light weight and strong load materials such as carbon fiber and foam. Because the cabin is used for pilot driving, the thermal insulation material is used on the cabin to prevent the damage caused by the low temperature at high altitude. Polyurethane foam used in some parts of cockpit shell. In the design of door locks and porthole, the aircraft uses polyurethane composite material, and some polyurethane foam outside the cockpit is used for battery insulation. At the same time, the body is also covered with silver coated raw materials, which can protect solar cell [4].

### *3.3 "Green Pioneer" Solar UAV*

The "Green Pioneer" solar UAV adopts the "composite wing" aerodynamic layout, which integrates the combined wing and tailless layout skillfully. It not only makes the structure of the aircraft more compact and lighter, but also greatly reduces the energy consumption, but also increases the aerodynamic characteristics of the wing. It has a wingspan of 7.5 m, a captain of 2.84 m and a cruise altitude of 5000m. It is powered by six small DC motors.

IOP Conf. Series: Materials Science and Engineering **608** (2019) 012016 doi:10.1088/1757-899X/608/1/012016

"Green Pioneer" adopted the "composite wing" layout, which is the first aerodynamic layout form in the world[5]. This structure can arrange solar panels on the upper and lower wing surfaces. The upper wing surface can receive solar radiation directly, and the lower wing surface can receive sunlight reflected by clouds and atmospheric particles, thus greatly increasing the area of solar radiation, as shown in Figure 3. Show. The layout of "composite wing" makes the load-bearing route of the whole aircraft structure shortest, reduces the weight needed to ensure the structural strength, and can effectively reduce energy consumption and increase useful load.

**Figure 3.** "Green Pioneer" Solar UAV.

*3.4 "MY-6" Solar UAV*

"MY-6" UAV is a long-endurance solar UAV developed by Northwest Polytechnic University. The whole fuselage is 1.2m long and the wingspan is 7m. The weight of the UAV is only 15kg, the maximum mission load is 5kg, and the flight altitude is 6,000 m. The duration in summer is 23 hours, while in winter it is 13 hours because of the weak sunshine. The aircraft adopts full-wing pneumatic layout, has ultra-light surface density structure, and has high efficiency of laying solar panels. The structure of aileron and spoiler is not designed on the wing, which can effectively increase the laying area of solar panels; the fuselage is designed as a large airfoil shape, so that the fuselage can generate part of lift and reduce energy consumption during flight; there is a vertical tail at each side of the fuselage and at the rear of the wing, which can play the role of the stabilizer; at the front edge of the left and right wings, there is a vertical tail at each side of the fuselage. There is a motor to power the UAV. The turning of UAV is mainly realized by differential rotation of the motors on both sides. There is an elevator at the tail of the fuselage, as shown in Figure 4. The connection between the wing and the fuselage is carbon pipe plug-and-pull, and the wing beam is carbon pipe. In terms of material, the whole aircraft is made of carbon fiber, which makes the aircraft have high strength, stiffness and light weight at the same time.

**Figure 4.** "MY-6" Solar UAV.

## 4. Major Challenges for Solar UAV

*4.1 Energy System*

Energy system is one of the key technologies in the design of solar UAV, which includes energy storage system, solar array and energy management and distribution system [6]. In order to improve the

IOP Conf. Series: Materials Science and Engineering **608** (2019) 012016 doi:10.1088/1757-899X/608/1/012016

performance of solar UAV, we can make breakthroughs in two aspects. On the one hand, we can improve the performance of a single panel, improve its energy conversion rate and reduce its weight. In the whole energy conversion process, except the conversion efficiency of solar cells is about 20%, the conversion efficiency of other parts is more than 70%[7]. As far as the current solar panels are concerned, they can be roughly divided into two categories: crystalline silicon solar cells and thin film solar cells. Most of the existing solar UAVs use crystalline silicon solar cells. With the development of solar cells, the thickness of crystalline silicon solar cells becomes thinner and lighter, but the welding process will also result in larger fragmentation rate. Although gallium arsenide thin film battery has high conversion efficiency and good radiation resistance, its application is limited to a large extent due to its high cost. On the other hand, better battery packaging technology is used to reduce energy loss. Because solar panels are brittle, the brittleness, impact resistance and disassembly of solar cells must be taken into account in battery packaging. The laying of solar array should take into account the deformation of the wing during flight, and adopt the layout form which is suitable for the bending of the wing. Therefore, improving the efficiency of energy system can effectively increase the flight time of solar UAV.

### *4.2 Structure Design and Pneumatic Layout*

Most of the solar UAVs designed at this stage only achieve the long-range goals, but the load is not outstanding, and the maximum load is only tens of kilograms. Therefore, reasonable structure design and aerodynamic layout are particularly important. A good design can greatly increase the load of UAV. At present, almost all solar UAVs adopt super-large aspect ratio mechanism, which will inevitably cause large deflection deformation during flight, which will cause very prominent aeroelastic problems. Therefore, in structural design, in addition to ensuring strength requirements, it should also ensure sufficient stiffness. Selection of light-weight and high-strength structural materials, rational structural layout design and fine processing of light and thin structural parts are important structural and technical problems that solar UAV must solve. At present, the aeroelastic calculation and analysis methods in the aviation field are not suitable for the non-linear aeroelastic problem of solar UAV. Therefore, new aeroelastic theories and methods must be developed.

### *4.3 Flight Control Technology*

Flight control technology is also one of the problems to be solved urgently, especially in the case of low-speed flight, the wing flutter caused by long-span flight altitude and the aeroelastic control of flexible wings. When the aircraft crosses the troposphere, it is easy to encounter gusts and atmospheric turbulence, which will cause wing tremor and cause aircraft disintegration. For example, “HP03” solar unmanned aerial vehicle is disintegrated and crashed in this situation.

## 5. Concluding remarks

Solar UAV has remarkable performance and broad development prospects. It is widely used in military and civilian fields. At present, many countries regard it as the existence of sub-satellite for weather prediction, reconnaissance and so on. Despite the continuous progress of science and technology, there are still many difficulties to overcome. From the structural analysis of the solar unmanned aerial vehicle which is more famous at home and abroad, it can be seen that the structural design can improve its performance to a great extent. Therefore, if there is a breakthrough in the aerodynamic layout and structure design, the development of solar UAV will enter a stage of rapid development.

## Reference

[1] Dong Xin 2016 Design optimization of composite wing structure for solar UAV *D. Xi'an University of Technology.*

[2] Zhang J, Zhang D H 2016 Essentials of configuration design of HALE solar-powered UAV *J. Acta Aeronautica et Astronautica Sinica.*

[3] NASA 2002 Heios prototype: the forerunner of 21st century solar-powered “atmospheric satellites” *R. Washington D.C.*

IOP Conf. Series: Materials Science and Engineering **608** (2019) 012016 doi:10.1088/1757-899X/608/1/012016

[4] Chen Chen, Wang Xihuai and Xiao Jianmei 2013 Modeling and Simulation of four-phase brushless DC motor system [J]. *Electronic Design Engineering* chapter 19 pp 51-52.

[5] Wu Jian 2002 Creating a "Green Pioneer": China's Solar Long-Term and High-Altitude UAV Program *J. International Outlook*.

[6] Hall D W, Watson D A and Tuttle R P 1985 Mission analysis of solar powered aircraft *R. Washington D.C.*

[7] Zhu X, Guo Z and Hou Z 2014 Solar-powered airplanes: A historical perspective and future challenges *J. Progress in Aerospace Sciences* chapter 71 pp 36-53

IOP Conf. Series: Materials Science and Engineering **608** (2019) 012017 doi:10.1088/1757-899X/608/1/012017

# Research on the Variation Analysis Model for CFRP Assembly

**Jianfeng Yu[1], Shuhan Huang[1*] and He Zhang[1]**

[1]School of Mechanical Engineering, Northwestern Polytechnical University, Xi'an, Shaanxi 710072, China

[*]Email: 15571221001@163.com

**Abstract.** The conventional error analysis model is based on Key Characteristics (KC) tree [1] and the nominal 3D assembly model. Before the actual execution of assembly process, the potential interferences and out-of-tolerance problems are simulated and predicted. Then the tolerance and assembly schemes of the parts are optimized accordingly. This method requires statistical deviation data of parts and it can only give a rough probability for mass production case. In this paper, the common CAT technologies, variation analysis and optimization method are presented with an example of wing box model in 3DCS. Based on KC tree and optimizing theories a model analysis method is proposed to improve the performance in the case of multi-material part assembly processes.

## 1. Introduction

A typical application process of CAT includes three main steps: identification of key characteristics [2], construction of deviation analysis models, and tolerance optimization.

### *1.1. Identification of key characteristics*

The manufacturing process of aircraft involves various aspects of technology and applications which involves many different concepts at each stage. Therefore, how to accurately identify the corresponding KCs from these complex data becomes an important issue. The identification methods are mainly the following.

**According to the assembly scheme**: Under the premise of a complete assembly scheme, characteristics which are used in locating and connection by parts are found out as the KCs. It normally starts from the lowest part level until the highest product KCs are found. Each step in the process also include analysis of the direction and route of error transmission to optimize the assembly scheme.

**Reasoning based on statistical data**: Starting from the error statistics in the manufacturing process, combined with the process requirements and assembly schemes, the source of error is reversed and the key characteristics are derived.

**The risk analysis method**: This method is based on the formula to determine the risk priority number (RPN) of the characteristics to be selected:

$$\mathrm{RPN}=S\times O\times D \tag{1}$$

In formula (1), *S* (Severity) is a subjective value used to measure the severity of the failure. *O* (Occurrence) represents the possibility of failure when using which is also a subjective value. *D* (Detection) represents the degree of difficulty in detecting potential failures before the product is completed. These three numerical evaluations range from 1 to 10. After evaluating all candidate, the high-risk characteristics will be chosen as KCs.

**The historical data analysis method**: This method is based on the statistical data of scrapped parts generated by the workshop and the analysis results of causes to obtain the KCs. With all data generated in the actual production process, its subjective factors are relatively low, and the results are reliable.

### *1.2. Establishing an deviation analysis model*

The purpose is to simplify the complex error behaviour and to optimize the part tolerance with the model. Researchers have proposed various models, such as the classics dimension chain model, analytical models based on deterministic locating, etc.

**The dimension chain model**: it is actually a highly simplified theoretical model because the amount of information that can be carried by each link is very small. In the process of extracting the dimension chain, most of the geometry characteristics of the part are neglected, so it is suitable for a simple structure. For complex one, the extraction process and calculation will be complicated. The result will also be too theoretical. In addition, for products that use a large number of fixtures in the assembly process, the dimension chain model also has limitations in analysing the introduction of fixtures' errors.

**Analytical model based on deterministic locating**: Compared with the dimension chain model, this model is closer to the actual assembly process. It can introduce the errors of the fixture well. The complex assembly can also be analysed and solved. However, because of its computational complexity [3], to solve this model relies more on computer programs. Sometimes the quality of the algorithm can even determine the analysis speed and the accuracy of the results. Facing the large-scale structures, the complexity of its locating will spend a lot of staff energy and time to complete the set-up process. The analysis and calculation process also has relatively high requirements for the hardware.

### *1.3. Tolerance optimization*

In fact, the difficulty of optimization process depends on many front-end factors. For example, a reasonable assembly scheme can allow a large room for tolerance optimization without having to compress the tolerance zone of some parts to very narrow, because the errors can be maximally eliminated during the assembly process. An excellent part design can also significantly reduce the sensitivity of the error transfer to the part geometry, thereby saving the cost of modifying the part shape and assembly method.

Generally iterative method is used to optimize tolerances. That is, starting from a presupposed set of tolerances, after the operation of repeated analysis and improvement, the set of tolerances will be approximated to the optimal solution.

## 2. Wing box assembly process and error analysis

"In the aircraft manufacturing process, the shape accuracy requirements are much higher than the assembly accuracy requirements" [4]. The appearance characteristics of aircraft are usually given priority, and the accumulated errors are eliminated through the connections of inner structure. Therefore, the assembly process also affects the transfer and cumulation of errors.

The actual assembly process can be expressed by formula (2) and Figure 1 shows the main structure of the wing box:

$$P = \{\text{step1},\ \ \text{step2},\ \ \text{step3},\ \ \text{step4}\} \tag{2}$$

P—process (wing box assembly).

IOP Conf. Series: Materials Science and Engineering **608** (2019) 012017 doi:10.1088/1757-899X/608/1/012017

Step 1: locating and assembly of the leading edge-front spar component [5].
Step 2: locating and assembly of the trailing edge-rear spar component.
Step 3: locating and assembly of the ribs.
Step 4: locating and assembly of the panels.

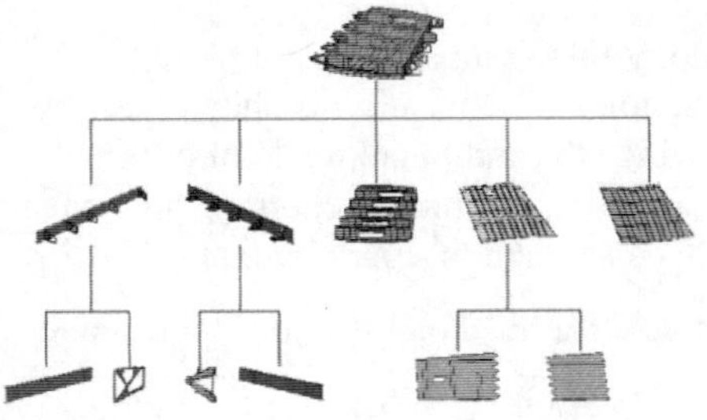

**Figure 1.** Structure of wing box assembly

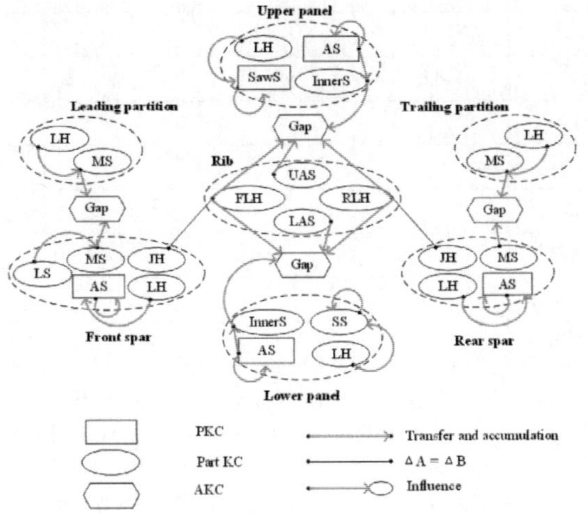


**Figure 2.** Transfer of KCs' own fluctuations

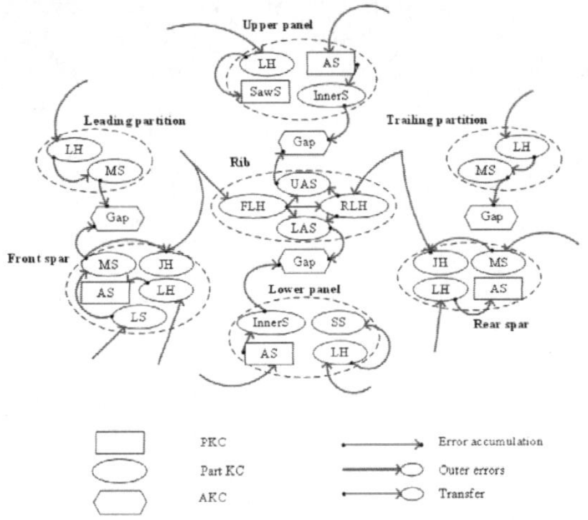


**Figure 3.** Introduction and delivery of locating errors

With analysing the detailed assembly process, a global error transfer routes can be drawn as Figure 2 which shows all parts, assemblies and KCs that appear in the process and how their deviations will transfer and accumulate. Besides, Figure 3 shows the introduction and delivery of locating errors. In the two figures: LH—Locating Hole. LS—Locating Surface. MS—Mating Surface. AS—Aerodynamic Surface. JH—Jig Hole. SS—Side Surface. SawS—Saw tooth Surface (they form the seam in upper panel).

## 3. Tolerance allocation technology based on 3DCS

Based on the detailed deconstruction and analysis in Chapter 2, this chapter will show the analysis result of the gap between metal wing ribs and panel made of composite material. To run the analysis in 3DCS, parts' preset tolerances, Move (assembly scheme), and Measure scheme are indispensable. Table 1 shows the original tolerance scheme of upper panel.

**Table 1.** Original Tolerance of Upper Panel /mm.

| KCs | Position | Size range | Range | Offset |
|---|---|---|---|---|
| LH 1 | Φ0.1 | 0.04 | | |
| LH 2 | Φ0.1 | 0.04 | | |
| LH 3 | Φ0.1 | 0.04 | | |
| SawS | | | 0.3 | 0 |
| UAS | | | 0.2 | 0 |
| 4-5InnerS | | | 0.223 | 0 |
| 5-6InnerS | | | 0.270 | 0 |
| 6-7InnerS | | | 0.316 | 0 |
| 7-8InnerS | | | 0.366 | 0 |

Due to design requirements, the skin is not uniform but gradually thickens from 4mm to 8mm. So the measurement points should be set in different thickness regions. For the connect position, a small size of gap is normally more acceptable than interference which causes the deformation of product shape. Figure 4 shows the results of conventional method which uses random variation to build the analysis model in four regions.

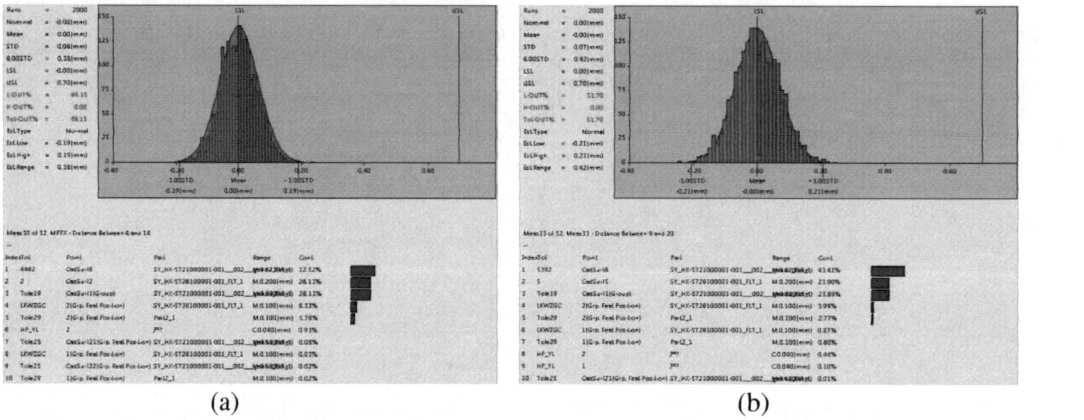

(a) (b)

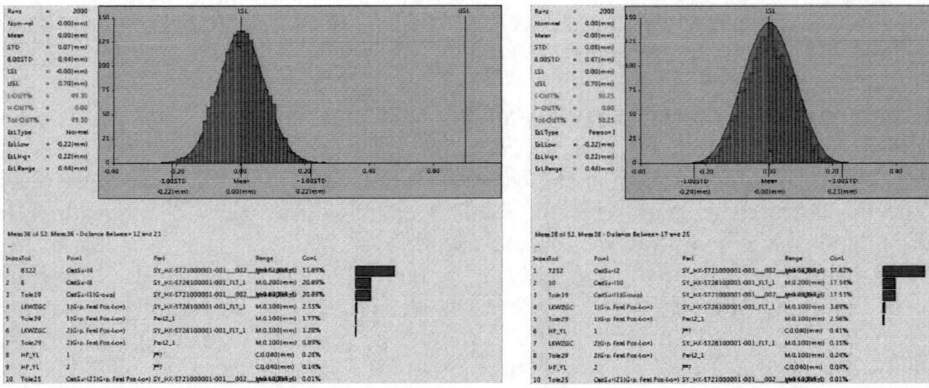

(c) (d)

**Figure 4.** Analysis results of gap in four different regions with original tolerance

The fluctuation range of the gap are: 0.38mm of 4-5mm region, 0.42mm of 5-6mm region, 0.44mm of 6-7mm region, 0.44mm of 7-8mm region, gradually growing larger with the skin thickening. This is because the deviation of skin is calculated at 5% of its nominal thickness, so the skin itself fluctuates more in thicker regions. In the figure, the unqualified ratio is all about 50%. Although the skin itself is the largest source of fluctuations, modifying the tolerance of the composite material skin is still not suitable, so the target turns to the aerodynamic surface of rib. After adjusting the tolerance zone of it inward by 0.1 mm, the new result is shown in Figure 5.

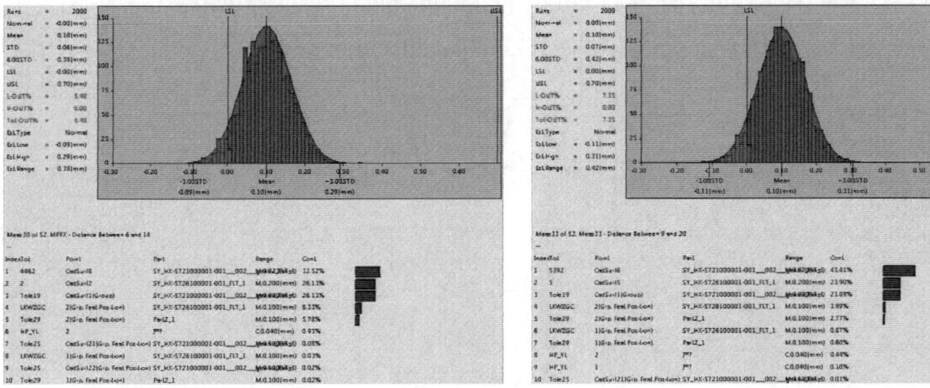

(a) (b)

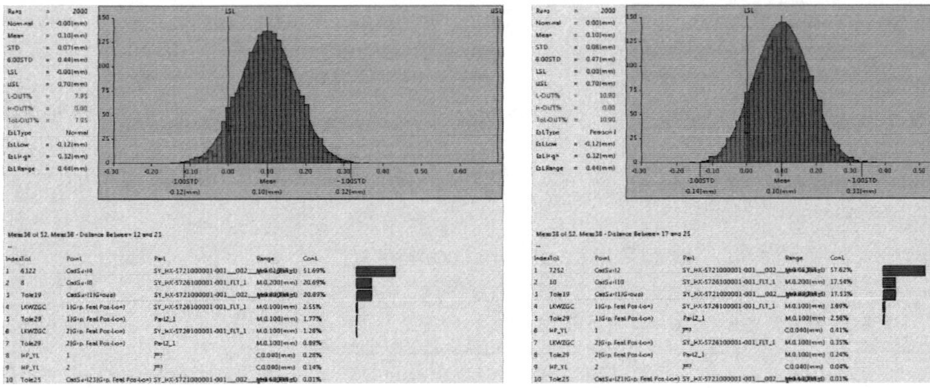

(c) (d)

**Figure 5.** Analysis results after optimization

The unqualified rate is respectively reduced to 5.4%, 7.35%, 7.95%, and 10.9%. The optimization effect is very obvious.

## 4. Variation analysis of multi-material parts assembly

The idea of multi-material assembly is to apply composite material to the secondary parts of the structure, but in important supporting frames, bearing structures and parts with high precision requirements. The machined metal parts are still used. Assembly made up of parts with different properties has brought a lot of problems to the assembly work.

Here proposes a new idea to improve the analysis method of the multi-materials assembly. The preset tolerance of the composite part was abandoned in the original analysis process. Set its true deviation directly. The goal is to transfer the error as much as possible to metal parts that can be optimized to a narrower tolerance zone. Figure 6 shows the difference between the new method and conventional one in error accumulation.

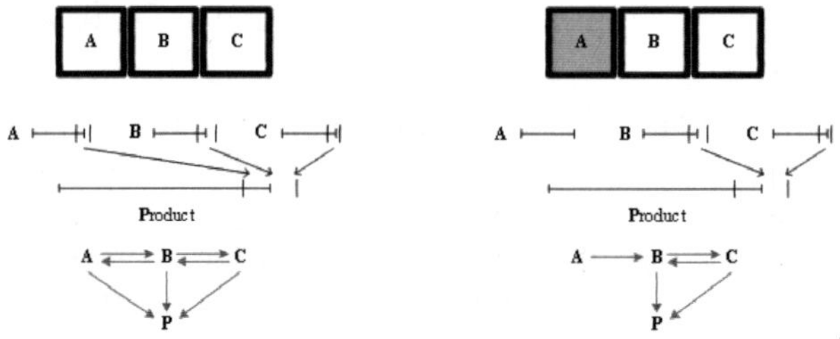


**Figure 6.** Full metal assembly and multi-materials assembly

For composite parts A with large deviations, the tolerance band is T and the deviation is ΔA.

$$\begin{cases} \Delta A \in (\min, \max) = T \\ P(\Delta A = \max) = 0 \\ |\max - \Delta A| = C \neq 0 \end{cases} \tag{3}$$

Formula (3) shows that after the deviation of A is set, there is a tiny blank space C between the deviation value and the original tolerance zone, and this interval will be allocated and merged into other parts that cooperate with A.

$$\begin{cases} A = C \times Con(partA) \\ B = C \times Con(partB) \\ \quad \cdots \end{cases} \tag{4}$$

Formula (4) shows that the allocation of the blank interval C is performed according to the contribution rate of other parts, and the assigned amount is added to the tolerance band of the corresponding part.

The new method is applied at the four representative measure points by adding the measured deviation value. To input the deviation data into the model, the data structure of compiled DEV file is adopted to bring deviation information into the 3DCS system. the actual measured corresponding deviation values are shown in Table 2.

**Table 2.** Actual Panel Variations.

| Area of skin | Variations/mm |
|---|---|
| 4-5mm | +0.048 |
| 5-6mm | -0.057 |
| 6-7mm | +0.012 |
| 7-8mm | -0.047 |

IOP Conf. Series: Materials Science and Engineering **608** (2019) 012017 doi:10.1088/1757-899X/608/1/012017

The corresponding aerodynamic surface of the wing ribs is still set to a tolerance of 0.2mm as the conventional method and a normal distribution is used to randomly select the deviation value. The Move and the four measurement points are the same as it in chapter 3. The new results are shown in Figure 7.

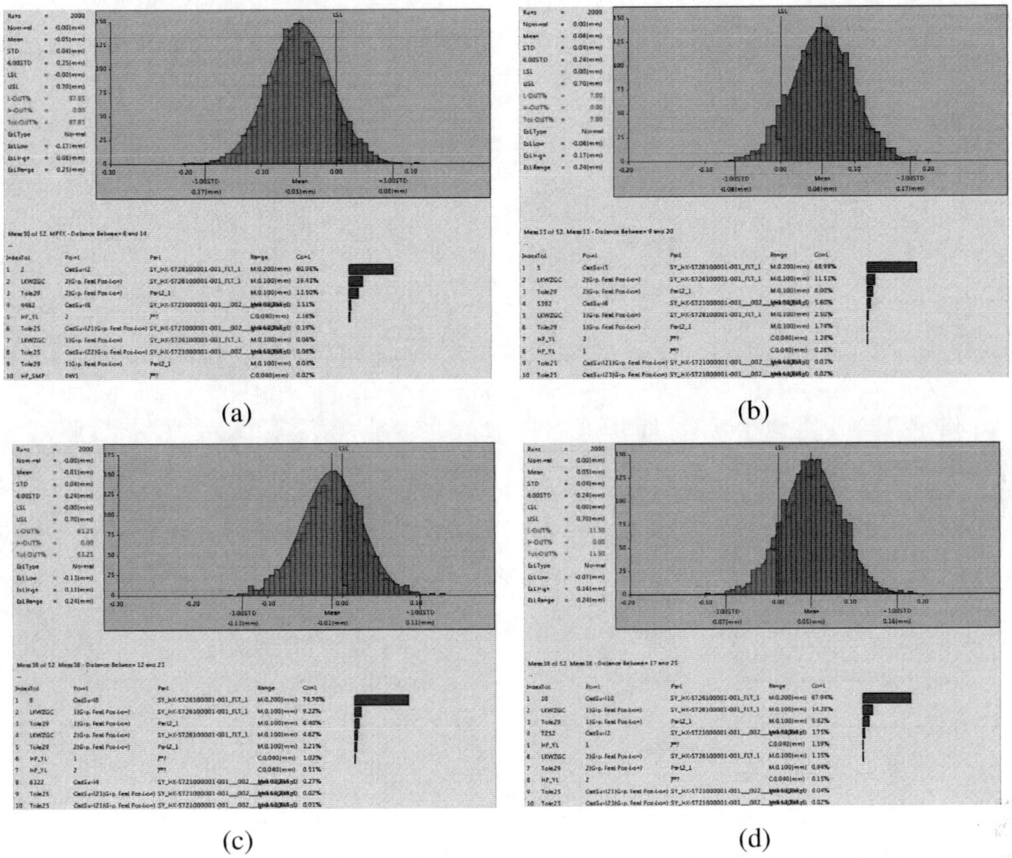

**Figure 7.** Analysis result of new method with actual variation involved

The analysis results show fluctuation range of this method: 0.25mm of 4-5mm region, 0.24mm of 5-6mm region, 0.24mm of 6-7mm region and 0.24mm of 7-8mm region. The range values of the four regions are very similar. At the same time, all the most contributing factors are the fluctuations of aerodynamic surfaces of the rib which are all above 60%. This shows that after the panel deviation is set to an actual constant value, its fluctuation is ignored and the critical analysis object is transferred to the rib. This is exactly what we expect.

In terms of the unqualified rate, the directions of deviation of the skin in the four areas are different. The out-of-tolerance regions also shows different proportions, and targeted analysis must also be carried out in formulating optimization plans.

In the 4-5mm area, due to the deflection of the skin towards the ribs, the wing ribs should be reduced in size accordingly, reducing its size by 0.15mm and the aerodynamic surface tolerance zone by 0.1mm.

In the 5-6mm area, since the skin deviates away from the ribs, it only need to reversely offset the aerodynamic surface tolerance zone by 0.05mm.

In the 6-7mm area, the skin shift is small and close to the nominal position, and the tolerance offset value should be larger to avoid interference, set the offset by 0.1mm.

The 7-8mm zone is similar to the 5-6mm zone. The skin deviates away from the ribs. It only needs to offset the tolerance zone of aerodynamic surface by 0.07mm.

According to the above optimization plan, the values used in analysis are optimized. Run the calculation and statistics again and the new results of the width of gap between skin and rib are as follows.

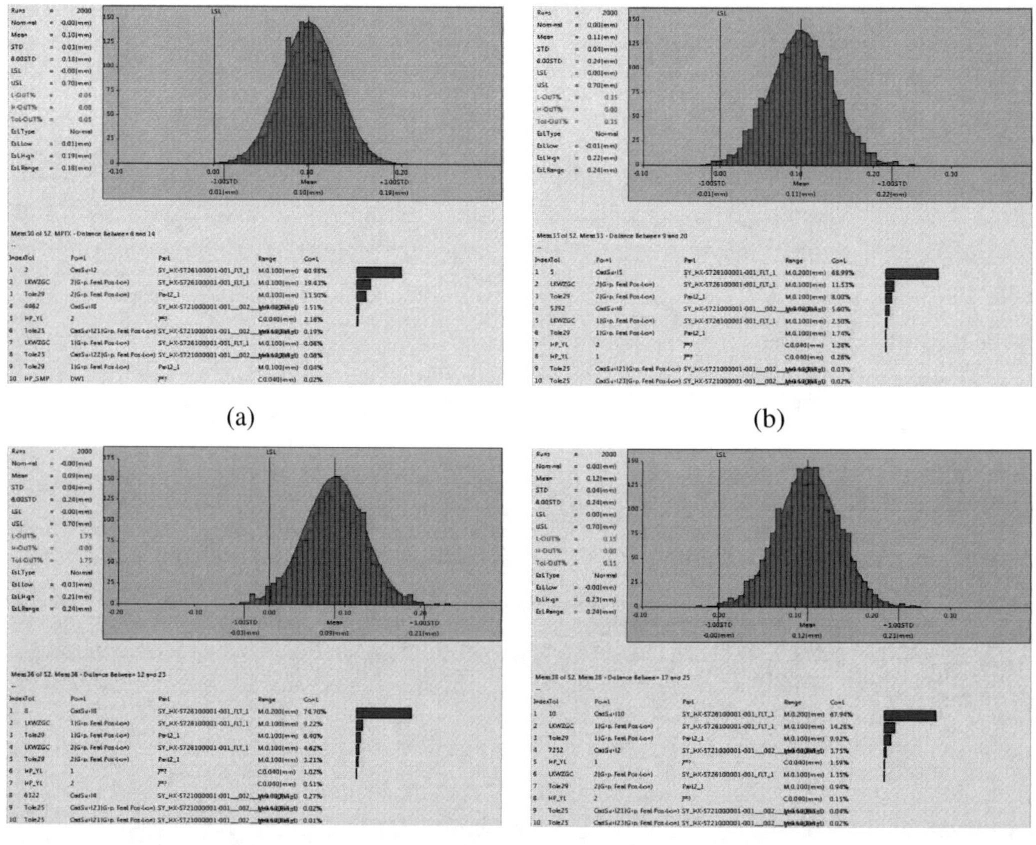

(a) (b) (c) (d)

**Figure 8.** Analysis result of new method after applying optimization

As shown in Figure 8, when using the optimized scheme, the unqualified rate at each measurement point is greatly improved and they are all less than 2%. In the example, by adding the actual measured deviation to the model, the error factors are concentrated on the rib that matches the skin. In order to eliminate these comprehensive errors, different targeted optimization schemes are also formed in regions with different deviations. The resulting tolerance scheme of the ribs can actually guide the corresponding characteristics in the manufacturing process to ensure that the machined ribs can be perfectly coordinated and matched with the specific skin during assembly. In the process of new product development and prototype assembly, this tolerance analysis method based on concessions for composite parts is more targeted and instructive than the conventional method of adding random deviations.

## References

[1] Chen Z H, Duo F Z, Tang X Q. 1000-6893, **33** (2012)
[2] Liu Z C, Zou J H, Fan Y Q. 1006-5911, **13** (2007)
[3] P. Franciosa, S. Gerbino, S. Patalano. *Variational modeling and assembly constraints in tolerance analysis of rigid part assemblie: Planar and cylindrical features* (Int J Adv Manuf Technol 2010)
[4] Feng Z M, Zou C, Liu J H. 1006-5911, 16 (2010)
[5] J. Muelaner, P.G. Maropoulos, *Integrated Dimensional Variation Management in the Digital Factory*. (7th International Conference on Digital Enterprise Technology, 2011)

# Human Interface Research of Civil Aircraft Cockpit Based on Touch Control Technology

**Wen Yang, Qin Sun, Xiaolong Gao, Dayong Dong and Xianchao Ma**

Shanghai Aircraft Design and Research Institute, Jinke Road 5188, Shanghai, China

Email: yangwen@comac.cc

**Abstract.** Touch screen technology is an important direction of the aircraft cockpit technology development. More and more touch screen test units are applied in the cockpit. However, most of the touch screen interface design is only a simple copy of the physical interface, which does not bring any benefits to the aircraft, but will cause some safety effects. Fundamentally speaking, these test units did not consider the impacts of flight crew operating procedures, human factors, flight scenes, and other factors on touch screen technology. This paper gives some advantages and disadvantages analysis for the touch control technology and provides a new concept of cockpit overhead control panel based on the touch control technology. The overhead touch control panel can not only effectively reduce the workload of the pilot, reduce the weight of the system, but also ensure the same flight safety with the physical panel.
**Keywords.** Touch Control, Human Factor, Workload, Hybrid Layout, Ergonomics Evaluation.

## 1. Introduction

With the development of civil transportation and science technology, it's an inexorable trend for civil aircraft cockpit stepping into tantalization, intelligentization and visualization. Touch technology gets a lot of attention in civil aircraft area due to its technical characteristics of high integration, simple control, and intuitive display. Cockpit products using touch technology have been researched and come into service in military aircraft and general aviation area. These products can provide the same function as traditional ones. However, considering human factor and high level regulations from flight crew operation, there are still many restrictions for touch technology products being applied in the civil aircraft cockpit. This paper analyzes the application characteristics of civil aircraft touch technology, provides application assumption of products using touch technology and proposes methods for cockpit ergonomics evaluation based on the current situation of touch technology. [1]

## 2. Human-machine Interface Development in Civil Aircraft Cockpit

The human-machine interface design in civil aircraft cockpit is a concentrated embodiment of the close combination of human and aircraft. The interface design can affect the operation and observation of pilots. Signals and messages from the cockpit during flight have an impact on judgments, decision, measurements, and safety. The cockpit is developing with the advance of technologies.

The analog instruments are widely used in the early civil aircraft cockpit to realize definite functions. It requires a strong capability for the flight crew to monitor the real-time state of these instruments. Figure 1 shows the dense physical instruments used in the cockpit of B747-200 which went into service in 1971.

**Figure 1.** Cockpit of B747-200. [2]

The B787 is a Dreamliner delivered by Boeing in 2011 and PFD/MFD was applied to replace most of the physical instruments. This cockpit was called "glass flight deck"[3] by Boeing and gradually accepted by pilots, which is shown in figure 2. It is simpler and safer for pilots to operate an aircraft because signals are integrated to display in "glass flight deck".

**Figure 2.** Cockpit of B787. [3]

B777X is about to be developed by Boeing in the future. Touch MFD will be used on B777X as shown in Figure 3. Meanwhile, Airbus A350 is setting out to research touch technology and try to use it on lower display and pedestal zone.

**Figure 3.** Cockpit of B777X (in researching).[4]

Touch screen technology is broadly used on G650 as shown in Figure 4. Thirteen touch screens integrated functions of multiple systems like flight regulation, communication and weather radar and so on to replace most of the switches. It is not a simple replacement of operation devices. The cockpit

IOP Conf. Series: Materials Science and Engineering **608** (2019) 012018 doi:10.1088/1757-899X/608/1/012018

is highly integrated and intelligentized and pilots can be completely immersed in the cockpit by the application of touch screens.

**Figure 4.** Cockpit of G650. [5]

In summary, civil aircraft experienced significant changes from instrument cockpit to glass cockpit by looking at the development history of b747-200, B787, B777X, and G650. Today, the glass cockpit is developing into touch cockpit. The cockpit is becoming more intelligent, lower workloads, and less amounts of the flight crew.

## 3. The feature of Touch Screen Technology

It is an obvious conclusion from the development of cockpit human-machine interface that touch cockpit is an inevitable trend of in the future. Touch screen technology is more intelligent and highly integrated which can effectively reduce the weight of the system and the workload of pilots. Furthermore, there are still have some difficulties in human factors of the touch screen technology, which impacts the pilot's human-machine interaction operation.

### *3.1. Digitization*

The development of digital technology is the precondition for the application of the touch screen in the aircraft cockpit. Whether the control interface adopts the touch technology is greatly affected by the aircraft terminal system. On the one hand, due to the high requirements of aircraft safety, commercial aircraft terminal system with high safety level tends to adopt the traditional physical connection and interface of these systems to realize by touch technology. On the other hand, some system functions must be completed without power on the aircraft which are also hard to apply touch technology. Therefore, the combination of touch control and physical control will be a major direction in cockpit human-machine interface design for civil aircraft currently.

### *3.2. Integration of Display and Control*

The integration of display and control is a major advantage of the development of touch screen technology. The integration of display and control means that a touch screen can display the information required by the user and direct operation can also be implemented in the corresponding area. The visual feedback (information change) generated by the operation can also be presented in the same area. The display area is completely separated from the control area in traditional cockpit thus more pilots efforts are required to switch between two different areas to obtain the status information of the aircraft. Therefore, the development of touch technology makes flight control more intuitive and efficient.

### *3.3. Intelligent*

Automated disposal procedure is an advantage by adopting the touch interface. The fundamental purpose of touch cockpit is to reduce the workload of pilots. The touch interface cannot make decisions and deal with tasks for pilots. However, with the development of intelligent technology, the machine behind control interface will automatically solve more tasks for pilots.

The fundamental purpose of touch cockpit is to reduce the workload of pilots. The touch interface cannot make decisions and deal with tasks for pilots. However, with the development of intelligent technology, the machine behind control interface will automatically solve more tasks for pilots. The pilot is a scarce resource in civil aviation and the number of the flight crew is a key factor affecting the profitability of an airline. The number of pilots on an aircraft has been reduced from five to two with the development of aircraft. With the development of touch screen intelligent technology, the single pilot is becoming possible.

### *3.4. Redefine of the working area*

The "cross area" in the civil aircraft cockpit is the main area for cockpit display and control. Control functions are mainly located at the overhead and central console area and display functions are concentrated at the instrument area. Glareshield area integrates display and control functions with FMCP as its pivotal equipment. Glareshield is the core area for pilots to perform forward-looking operations and gain external information faster while operating the equipment located in this area. More and more aircraft are now equipped with head-up displays (HUD), which proves that the concept of forward-looking operation will be the focus of cockpit development in the future.

The development of touch screen technology has greatly reduced the traditional control and display interface and makes more display and control functions to be concentrated in the forward view area possible. The cockpit layout will be changed dramatically.

### *3.5. Human Factors*

The application of touch screen technology has enormous advantages in the civil aircraft cockpit. However, some defects and problems need to be paid attention to, especially human factors, which will be the key factor in the development of touch screen technology.[6]

*3.5.1 Misoperation.* Civil aircraft cockpit is a narrow space and the aircraft can enter into turbulence during the flight. As a cockpit control interface, the touch screen is easier to be triggered by mistake compared with traditional physical control devices, which is an issue that needs to be evaluated. Based on the misoperation of the touch screen, there are mainly the following considerations:

- Touch Screen Activation. The touch screen should be in a dormant state when it has not been used for a long time. The touch screen can display normally but the operation function is in a dormant state. When the operation is required, the touch screen control function should be activated in some way.
- The Logic of Operation Response. Touch screen button should be activated when the finger leaves the screen. That means a certain area can be activated only when the operator's finger leaves. When a pilot accidentally touches the screen or clicks on the wrong area, he can move his finger to another area on the screen and then lift it up thus the corresponding control function of the wrong area will not be activated.
- Operation confirmation. Touch controls should be designed to provide a confirmation step before activating a function to prevent misoperation. It is also necessary to balance the confirmation steps with the increased pilot workload. The operation confirmation function should be set for the functions with high safety level on the aircraft which will cause irreversible safety impact on the aircraft thus needed to be confirmed twice before triggering.

*3.5.2 Operation Feedback.* A certain amount of operating force is usually required to confirm the operating instructions on physical devices. For example, the key operation is accompanied by feedback such as sound and force, and the joystick will transmit a reaction force feedback to the pilot when it is operated. However, the touch screen is a static plane which is difficult to induce the pilot's manipulation perception. Although operation feedback can be generated by means of sound and vibration, it is still unable to establish a perfect information feedback mechanism. The design of appropriate information feedback mechanism of touch screen technology used in the complex civil aircraft cockpit is one of the main contents of its research.

*3.5.3 Accuracy of Display.* The design of the display interface should inform the pilot which parts are touchable and which are not in a very direct and clear manner. If a certain position on the interface is thought to be touched to achieve a certain function but corresponding feedback is not received after clicked, it will frustrate the pilot's operation and affect the pilot's mood and normal operation.
The effective touch function area on the display screen should be clearly defined to ensure that the pilot can easily identify the effective control function area. Finger movement within a predetermined valid area shall not inadvertently operate other valid areas.

*3.5.4 Operational Stability.* The touch screen operation usually requires a pilot's arm to be suspended which can easily cause fatigue [7]. Traditional physical devices can provide operating fulcrums for the pilot, which is helpful for the pilot to select the target accurately during the flight. Without providing fulcrums, touch screen operation has higher requirements for pilots to accurately select the target which results in increased pilot fatigue.

*3.5.5 Operational Efficiency.* Touch screens used in civilian aircraft can reduce weight and save space. However, operation pages are overlaid which makes pilots spend more steps to complete triggering. It is particularly important for the operation which affects the flight safety and processing urgently needs the pilot to carry on. For this kind of operation, pages can be set by touch screen logic to quickly jump out, so that the pilot can efficiently handle the relevant operation. From the analysis of the operation mode of the display control device, the touch screen device is more suitable for the operation that can be completed by clicking, such as continuous input, click, etc., which is better than the traditional input inefficiency. But in precision adjustment or continuous adjustment aspect, the superiority of the touch screen is not obvious, even inferior to the traditional physical input equipment. Therefore, the touch screen should be applied for input operation, such as MCDU on the central console and buttons in each area to replace traditional control equipment. Devices like handles, knobs, and joysticks should be better not using the touch screen.

*3.5.6 Cockpit Commonality.* There is often a process involved in the acceptance of new technology for both airlines and pilots. More training is needed for pilots who are familiar with traditional aircraft models to learn new technologies which mean airlines need to spend more operating costs. Therefore, more common issues of cockpit touch screen should also be considered to reduce extra costs in the process of operation.

*3.5.7 Sunlight readability.* The display information on the touch screen should be readable under all predictable lighting conditions. Lighting conditions to be evaluated should include at least direct sunlight to the display screen, dusk, night flight through clouds, thunderstorms and other flight scenes. The glare in a variety of expected flight scenes should be paying attention to due to the widespread use of touch screens in the cockpit. Design criteria or specification of the display information itself, such as the design specification of character size, key size, and key color should also be considered while evaluating the flight scene. It is necessary to conduct tests and analysis and research on touch screen devices to determine specific design requirements while researching in the touch screen.[8]

## 4. Application of Cockpit Touch Screen

The purpose of the cockpit touch screen technology application is reducing the workload of flight crew under safety conditions. At present, the touch screen technology application is still in the primary stage, because of the low level automation in the cockpit. There are some considerations of the application touch screen technology, the details as follows:

- The hybrid layout of the touch screen interface and the physical interface. According to the advantage of these two technologies, it is better to provide a hybrid layout interface in the cockpit. The physical interface is preferred to perform the high safety, fast operation, and continuous holding functions. The touch interface is preferred to perform the low safety and single point touch functions.
- Pay more attention to reduce the pilot workload before the flight. There is a better environment in the cockpit before the flight, and pilots have a lot of work to deal with. The touch screen technology can increase the working efficiency before the flight, and improve the dispatch rate of the aircraft.
- No paper concept. No paper concept makes more space in the cockpit and more efficient of data updating and query. The touch screen technology provides a better condition for the no paper concept.
- Smart guidance. In the flight, pilots may not have enough time to deal with the failure problems in the emergency condition. The application of touch screen technology can guide pilots to make the choices according to the AFM. The pilots can allow the computer to deal with some actions based on the choices, but the final decisions should be made by pilots based on the safety consideration.

Based on the above considerations, this chapter gives a proposal of the touch screen technology applications on the current technology.

### *4.1. Graphical Flight Plan*

The traditional flight plan receives the text date from the keyboard. The pilot needs to remember the number of the airport, and more time to input it. The graphical flight plan can reduce the reaction time of in pilots' mind, reduce the workload of the flight crew, and improve flight safety. The graphical flight plan with touch screen has some other advantages.[9] Pilots can quickly change flight plans on the touch screen, get the fight plan from the ground automatic, zoom in or out the flight path on the touch screen in order to get the overview of the path. In summary, the graphical flight plan with the touch screen can improve the efficiency of pilots.

### *4.2. Electronic Flight Bag (EFB)*

As an electronic information management and display system mainly used in the cockpit, EFB with touch screen technology can display a variety of aircraft information data instead of the traditional paper data, and carry out the basic aircraft performance calculations. The EFB installation can be fixed or portable.[10] The main contents displayed in the EFB include navigation charts, meteorological charts, flight manuals and logs, passenger lists, electronic maintenance manuals, electronic checklist, and other related information. The pilots can query manuals and data more quickly and effectively through the EFB with touch screen technology. While improving flight safety margin, an aircraft can save more than ten kilograms of flight data paper a year.

### *4.3. Traditional Overhead Control Panel*

IOP Conf. Series: Materials Science and Engineering **608** (2019) 012018 doi:10.1088/1757-899X/608/1/012018

As we know, most of the control commands come from the overhead control panel in the cockpit. The traditional cockpit overhead control panel contains nearly 200 control devices. There are so many factors of control devices arrangement need to be considered.

- Golden Area of Overhead Control Panel. Generally speaking, the traditional cockpit overhead control panel is divided into three columns, the middle column area which can be easily operated by the pilot and co-pilot. According to the aircraft flight manual, the important control devices which need to be operated by the pilot and co-pilot are generally concentrated in the middle column. On the other hand, the middle area of the control panel is an ergonomics comfortable area for pilot operation. The power supply, fuel, and air condition system control devices with high operating frequency are generally concentrated in this area. The front area of the cockpit overhead control panel is the ergonomics uncomfortable area for pilot operation. It is difficult to carry on the "press" action in this area. Most of the toggle switches should be arranged in this area. Thus, the golden area of the overhead control panel is concentrated in the red rectangle area of Figure 5.

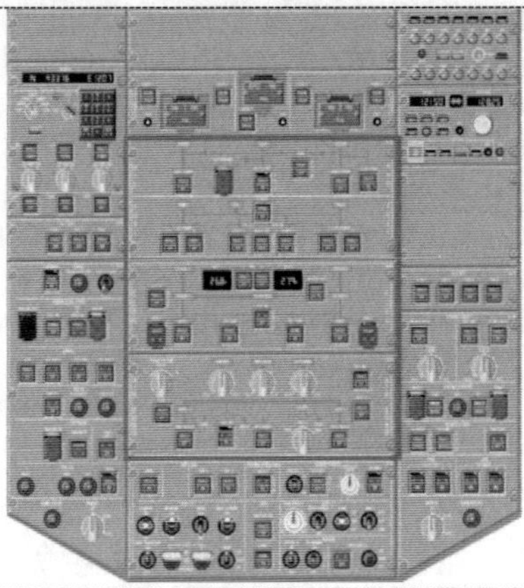

**Figure 5.** A320 Cockpit Overhead Control Panel. [11]

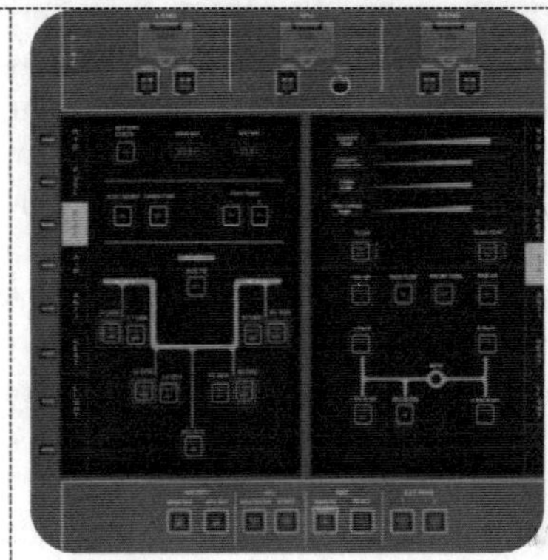

**Figure 6.** Cockpit Overhead Touch Control Panel.

- Operation Habit. The layout of traditional overhead control panel devices needs to consider the operation sequence according to the electronic checklist. So that the operation sequence should conform to a certain rule in the overhead control panel. In this way, the arrangement of overhead control panel devices should conform to the pilot operating habit, in order to reduce the workload of the pilot.
- Mistake Operation. The devices with similar shape and redundant design in system logic should be arranged separately in a distance to prevent mistake operation under the heavy workload in the cockpit.

In summary, there are still some limitations for the layout of the traditional overhead control devices, and it will make some difficult for the engineering design. Therefore, before the application of touch screen technology in the cockpit, it is necessary to make innovations in the layout of overhead control devices so as to take the advantages of touch screen technology. [12]

### *4.4. Overhead Touch Control Panel*

This chapter gives a proposal of the hybrid overhead control panel, which combines physical and touched control functions, as shown in Figure 6. There are the following characteristics for the overhead control panel with touch screen.

IOP Conf. Series: Materials Science and Engineering **608** (2019) 012018 doi:10.1088/1757-899X/608/1/012018

*4.4.1 Hybrid Layout.* The overhead control panel adopts a hybrid layout of physical devices and touch screen, which takes advantage of the touch screen and also makes the consideration of the aircraft safety.

- The hybrid layout consists of two dual-redundant touch screens, which provide the control interface for the pilot and co-pilot respectively, which not only ensures the safety of the system but also meets the ergonomic operation of the pilot and co-pilot.
- The physical devices such as engine fire controls and power system controls should be consistent with the traditional controls. Because these control functions failure will make a big impact on aircraft safety. The pilots need high-reliability controls for these function. On the other hand, the pilots still need to switch the aircraft power on if the touch screen loses power, so it is better to use mechanical controls for the power system.
- In order to achieve better ergonomics, the important and low-frequency physical devices are located above the touch screen, and the important and high-frequency devices are located below the touch screen.
- The main menu inside the touch screen adopts the longitudinal menu, which is close to the pilot and co-pilot. In this way, the longitudinal menu not only conforms to the pilots' operating habits but also meets the ergonomic requirements.

*4.4.2 Smart Warning System.* The overhead touch control panel can push the corresponding warning information automatic. There are two ways for pilots to deal with warning procedures: in the emergency situations, the lower risk warnings are automatically processed by system CPU; For the higher risk warnings, pilots can operate step by step according to the pushed information, and display the results of the operations on the touch screen. The smart warning system greatly reduces the workload of the pilots in emergency situations. At the same time, there are two columns of push buttons with annunciators, which are located on the left and right sides of the touch screens. The pilots can quickly enter into the operation page of the touch screen with pressing the corresponding push button when the warning occurs.

*4.4.3 Programmed Electronic Checklist.* In the traditional aircraft operation procedures, pilots need to make a step-by-step operation inspection according to the paper operation procedures. The electronic checklist is loaded in the touch screen, and the pilots can carry on the checklist automatic or semi-automatic according to their needs. In the fully automatic mode, the checklist would be done with one key operation, the pilots only need to monitor the result after the operation. In the semi-automatic mode, the checklist would be done by the pilot's step by step operation according to the recommends of the touch screen.

*4.4.4 Integration of Display and Control.* The touch screen combines displayed information and controls on the same page. The control results of the system can be seen, which greatly reduces the workload of pilots.

*4.4.5 Operation with Holding.* In the middle of the two touch screens, a grip groove is designed for pilots to grip when they operate with one hand. When the pilot's hand is in the holding status, the control finger can cover the whole touch screen control panel because of the suitable dimension of the touch screen. In this way, the design feature can greatly reduce the fatigue of pilots and improves control stability.

## 5. Ergonomics Evaluation

IOP Conf. Series: Materials Science and Engineering **608** (2019) 012018 doi:10.1088/1757-899X/608/1/012018

For the installed system and equipment, it shall be designed to ensure that the intended functions can be performed under any foreseeable operating condition. [13] The touch screen panel shall be involved in the closed-loop of the functions for each system. It can be judged that the touch screen panels meet the aircraft functional requirements by the ground function test of each system.

On the other hand, as a new type of human interface, the touch screen panel shall be performed by ergonomics evaluation in the cockpit. There are two kinds of subject evaluation method were used for the human factor evaluation of the flight deck HMI. One is the static evaluation based on the flight deck interface and description in JDP phase. The other method is the dynamic evaluation based on the typical scenario tasks in the verification phase.[14] The dynamic evaluation is very important for the touch screen control panel, which needs a specific platform with part or full flight deck system functions to perform the evaluation.

This chapter suggests two kinds of dynamic evaluations. One is the evaluation based on the Simulation Cockpit. In the Simulation Cockpit evaluation process, it needs to define all kinds of scenarios in the flight. The engineers can evaluate the HMI to make sure all the functions of the touch screen panel to fulfill with the aircraft requirements in the different scenarios.[15] The other evaluation based on real aircraft. The pilot needs to give the evaluation of the touch screen panel in the normal flight. The evaluations do not need to cover the full functions. The pilot can give the conclusion that whether the touch screen panel is high efficiency in the flight.

## 6. Conclusion

At present, there are some big changes for the development of civil aircraft cockpit. In order to reduce the workload of the flight crew, increase the efficiency of the flight crew, improve the flight safety, and reduce the airline operating costs, many new ideas and technologies of the cockpit are created by the engineers. Such as the concept of the single flight crew, the application of touch screen control technology, and the active side stick. It can be found that the development of aircraft is always in the direction of higher intelligence through these new technologies. Therefore touch screen technology will be widely used in civil aircraft in the next 15 years, although there are still some ergonomics problems in the current applications.

## Reference


[1] Jin-ling CHENG, Jun-hui YE, Lin-fang YAN. 2017 *Application of Touch-Screen Control in Future Civil Aircraft Cockpit* (Electronics Optics & Control)

[2] *http://pic.carnoc.com/file/160303/16030311570178.jpg*

[3] Civil Aviation Resource Network. Boeing Microrecord: Revolution of Aircraft Cockpit Technology. 2016 *http://news.carnoc.com/list/338/338229.html*

[4] *https://www.bilibili.com/video/av16316060/*

[5] *http://www.gulfstream.com/aircraft/gulfstream-g650er#*

[6] Cardosi K M. 1995 *Human factors in the design and evaluation of air traffic control systems* (Federal Aviation Administration)

[7] Noyes J M, Starr A F. 2007 *A comparison of speech input and touch screen for executing checklists in an avionics application.* (The International Journal of Aviation Psychology 17(3):) p299-315

[8] Wen Yang. 2016 *The Research of Commercial Aircraft Cockpit Control Panel Based on the Quite-Dark Cockpit* (Fudan University)

[9] Gert Stuyven, Barco NV. 2012 *Concept for an Avionics Multi Touch Flight Deck* ( SAE International by Aviation Industry Development Research)

[10] Zhenling XU, Cili JIA. 2011 *Portable Entertainment Devices Improve IFE Innovations* ( International Conference on Information, Services and Management Engineering)

[11] *http://www.sohu.com/a/121129550_465912*

IOP Conf. Series: Materials Science and Engineering **608** (2019) 012018 doi:10.1088/1757-899X/608/1/012018

[12] Alapetite A, Fogh R, Zammit-Mangion, et al. 2012 *Direct tactile manipulation of the flight plan in a modern aircraft cockpit* (Proceeding of HCI Aero'2012. Brussels: International Conference on Human Computer Interaction in Aerospace) p 12-14
[13] Civil Aviation Administration of China *CCAR 25.1309(a)*
[14] Dayong Dong, Baofeng Li, Haiyan Liu, Wenjun Dong, Hongtao Liu, Zhefeng Jin. 2011 *An analysis and evaluation procedure in civil aircraft flight deck design*
[15] Dong D Y, Yu J H, Li B F. 2016 *Airworthiness compliance certification technology of civil aircraft flight deck human factor* (Acta Aeronautica et Astronautica Sinica 17(1):) p310-316

IOP Conf. Series: Materials Science and Engineering **608** (2019) 012019 doi:10.1088/1757-899X/608/1/012019

# A CAS Message Simulation Module Used in Civil Aircraft for Air Management System Failure Diagnosis

**Xi Haiyan**

Shanghai Aircraft Design And Research Institute

Email: xihaiyan@comac.cc

**Abstract.** Air management system is one of the most important airborne systems on civil aircraft, which is responsible for controlling aircraft bleed air, cabin pressure control, wing anti-icing, cabin ventilation and temperature control. And the air management control system is the brain and neural network of the entire air management system. The air management control system control the various subsystems of the air management system and control the air management system to work together with other interfaced systems. Air management system control logic and interface signals are the key points and most difficulty parts in air management system control. For this highly complex system, failure diagnosis is most important and difficult part in ground test and flight test. Usually faults are reflected on the CAS message and synoptic pages as the alarms. However, the causes of such alarms may be varied. The traditional trouble shooting approach is to check the control specification documents which are complicated and hard to find the root cause. If there is a CAS message simulation module database that can be quickly traced back to the trigger logic of any one of the alarms, and the signal relationship between the alarms is displayed, the control logic of the system is clearly displayed, the failure diagnosis would became simple and intuitive. Based on the above design intention with referring to control specification, this paper created a set of air management system alarm models by using Matlab/Simulink. These alarm models are used for system troubleshooting and design's optimization. Finally, an application of cabin pressure control alarm with both trouble shooting and optimization is introduced.

## 1. Introduction

The size and complexity of embedded software in automotive systems has been increasing rapidly[1]. The air management system is one of the most important airborne systems on a civil aircraft. It is responsible for controlling aircraft bleed air, cabin pressure control, wing anti-icing, cabin ventilation and temperature control. The air management control system control the whole air management system and coordinate the sub systems to work together. The control system is also responsible for communicating with other interfaced system. Therefore, to realize the above control functions, the air management control system is allocated several thousand requirements. For this high complex system, to analysis the system's failures in the ground test or flight test is one of most important and difficult parts in civil aircraft design. Normally such faults are usually reflected on the CAS messages and synoptic pages. However, the causes of such alarms may be varied. The traditional trouble shooting approach is to check the control specification documents which thousands of pages which are complicated and hard to find the root cause. If there is a CAS message simulation module database that can be quickly traced back to the trigger logic of any one of the alarms, and the signal relationship between the alarms is displayed, the control logic of the system is clearly displayed, the failure diagnosis would became simple and intuitive. Based on the above design intention with referring to control specification, this paper creates a set of air management system alarm models and uses this model for the trouble shooting and system's optimization.

## 2. CAS Message Model Design Strategy

### *2.1. System architecture*

There is also a greater need to handle system-level control strategies, via communication networks and command software[2].The Air Management System Integrated Control System is typically equipped with two Integrated Air Management System Controllers (IASCs), each containing two control channels (channel A and B) and one safety channel (channel S). For different system control requirements, different backups are used between the control channels to ensure that when one control channel fails, the other control channel still has full control capability. The safety channel is used to monitor and protect the functions of the system with high security.

### *2.2. CAS message model*

According to the air management system control logic of a civil aircraft, and refer to[3][4], based on MATLAB/Simulink R2016, a set of air management system alarm model is created by using Simulink function. The model takes the civil aircraft air management system related controlled equipment, sensors and external signals as input. Based on the control logic described in the control specification, the result of directly triggering the CAS message is the output.

Take a FAULT alarm with bleed air failure as an example to introduce a model construction process. The ventilating failure FAULT alarm, the source of the fault includes temperature anomaly, pressure anomaly, equipment anomaly and sensor anomaly. The models are shown in Figure 1:

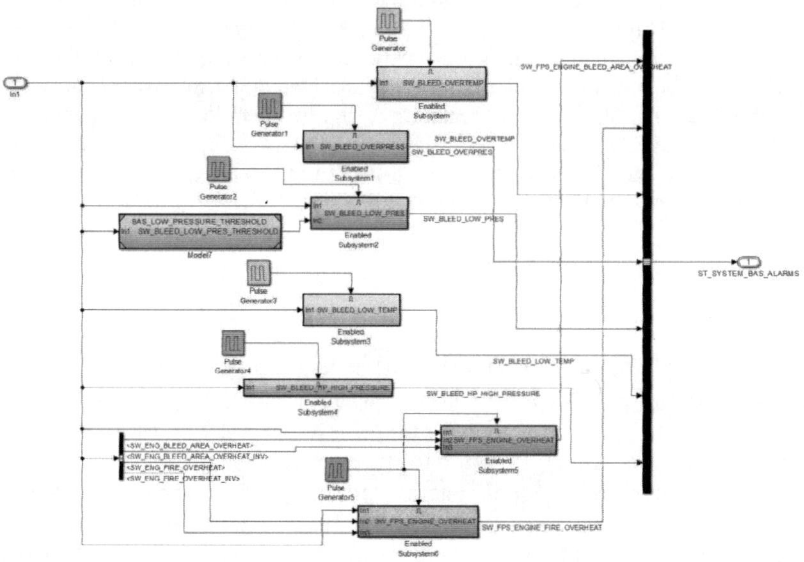

**Figure 1.** Bleed air system alarm model

## 3. Application

### *3.1. Failure Description*

The air bleed system starts up through the APU bleed air when aircraft is on the ground, and the cabin pressure control system uses the automatic mode. In the case of single PACK mode, the cabin pressure increased abnormally, and the CABIN PRESSURE DEGRADED alarm message appears. At the same time, we check the two outflow valve states that when the double PACK is closed, the outflow valve 1 is opened, the outflow valve 2 is closed, and the cockpit and cabin pressure are normal; after the APU started up, when the one-side PACK is opened, The outflow valve 1 is adjusted from the full open position to the full closed position, and the outflow valve 2 is kept closed. The outflow valve 1 is shown in figure 2 and the outflow valve 2 is shown in figure 3.

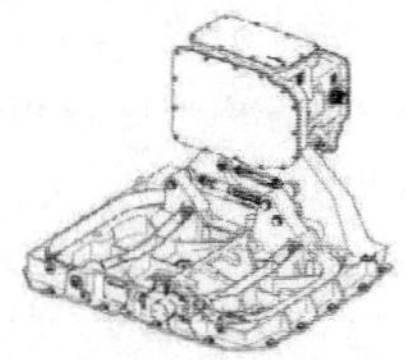

**Figure 2.** Outflow Valve 1

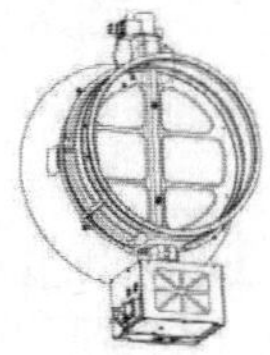

**Figure 3.** Outflow Valve 2

### *3.2. Failure Diagnosis*

Basing the model on test data could be problematical when testing becomes impractical with many years lifetime tests[5]. After analyzing the alarm simulation model, it can be found that the alarm of cabin pressure degraded, which is triggered by the failure of the outflow valve 2's automatic control, and the cabin pressure degraded logic module is shown in Figure 4.

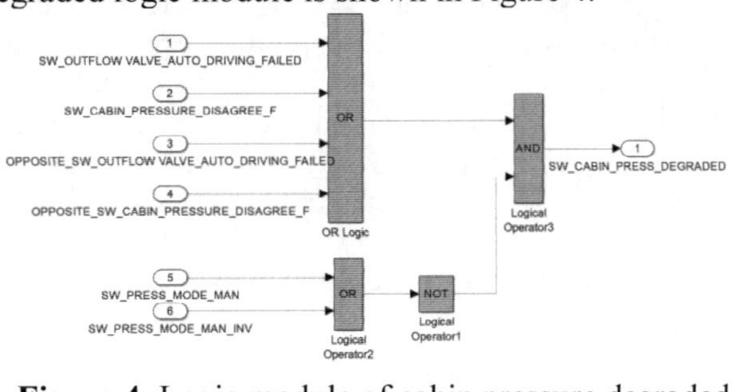


**Figure 4.** Logic module of cabin pressure degraded

There are three failure conditions for the outflow valve 2's automatic control failure. The first is the outflow valve 2's internal failure, the second is the command failure from the controller to outflow valve2, and the third one is the communication failure between controller and the outflow valve 2. These three failure conditions are shown in Figure 5.

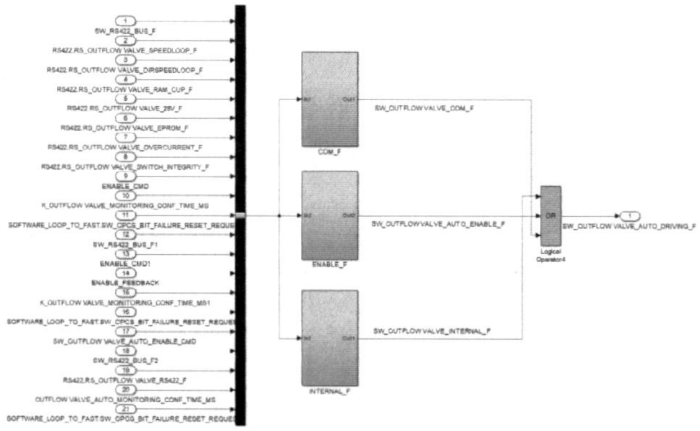

**Figure 5.** Logic module of outflow valve automatic control failure

From Figure 5, it can be seen that the alarm model can quickly analyze the potential fault source and locate the fault signal. By using failure logic module, all 21 potential fault sources can be easily sorted. To check each signal's correctness is the trouble shooting solution. Finally, we found the root cause of the outflow valve 2 auto mode fault is the item 9 in figure 6, enable command failure. One port of item 9's connector is out of connection, after fixing the connection, the outflow valve 2 works under the control.

*3.3. Logic optimization*

In this civil aircraft, the cabin pressure control system has two control valves, outflow valve 1 and out flow valve 2, when one outflow valve is working, the other one is stand-by.

In current design logic, in the case where the double PACK worked, the outflow valve 1 is normally opened, the outflow valve 2 is closed, the cockpit and cabin pressure are normal, while if the outflow valve 1 is failed, the outflow valve 2 will take over. but in the case where the single PACK automatic mode, the outflow valve 2 need to be worked and the outflow valve 1 is closed, but in case of outflow valve 2 is failed, the outflow valve 1 will not take over. Both of outflow valves closed position causes abnormal pressure in the cabin, and there is no FAULT CAS message to inform the pilot.

The current design logic may cause to this single point of failure, and it needs to be optimized. By compared with current cabin pressure alarm module, the optimization is shown in table 1.

**Table 1.** Optimization of outflow valve automatic control failure alarm

| CAS message | Before Optimization | After Optimization |
|---|---|---|
| FAULT | The Cabin Press Fault CAS message will be triggered when the following three items are gathered<br>1. The outflow valve 1 drive failure or exhaust valve 1 pressure sensor is disagree;<br>2. The outflow valve 2 drive fails or the outflow valve 2 pressure sensor is disagree;<br>3. Manual mode is not activated. | The optimization increases control channel's and single pack mode's identification.<br>1.If 1A channel which control the outflow valve2 works, then see if it is a single PACK. If both are yes, confirm whether this channel is invalid. If it's invalid, and the system is in the automatic mode, it will report FAULT;<br>2.If the control is not in the 1A channel, then see if it is a single |

| CAS message | Before Optimization | After Optimization |
|---|---|---|
| | | PACK, and if so, confirm whether the opposite channel is invalid, if it fails, and the system is in the automatic mode, it will report FAULT;<br>3.If the system is not in single PACK mode, it still uses the previous decision logic. |
| DEGRADED | The Cabin Press Degraded CAS message will be triggered if outflow valve 1 drive failure or outflow valve 1 pressure sensor is disagree or outflow valve 2 drive failure or outflow valve 2 pressure sensor is disagree. | Keep original design. Because when the FAULT message is triggered, the Degraded message will be inhibited automatically. |

The logic of the cabin pressure failure alarm before optimization is shown in Figure 6:

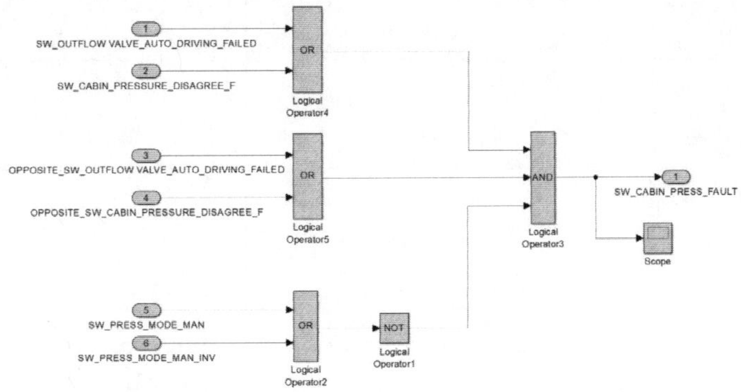


**Figure 6.** Logic module of cabin pressure fault alarm

Considering the optimization mentioned in table 1, the modified cabin pressure fault alarm increases the consideration the both single PACK mode and channel's identification. The optimized alarm logic is shown in Figure 7:

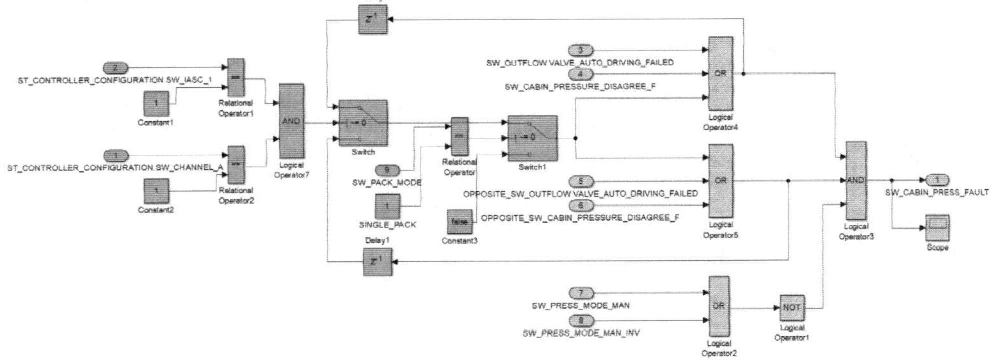


**Figure 7.** Optimized logic module of cabin pressure fault alarm

After the logic modification is completed, we have tested the failure test for three different working conditions. The test results are shown in Table 2:

**Table 2.** Verification results for cabin pressure failure alarm's optimization

| Failure mode | Output before | Output after |
|---|---|---|
| 1A channel control, single PACK mode; At this time, the out valve 2 is disabled. | 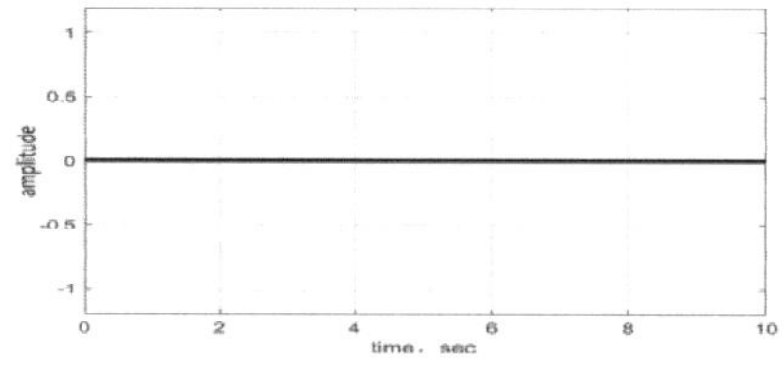 | 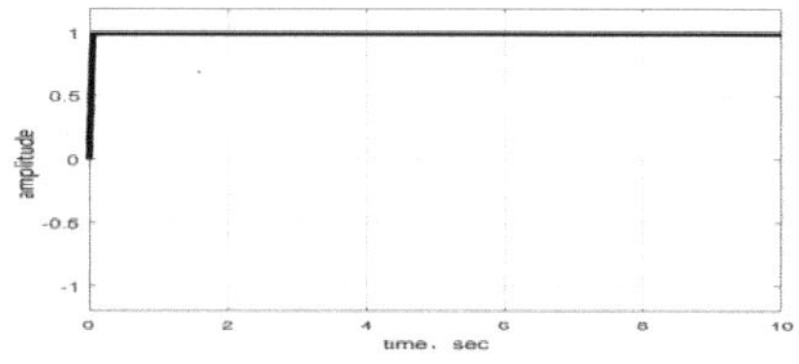 |
| 2A channel control, single PACK mode; At this time, the outflow valve 2 is disabled. | 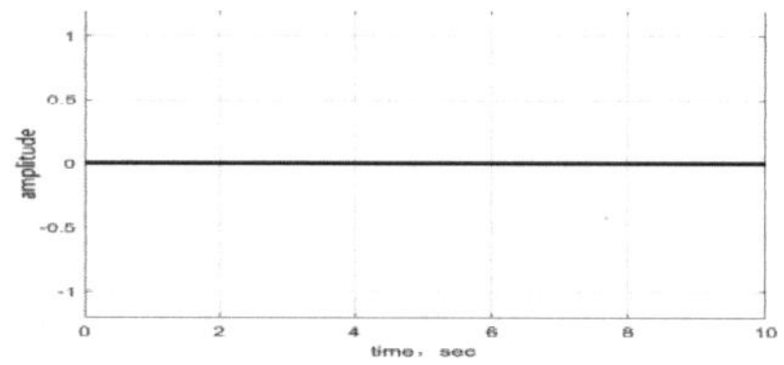 | 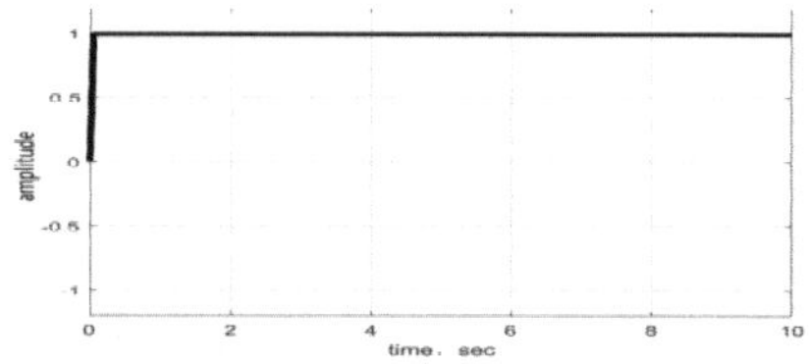 |
| Not in single PACK mode, the outflow valve 2 is disabled. | 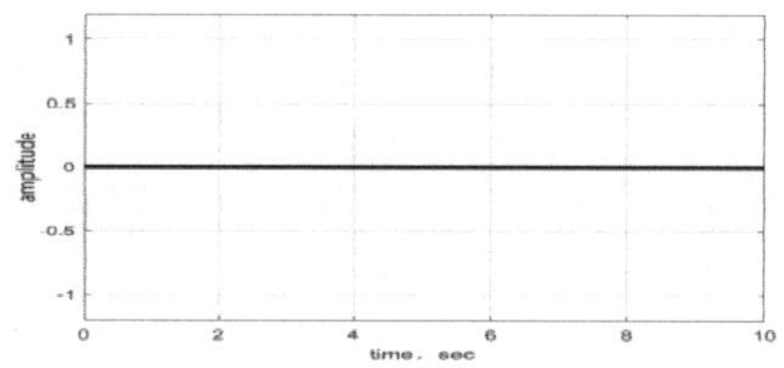 | 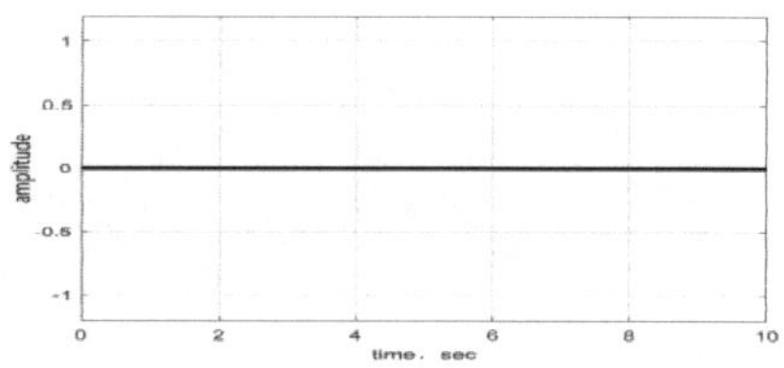 |
| Note: When oscilloscope's output equals 0, it means not failure; When oscilloscope's output equals 1, it means failure. | | |

Through the oscilloscope output, it can be seen that in order to verify the validity of the model, first test case is the 1A channel control, the single PACK automatic mode, and the failure of the outflow valve 2 at this time, the left picture shows the oscilloscope when the model is not modified. The output result shows that the output value is 0, that is, the system does not report FAULT; the right picture shows the output of the oscilloscope after the model is optimized. It can be seen that after a 50ms pulse, the output value is 1, that is, the system reports FAULT message according to the optimized requirements. The second case is the 2A channel control and other conditions the same as the first case, we get the same results. Finally, we cancel the single PACK mode. At this time, the outflow valve 2 is invalid. It can be seen that the modified logic is consistent with the report result of the logic before the optimization which is just our expectation. that is, it is a report, which is in line with our design expectations.

## 4. Summary

This paper first introduces the air management system and its high complexity. Then a developed simulink-based alarm model for the air management system function is introduced. Based on this alarm modle, the paper analyzes and locates the single PACK mode failure fault of the cabin pressure control system by using the alarm model. With the support of the model, the fault is solved. After further analysis of this fault and current design logic, it is confirmed that the current logic needs

IOP Conf. Series: Materials Science and Engineering **608** (2019) 012019 doi:10.1088/1757-899X/608/1/012019

further optimization. Finally, the paper analyzes the valve control logic of the cabin pressure control system in detail, and clarifies the optimization direction. Based on the existing logic, the failure judgment of the single PACK mode is considered, and the optimized logic improves the reliability of the system.

**References**

[1] Srihari Sukumaran, 2011, *Applying Software Dependence Analysis for Automotive Embedded Software Andersen, SAE International*

[2] Allen, J., 2012, *Simulation and Test systems for validation of Electric Drive and Battery Management Systems, SAE International*

[3] Wu, Luochang, 2012, *On monitoring and testing system of the hydropower unit vibration based on virtual instrument technology , ICMST 2011*

[4] Ma You-Jie, 2011, *Simulation and parameters tuning of second order ADRC controller in SIMULINK, ICMST 2011*

[5] Nabil Mohammed Hammad, 2010, *Vehicle Valve Regulated Lead Acid Bettery Modeling and Fault Diagnosis, SAE International*

# Corrosion Detection for Large Steel Structure base on UAV Integrated with Image Processing System

**Qingcheng Chen[1*], Xiang Wen[1], Shujie Lu[2] and Dongxiao Sun[2]**

[1] Shanghai Institute of Special Equipment Inspection & Technical Research,
[2] Shanghai Jiao Tong University, Shanghai, China

[*]Corresponding author e-mail: chenqczju@163.com

**Abstract**. Large cranes and other outdoor large steel structures need to be checked regularly to ensure their safety, and corrosion detection is one of the most important inspection item. Traditional manual inspection has many disadvantages including high labour intensity, low efficiency and low detection accuracy. In this paper, a novel corrosion detection method based on UAV intelligent image recognition is proposed. The six-rotor UAV can collect the photos of the steel structure device safely and reliably. With the intelligent corrosion recognition algorithm proposed in this paper, infection such as background noise can be eliminated, and the corrosion area in the image can be identified automatically with high precision. A test device has been built and the results show that the device and method can identify the corrosion area effectively.

## 1. Introduction

Cranes implemented in ports and other places are always exposed in salty sea wind environment and working with heavy loads. Their metal materials are easily to be corroded, and cracks are generated. This may bring big risks to lifting operation, and also very harmful for the on-site working staff and the mission loads. Detecting the corrosion and cracks situation is one of the most important routine work for safety production.

In the traditional detection method, engineers should climb up to the cranes and check if there is any unaccepted corrosion or cracks. Auxiliary detection equipment is always needed for the large steel structure, which makes the detection process complicated, time consuming, difficult and often dangerous for the engineers. Also, there are always many hard-to-reach positions with limited manual detection and make the detection less effective. A new detection method with UAV(unmanned aerial vehicle) comes into use recently, with many advantages including higher efficiency, safe for inspectors and higher accuracy [1-4].

In 2016, an UAV was used in a crane structural detection [5], during the 20-minute flight, multiple sets of high-altitude pictures were shot by a CCD camera, then these pictures are transferred to the ground station and post-processing were performed. Since there was no ultrasonic ranging module in this detection platform, and it is difficult to quantify the defects in the post-processing. In 2017, another UAV was used in the disposal of special equipment and hazardous chemicals [6]. The aerial photography and infrared imaging technologies were used to test and evaluate the safety status of equipment. Also, UAVs were used to inspect large-scale playground facilities and other equipment to detect high-altitude corners that cannot be reached by inspectors [7-14].

Although the UAV detection method mentioned above improves safety and reduces labor intensity compared with manual detection method, there are also some problems to be solved. The UAV is only used to take pictures during the detection process, and the actual corrosion check is mainly finished by the experienced engineers after the detection. The results cannot be achieved in the real-time way, and some corrosion areas needs to be confirmed repeatedly, therefore, experienced engineers are still needed.

In this paper, an automatic corrosion detection device based on real-time image processing technology of UAV is designed, with the corrosion image real-time processing technology, the labor intensity of corrosion detection can be further reduced, the accuracy and effectiveness of detection can be improved.

## 2. Structure of the corrosion detection system

The structure of the corrosion detection system is shown in Figure 1, which is mainly composed of a six-axis UAV platform, image sensor modules and a ground station.

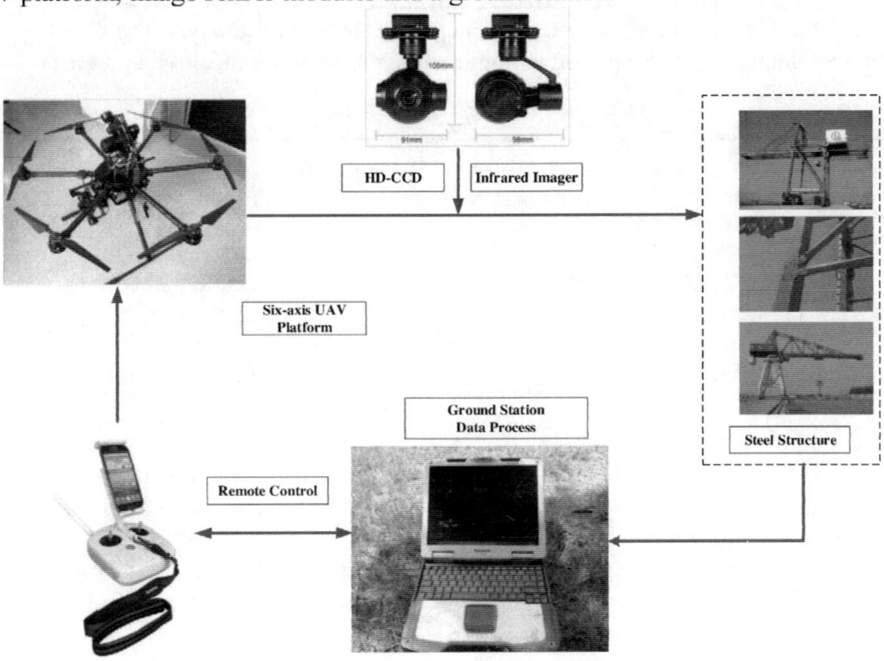


**Figure 1.** the structure of corrosion detection system

(1) six-axis UAV platform

According to the requirement of special equipment inspection, the UAV intelligent detection system has been improved. The platform is designed with the features of intelligent obstacle avoidance, positioning, stable hover and other flight characteristics when used in the unstructured special equipment detection environment. The multi-axis UAV can take off and land vertically at any place, and has low requirements on the working place, also, its simple mechanical structure is easy to maintain. Therefore, a six-axis UAV is used in the platform, some technical data is shown in Table 1.

By carrying a variety of sensor modules such as visual sensing, ultrasonic sensing and laser sensing, the platform is based on the multi-sensor information fusion system. The corresponding rules are designed and the information obtained by sensors is complemented and processed to improve the intelligence of the whole system.

**Table 1** Technical data of the six-axis UAV platform

| Item | Value |
|---|---|
| Flight mass/kg | 8.5 |
| Wheelbase/mm | 800 |
| Duration of flight/min | 20 |
| Classification of wind resistance | 5 |
| Blade diameter /cm | 15 |

(2) Image sensor modules

The UAV cradle system is equipped with a high-definition CCD camera and an infrared video camera, some technical data is shown in Table 2. The focal length is adjustable and high-definition image data can be accessed instantly in the stable flight. Also, a machine vision processing unit is embedded in this system. the potential structural failure of steel structure can be suspected and recorded, and this information can be shown on the station system. After the analyzing of image data, it is convenient for the inspectors to locate structural defects and analyze the fatigue and powerful connection parts effectively, making the evaluation and testing more intuitive and clear.

**Table 2** Technical data of the CCD camera

| Item | Value |
|---|---|
| Size/mm | 105*95*98 |
| Working temperature/°C | -10～50 |
| Pitch control Angle /° | ±90, accuracy±0.02 |
| Roll control Angle /° | ±45, accuracy±0.02 |
| Pointing control Angle/° | ±150, accuracy±0.03 |
| Mass/g | 400 |
| Optical zoom /mm | F4.9～49 (10X zoom) |

(3) Ground station and remote control

A ground monitoring and controlling platform system is designed to enable the UAV platform system communication. During the flight period, the flight status data including GPS position, flight direction, the battery level, can be transmitted in a real-time and reliable way. The station can display a variety of flight data, capture the image data and process these data. The data processing system can analyze and process a large number of image point cloud data. The suspicious structural defects can be identified and reported early.

## 3. Automatic corrosion detection algorithm

(1) Conventional image processing methods

Using conventional image processing methods, it is possible to obtain acceptable detection results on pictures with strong contrast between background and corrosion. However, from the HSV color space shown in Figure 2, the contrast between the background of the object and the corrosion color is not obvious. At the same time, some photos are affected by the sunlight, which leads to unsatisfied results when using this conventional processing method, as shown in Figure 3.

**Figure 2.** HSV colour space conversion

**Figure 3.** Process results with conventional method

(2) Image processing method based on deep learning theory

A new image processing method based on deep learning theory is proposed for the corrosion area detection. This framework can be built in an efficient C++ language with built-in Python and MATLAB interfaces. It has been widely used in a range of applications in video or image processing.

The target detection neural network uses Faster RCNN, which is based on R-CNN and was first proposed in 2015. The innovation is the network integrated with feature extraction, candidate frame selection, border regression and classification. In order to obtain a better result, VGG16 model is designed, as shown in Figure 4. The detection accuracy and detection efficiency can be effectively improved with this network. Therefore, this method is improved and applied in the image processing for corrosion detection.

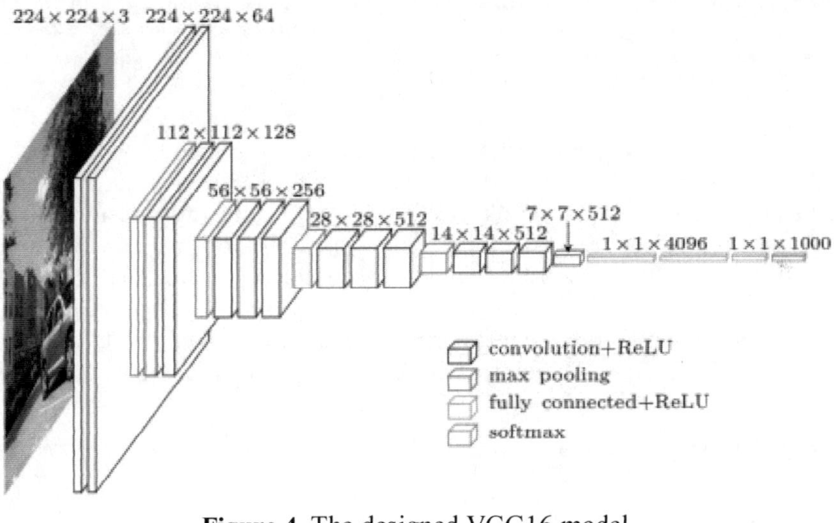


**Figure 4.** The designed VGG16 model

## 4. Experimental result and analysis

(1) Experimental setup and data preparation

Since a lot of images are needed for the neural network training, a long-time data collection should be conducted firstly. With the UAV platform as mentioned above, more than 100 pictures were captured during a one-hour flight around a frame crane. On these pictures, a lot of dust areas can be seen. Also, the distance information was obtained by the radar sensor when taking the pictures.

After kicking off some repeated pictures and unaccepted pictures, 63 pictures were chosen as the train objects.

(2) Neural network training

With the help of the LabelImg-master® software, the corrosion area data was scaled as shown in Figure 5. Both the corrosion area and its location information were extracted for every picture. After this treatment, different small corrosion area pictures are generated as shown in Figure 6.

IOP Conf. Series: Materials Science and Engineering **608** (2019) 012020 doi:10.1088/1757-899X/608/1/012020

**Figure 5.** Data scaling

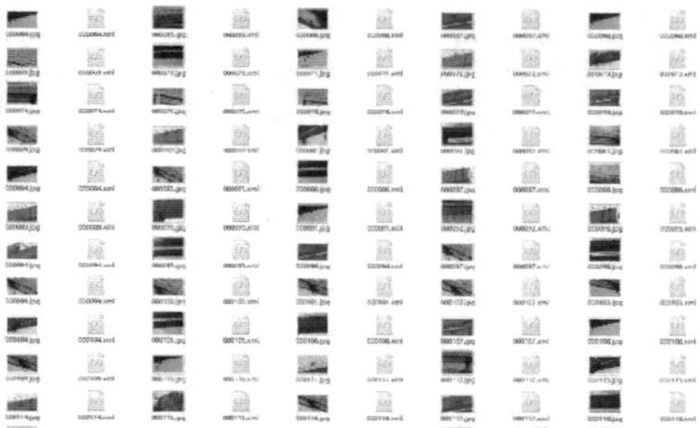

**Figure 6.** small corrosion area and its location xml-format files

Then with these small corrosion area pictures and xml-format files, an auto neural network training process was carried out. In order to get a better and fast result, a pre-training model was loaded, some of the training parameters were configured as shown in Figure 7.

```
train_net: "/home/dd/py-faster-rcnn/models/pascal_voc/VGG16/faster_rcnn_end2end/train.prototxt"
base_lr: 0.001
lr_policy: "step"
gamma: 0.1
stepsize: 50000
display: 20
average_loss: 100
# iter_size: 1
momentum: 0.9
weight_decay: 0.0005
# We disable standard caffe solver snapshotting and implement our own snapshot
# function
snapshot: 0
# We still use the snapshot prefix, though
snapshot_prefix: "vgg16_faster_rcnn"
iter_size: 2
```

**Figure 7.** Training parameters configuration

(3) Results and analysis

After using some pictures and more 20k training cycles. The learning results can be verified with the rest pictures which have not been used for training. The corrosion area can be marked with red box and labelled automatically, the "corrosion" indicates the corrosion area and the number means the probabilities.

From the picture shown in Figure 8, most corrosion areas are marked and labelled correctly, almost no omission or wrong marking. Therefore, this corrosion detection method can be proved to be effective and can be used in a real-time mode.

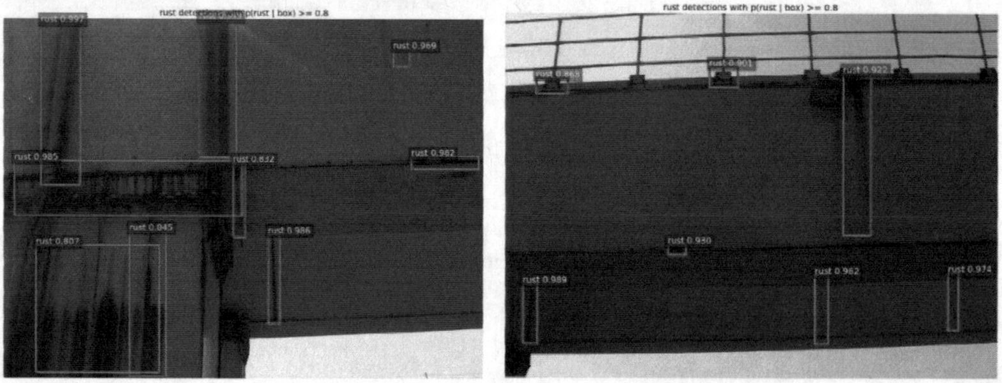

**Figure 8.** Corrosion detection experimental results

## 5. Conclusion

In this paper, a novel corrosion detection method based on UAV intelligent image recognition is proposed. Firstly, a corrosion detection system is designed, which is mainly composed of a six-axis UAV platform, image sensor modules and a ground station. The experiment has proved that this system is very solid and reliable for taking clear pictures and collecting other information of the large steel structure. Then, a new image processing method based on deep learning theory is proposed for the corrosion area detection. With this method, the experimental results show that the detection accuracy and detection efficiency can be effectively improved after the deep neural network training. Finally, since this method can be programed and applied in the ground station, the corrosion detection process can be carried out automatically, which will be a great progress for intelligent detection.

## References

[1] Morgenthal G, Hallermann N. Quality assessment of unmanned aerial vehicle (UAV) based visual inspection of structures[J]. Advances in Structural Engineering, 2014, 17(3): 289-302.

[2] Hallermann N, Morgenthal G. The application of unmanned aerial vehicles for the inspection of structures[J]. Proc. of PLSE, 2012: 1085-1095.

[3] Belmonte N, Staulo S, Fiorot S, et al. Fuel cell powered octocopter for inspection of mobile cranes: Design, cost analysis and environmental impacts[J]. Applied energy, 2018, 215: 556-565.

[4] De Melo R R S, Costa D B, Álvares J S, et al. Applicability of unmanned aerial system (UAS) for safety inspection on construction sites[J]. Safety science, 2017, 98: 174-185.

[5] ZHOU Qian-fei, et.al, Research on UAV Intelligent Visual Inspection System for Large Crane Metal Structure, IInternal CCombustion Engine & Parts, 2018, 271(19):230-234.

[6] Ling Zhangwei, et.al, Design and Application of Special Equipment UAV Detection System, Chemical management, 2018, No.487(16):77-78.

[7] ZHOU Jian, et.al, Research on the Online Monitoring System of Quadrotor UAV Based on STM32, Electronic Product Reliability And Environmental Testing,, 2018(1).

[8] Wang Kailei, et,al, Crane metal structural damage detection method based on UVA platform, Hoisting and Conveying Machinery, 2018(10).

[9] Moranduzzo T, Melgani F. Monitoring structural damages in big industrial plants with UAV images[C]//2014 IEEE Geoscience and Remote Sensing Symposium. IEEE, 2014: 4950-4953.

[10] Phung M D, Quach C H, Dinh T H, et al. Enhanced discrete particle swarm optimization path planning for UAV vision-based surface inspection[J]. Automation in Construction, 2017, 81: 25-33.

[11] Kim H, Sim S H, Cho S, et al. Unmanned aerial vehicle (UAV)-powered concrete crack detection based on digital image processing[C]//International Conference on Advances in Experimental Structural Engineering. 2015.

[12] Yeum C M, Dyke S J. Vision‐based automated crack detection for bridge inspection[J]. Computer‐Aided Civil and Infrastructure Engineering, 2015, 30(10): 759-770.

[13] Ortiz A, Bonnin-Pascual F, Garcia-Fidalgo E. Vision-based corrosion detection assisted by a micro-aerial vehicle in a vessel inspection application[J]. Sensors, 2016, 16(12): 2118.

[14] Li X, Yang Q, Chen Z, et al. Visible defects detection based on UAV-based inspection in large-scale photovoltaic systems[J]. IET Renewable Power Generation, 2017, 11(10): 1234-1244.

IOP Conf. Series: Materials Science and Engineering **608** (2019) 012021 doi:10.1088/1757-899X/608/1/012021

# Mechanism and Characteristics of High-Frequency Pulsed Jet Circulation Control

**Ren Zhan, Deng-cheng Zhang, Yan-hua Zhang, Wu-ji Zheng and Yi-wen Li**

Aeronautic Engineering Technology College, Airforce Engineering University, 710038 Xi'an, China

Corresponding author: Dengcheng Zhang@dengcheng_zhang@sina.com

**Abstract**: Conventional circulation control technology presents problems in certain situations. Pulsed jet circulation control appears to be a feasible alternative to solve such problems, especially for the high-frequency pulsed jet (HFPJ). To acquire the aerodynamic characteristics of the HFPJ and reveal the mechanism behind it, numerical simulation is conducted using a verified computational fluid dynamics (CFD) method. The action mode of the pulsed jet characterized by equilibrium and oscillation is proposed according to the analysis of a dynamic procedure. Lift and drag characteristics are obtained, defining the critical frequency to demarcate high and low frequencies. An oscillation function is also established to illustrate the difference between the characteristics of the HFPJ and a low-frequency pulsed jet (LFPJ) by analyzing the simulation results. The pressure distribution and flow field in the simulation results explain the stall characteristic of the HFPJ. Series results indicate the HFPJ demonstrates superior lift augmentation and drag reduction compared to the LFPJ and steady jets.

## 1 Introduction

Circulation control technology based on the Coanda effect[1] has been widely studied since the 1930s. Circulation control airfoil (CCA) [2] consists of conventional airfoil and a circular trailing edge [3]. First, a plenum is opened in the wing interior while a small slot is established on the trailing edge of the upper wing. The jet sheet (i.e., wall jet) is ejected from the slot through the inner tube tangent to the trailing edge. The jet sheet mixed with the external flow circles the trailing edge under the action of pressure and centrifugal force with delayed boundary layer separation. The effect produces substantial lift augmentation in the airfoil, with the greatest incremental lift efficiency ($\Delta C_l/\Delta C_u$) exceeding 80 and reducing the landing speed of the plane by approximately 50%[4]. Relevant research has been carried out by Englar and others. The medium height slot, low momentum coefficient, and small angle of attack (AOA) are conducive to lift augmentation[1][5][6]. The United States Navy utilized the A-6A CCW/ short take-off and landing (STOL) aircraft to conduct flight tests to verify STOL on circulation control technology [7][8]. In addition, series tests were developed by colleges and universities including West Virginia University[9–12].

Circulation control technology offers advantages in lift augmentation and noise reduction but also has inherent disadvantages: the sharp increment of induced drag increases the drag coefficient as the AOA grows, potentially leading to an abrupt stall on the wing[13]. A higher mass flow rate of the steady jet originating from the engine increases corresponding drag[14][15]. The emergence of circulation control technology in the pulsed jet alleviates some adverse effects of conventional jets. Meanwhile, pulse-jet circulation control achieves a similar or identical lift augmentation effect with a smaller mass flow rate[16]. Although experimental and numerical studies have been carried out by Oyler and Schatz et al[17][18], study conditions were limited to low frequency, low Mach number (0.1Ma), and an AOA equal to 0. Drastic fluctuation in aerodynamic characteristics occurs under these conditions; thus, the advantages of the pulsed jet cannot be fully realized. Compared with no blowing circulation control technology, the lift characteristics are greatly improved by using a steady jet or low-frequency pulsed jet (LFPJ), although better drag characteristics are hard to achieve simultaneously. The high-frequency pulsed jet (HFPJ) has therefore been studied systematically as the basis for predecessors to eliminate defects.

In this paper, the critical frequency for distinguishing the HFPJ and LFPJ is presented according to the frequency characteristics of the pulsed jet. Preliminary research on the characteristics and mechanism of the HFPJ has been conducted. Results show that drag characteristics are greatly improved, an advantage that is otherwise unattainable in a traditional high-lift system or circulation control using a steady jet. The conclusions provide theoretical evidence for continued research on aircraft performance when applying the circulation control technique using the HFPJ.

**2 Introduction of Numerical Simulation Method**

The following case was simulated using ANSYS CFX. The main control equation is an unsteady compressible Reynolds average N-S equation. SST[19] was utilized as the turbulence model, and the finite element method (FEM) was adopted to discretize the equation. The time step was set at 0.01*T*(period of the jet sheet). Mach number of the flow was set to 0.1 with the total pressure 1 atm. The supercritical airfoil indicated in Fig. 1 was adopted to generate the applied CCA as the baseline.

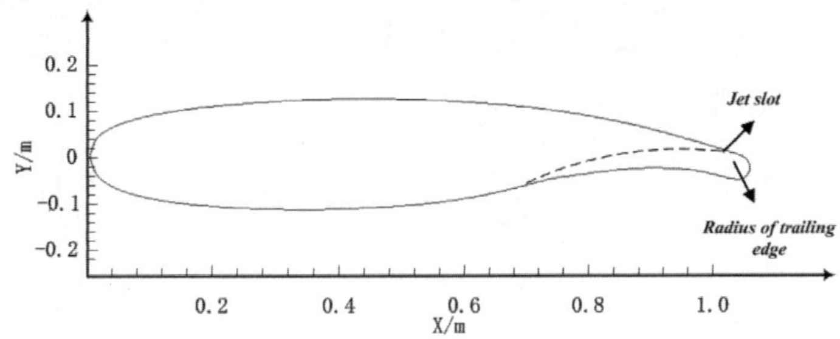


**Figure 1.** The applied CCA.

The impulse function of $v_{jet}(t)$ is defined as follows:

$$v_{jet}(t)=\begin{cases} 2v_{jet-av}\cdots\cdots\left(t\in\left(\dfrac{k}{f},(2k+1)\dfrac{1}{2f}\right)\right) \\ 0\cdots\cdots\left(t\in\left((2k+1)\dfrac{1}{2f},(k+1)\dfrac{1}{f}\right)\right) \end{cases} \quad (1)$$

where $v_{jet-av}$ is average jet velocity. $C_{\mu-av}$ is defined similarly to $v_{jet-av}$.

To accurately simulate the complicated pulse flow field around the Coanda surface, the proposed method are as follows: considering the complicated flow field near the slot, airfoil rear (under a large AOA), leading edge, and trailing edge, the mesh of those areas is densified to confirm the precision of the calculation distinguished by different methods of increasing nodes at the edges. The height of the first layer grid was set to $1.0\times10^{-6}$ m such that $y^+<1$.The model included 339410 cells as shown in Fig.2. Different structural grid scales (G1–4) were adopted to verify grid independence with 210860, 276740, 324972, and 411800 cells, respectively.

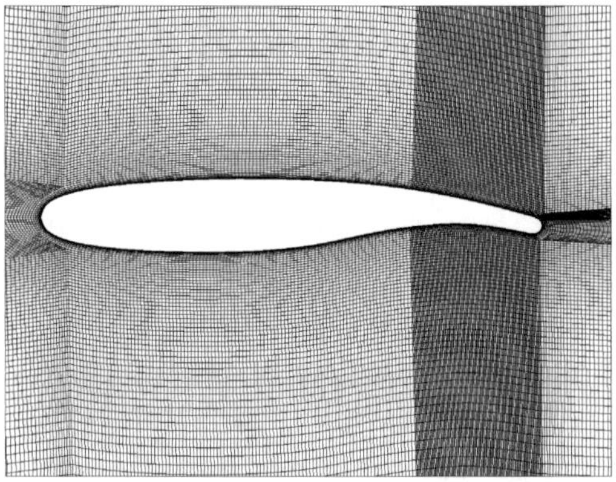

**Figure 2.** Adopted computational grid with 339410 cells.

The CCA from the reference[20] was used to verify the calculation method. The calculated results were consistent with experimental data from the reference[20], depicted in Fig. 3 for G1–4, indicating that the CFD calculation method is appropriate for numerical simulation of the HFPJ in the subsequent chapters. The time step was set at 0.05$T$ for each cycle.

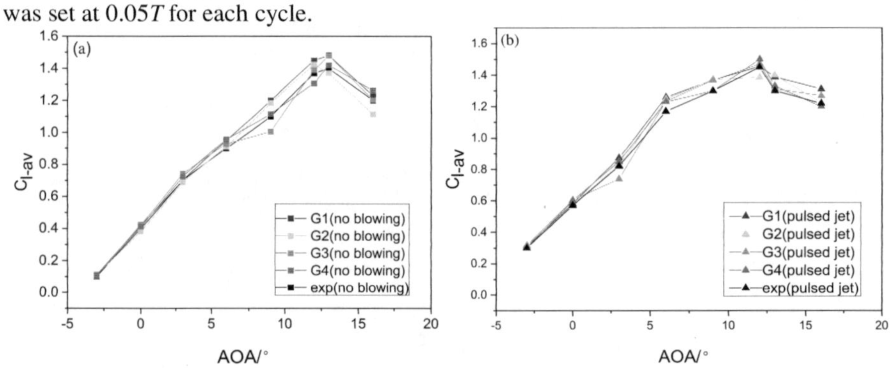


**Figure 3.** Validation of CFD method: (a)no blowing; (b) pulsed jet

## 3 Results and Discussion

### *3.1 Action Mode of Pulsed Jet Circulation Control*

The composition of the pulsed jet is twofold, containing jet-effective and jet-ineffective parts for any frequency of the pulsed jet $f$. The function is the acquisition of equilibrium and periodic variation of circulation caused by periodic changes in the flow field. The mechanism of the HFPJ circulation control is illustrated in Chapter 3;and the flow field around the trailing edge of the airfoil at two frequencies (10Hz) is displayed in Fig. 4.

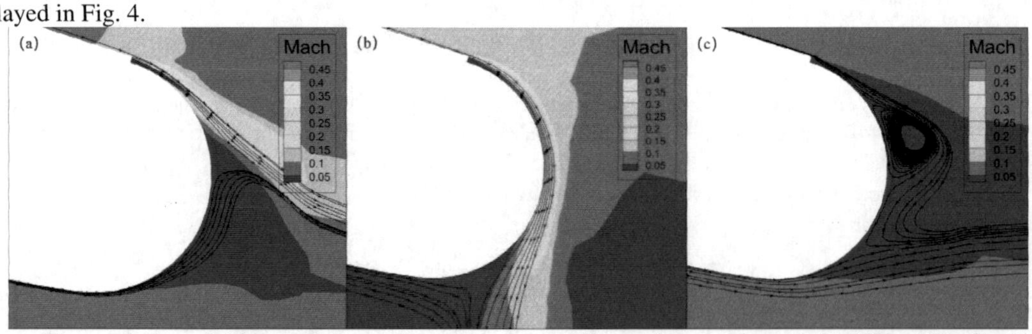


**Figure 4.** Flow field around the trailing edge at two different frequencies: (a) $f$ = 10Hz ($t/T$→0); (b) $f$ = 10Hz ($t/T$ =0.5); (c) $f$ = 10Hz ($t/T$→1)

IOP Conf. Series: Materials Science and Engineering **608** (2019) 012021 doi:10.1088/1757-899X/608/1/012021

In the simulation, before acquiring the equilibrium, the jet from the slot on the trailing edge began to function at $t$ = 0. The consequent input momentum accelerated the boundary layer flow, deflecting the streamline encircling the Coanda surface. The delay in separation produced more circulation around the airfoil along with lift ($C_l$) and drag ($C_d$). Circulation around the airfoil peaked at $t = 0.5T$. The jet sheet disappeared when $t > 0.5T$, and the flow continued to move around the Coanda surface at a short distance due to inertia. The current separation point could not be maintained due to the high pressure drag with no input momentum. At the same time, separation was enhanced and upper flow started to decelerate along with a circulation destruction that reduced $C_l$. $C_d$ declined as the static pressure increased and the flow decelerated, concluding the entire process cycle. The flow field was changed despite the jet being stopped (streamline deflection) compared to the initial condition ($t$ = 0). As such, the increment of circulation was positive in a single cycle. If the jet sheet was re-imposed before the increment decreased to 0, circulation restarted with a micro-increment of circulation in every cycle before the equilibrium. Accumulation in a cycle changed to 0 under the equilibrium. After completion of the above process, periodic oscillation of the circulation occurred within a specific frequency in a certain range. This trend also held true for circulation, $C_l$, and $C_d$. Meanwhile, number of experienced cycles before the equilibrium $n$ depends on the frequency of the pulsed jet, illustrated in Fig. 5.

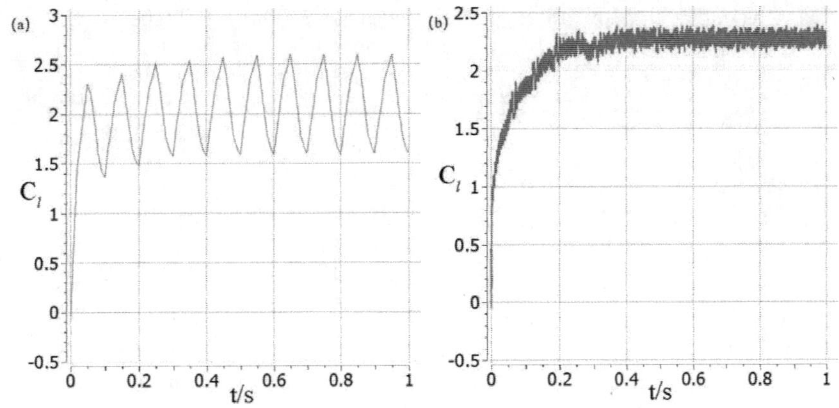


**Figure 5.** Procedure of reaching equilibrium: (a) $f$ = 10Hz;(b) $f$ = 200Hz.

The difference in aerodynamic characteristics between un-equilibrium and equilibrium indicated in Fig. 5 is related to the difference in the flow field around the rear lower surface of the CCA, except for the distinction of the upper-surface flow field as the main reason. A high-velocity region on the upper surface flow also expanded as time passed due to the gradually delayed boundary separation. Moreover lower surface flow decelerated in the process of approaching the equilibrium, a trend induced by the high-static-pressure area located on the lower part of the trailing edge that provides a high pressure gradient as shown in Fig.6. The high pressure gradient disappeared when equilibrium was achieved.

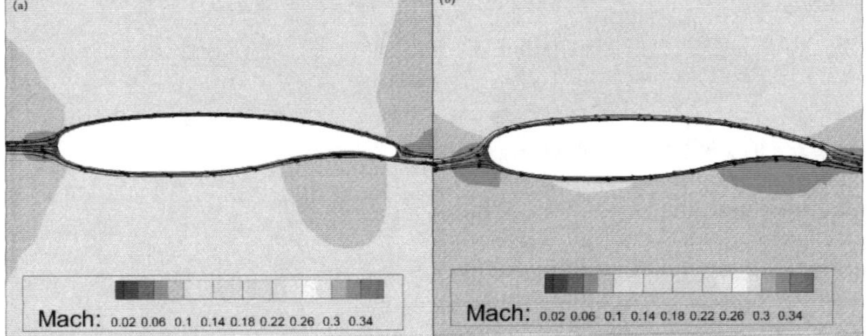


**Figure 6.** Flow field of CCA lower surface (200Hz, $t/T$ = 0.5): (a) un-equilibrium;(b) equilibrium

### *3.2 Aerodynamic Characteristics and Mechanism of HFPJ Circulation Control*

Lift and drag characteristics are profoundly affected by the frequency of the pulsed jet. Therefore, it is necessary to study the frequency characteristic of the pulsed jet to identify the optimal frequency range. Fig. 7 shows the curve of lift and drag characteristics that shifted in line with the frequency at a certain $C_{\mu\text{-}av}$. Average lift coefficient $C_{l\text{-}av}$ and average drag coefficient $C_{d\text{-av}}$ are defined as follows:

$$C_{l-av} = \frac{1}{T}\int_0^T C_l(t)dt \tag{2}$$

$$C_{d-av} = \frac{1}{T}\int_0^T C_d(t)dt \tag{3}$$

Fig. 7(a) shows that average lift coefficient increment$\Delta C_{l\text{-}av}$ diminished first and grew later as the frequency increased. It tended to be equal to that of the steady jet and remained constant once the frequency exceeded 300Hz. Fig. 7(b) shows that $C_{d\text{-av}}$ diminished to a specific value at a higher frequency and then stayed constant. Above all, the characteristic showed significant differences between the HFPJ and no blowing or LFPJ. The$\Delta C_{l\text{-}av}$ of the HFPJ appeared to remain constant when the frequency was above a specific value. The HFPJ is supposed to be characterized by this unique feature. Thus, in the demarcating method, the critical frequency of circulation control using the pulsed jet is defined as the point at which $C_{l\text{-av}}$ remains constant, and the high/low frequency is higher/lower than that. The results indicate that the respective effects of lift augmentation and drag reduction were improved by using the HFPJ.

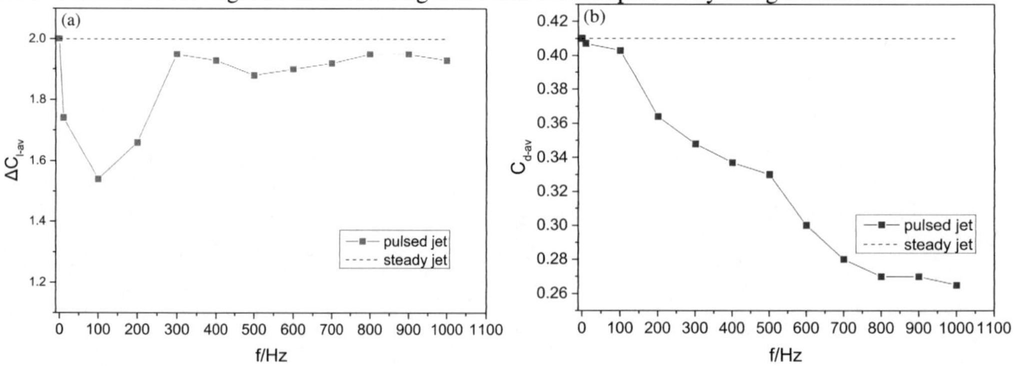


**Figure 7.** Frequency characteristic curve: (a) average incremental lift coefficient; (b) average drag coefficient

According to the action mode of the pulsed jet, circulation appeared to increase and decrease in two different parts. To reveal the mechanism of lift and drag characteristics of the HFPJ, a model of the accumulating circulation process from the initial state to the equilibrium was established. Moreover, an oscillation model was proposed to explain the rule of circulation at the equilibrium of $C_l$. Assume that a series of equilibria are distinguished by frequencies.

Periodic oscillating function of the circulation at equilibrium $\Gamma(t)$ is defined as follows:

$$\Gamma(t) = \Gamma_M + \mathrm{F}(t) \tag{4}$$

where F(t) is negative circulation function of the fluctuating value

The relationship between $T_F$ and $f$ is expressed as

$$T_F = \frac{1}{f} \tag{5}$$

where $T_F$ is period of F($t$)

According to the oscillation model, the different aerodynamic equilibrium states differentiate the time characteristics of circulation ($\Delta C_l$). The distinction between peak of $\Gamma_M$ and trough of $\Gamma_m$ symbolizes the change in the equilibrium point at different frequencies. $\Gamma_M$ and $\Gamma_m$ collectively determine $\Delta C_{l\text{-}av}$ as shown in Fig. 8. $\Gamma_M$ dominates $\Delta C_{l\text{-}av}$ when $f < 200$Hz and nearly reaches a constant value above $f =$ 300Hz, whereas $\Gamma_m$ dominates $\Delta C_{l\text{-}av}$ when $f > 200$Hz. This conclusion is exemplified in Fig. 8, and the rule of $\Gamma_M$ and $\Gamma_m$ is as follows: under the influence of the combination of $\Gamma_M$ and $\Gamma_m$, the $\Delta C_{l\text{-}av}$ might reach an extreme value as the frequency increases, as depicted in Fig. 8. Variation in $\Delta C_{l\text{-}av}$ depends on $\Gamma_M$ and $\Gamma_m$ in two respective situations ($f > 200$Hz and $f < 200$Hz).

The delay of separation, or the position of the separation point, represents the value of $\Gamma_M$. Conada turning angle, $\theta_{jet}$ (transient value; the variable changes over time) is recognized as the measurement of the displacement of the separation point along the Coanda surface. Given that the pressure gradient rises as $\theta_{jet}$ increases, it becomes harder for the jet sheet to overcome the larger gradient with a lower $\theta_{jet}$ increasing rate $\frac{d\theta_{jet}}{dt}$ until it declines to 0. This process facilitates circulation accumulation to focus on the time before $\frac{d\theta_{jet}}{dt}$ starts to decline sharply. In the simulation, $\theta_{jet-m}$ at different frequencies demonstrated negligible change based on frequency over 100Hz: maximum $\frac{d\theta_{jet}}{dt}$ (t/T = 0.5) of pulsed jet $\frac{d\theta_{jet-M}}{df}$ remained virtually constant if the corresponding jet-effective time was close to that mentioned above, such that $\Gamma_M$ changed little at frequencies over 300Hz as well.

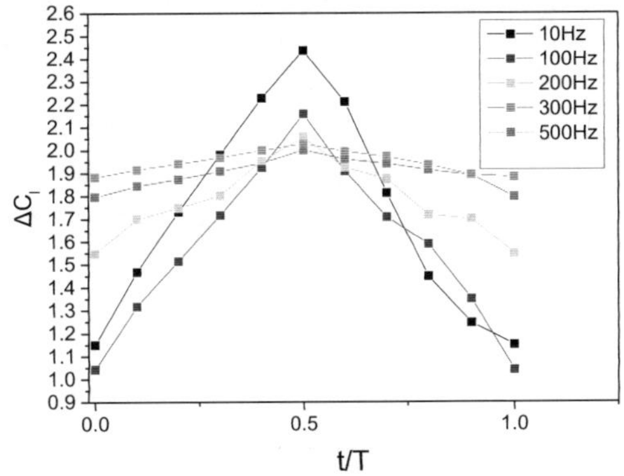


**Figure 8.** Time characteristics of lift coefficient increment.

The $\Gamma_m$ achieved higher values at high frequencies according to Fig 8. As shown in Fig. 9, the boundary layer separation on the Coanda surface consisted of successive deceleration and detachment from to the flow. Deceleration, or moving backward from the separation point, occurred more quickly than detachment. This process operates as follows: after jet termination, the rest momentum of the jet transfers from the fluid region (closer to the slot) to the separation point. Thus, the position of the separation point can be maintained or changed slightly if the jet-ineffective time is sufficiently brief. On the other hand, centrifugal force declines as the flow decelerates, contributing to attachment on the Coanda surface of the flow. A regenerative jet sheet in the next cycle replenishes momentum to the boundary layer before the rest momentum in the current cycle can be counteracted by the impulse of the pressure gradient and frictional drag if the frequency is high enough. Consequently, boundary separation is better at high frequencies when $t/T \rightarrow 1$. This phenomenon explains why $\Delta C_l$ at $f =$ 100Hz was much lower than that at $f =$ 500Hz as shown in Fig. 9 (i.e., because deceleration started and continued for a period of time).

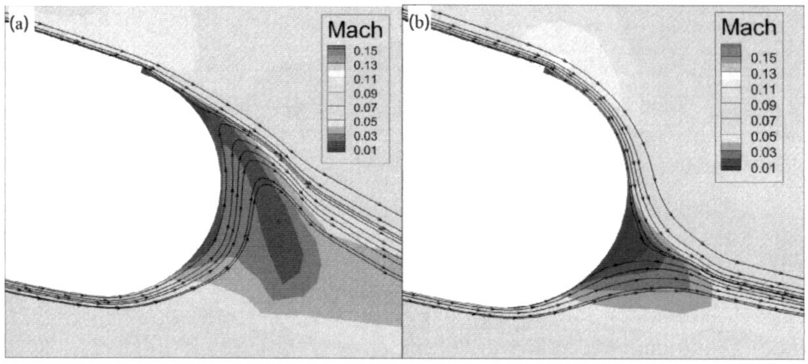


**Figure 9.** Flow field around the trailing edge at 10Hz and 500Hz ($t/T \to 1$): (a) $f$ = 100Hz; (b) $f$ = 500Hz

As for drag, the reduced jet-effective time resulted in displaced separation up along the Coanda surface as the frequency increased, leading to an increase in static pressure at the trailing edge and reducing the pressure drag.

In terms of the action mode of the pulsed jet, the corresponding shorter effective and ineffective time for the HFPJ resulted in relatively high $\Gamma_m$, along with $\Gamma_M$ nearly equal to that of LFPJ. Higher $\Delta C_{l\text{-}av}$ is attributed to $\Gamma_m$ and $\Gamma_M$ at high frequencies. Compared to the LFPJ and steady jet, the HFPJ reduced drag via undermined pressure drag at the equilibrium. The distinction of the pressure distribution at the trailing edge between the HFPJ and LFPJ contributed to refined drag characteristics due to pressure drag reduction.

### *3.3 Stall Characteristics of High-frequency Pulsed Jet*

Under conditions of different respective circulation controls, namely steady jet, the HFPJ at either 200Hz or 500Hz, and no blowing with $c_{\mu-av}$ = 0.02, the curve of changes in the lift-drag coefficient with AOA was obtained through numerical simulation; see Fig 11. The stall AOA (10°) under steady-jet circulation control was relatively low compared to under a no-blowing condition at small AOA. $C_{d\text{-}av}$ increased slowly and showed no sharp increase above 20°. The stall AOA (15°) under conditions of the HFPJ jet and no blowing was nearly equal: $C_{l\text{-}av}$ declined slowly and maintained a higher value as HFPJ was imposed when AOA exceeded the stall AOA. The drag characteristic was improved by HFPJ circulation control as shown in Fig. 10(b); that is, $C_{d\text{-}av}$ was less than AOA when applying HFPJ circulation control.

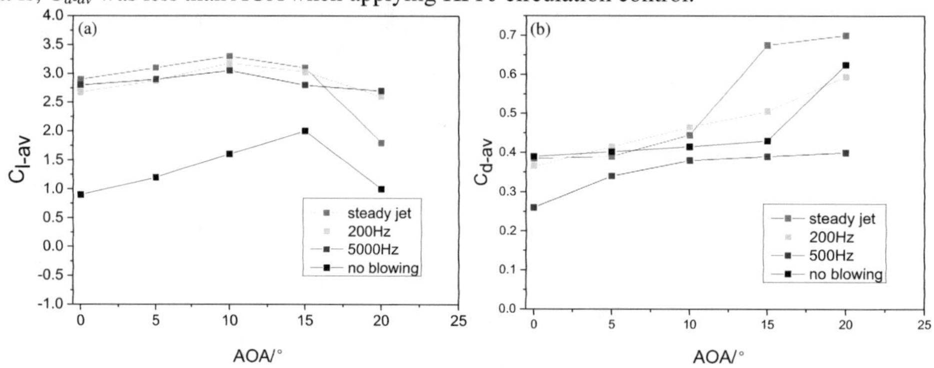


**Figure 10.** Lift-drag characteristic curve: (a) lift characteristic curve; (b) drag characteristic curve

With the flow field distribution of the steady jet and HFPJ at a certain time with no blowing, respectively, the time of the simulated flow field represents when the transient lift and average lift are equal. As shown in Fig. 11, stall became most serious when no jet circulation control was applied, but it was greatly alleviated by the HFPJ with deflection of streamlines leading to lower particular AOA and elimination of vortex.

IOP Conf. Series: Materials Science and Engineering **608** (2019) 012021 doi:10.1088/1757-899X/608/1/012021

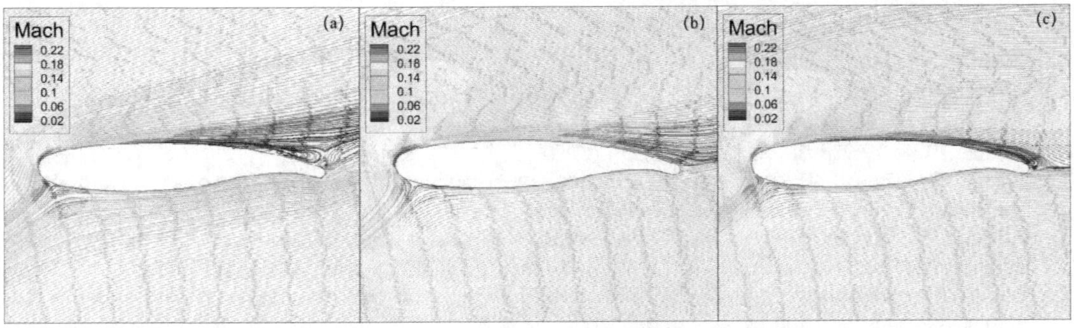


**Figure 11.** Flow around airfoil at 20° :(a) no blowing;(b) steady jet; (c) high-frequency pulsed jet

Fig. 12 illustrates the pressure diagrams of the trailing edge when AOA = 20° with different forms of circulation control. All conditions were set at the same reference pressure to determine a conclusion by comparison. The high-pressure region expanded clearly with the HFPJ compared to the steady jet and no blowing, leading to improved drag characteristic of the HFPJ as depicted in Fig. 11: the HFPJ increased the area of the high-pressure region around the rear of the airfoil and thus reduced pressure drag. In addition, frictional drag was nearly invariable because it was unaffected by variation in the small AOA; hence, the drag became lower overall as the HFPJ was imposed. The frictional drag affected the total drag chiefly by increasing the drag along with the AOA.

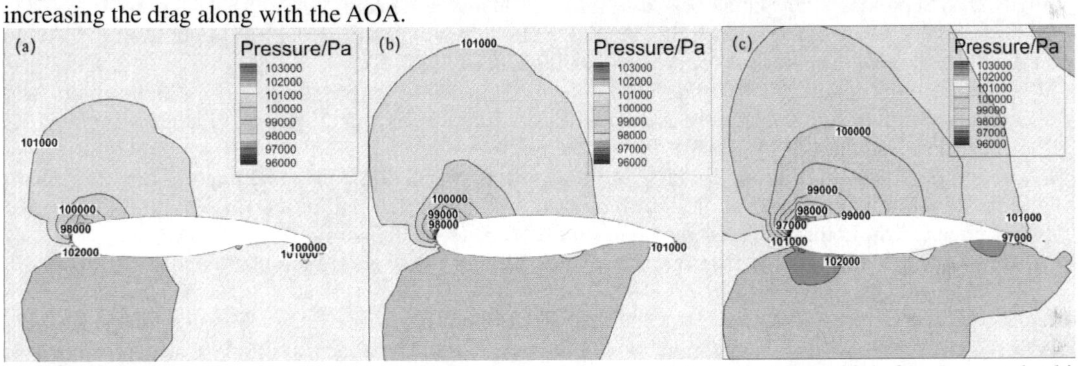


**Figure 12.** Pressure counter of airfoil at 20 ° : (a) no blowing; (b) steady jet; (c) high-frequency pulsed jet

## 4 Conclusion

In this study, the action mode of the pulsed jet is described in a procedure to reach and maintain the equilibrium of the circulation around the airfoil using the given mechanism. Lift characteristics of the pulsed jet identified the critical frequency (300Hz in the simulated case) to confirm the high or low-frequency range. The HFPJ circulation control exhibits particular advantages in lift augmentation and drag reduction by oscillation model: the HFPJ increases the maximum and minimum value of circulation in a cycle by modifying the boundary layer separation condition. Pressure drag is reduced by narrowing the region of low static pressure around the trailing edge. For the stall characteristic, the HFPJ expands the high-pressure region and reduces effective AOA, thereby increasing the stall AOA while reducing pressure drag. The simulation results must be verified by experiments but provide a reference for future studies.

## References

[1] Englar RJ., and Huson G. Development of advanced circulation control wing high-lift airfoils. Journal of Aircraft 2012;21:476-483.

[2] Nielsen J. and Biggers J. Recent progress in circulation control aerodynamics. In: 25th AIAA Aerospace Sciences Meeting, Aerospace Sciences Meetings, Reno, USA, 24 March - 26 March 1987, AIAA 1987-1. Reston: AIAA.

[3] Zhu Z and Zongcheng W. Study of the circulation control technology. Acta Aeronautica Et Astronautica Sinica 2016; 37:411-428.

[4] Englar RJ and Williams RM. Test techniques for high-lift, two-dimensional airfoils with boundary layer and circulation control for application to rotary wing aircraft. Canadian Aeronautics and Space Journal 1975; 19:93-108.

[5] Zhang SY, Zhong YC, Peng-Zhe FU, et al. Numerical investigation of dual-radius in circulation control airfoil aircraft design. Aircraft Design 2011;31:1-6.

[6] Abramson J, Rogers E. High-speed characteristics of circulation control airfoils. In: 21st Aerospace Sciences Meeting, Reno, USA, 10 January - 13 January 1983 ,paper no. 83-0265. Reston: AIAA.

[7] Englar RJ, Hemmerly RA, Taylor DW, et al. Design of the circulation control wing STOL demonstrator aircraft. Journal of Aircraft 2015; 18:51-58.

[8] Pugliese A. and Englar R. Flight testing the circulation control wing. In: Aircraft Systems and Technology Meeting, Aircraft Design and Technology Meeting, New York, USA, 20 August - 22 August 1979, AIAA 1979-1971.Restion: AIAA.

[9] Broniatowski M. Circulation controlled STOL wing optimization. Journal of Aircraft 1984; 21:128-134.

[10] Sparks R, Michie S, and Crowther W. Development of an integrated circulation control Fluidic thrust vectoring flight test demonstrator. Texas Heart Institute Journal 2005; 32:194-197.

[11] Fielding J, Lawson C, Rui MP, et al. Design, build and flight of The DEMON demonstrator UAV. In: 11th AIAA Aviation Technology, Integration, and Operations (ATIO) Conference, Virginia Beach, USA, 20 September - 22 September 2011, AIAA 2011-6963. Reston: AIAA.

[12] Buonanno A, Cook MV, Erbsloeh SD. Flight Dynamic Simulation of a Flapless Flight Control UAV. In: 25th International Congress of the Aeronautical Sciences, 2006:1-11.

[13] Yaros SF, Sexstone MG, Huebner LD, et al. Synergistic airframe-propulsion interactions and integrations: a white paper prepared by the 1996-1997. Technical Report, Langley Aeronautics Technical Committee, USA, April 1998.

[14] Gregory J., Sally V., Anthony W, et al. An active flow circulation controlled flap concept for general aviation aircraft applications. In: 1st Flow Control Conference, Fluid Dynamics and Co-located Conferences, St. Louis, USA, 24 June - 26 June 2002, AIAA 2002-3157. Reston: AIAA.

[15] Kind RJ and Maull DJ. An experimental investigation of a low-Speed circulation-controlled aerofoil. Aeronautical Quarterly 1968; 19:170-182.

[16] Jones G and Englar R. Advances In Pneumatic Controlled High Lift Systems Through Pulsed Blowing. In: 21st AIAA Applied Aerodynamics Conference, Fluid Dynamics and Co-located Conferences, Orlando, USA, 23 June - 26 June 2003, AIAA 2003-3411.Reston: AIAA.

[17] Oyler TE and Palmer WE. Exploratory Investigation of Pulse Blowing for Boundary Layer Control.1st ed. 1972.

[18] Schatz M. and Thiele F. Numerical study of high-lift flow with separation control by periodic excitation. In: 39th Aerospace Sciences Meeting and Exhibit, Aerospace Sciences Meetings. Reno, USA, 08 January - 11 January 2001,AIAA 2001-296. Reston :AIAA.

[19] Rumsey CL and Nishino T. Numerical study comparing RANS and LES approaches on a circulation control airfoil. International Journal of Heat & Fluid Flow, 2011; 32:847-864.

[20] Shah N, Chi W, Kontis K. Active Flow Control Using Steady and Pulsed Blowing at Subsonic Speeds. In: 46th AIAA Aerospace Sciences Meeting and Exhibit, Reno, USA, 07 January - 10 January 2008,AIAA 2008-742.Reston:AIAA

IOP Conf. Series: Materials Science and Engineering **608** (2019) 012022 doi:10.1088/1757-899X/608/1/012022

# Generation Method of Air Defense Missile Weapon Interception Scheme

**Song Guibao, Song Jiaming, Jia Runa, Jiang Zijie and Liu Zhan**

Naval Aviation University, Yantai 264001, Shandong, China

15098521852@163.com

**Abstract**: The research on the generation method of the anti-aircraft missile interception scheme before the war can lay the foundation for the rational allocation of the next air defense missile. In this paper, the clustering algorithm based on fuzzy C-means is used to cluster anti-ship missiles. Based on this, the number of anti-aircraft missiles launched during the first interception is determined. Then, by analyzing the depth of the launching area and the interception process, the number of interception of each layer of air defense missiles in the air defense interception operation is calculated. Finally, the anti-aircraft interception scheme that can achieve the expected damage probability of anti-ship missiles can be solved. The case is verified by the case that the calculation efficiency is relatively high and has certain application value.

## 1. Preface

At present, the three-layer air defense interception with air defense missiles as the main defensive weapon is the main method of surface warship anti-aircraft anti-missile operations. At the same time, the use of universal vertical launchers on ship platforms has changed the way missiles are deployed, and surface ships can flexibly configure ship-borne missiles based on operational requirements. However, there is no specific study on the configuration methods of various types of ship-borne missile weapons before the war. Therefore, this paper aims to determine the generation method of the pre-war air defense missile interception scheme by appropriately simplifying the main links of air defense and anti-missile warfare, thus laying a foundation for the rational allocation of the next air defense missile [1, 2].

## 2. Problem analysis and solution process

### *2.1. Solution process of formation air defense missile interception scheme*

(1) Firstly, cluster analysis of anti-ship missiles. In air defense operations, different types of anti-ship missiles have different penetration capabilities and damage capabilities, and different interception schemes must be adopted. In the multi-layer air defense interception operation, the first interception method is particularly important. Based on the penetration capability index and the damage capability index as the reference variables, the anti-ship missiles are clustered and the cluster center points are determined. Based on this, the number of anti-aircraft missile launches during the first interception is determined[3,4]

(2) Calculation of the number of interceptions. The boundary of the three-layer interception area is divided based on the maximum tracking distance of the radar and the maximum range of the far,

IOP Conf. Series: Materials Science and Engineering **608** (2019) 012022 doi:10.1088/1757-899X/608/1/012022

medium and short-range anti-aircraft missiles. The calculation of the depth of the launch area is appropriately simplified and the maximum number of intercepts at each layer is calculated and calculated.

(3) The air defense missile interception plan is determined. Based on the maximum interception times of the long-, medium-, and near-three-layer air defense interception operations and the interception probability of air defense missiles, various air defense missile interception schemes for the expected interception probability of incoming missiles are calculated[5].

### *2.2. Anti-ship missile clustering analysis*

On the one hand, when launching anti-aircraft missiles to intercept anti-ship missiles, the first interception effect determines the success or failure of interception operations to a certain extent. The greater the threat of incoming missiles, the greater the number of ship-to-air missiles launched during the first interception. . On the other hand, the different target characteristics of different incoming missiles increase the difficulty of determining the air defense missile interception scheme before the war. Therefore, cluster analysis method is used to comprehensively analyze different incoming targets according to the characteristic indicators, determine the cluster center, and then determine the number of anti-aircraft missile launches during the first interception according to the cluster center feature values.

*2.2.1. Cluster Analysis Based on Fuzzy C-Means.* Fuzzy clustering is an important component in the field of pattern recognition. Fuzzy C-means clustering is currently the more common method. It clusters the data by calculating the geometric closeness, so that the target achieves aggregation based on the main features. The general idea of its calculation is: The number of samples is divided into c categories, denoted as , and the cluster center of each category is expressed as , and we use the degree of membership of the sample pair to satisfy the condition. Express the objective function as:

$$J_b(v) = \sum_{i=1}^{n} \sum_{k=1}^{c} (\mu_{ik})^b)^2 \tag{1}$$

Clustering Analysis Based on Simulated Annealing Ant Colony Algorithm

The simulated annealing ant colony algorithm generally adopts the calculation process of ant colony algorithm, but updates the population by simulated annealing algorithm to avoid falling into local optimum[6,7]. The flow chart of fuzzy C-means clustering algorithm based on simulated annealing ant colony algorithm is shown in Fig. 1.

Initialization parameters including population size sizepop, Maximum evolution MAXGEN, Cross probability $P_c$ And mutation probability $P_m$ ,Annealing initial temperature $T_0$ And termination temperature$T_{end}$And temperature cooling factor k.

The selected genetic algorithm is encoded as follows: using binary encoding, Each chromosome consists of c cluster centers.thereby m chromosomes constitute the sample variables to be optimized. Each variable is binary coded with k bits. Fitness functionTarget function$J_b$,and

$$\mathrm{FintV} = \mathrm{ran\ king}(J_b) \tag{2}$$

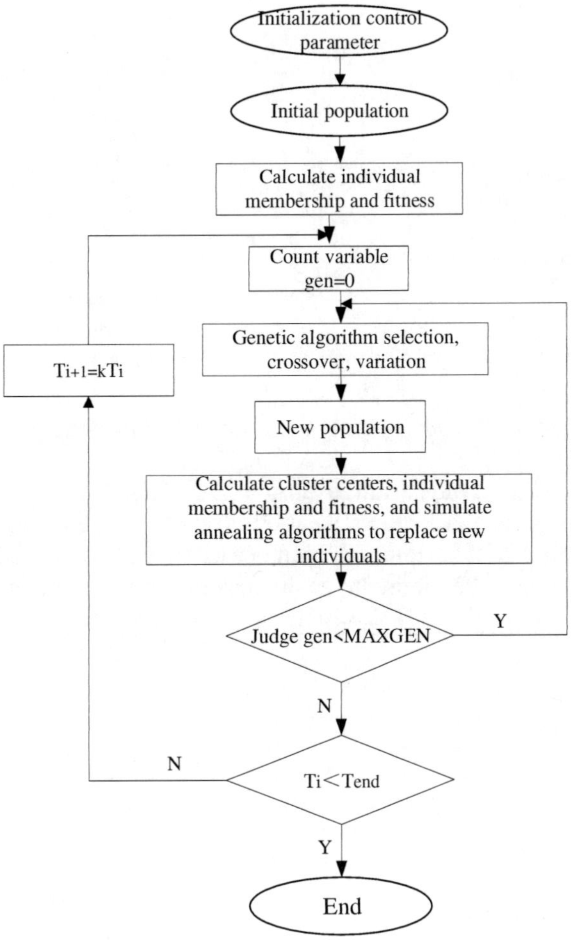


**Figure 1.** Flow chart of clustering algorithm

*2.2.2. Inbound missile clustering analysis process.*

(1) MATLAB solving process of simulated annealing ant colony algorithmGenerally speaking, the faster the anti-ship missile's penetration speed, the stronger the anti-interference ability, the lower the penetration height, and the stronger the penetration capability. At the same time, the stronger the target detection and selection ability, the greater the lethality of the warhead, and the greater the damage it poses to our ship platform [8,9]. The penetration capability index and the damage capability index of different anti-ship missiles are characterized by the value of interval [0,1]. These two characteristic indicators reflect the threat degree of the incoming missile and are used to determine the missile launch during the first interception. Quantity.

Because of the difficulty in obtaining samples, MATLAB randomly generated the penetration capability index and the damage capability index of 400 anti-ship missiles. Since the sample is too large, it will not be expanded here.

Set the maximum number of iterations to 20, The termination tolerance of the objective function is e-6, and the number of categories is 4; Simulated annealing algorithm parameters: k=0.8,$T_0$=100, $T_{end}$=1;Genetic algorithm parameters: size pop=10;MAXGEN=10, $P_c$ =0.7,$P_m$=0.01.

According to the above method, the cluster analysis is performed. The results are as follows: the termination temperature is 0.922, the objective function Jb=3.4566, and finally the coordinates of the four cluster centers are (0.17, 0.41), (0.72, 0.18), (0.55, 0.55) and 0.57, 0.81), clustering diagram is as Fig 2 shows:

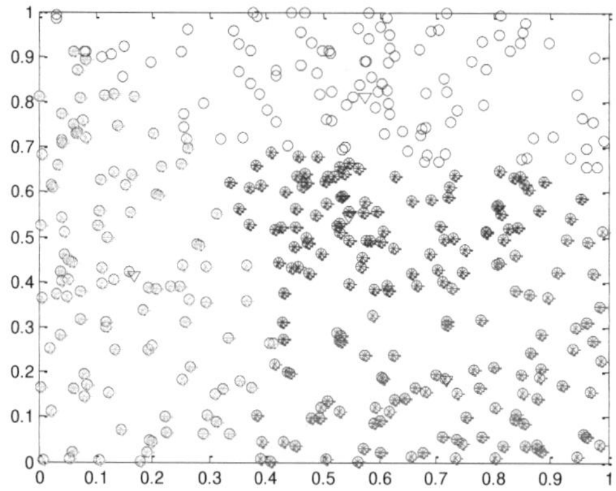


**Figure 2.** Cluster analysis of incoming missiles

(2) Analysis and application of target clustering results

A comprehensive analysis of the penetration capability index and the damage index of the obtained cluster center point is performed to determine the number of air defense missile launches for the first interception of the incoming missile belonging to each cluster center. As shown in Table 1:

**Table1**. The first missile salvo number table

| Cluster center | Number of first-shot missiles |
|---|---|
| (0.17,0.41) | ⩾1 |
| (0.72,0.18) | ⩾2 |
| (0.55,0.55) | ⩾2 |
| (0.57,0.81) | ⩾3 |

## 2.3. Formation process of formation air defense *missile interception plan*

### 2.3.1. *Target intercept count calculation.*

(1)The kill zone of an air defense missile against an air target is an area composed of points of an air defense missile that can produce an expected damage effect after the target encounters, and the area is usually an irregular three-dimensional area. The kill zone only describes the encounter between the anti-aircraft missile and the incoming target. To make the decision on the timing of the missile launch, it is necessary to calculate the launch zone. The launching area is the area that can be used to make the missile and the target of the air-defense missile launched by the target in the killing zone. The near and far boundaries of the kill zone are usually determined by the missile's comprehensive tactical indicators and ship platform indicators. Its calculation formula is [10]:

$$D_{fy} = \sqrt{{D_{sy}}^2 + (V_m \cdot t_{dy})^2 + 2V_m \cdot t_{dy}\sqrt{{D_{sy}}^2 - H^2 - P^2}} \tag{3}$$

$$D_{fj} = \sqrt{{D_{sj}}^2 + (V_m \cdot t_{dj})^2 + 2V_m \cdot t_{dj}\sqrt{{D_{sj}}^2 - H^2 - P^2}} \tag{4}$$

Wherein,$D_{fj} and D_{fy}$ To indicate the near and far boundaries of the launch area; $D_{sj}$ and $D_{sy}$ To indicate the near and far boundaries of the kill zone; $V_m$ Indicates the target flight speed; $t_{dy}$ Indicates the time required for an air defense missile to reach the far side of the kill zone; $t_{dj}$ Means the time required for the first anti-aircraft missile to reach the vicinity of the kill zone, $H$ Indicates the target height, $P$indicating the target route shortcut. The depth of the launch area refers to the distance between the far and near points of the launch area. Used$h_f$ to indicate The period of the target through the depth of the launch area is the period of time during which our missile can be effectively intercepted. To highlight the main contradictions, make the following assumptions:

IOP Conf. Series: Materials Science and Engineering **608** (2019) 012022 doi:10.1088/1757-899X/608/1/012022

a. Treat our surface ships and their formations as points, and do not distinguish between the defense angles of the attacked ships and other protective ships.

b. The incoming target passes straight through the speed in our combat airspace at a uniform speed, and the route shortcut is zero. Our air defense missiles pass at a constant speed$V_d$ along the direction of the missile.

c. Determine the boundary of the far-end air defense intercept zone according to the radar tracking distance and the smaller value of the maximum range of the long-range missile, and determine the boundary of the middle intercept zone and the near-layer intercept zone according to the maximum range of the medium-range and short-range air defense missiles.

d. It can always intercept the incoming target at a certain height, and approximate the depth of the launching area to the horizontal depth of the launching area. In order to meet the defensive operational requirements of warship formation, the requirements for radar detection distance [11,12] are:

$$R_j \geq d_{sy} + V_t(t_{fy} + t_{xs}) \tag{5}$$

In the formula, $d_{sy}$ indicates the horizontal distance of the far zone of the killing zone, and $V_t$indicates the horizontal speed of the target, $t_{fy}$indicating the time required for the air defense missile to fly to the far side of the killing zone, $t_{xs}$indicating the reaction time of the air defense missile.

The depth of the launch area $h_f$is simply calculated as:

$$h_f = d_{sy} - d_{sj} + V_m \cdot t_{dy} - V_m \cdot t_{dj} \tag{6}$$

$d_{sj}$Indicates the horizontal distance of the vicinity of the killing zone. The other letters have the same meaning as above.

*2.3.2 Calculation method for interception times of each layer.* Regardless of the flight altitude, the three-layer air defense interception area is divided according to the range of the long-range, medium-range and short-range anti-aircraft missiles, as shown in Fig. 3. Determine the near and far boundaries of each air defense interception zone. The interception number of incoming missiles in each interception zone is calculated as follows: Set the first encounter between the incoming missile and our missile$P_0$, The distance is$D_0$ , by the title, $D_0 = D_{sy}$;The first encounter point is$p_{i-1}$ , the distance is$D_{i-1}$ , and the following conditions are met:

$$D_i = D_{i-1} - V_m \cdot t_i \tag{7}$$

Where$t_i = t_p + t_{xs}+t_{di}$, $t_p$indicating the time of assessment, $t_{xs}$indicating the response time of the air defense missile, $t_{di}$indicating the time required for our air defense missile to fly to the point of encounter, and

$$t_{di} = \frac{D_i}{V_d} \tag{8}$$

When$D_i$<$D_{sj}$ he layer interception ends.

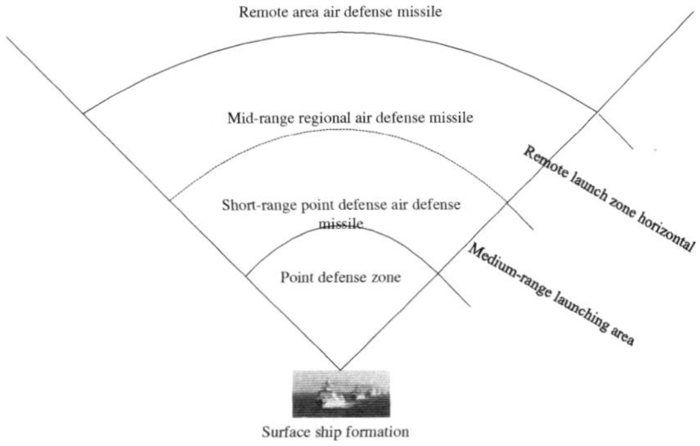


**Figure 3**. Formation defense zone division

*2.3.3 Interception scheme.*

(1) Calculation method of target intercept probability. In the interception of threat targets, the principle of early defense and time reduction should be grasped. When remote interception is performed, the remote interception is preferred, and the earlier the target is successfully intercepted, the greater the probability of success. Set $P_1, P_2, P_3$indicate the intercept probability of long-range, medium-range, and short-range missiles, respectively. Representing the number of launches of long-range, medium-range, and short-range missiles, $x_1, x_2$ and$x_3$ respectivelythe formula for calculating ,the intercept probability of incoming threat targets is:

$$P_L = 1 - (1 - P_1)^{x_1}(1 - P_2)^{x_2}(1 - P_3)^{x_3} \quad (9)$$

(2) Anti-aircraft missile interception scheme under expected intercept probability. First, the number of missiles launched during the first interception is determined according to the clustering center to which the incoming anti-ship missile belongs. According to general operational principles, additional air defense intercepts usually only launch one air defense missile. Then, according to the farthest detection distance of the radar, the maximum range of the long-range, medium-range and short-range anti-aircraft missiles, the boundary of the three-layer air defense interception area is determined; according to the above attack, the number of interception levels of each layer is determined. Finally, based on the interception times and interception probabilities of each layer of missiles, a simple arrangement is made to determine the anti-aircraft interception scheme that can achieve the expected interception probability of anti-ship missiles.

## 3. Case Analysis

We formed a fleet for the destroyer, and the enemy was two fighters; the enemy attacked us and we intercepted the missile. The enemy fighters mounted the air-to-ship missiles when they were formed from our formation with a maximum range of 120km and a height of 5km, and their penetration speed was 2 Mach. The maximum tracking distance of our destroyers for anti-ship missiles is 240km. Our missile types include: regional ship-to-air missile No. 1 with a maximum range of 180km, an average flight speed of Mach 3, a single hit probability of 0.5, regional ship-to-air missile No. 2 with a maximum range of 32km, an average flight speed of 2 Mach, and a single hit probability of 0.7. The short-range defense ship-to-air missile No. 3 has a maximum range of 8km, an average flight speed of Mach 1.5, and a single-shot probability of 0.8 for anti-ship missiles. It is known that the altitude of the ship-to-air missile is within the range of our radar tracking and missile flight altitude. Our launch response time is 5s and the damage evaluation time is 5s. Solve the feasible interception scheme when the ship-to-air missile reaches the intercept probability of more than 85%. The solution process is as follows:

(1)Determine the type of threat target to which the ship-to-air missile belongs. The missile belongs to the cluster center (0.17, 0.41). The penetration capability is poor and the damage capability is general. Therefore, the plan to select the missile single-shot is intercepted for the first time.

(2) The farthest killing distance of our missile is 120km, the radar tracking distance is 240km, and the long-range reference formula of the launching area (5) is 150km<240km, so the radar can guide the missile when the incoming missile distance is 150km. According to the missile range, the three-layer interception area is simply determined, so that the three-layer air defense interception airspace is 32-120km for the far-end interception area, 8-32km for the middle-level interception area, and 1.5-8km for the short-range interception area.

(3) Taking the calculation of the number of intercepts in the far layer as an example: when the target enters the far boundary of the launching area, the first interception is performed, and a long-range air defense missile is launched. At this time, the first encounter point distance is 120km; the second intercepting distance is Calculate the reference formula (8), the second encounter point distance is 68km; the third intercept distance is 30.6km <32km, so the far layer can only intercept twice. It is also calculated that the middle layer can be intercepted once, and the short range can be intercepted once. Refer to formula (4) and perform a simple arrangement to obtain three interception schemes as shown in Table 2. According to the economic principle, instead of option one, there are

two interception schemes for the ship-to-air missile. The scheme also conforms to the actual fire interception process in the general combat process, and verifies the practicability of the algorithm.

**Table 2**. Plan of three-layer air defense

| Interception process | Distal interception | Middle interception | Short-range interception |
|---|---|---|---|
| Interception scheme one | 2 | 1 | 0 |
| Interception scheme II | 1 | 1 | 0 |
| Interception scheme three | 0 | 1 | 1 |

## 4. Conclusion

By analyzing the air-to-air interception process of ship-to-air missiles, this paper determines the interception schemes of various ship-to-air missiles, which can provide a selection interval for the decision-making configuration of ship-to-air missiles. The next step is to further refine the analysis process to better match the actual operational situation.

## References

[1] Zhai Xudong, Wang Yongchun. Theory and application of ship formation formation collaborative air defense mission planning [M]. Beijing: National Defence Industry Press, 2013: 56-88.

[2] Jiang Daohong, Liu Huimin. Introduction to Combat Coordination [M]. Beijing: National Defense University Press, 2004: 44-46.

[3] Li Jian. Research on dynamic firepower distribution and effectiveness evaluation of weapon targets [D]. Nanchang Aviation University, 2014: 28-44.

[4] Dong Chaoyang, Lu Yao, Wang Qing. Improved genetic algorithm for solving firepower distribution problem[J].Ordnance Engineering Journal,2016,1(01):97-99.

[5] Zhang Zuiliang. Military Operations Research [M]. Beijing: Military Science Press, 1993: 100-132

[6] LIU Ai-jun, YANG Yu, LI Fei, XING Qing-song, LU Hui, ZHANG Yi-dong.Research and application of chaotic simulated annealing particle swarm optimization algorithm[J].Journal of Zhejiang University(Engineering Science),2013,47(10):1722- 1730.

[7] Tong Yue. Research on transportation resource allocation problem and algorithm in urban cargo distribution [D]. Tsinghua University, 2013: 39-50.

[8] Wang Wei,Shi Hongquan,Wang Lei,Li Tao.A Study on Air Target Threat Assessment of Ship Formation Cooperative Operation[J].Modern Defence Technology,2014,42(06):79-83+110.

[9] Li Jia. Research on mission formation and air defense threat assessment model based on Bayesian network [D]. Huazhong University of Science and Technology, 2016: 21-30

[10] REN Yiguang, SHA Jichang, JING Tao. Modeling and Simulation of Interception Times of Ship-to-Air Missile Anti-missile Combat[J].Fire Control & Command Control, 2009, 34(10): 108-110+114.

[11] JING Fabiao, Kang Xiaoyu. Model of interception frequency of formation air defense and antimissile warfare[J].Ship Science and Technology,2011,33(03):120-122.

[12] Lou Shouchun. Analysis of surface-to-air missile weapon system [M]. Beijing: National Defense Industry Press. 2013: 72-73.

IOP Conf. Series: Materials Science and Engineering **608** (2019) 012023 doi:10.1088/1757-899X/608/1/012023

# Uncertainty Analysis of Squeeze-Type Regulation Method for Equal Expulsion of Parallel Tanks

**W J Yin[1,*], Z Lin[1] and M J Wang[1]**

[1]Beijing institute of Control Engineering, 100094 Beijing, China

[*]yinwenjuan1221@126.com

**Abstract.** Focused on the problem of unbalanced propellant consumption in parallel tanks of propulsion system, a squeeze-type regulation method based on the gas law method was discussed in the paper. Taking the tank system as the research object, the thermodynamic models under isothermal, adiabatic and other conditions were given and the factors affecting the uncertainty of the model were analysed. The results show that the measurement uncertainty of the model mainly comes from the pressure measurement accuracy and the volume measurement accuracy. The model was applied to different cases and the influence law of different tank parameters on the model uncertainty was studied. It is found that under the premise of the fixed pressure difference of the parallel tanks, the uncertainty of the regulation model is almost linear with the regulation amount. In addition, the relative uncertainty of the model can be reduced effectively by improving the pressure difference of the parallel tanks. The accuracy of the regulation model can reach to 7.7%, which can meet the requirements for on-orbit use.

## 1. Introduction

Considering the bearing capacity and configuration, medium and large spacecrafts generally adopt the bipropellant propulsion systems with four-tank design [1~4]. Generally, the tanks are symmetrically arranged in the satellite, and the same kind of propellant is used in parallel to ensure that the center of the propellant is on the central axis of the satellite. When the propellant expulsion difference of the parallel tanks is large, the deviation of the remaining propellant in the tanks is also too large, which causes the satellite centroid to be skewed and the disturbance torque to become larger, which makes the attitude control thrusters work frequently and the propellant utilization efficiency decrease.

The expulsion of the parallel tanks is generally adjusted by the flow resistance matching method [5~6], but the requirements of simulation accuracy and ground test are high. Controlling propellant flow rate of parallel tanks through the switching strategy of the latch valve during the orbit transfer phase can also effectively solve the problem of the unbalanced propellant consumption in parallel tanks, but the implementation is difficult and may affect the performance of the engine. There is another way to squeeze the propellant from the tank with more remaining amount to the tank with less remaining amount by adjusting the pressure of the tanks after the end of the orbit transfer phase. The squeeze-type regulation method is simple and has strong operability.

In this paper, the squeeze-type regulation method for the equal expulsion of parallel tanks is described. The thermodynamic models under isothermal, adiabatic and other conditions were given. The model uncertainty was analyzed from the aspects of environment and measurement parameters.

The influence law of different parameters on the model uncertainty was studied, and the reference to the specific implementation of the regulation method is provided.

## 2. Description of the model

### *2.1. Overview of regulation process*

The principle of the squeeze-type equal expulsion regulation system is shown in Figure 1. The system consists of the following parts:

Helium tank: High-pressure helium is stored in the helium tank and flows into the propellant tanks through the pressure regulator.

Latch valve: The latch valves are arranged upstream and downstream of the propellant tanks to control the gas inflow and the propellant outflow.

Propellant tank: The propellant tank is used to store propellant, which is filled with helium to squeeze the propellant to the engine or thrusters inlet.

Pressure transducer: The pressure transducer is used to monitor the pressure change of the propellant tank. The accuracy of the pressure transducer directly affects the strategy and final accuracy of the regulation method.

Temperature sensor: The temperature sensor is placed on the wall of the propellant tank to monitor the temperature change.

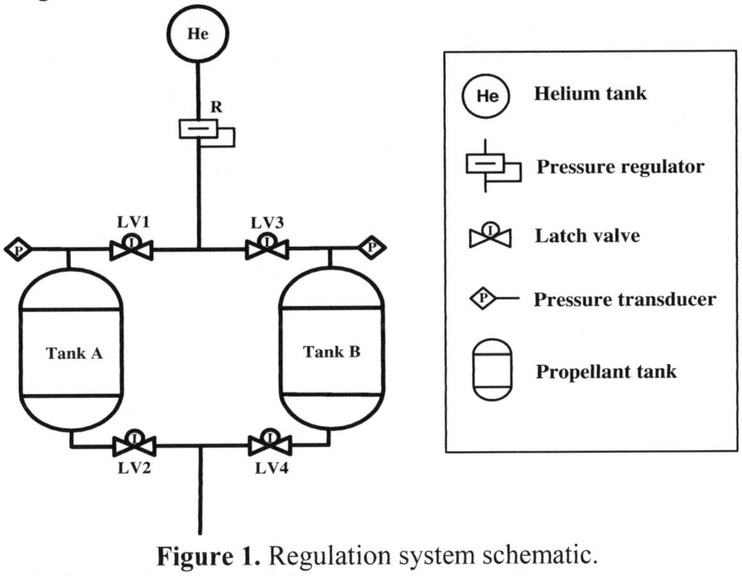


**Figure 1.** Regulation system schematic.

The steps of regulation method are as follows:

1) At the end of the orbital transfer, close the downstream latch valves LV2 or LV4 to prevent the propellant flow between the two parallel tanks. The pressure and temperature of the tank are recorded, and the volume of the gas in the tank A and tank B can be expressed as

$$V_{gAi} = V_t - \frac{m_{Ai}}{\rho_A}, \quad V_{gBi} = V_t - \frac{m_{Bi}}{\rho_B} \tag{1}$$

Where, $V_t$ is the nominal volume of the propellant tank, $m_{Ai}$ and $m_{Bi}$ are the propellant remaining of tank A and tank B. $\rho_A$ and $\rho_B$ are the density of tank A and tank B.

2) By controlling the latch valve LV1 and LV3, the pressure of the tank A and the tank B is conditioned to a predetermined pressure which is recorded as $P_{Ai}$ and $P_{Bi}$, the pressure of the helium in the tank A and tank B can be calculated as follows:

$$P_{gAi} = P_{Ai} - P_{vAi}, \quad P_{gBi} = P_{Bi} - P_{vBi} \tag{2}$$

Where, $P_{vAi}$ and $P_{vBi}$ are the saturated vapor pressure in the tank A and tank B.

3) Open the downstream latch valves LV2 and LV4, and pressure difference will repel the propellant from one tank into another tank. While the pressure of parallel tanks reaches equilibrium, the regulation process is completed and the adjusted volume $\Delta V_1$ is obtained.

*2.2. Calculation model*

Assuming that the gas is ideal and the tank system is in thermodynamic equilibrium state before and after the regulation, ignoring the influence of propellant density and expecting that some propellant in tank B will be squeezed into tank A. Various thermodynamic models are discussed and established under different assumptions in the following.

1) Isothermal model

Assuming the regulation amount is small or the heat exchange is timely in the regulation process, the gas temperature in the propellant tank is kept consistent. Ignoring the saturated vapor pressure, the governing equation under the assumption of isothermal is given by

$$P_{Bi} = P_{Ai}\left(1+\frac{\Delta V_1}{V_{gBi}}\right)\left(1-\frac{\Delta V_1}{V_{gAi}}\right)^{-1} \tag{3}$$

2) Adiabatic model

Assuming that the entire regulation process takes a short time, the temperature change caused by the volume change of the gas is dominant, and the heat exchange between the gas and objects such as tank wall,propellant and external environment is negligible. The regulation process is adiabatic. Ignoring the saturated vapor pressure, the governing equation is given by

$$P_{Bi} = P_{Ai}\left(1+\gamma\frac{\Delta V_1}{V_{gBi}}\right)\left(1-\gamma\frac{\Delta V_1}{V_{gAi}}\right)^{-1} \tag{4}$$

Where, $\gamma$ is the adiabatic coefficient of helium.

3) non-isothermal and non-adiabatic model

Assuming that the propellant tank pressure change during the regulation process is small and the propellant temperature is almost unchanged, the governing equation is given by

$$\begin{cases} P_{Bi} = \left( \left(P_{Ai} - P_{vAi}\right)\left(1+\frac{\Delta T_{gA}}{T_{gAi}}\right)\left(1-\frac{\Delta V_1}{V_{gAi}}\right)^{-1} + P_{vAe} - P_{vBe} \right)\left(1+\frac{\Delta V_1}{V_{gBi}}\right)\left(1+\frac{\Delta T_{gB}}{T_{gBi}}\right)^{-1} + P_{vBi} \\ \Delta T_{gA} \approx (\gamma-1)T_{gAi}\left(\frac{\Delta Q_A}{P_{gAi}V_{gAi}} + \frac{\Delta V_1}{V_{gAi}}\right) \\ \Delta T_{gB} \approx (\gamma-1)T_{gBi}\left(\frac{\Delta Q_B}{P_{gBi}V_{gBi}} - \frac{\Delta V_1}{V_{gBi}}\right) \\ P_{vAe} \approx P_{vAi}\ ,\quad P_{vBe} \approx P_{vBi} \end{cases} \tag{5}$$

Where, $P_{vAe}$, $P_{vBe}$ are the saturated vapor pressure after the regulation, $T_{gAi}$, $T_{gBi}$ are the gas temperature before the regulation. $\Delta Q_A$, $\Delta Q_B$ are the gas thermal energy change in tank A and tank B. $\Delta T_{gA}$, $\Delta T_{gB}$ are the temperature change of gas in tank A and tank B.

The calculation of the non-isothermal and non-adiabatic model is complicated, it is not only difficult to quantitatively analyze the heat exchange behavior of the tanks, but also unsuitable for the rapid implementation of the active regulation on the orbit. The isothermal model and the adiabatic model are simple to calculate. Although the assumption of gas state is different from the actual regulation process, the model can be corrected by ground test or previous on-orbit regulation data.

## 3. Error analysis

The calculation model includes effects of the pressure measurement, volume measurement, temperature measurement and the non-ideal gas effects (compressibility) on the regulation accuracy..

For ideal gas, assuming that the propellant temperature changes are negligible, and introducing the concept of polytropic process, the following form of the governing equation can be obtained:

$$\Delta V_1 = \left(P_{Bi} - P_{Ai}\right)\left(n_B\left(P_{Ai} - P_{vAi}\right)\left(V_t - \frac{m_{Bi}}{\rho_B}\right)^{-1} + n_A\left(P_{Bi} - P_{vBi}\right)\left(V_t - \frac{m_{Ai}}{\rho_A}\right)^{-1}\right)^{-1} \quad (6)$$

$n_A$, $n_B$ are the polytropic exponent of tank A and tank B, respectively.

Using conventional techniques, the uncertainty in $\Delta V_l$ can be expressed in terms of the measurement uncertainties of $P_{Ai}$, $P_{Bi}$, $P_{vAi}$, $P_{vBi}$, $m_{Ai}$, $m_{Bi}$ and $V_t$ et al.

$$U_{\Delta V_1}^2 = \left(\frac{\partial \Delta V_1}{\partial P_{Bi}} U_P\right)^2 + \left(\frac{\partial \Delta V_1}{\partial P_{Ai}} U_P\right)^2 + \left(\frac{\partial \Delta V_1}{\partial P_{vBi}} U_{P_v}\right)^2 + \left(\frac{\partial \Delta V_1}{\partial P_{vAi}} U_{P_v}\right)^2 + \left(\frac{\partial \Delta V_1}{\partial V_t} U_{V_t}\right)^2 + \left(\frac{\partial \Delta V_1}{\partial m_{Ai}} U_m\right)^2 + \left(\frac{\partial \Delta V_1}{\partial m_{Bi}} U_m\right)^2 \quad (7)$$

Where $U$ is the uncertainty of a measured or computed value.

Obviously, while $n_A = n_B$, the relative uncertainty is independent of the polytropic exponent. Therefore, using Equation(7) for uncertainty analysis is comprehensive and feasible.

The uncertainty analysis can determine the relative importance of each of the sources of uncertainty. This is useful for assessing the feasibility of making improvements to the regulation. The error distribution function UPC is defined as:

$$\mathrm{UPC}(x_i) = \left|\frac{U(x_i)}{U_{\Delta V_1}}\right| \times 100 \quad (8)$$

Where, $x_i$ is the source of each uncertainty, $U(x_i)$ is the uncertainty component associated with $x_i$.

### *3.1. Measurement values*

The sample set of readings used for the uncertainty analysis is given in Table 1. Taking NTO propellant tanks as an example.

**Table 1.** Sample set of various parameters.

| Symbol | Value | Symbol | Value |
|---|---|---|---|
| $P_{Bi}$ | 1.65MPa | $U_P$ | ±(0.01~0.1)%FS |
| $P_{Ai}$ | 1.60MPa | $U_{P_v}$ | ±0.005MPa |
| $m_{Bi}$ | 320Kg | $U_{V_t}$ | ±0.05% |
| $m_{Ai}$ | 300Kg | $U_m$ | ±0.5% |

### *3.2. Results and Discussion*

Figure 2 shows the quantitative variation in UPC with the uncertainty in the propellant tank pressure sensor varies from 0.01% FS to 0.1% FS. It can be noted that the saturated vapor pressure error has little effect on the model accuracy. Correspondingly, the gas volume measurement accuracy and the pressure sensor accuracy have a significant influence on the overall model accuracy. Therefore, the accuracy of the regulation method can be improved effectively by using high-precision pressure sensors and adopting high-precision propellant gauging techniques.

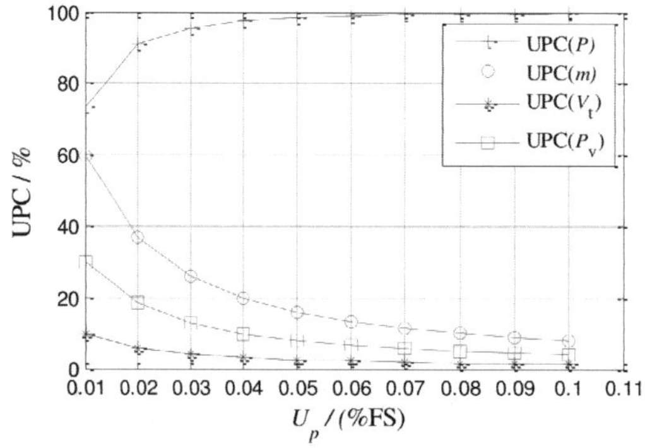


**Figure 2.** Variation in UPC with uncertainty in tank pressure sensor

## 4. Effect of tank parameters on model uncertainty

During the ground test, it was found that the actual regulation process was between the isothermal process and adiabatic process. In order to analysis the effect of tank parameters on the relationship between model uncertainty and regulation amount, set $n_A = n_B = 1.35$ and calculate the model uncertainty under different assumptions and tank parameters (tank pressure varies from 1.4MPa to 1.8MPa, propellant remaining varies from 200kg~800kg).

Assuming $\Delta m = m_{Bi} - m_{Ai} = 20\text{kg}$ , the relationships between the model uncertainty and the regulation amount with different parallel tanks pressure differences are shown in Figure 3. It can be observed that the difference between the maximum uncertainty and the minimum uncertainty of the model is small with different values of $\Delta P$ . The model uncertainty for the same amount of regulation increases as the pressure difference of the parallel tanks decreases. Thus, it can be concluded that relative uncertainty of the regulation model can be reduced by increasing the pressure difference of the parallel tanks during the on-orbit implementation.

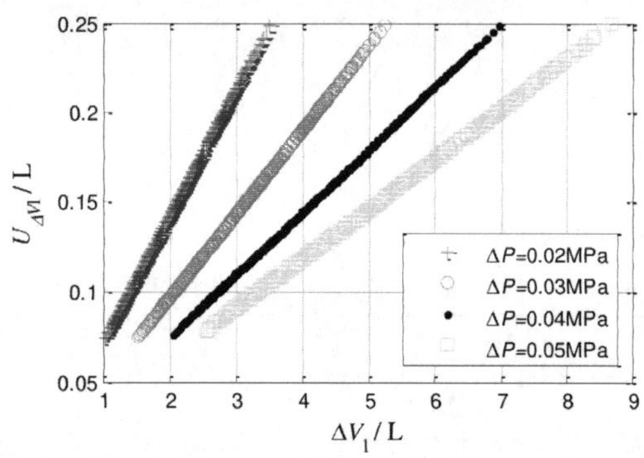


**Figure 3.** Variation of $U_{\Delta V_l}$ for different tank pressure difference

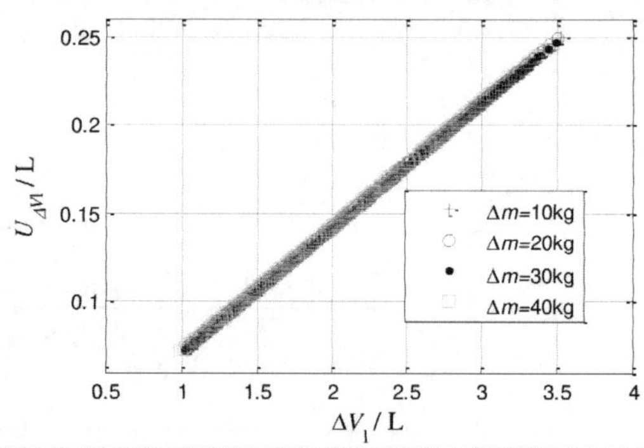


**Figure 4.** Variation in $U_{\Delta V_l}$ for different propellant remaining difference.

Assuming $\Delta P = P_{Bi} - P_{Ai} = 0.02\text{MPa}$ , the relationships between the model uncertainty and the regulation amount under different propellant remaining differences are shown in Figure 4. It can be observed that several curves are almost completely coincident, thus, the difference in tank propellant remaining has little effect on the relationship between the model uncertainty and the regulation amount.

Through the above analysis, it can be noted that the smaller the pressure difference of the parallel tanks before the regulation, the larger the relative error of the regulation amount. Generally, while the performance of equal expulsion of parallel tanks is less than one percent on-orbit, it can be accepted. Combining with the analysis above, the maximum relative uncertainty of the model can be calculated as shown in Figure 5. While the propellant remaining measurement accuracy is 0.5% and the pressure sensor measurement accuracy is 0.1%, the maximum relative uncertainty of the model is about 7.7%, which can meet the requirements for on-orbit use.

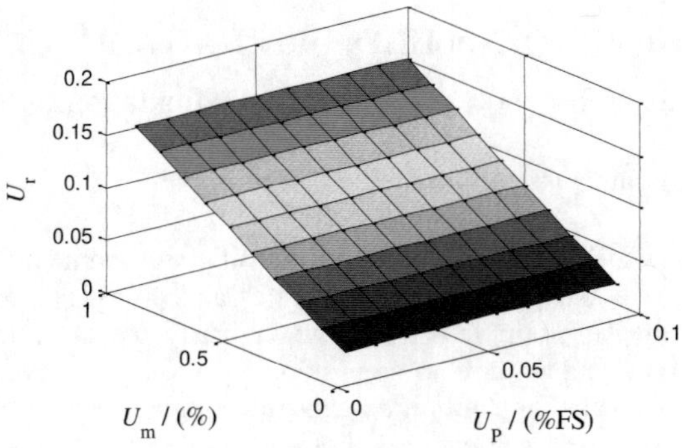


**Figure 5.** $U_r$ vs $U_P$ and $U_m$

## 5. Conclusions

In this paper, the calculation models under various conditions of the squeeze-type regulation method for the equal expulsion of parallel tanks are proposed and the error of the regulation model is analysed. The conclusions are as follows:

1) The accuracy of the model mainly depends on the volume measurement accuracy and pressure measurement accuracy of helium in the tanks.

2) Under the premise that the pressure differential of the parallel tanks is fixed, the model uncertainty is almost linear with the regulation amount. In addition, increasing the pressure differential of the parallel tanks can effectively reduce the relative error of the regulation model.

3) The maximum relative uncertainty of the model is about 7.7%(while $U_m$=0.5%, $U_P$=0.1%), which can meet the requirements for on-orbit use.

## References

[1] Immich H, Langel G, Munding G. Satellite unified bipropellant propulsion system experience and improvements[R]. AIAA 89-2506

[2] Poussin J F, Berger G. Eurostar E3000 Three-year Flight Experience and Perspective [R]. AIAA 2007-3124

[3] Bouhia A, Billand M, Roulet M. Implementation of A Large Multi-Beam Mission on Eurostar E3000[R]. AIAA 2008-5447

[4] Purohit G P, Smolko J W, Ellison J R. Propellant management device performance during an off-nominal transfer orbit mission [R]. AIAA 98-3168

[5] ZHANG Y H. Equalizing expulsion and control of parallel tanks in propulsion system [J]. Journal of Rocket Propulsion, 2013, 39(3):67-71

[6] Yuan L, Wang S, Lian R Z. Progress of optimization for bipropellant uniform propulsion system [J]. Journal of Rocket Propulsion, 2014, 40(6): 8-12

# An Improved Regression Analysing Method for Multi-position Calibration Test of Gyrowheel System

**Kai CUI, Yuyu ZHAO, Xin HUO, Hui ZHAO and Yu YAO***

Harbin Institute of Technology, Control and Simulation Center, 150001 Harbin, China

*Corresponding author: yaoyu@hit.edu.cn

**Abstract.** Gyrowheel is an innovative rate sensing and momentum management instrument. It could measure two-axes external angular rates and provide three-axes control torques at the same time. In order to improve the rate sensing performance, calibration tests are necessary to identify error terms and acquire compensation equations. However, there are too many possible error terms which make the calibration tests too complex to carry out. Considering that some of the error terms might be neglected without decreasing the compensation accuracy, proper test procedure to identify and neglect insignificant error terms is of great help to simplifying calibration tests and compensation equations. In this paper, a method to figure out and exclude insignificant error terms based on variance analysis is proposed. Test result shows that this method can simplified the form of the compensation equations without decreasing the compensation accuracy.

## 1. Introduction

Gyrowheel is an innovative attitude determination and control instrument, it provides control torques about the whole three axes and measures the angular rates about two axes of a spacecraft at the same time. Because of the multi-role ability of Gyrowheel, it can remarkably reduce the size, mass and power of attitude control system of a spacecraft, leading to the reduction of the ultimate costs for the spacecraft[1-4].

When acting as an angular rate sensor, Gyrowheel system operates like a Dynamic Tuned Gyroscope (DTG). Because of the imperfections of the instrument, angular rate output could be affected by gravity acceleration. Multi-position calibration tests have been widely used to identify and eliminate such errors[5-7]. These calibration tests focus on how to identify different error terms related to gravity acceleration input. However, some of these error terms may have comparingly slight influence on the accuracy of compensation equations. Considering the complexity of tilt-condition tests for Gyrowheel system, proper method to simplify the procedure of multi-position tests is required.

In this paper, the resource of the error terms of Gyrowheel system about gravity acceleration is analysed. A method to identify these error terms based on multi-position tests is studied, and the way to figure out insignificant error terms is proposed. Test results show that this process can help to neglect insignificant error terms effectively without decreasing the compensation accuracy.

## 2. Rate Sensing Principle of the Gyrowheel System

Figure.1 shows a cut-away view of the Gyrowheel. The basic structure of Gyrowheel system is similar to a DTG, except that its rotor is able to tilt in a range of 5 degrees and change the spinning speed. The spinning axis of the rotor is defined as the z axis and x and y axes are the two sensitive axes. The rotor spins along the z-axis and torque coils could apply control torques along x and y axes. By altering the tilt angle and spinning speed of the rotor, Gyrowheel could provide three-axes control torques. And when there is angular rate input along x or y axis, the currents of the torquers would be altered to maintain the tile angle. As a result, the angular rete input could be measured by processing the current data of two-axes torquers.

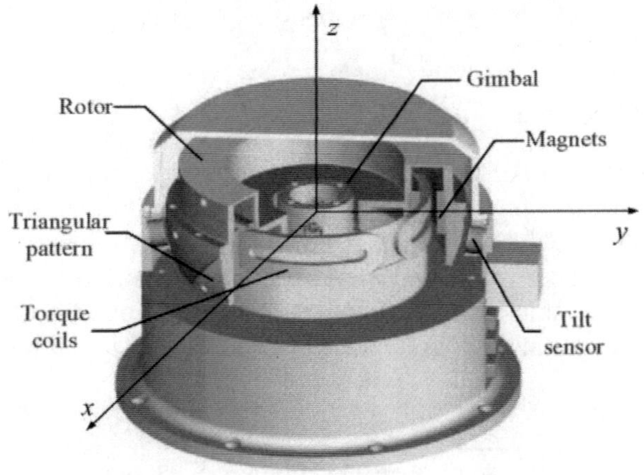


**Figure 1.** Cut-away view of the Gyrowheel.

The rate sensing equations can be written as (1) and (2) after reasonable simplification. Where $\omega_{cX}$ , $\omega_{cY}$ represent angular input along two axes. $\Phi_x$ , $\Phi_y$ are tilt angles of the rotor, $\omega_z$ is the spinning speed and $i_x$ , $i_y$ represent the torque coils current. $k_{scale,x}$, $k_{scale,y}$ are the scale factors and $I_{t,x}$, $I_{t,y}$, $c_{g,x}$, $c_{g,y}$ could be just regarded as factors determined by parameters of the prototype.

$$\omega_{cX} = \frac{1}{\omega_z}[k_{scale,x} i_x - c_{g,y}\dot{\Phi}_y - K_{dyn,y}\Phi_y] - I_{t,x}\dot{\Phi}_x + 2c_{g,x}\Phi_x \tag{1}$$

$$\omega_{cY} = \frac{1}{\omega_z}[k_{scale,y} i_y - c_{g,x}\dot{\Phi}_x - K_{dyn,x}\Phi_x] - I_{t,y}\dot{\Phi}_y - 2c_{g,y}\Phi_y \tag{2}$$

Design limitations and constructional deficiencies could lead to disturb torques which produce errors when measuring angular rates. The negative effect of these disturb torques might be reduced by compensation equations which could be obtained by implementing calibration tests. Similar to the DTG, the drifting errors could be categorized as: g insensitive terms, g sensitive terms and g-squared sensitive terms. It has been widely accepted that the g-squared sensitive terms are in most cases negligible. However, in order to function as a momentum actuator, the Gyrowheel have significantly larger rotor mass and inertia compared to DTGs. That leads to comparingly larger g-squared sensitive terms which may not be neglected entirely. Thinking of all these error terms, the Gyrowheel error model for $x$ axis can be given as (3), and equations for $y$ axis have the same form.

$$\begin{aligned} \omega_{dX} &= D(x)_0 + D(x)_x g_x + D(x)_y g_y + D(x)_z g_z \\ &+ D(x)_{xy} g_x g_y + D(x)_{yz} g_y g_z + D(x)_{xz} g_x g_z \\ &+ D(x)_{xx} g_x^2 + D(x)_{zz} g_z^2 \end{aligned} \tag{3}$$

Where $\omega_{dX}$ is drifting angular rates along the two sensitive axes. $D(x)_0$, represents the g insensitive error terms, $D(x)_i$ , $i = x,y,z$ are the g sensitive error coefficients, $D(x)_{ij}$, $ij = xx,zz,xy,yz,xz$, are all the possible g-squared sensitive error coefficients, and $g_x,g_y,g_z$ represent components of gravity vector along the three axes.

In the early process of the calibration tests, the Gyrowheel is actualized based on the null tilt condition in order to maintain a relatively high accuracy. The research in this paper is also done under this condition. In the null tilt condition, after incorporating error terms into the rate sensing equation, the equation becomes (4), which is the original form of the compensation equation for $x$ axis.

The model listed above considered every possible g sensitive and g-squared sensitive error terms. However, the importance of each error terms is different and some terms may be negligible. The final equations used to compensate the error terms may have a simpler form. A method to work out the insignificant error terms based on multi-position tests will be discussed in following sections.

$$\begin{aligned}\omega_{cX} &= k_x i_x - D(x)_0 - D(x)_x g_x - D(x)_y g_y - D(x)_z g_z \\ &-D(x)_{xy} g_x g_y - D(x)_{yz} g_y g_z - D(x)_{xz} g_x g_z \\ &-D(x)_{xx} g_x^2 - D(x)_{zz} g_z^2\end{aligned} \tag{4}$$

## 3. Multi-Position Test

In order to identify different error coefficients in (4), multi-position tests should be carried out. Multi-position tests use a two-axes motion table to provide precise position reference. The rotation rate of the earth and components of gravity vector are regarded as nominal inputs of the Gyrowheel. Before carrying out a test, the Gyrowheel y axis and z axis are aligned with the table elevation and azimuth axes respectively. The motion table is implemented that its elevation axis is in the west direction and azimuth axis directs upward, as shown in Figure.2.

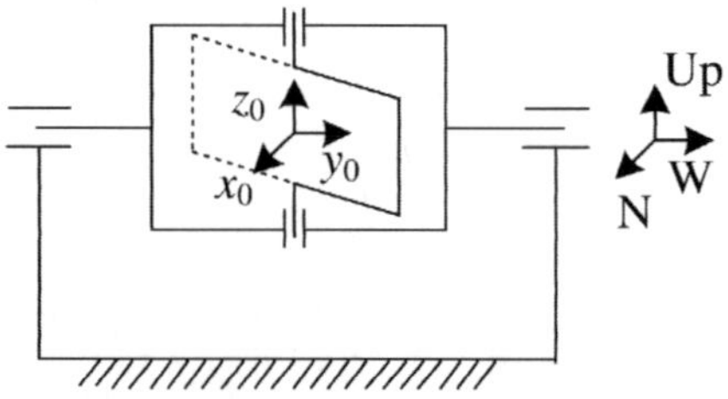


**Figure 2.** Diagram of the two-axis motion table.

Defining that $\psi$, $\Phi$ are the azimuth and elevation angles respectively, $\omega_e$ is the earth's rotation rate, and $\lambda$ is the latitude, the gravitational acceleration components and the earth rate component can be expressed as:

$$\begin{cases} g_x = g\sin\Phi\cos\psi \\ g_y = -g\sin\Phi\sin\psi \\ \omega_{cX} = \omega_e\cos\psi\cos(\lambda+\Phi) \\ \omega_{cY} = -\omega_e\sin\psi\cos(\lambda+\Phi) \end{cases} \tag{5}$$

Considering that mathematical expressions for the two axes are similar, now take equation of x axis for analyse purpose. (4) could be rewritten in a matrix form:

$$\mathbf{G}_X^T \mathbf{D(x)} = k_x i_x - \omega_{cX} \tag{6}$$

Where

$$\mathbf{G}_X = [1 \quad g_x \quad \cdots \quad g_z^2]^T \tag{7}$$

$$\mathbf{D(x)} = [D(x)_0 \quad D(x)_x \quad \cdots \quad D(x)_{zz}]^T \tag{8}$$

$$\boldsymbol{Y} = \boldsymbol{A\beta} + \boldsymbol{\eta} \tag{9}$$

Equations in the form of (6) could be obtained at each position by changing the azimuth and elevation angle of the motion table. Different azimuth and elevation angles lead to different $\boldsymbol{G_X}$ and different torquer current data. Considering that random error is unescapable when carrying out the test, the equation of data is in the form of (9).Where $\boldsymbol{A}$ is the structural matrix of the regression equations. It's determined by how the positions are configured. Elements in vector $\boldsymbol{Y}$ are drifting angular rates. The regression method finds a specific $\boldsymbol{\beta}$, which minimize the variance of $\boldsymbol{\eta}$. This can be done by solving the regression equation (10).

IOP Conf. Series: Materials Science and Engineering **608** (2019) 012024 doi:10.1088/1757-899X/608/1/012024

$$A^T A\hat{\beta} = A^T Y \tag{10}$$

## 4. Regression Analysing Method

Except for identifying error coefficients, multi-position tests could also be used to figure out the importance of different error terms. Considering that different error coefficients might have different dimensions, it's not reasonable to evaluate importance by simply comparing magnitudes of different coefficients. Instead, some variance analysis methods should be applied. Because of the similarity of two axes, analysis is done with x axis.

Each value in $\boldsymbol{Y}$ in (10) is acquired from a certain position. Suppose that:

$$\hat{Y} = [\hat{y}_1 \quad \hat{y}_2 \quad \cdots \quad \hat{y}_N]^T \tag{11}$$

Where

$$\hat{Y} = A\hat{\beta} \tag{12}$$

Regression square sum could be calculated by:

$$U_X = \sum_{i=1}^{N} (\hat{y}_i - \bar{y})^2 \tag{13}$$

Where $\bar{y}$ is the average value of vector $\boldsymbol{Y}$. $U_X$ evaluates the total effect of different error terms on the torquer current. The more error terms are included, the larger the regression square sum would be. If one of the error terms is excluded, the regression square sum could only decrease. The larger the regression square sum decreases after an error term is excluded, the more important that specific term is. The reduced value after excluding a specific error term is called partial regression square sum. Suppose that $\hat{\beta}_i$ is the number $i$ value of vector $\hat{\beta}$ and the partial regression square sum of it is $P_i$, it could be proven that:

$$P_i = \frac{\hat{\beta}_i}{c_{ii}} \tag{14}$$

Where $c_{ii}$ is the number $i$ diagonal elements of matrix $A^T A$.

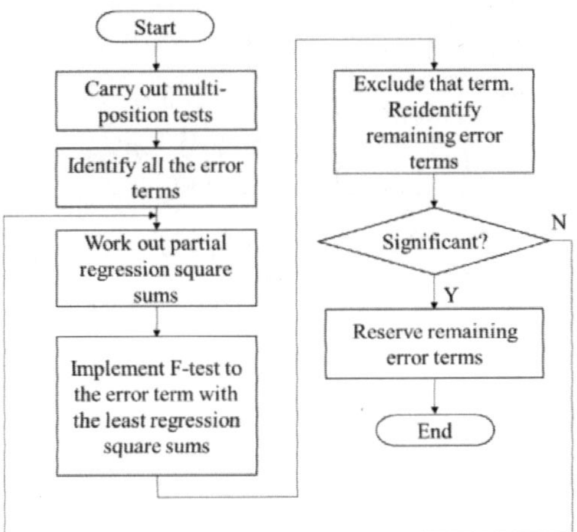


**Figure 3.** Procedure to exclude insignificant error terms.

The process of excluding insignificant error terms could be expressed by the flowchart in Figure.3. In which F-test is implemented by calculating the statistic quantity:

$$F_i = \frac{P_i}{Q/(N-M-1)} \qquad Q = \sum_{i=1}^{N} (y_i - \hat{y}_i)^2 \tag{15}$$

and comparing it with values in F distribution table.

To evaluate whether this process decreases the accuracy, compensation accuracy could be worked out as $\sigma$, which is calculated by (16).

IOP Conf. Series: Materials Science and Engineering **608** (2019) 012024 doi:10.1088/1757-899X/608/1/012024

$$\sigma = \sqrt{\frac{Q}{N-M-1}} \tag{16}$$

This value is the residual standard deviation of the compensation equation, and could be used to evaluate the error of the compensation equation. The smaller the $\sigma$, the higher the accuracy of the compensation equation.

## 5. Test Results

A series of 24-position tests have been carried out. The relationship between drifting angular rates and test positions is shown in Figure.4. The identification of all the possible term are listed in Table.1. And the compensation accuracy values of the equations are listed in (17) and (18).

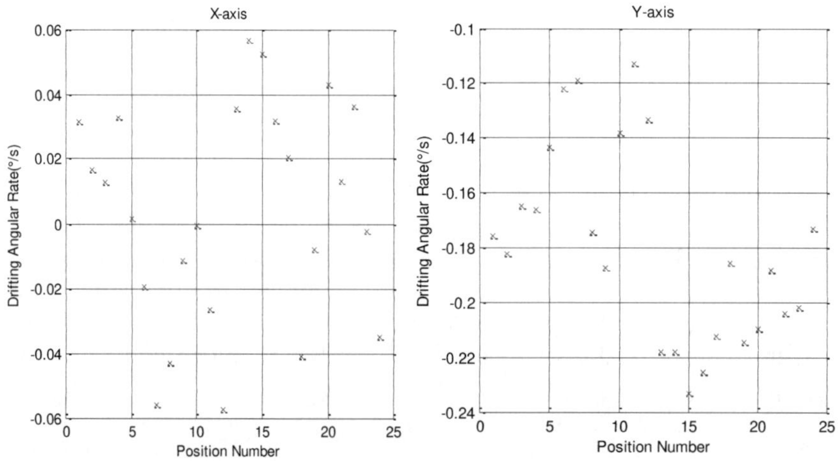


**Figure 4.** Drifting angular rate of two axes in different positions.

**Table 1.** Identification of all the possible error terms.

| x-axis Error Coefficients | Values | y-axis Error Coefficients | Values |
|---|---|---|---|
| $D(x)_0$(° /s) | -0.0272 | $D(y)_0$(° /s) | -0.1599 |
| $D(x)_x$((° /s)/g) | -0.0365 | $D(y)_x$((° /s)/g) | 0.0165 |
| $D(x)_y$((° /s)/g) | 0.0054 | $D(y)_y$((° /s)/g) | 0.0200 |
| $D(x)_z$((° /s)/g) | -0.0067 | $D(y)_z$((° /s)/g) | 0.0138 |
| $D(x)_{xy}$((° /s)/g$^2$) | 0.0098 | $D(y)_{xy}$((° /s)/g$^2$) | -0.0144 |
| $D(x)_{yz}$((° /s)/g$^2$) | -0.0006 | $D(y)_{yz}$((° /s)/g$^2$) | -0.0035 |
| $D(x)_{xz}$((° /s)/g$^2$) | -0.0217 | $D(y)_{xz}$((° /s)/g$^2$) | -0.0078 |
| $D(x)_{xx}$((° /s)/g$^2$) | 0.0004 | $D(y)_{yy}$((° /s)/g$^2$) | 0.0189 |
| $D(x)_{zz}$((° /s)/g$^2$) | 0.0606 | $D(y)_{zz}$((° /s)/g$^2$) | -0.0477 |

$$\sigma_X = 0.0131^\circ/\mathrm{s} \tag{17}$$

$$\sigma_Y = 0.0253^\circ/\mathrm{s} \tag{18}$$

After implementing the process of excluding insignificant error terms the remaining error terms are listed in Table.2. And the compensation accuracy values after simplification are in (19) and (20).

IOP Conf. Series: Materials Science and Engineering **608** (2019) 012024 doi:10.1088/1757-899X/608/1/012024

$$\sigma_X = 0.0123\,^\circ/\mathrm{s} \tag{19}$$

$$\sigma_Y = 0.0245\,^\circ/\mathrm{s} \tag{20}$$

**Table 2.** Remaining error terms.

| x-axis Error Coefficients | Values | y-axis Error Coefficients | Values |
|---|---|---|---|
| $D(x)_0$(° /s) | -0.0269 | $D(y)_0$(° /s) | -0.155 |
| $D(x)_x$((° /s)/g) | -0.0365 | $D(y)_y$((° /s)/g) | 0.0200 |
| $D(x)_z$((° /s)/g) | -0.0067 | $D(y)_z$((° /s)/g) | 0.0138 |
| $D(x)_{xx}$((° /s)/$g^2$) | -0.0218 | $D(y)_{zz}$((° /s)/$g^2$) | -0.0571 |
| $D(x)_{zz}$((° /s)/$g^2$) | 0.0604 | | |

After excluding these insignificant error terms, the compensation accuracy doesn't decrease, which means the simplification of the compensation equations is reasonable.

## 6. Conclusion

In this paper, an improved regression analysing method for Gyrowheel system has been proposed. Based on Multi-position tests, variance analysis is used to exclude insignificant error terms. Partial regression square sums are worked out to determine the importance of error terms and F-tests are carried out to determine whether the insignificant error term is negligible. Test results indicated that this method can significantly simplify the form of the compensation equations and further test procedure while at the same time maintaining the compensation accuracy. This will be of good help to Gyrowheel calibration tests, especially when tests for tilt condition are implemented.

**Acknowledgement**

This research is supported by the National Nature Science Foundation of China (NNSF) under Grant No. 61427809 and 61773138.

## References

[1] G. Tyc, D.A. Staley 1999 Gyrowheel™-An Innovative New Actuator/Sensor for 3-axis Spacecraft Attitude Control *Proceedings of AIAA SPACE 2009 Conference on Small Satellites*
[2] Y. R. Gang 2006 Rotor kinematic analysis of Gyrowheel *Proceedings of the 12th Space and Motion Control Technology Conference* 528-531
[3] B. Liu, J. C. Fang, G. Liu 2011 Design of a Magnetically Suspended Gyrowheel and Analysis of Key Technologies *Acta Aeronautica et Astronautica Sinica* **32.8** 1478-1487
[4] B. G. Liang 2013 Research on the measurement and control of the attitude of the spacecraft by Gyrowheel *Harbin Institute of Technology*
[5] American National Standards Institute 2005 IEEE Specification Format Guide and Test Procedure for Two-Degree-of-Freedom Dynamically Tuned Gyros *ANSI/IEEE std* 813-1988
[6] J. M. Hall 2008 Calibration of an Innovative Rate Sensing /momentum Management Instrument for De-tuned Operation and Temperature Effects *Carleton University Ottawa*
[7] Y. Y. Zhao 2016 An improved multi-position calibration method for the Gyrowheel based on D-optimal theory *Control Conference IEEE* 5335-5339

# Mission Capability Assessment of 3D Printing Cubesats

**Zhiyong Chen**[1, *] **and Nickolay Zosimovych**[1]

[1]Interlligent Manufacturing Key Laboratory of Ministry of Education, Shantou University, Shantou, China

[*]17zychen5@stu.edu.cn

**Abstract:** With the successful development of integrated technologies, many spacecraft subsystems have been continuously miniaturized, and CubeSats have gradually become the main executors of space science exploration missions. It is a new challenge to combine the CubeSat design with 3D printing. Compared with traditional manufacturing through machining, 3D printing technology has several advantages including short period of manufacturing, high accuracy in manufacturing small parts and low cost. The research task of this paper is a LEO (Low Earth Orbit) CubeSat mission, with a maximum acceleration of 5 g during the launch process, the internal operating temperature range of the CubeSat is from 0 to 40 °C, external temperature from -80 to 100 °C. The environmental factors were fully considered in the CubeSat design process, the impact load received during the CubeSat launch process and the working environment after reaching orbit were simulated and analyzed by ANSYS after the design parameters were obtained to verify the feasibility of the design.
**Key words:** CubeSat, 3D printing, ANSYS

## 1. Introduction

This paper introduces the design and manufacture of CubeSat platform based on 3D printing technology. The great potential in space exploration of CubeSat can be seen by the recent explosion of micro-satellite missions. Such as the biological science micro-satellite *Biosentinel* [1], an 3U experimental CubeSat *STU-2* [2], and night vision observation satellite *Athenoxat-1* [3]. Short research cycle and low cost are the main advantages of the CubeSat, however, the CubeSat structure usually made by traditional mechanical processing methods. Such processing methods are costly and require a long cycle, which is against the CubeSat's design intention. With the development of manufacturing technology, 3D printing technology gradually replaced many traditional processing methods. A 3D printing CubeSat platform realized within the activities of the Space Robotics Laboratory and V-Lab of the II Faculty of Engineering of Bologna University in collaboration with the Aerospace System Laboratory of the University of Rome "La Sapienza" [4]. Using rapid prototyping technology for manufacturing has several advantages including high fast implementation, accuracy in manufacturing small parts and low cost. In addition, this technology is very useful for the manufacture of small satellites, because the accuracy of the details can be achieved, which is quite difficult and expensive by using traditional technology.

The main content of this research is a satellite in the shape of a cube with side length of 10 cm. Each cube cell is called a unit, weighs about 1kg, and the CubeSat is named based on the number of units (2U, 3U, ...). The CubeSat is made by 3D printing technology, using PLA as the material. The research is based on the LEO mission, with a maximum acceleration of 5 g during the launch process, the internal operating temperature range of the CubeSat is from 0 to 40 °C, external temperature from

-80 to 100 °C. These factors should be fully considered during the CubeSat design process, using ANSYS to do the environmental simulation and analysis to verify the feasibility of the design.

## 2. Methodology

*2.1 Structure*
The framework structure for a single CubeSat provides enough internal workspace for the hardware required to run the CubeSat. Although there are various CubeSat structure designs, several consistent design guidelines can be found by comparing these CubeSats.

The CubeSat frame is usually designed as a cube with a side length of 100 mm, square columns (8.5 mm side and 113.5 mm height), placed at four parallel corners. The CubeSat is usually made of aluminum, because it has characteristics such as ease of machining, high-strength-to-weight ratio, lightweight and low cost [5]. In this paper, the 3D printing technology is combined with the structural framework design, and the applicability of the technology in the final product manufacturing has been discussed. PLA (polylactic acid) is selected as the material. This new material has the characteristics of lightweight and high-strength-to-weight ratio, and has the advantages of rapid realization, high accuracy and low cost of 3D printing technology [6].

Compared with traditional metal processing technology, the design scheme combining PLA and 3D printing technology is much more advanced. Traditional manufacturing techniques typically remove materials, while 3D printing typically adds them layer by layer. The technique is also known as FDM (Fused Deposition Modeling), in which the material is fused into a very thin wire and deposited in layers. The adoption of this technology changes the concept of primary and secondary structure in the traditional design process, because the whole structure can be produced at the same time, which not only reduces the number of parts, reduces the need for screws and adhesion, but also improves the stability of the overall structure. One disadvantage of this technique is that the material is not isotropic. On the other hand, the direction of material deposition needs to be optimized to be consistent with the priority direction of maximum load. After the completion of production, the products also need to be processed to achieve the desired effect, such as surface grinding [7].

*2.2 Power subsystem*
CubeSat runs on two main sources of power: secondary batteries and solar panels. The architecture of the power subsystem is very simple. The solar panel is connected directly to the battery through a blocking diode. The power subsystem of the CubeSat can be customized according to the operational requirements of each subsystem, including propulsion, control and data handling (CDH), attitude determination and control system (ADCS), payload, etc. [8]. The input power required by the subsystem can be determined by the next equation:

$$P_{\mathrm{req}} = P_{in} / \eta_{pwr} , \tag{1}$$

where $P_{in}$ is the input power of the subsystem, $\eta_{pwr}$ is the utilization efficiency. This equation gives the required power for all subsystems, the sum of these is the total power $P_{tol}$ required by the CubeSat. This total power then needs to be distributed to each electrical component by the distribution unit (PDU), which is powered by a power processing unit (PPU) connected to a high-voltage electronics assembly (HVEA). The power required by PDU can be determined by equation:

$$P_{PDU} = P_{tot} / \eta_{PDU} , \tag{2}$$

where $P_{PDU}$ is the total power required by PDU and $\eta_{PDU}$ is its efficiency. The power required by PPU and HEVA can also be calculated according to the respective efficiency of components in a similar way. It should be noted that the correct power budget should be achieved in a certain order (PDU, PPU, and then HEVA). The mass of wiring and related constraints usually accounts for 10-25% of the whole power subsystem; wiring efficiency (about 95%) is also traditionally used to estimate the power required for a PPU. It is also common to add a power margin of 10-20% to the power budget [9]. The resulting value is the amount of power the solar array needs to provide.

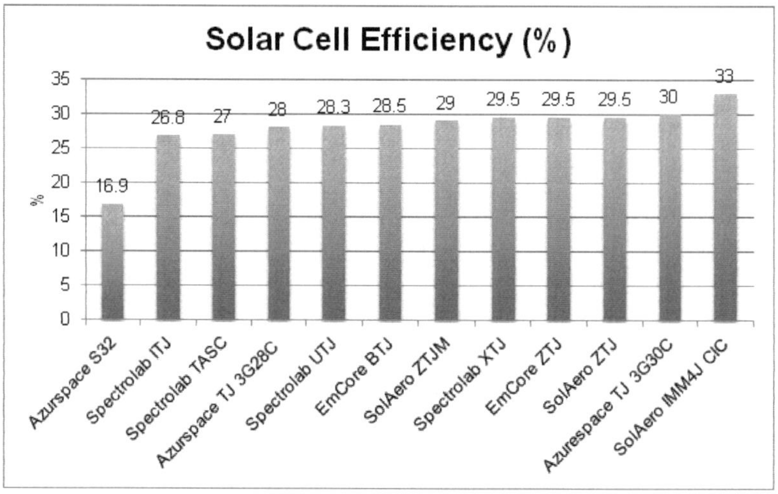


**Figure 1.** Solar Cell Efficiency [5].

Suitable solar cell arrays must be used to provide the required power for the CubeSat. To guarantee the power supply during the CubeSat's mission cycle, the state of the solar array needs to be considered not only at the beginning-of-life (BOL), but also at the end-of-life (EOL) phase. The degradation coefficient $D$ is introduced here, which can be defined by equation:

$$D=(1-D_a)^L, \tag{3}$$

where $D_a$ is the approximate degradation rate of the solar array (%), and $L$ is the duration of the mission (year). Figure 1 shows the typical photovoltaic solar cells, they have an efficiency range of 12-25% at the beginning-of-life (BOL), and there are a few new commercial photovoltaic solar cells designs that can achieve up to 30% efficiency [5]. Therefore, the efficiency of these solar cells at the end-of-life (EOL) can be calculated by equation:

$$\eta_{EOL}=\eta_{BOL}D, \tag{4}$$

where $\eta_{BOL}$ and $\eta_{EOL}$ respectively represent the efficiency of the solar cell array at the beginning-of-life (BOL) and at the end-of-life (EOL). Other parameters of the solar cell array, such as power density $P$ and specific power $\rho$ , they can be calculated by equation (4), and only need to be replaced $\eta_{EOL}$ with $P_{EOL}$ or $\rho_{EOL}$ , and at the same time replaced $\eta_{BOL}$ with $P_{BOL}$ or $\rho_{BOL}$ . From these parameters, we can estimate the mass $m_{SA}$ and expansion area $A_{SA}$ of the solar cell array:

$$m_{SA}=\frac{P_{tot}}{\rho_{EOL}}, \tag{5}$$

$$A_{SA}=\frac{P_{tot}}{I_{S,\min}\eta_{EOL}\cos(\theta_S)}, \tag{6}$$

where $\theta_S$ is the incidence angle of sunlight, $I_{S,\min}$ is the local minimum solar irradiation intensity. If the solar array is a mobile deployment system, with gimbal and extra support bracket, etc., additional quality budget needs to be considered. The power available at the beginning-of-life (BOL) and at the end-of-life (EOL) of a solar array can be calculated by next equation:

$$P_{SA}=A_{SA}\eta I_{S,\min}\cos(\theta_s) \tag{7}$$

For CubeSat, it is vital to keep the system running when it is not exposed to sunlight, and this part of power is provided by the battery. The mass of the battery can be estimated by equation:

$$m_b = \frac{P_{b,req} t_b}{\rho_b \eta_b d}, \tag{8}$$

where $P_{b,req}$ is the power the battery needs to provide, $t_b$ is the time for the battery to provide power (hours), $\rho_b$ is the specific power of the battery, $\eta_b$ is the utilization efficiency of the power provided by the battery, and $d$ represents the discharge depth of the battery. The charging rate of the battery is considered to ensure that the CubeSat can fully charge the battery during each orbit day. The charging time of the battery can be calculated by next equation:

$$t_c = \frac{P_{b,req} t_b}{V_b \eta_b d C_r}, \tag{9}$$

where $V_b$ is the battery output voltage and $C_r$ is the charge rate of the batteries for a given charger.

*2.3 Other Subsystems*

Although not the primary focus of this research, a complete CubeSat design requires thermal subsystem, the communication system, ADCS, and CDH subsystems. Due to the high surface area-to-volume ratio, the thermal system design requirements of the CubeSat should be the highest of all subsystems during the design process. The normal operating temperature range of internal components is from 0 to 40 °C, and the solar cell array can run between -80 and 100 °C [9]. Specific temperature depends on the different system of survival, but usually only extending ±10 °C from the working temperature range. The primary function of the communication system is to provide the signal connections to ground stations on the earth. The CubeSat consists of the satellite's onboard antennas, radios and data circuit boards. In addition, to ensure that the solar arrays point accurately to the sun during orbit, a three-axis stability system is required, the power requirements and mass of the electronic package must also be considered. The CDH subsystem acts as the CubeSat's brain, receiving telemetry information and controlling the CubeSat's actions when no human operator is present. Finally, a comprehensive set of autonomous navigation software is also a necessary condition for the successful operation of spacecraft.

## 3. Parametric results for CubeSat

*3.1 CubeSat mission Overview*

The CubeSat under development will be able to test the effectiveness of structures manufactured using rapid prototyping technology in an orbital environment and will be able to perform performance tests on the camera payload to evaluate its reliability and endurance. This camera is a commercial camera, using COTS optical devices, which can adapt to different observation tasks by changing the system's field of view and resolution.

*3.2 Structure*

Depending on the design specification and the material selected, the CubeSat space frame usually has a mass of no more than 200 g. To meet the related design requirements, such as the installation of related subsystems, additional braces and brackets may be required, the quality varies from 5-50 g. The mounting of the brackets completes the frame structure and provides CubeSat hardware and payload mounting points [10].

The outer panel is usually made of metal or fiberglass for the circuit board, with a mass of about 50 g per 10x10 cm of coverage area. These panels are mounted on the CubeSat frame surface to shield the internal hardware and provide mounting points for the expandable solar array as fixpoints.

Except the design considerations mentioned above, during the structural design process, 15-25 %

mass allowance should be reserved for components that may need to be installed.

*3.3 Power subsystem*

The mass of the gimbal and mounting hardware for the four solar arrays is approximately 40 % of the total mass of the solar array, resulting in a mass of about 60 g for each array and a power demand of about 2W. The CubeSat uses Spectrolab UTJ (ultra - triple junction solar cells), it can provide the power density of more than 380 W/m$^2$, 450 W/kg of specific power and efficiency of about 28 % at BOL [11]. A battery performance degradation of 7.8%, which included the assembly, design, and the storage-time losses, was assumed to happen before launch with an annual performance degradation of 0.5% during the task [11]. The view factor to the sun was approximated as 0.9, and assuming that the solar cell packing fraction is 0.9 [12], the average solar incidence angle was assumed as 30 deg (a conservative estimate), these values can identify the solar array area of 0.078 m$^2$, the total quality of solar array is about 124 g. In addition, CubeSat's PPU has a mass of about 50 g with an efficiency of 92 %, while PDU has a mass of about 35 g with an efficiency of 85 %. These values are very conservative estimates for the power subsystem. The mass of HVEA is approximately 45 g, and the efficiency is 97 %. These values refer to the mass quoted by the manufacturer of spacecraft electronics [5]. Table 1 shows the mass equipment list (MEL) of the CubeSat.

**Table 1.** MEL for the 1U CubeSat

| Subsystem | Component | Mass, kg | No. |
|---|---|---|---|
| Structure | | 0.652 | |
| | Frame | 0.200 | 1 |
| | Brackets | 0.005 | 8 |
| | Braces | 0.020 | 4 |
| | 10×10cm exterior panel | 0.050 | 6 |
| | Solar array gimbal and mounting system | 0.024 | 4 |
| Power | | | |
| | Solar arrays | 0.035 | 4 |
| | Battery | 0.178 | 1 |
| | Battery charger | 0.050 | 1 |
| | PPU | 0.050 | 1 |
| | PDU | 0.035 | 1 |
| | HVEA | 0.050 | 1 |
| CDH | | | |
| | CDH board | 0.050 | 1 |
| | EPS board | 0.100 | 1 |
| Bus Dry Mass | | | |
| | Payload (assumed) | 0.350 | 1 |
| LEO mission wet mass | | 1.719 | |

## 4. Simulation

Due to the unique nature of the space environment, CubeSat must perform a series of simulation tests before it can operate in the orbit. The CubeSat structure was analyzed by means of finite element analysis software ANSYS. In addition, random vibration analysis was performed to ensure that the CubeSat structure can withstand the impact load during the launch process. The feasibility of the CubeSat structure is verified by these results.

*4.1 Static Structural*

IOP Conf. Series: Materials Science and Engineering **608** (2019) 012025 doi:10.1088/1757-899X/608/1/012025

The CubeSat structure is validated by the numerical experiment. During launch process, CubeSat will be fixed inside the P-Pod, and the corresponding structural constraints should be added to the numerical model. In addition, the maximum acceleration impact during the launch process should also be considered. Static Structural module of ANSYS is used for calculation and analysis, the results show that the maximum stress of CubeSat Structure is 8.06 MPa, lower than the PLA yield strength of 40 Mpa. Figure 2 is the Von mises stress diagram of the CubeSat structure.

**Figure 2.** The Von mises stress diagram of the CubeSat structure.

*4.2 Thermal stress*

Temperature difference is a major characteristic of the space environment, CubeSat running around LEO will go through a temperature change of 100 °C. The simulation model of steady state temperature through numerical analysis software, set the conditions for internal temperature range from 0 to 40 °C, external temperature range from -80 to 100 °C. CubeSat structure should be able to resist a sudden change of temperature (circulating temperature between 20 to 60 °C), so the thermal shock test is also needed. According to the experimental results, the maximum value of Thermal strain is only 0.009, lower than 0.01, which is acceptable. Figure 3 is the thermal strain diagram of the structure.

The effects of radiation on PLA material can be neglected [13], the PLA material can withstand years of total dose of radiation to $10^9$ rad (SI), and the absorbed radiation dose of CubeSat run a year is far lower than this value in LEO (less than $10^7$ rad, height 2000 kilometers [14]). As for the direct sunlight, it should be considered that the structure of the PLA material is almost completely covered by the solar panels.

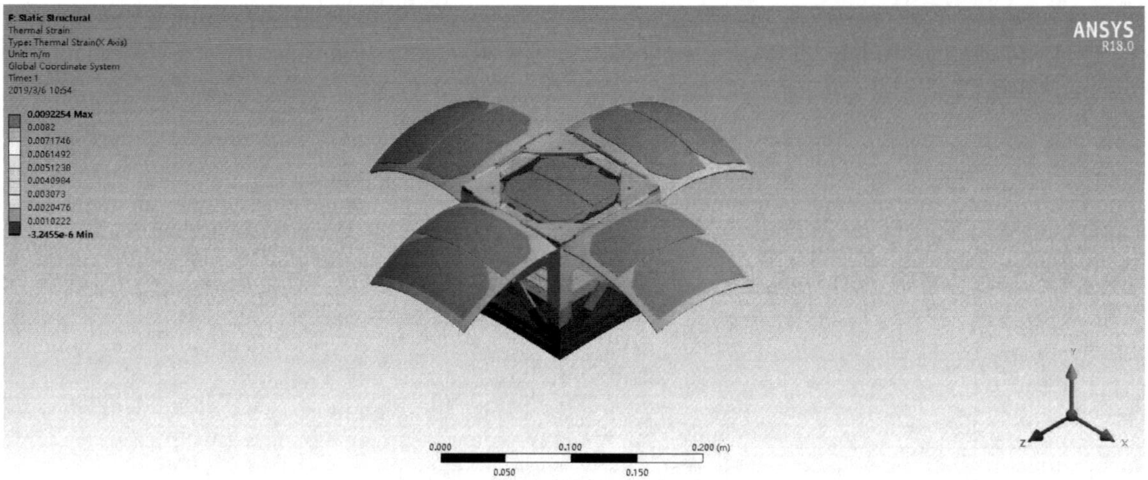


**Figure 3.** The thermal strain diagram of the CubeSat structure.

*4.3 Random Vibration*

To make the CubeSat structure conform to the emission conditions, random vibration simulation experiments are required. During the experiment, typical vibration characteristics of launch environment were simulated. NASA GEV qualification and acceptance were used as reference, and the specific parameters are shown in Table 2. The specific contents of the experiment include “Harmonic Response” and “Random Vibration”. Two identical harmonic response were performed before and after the random vibration test to assess the degree of structural degradation that may result from the launch load.

**Table 2.** NASA GEV qualification and acceptance [5].

| | Qualification | Acceptance |
|---|---|---|
| Frequency (Hz) | ASD ($g^2$/HZ) | ASD ($g^2$/HZ) |
| 20 | .026 | .013 |
| 20-80 | +6 dB/oct | +6 dB/oct |
| 80-500 | .16 | .08 |
| 500-2000 | -6 dB/oct | -6 dB/oct |
| 2000 | .026 | .013 |
| Overall | 14.1 Grms | 10.0 Grms |

Figure 4 shows the response curve of the two Harmonic response tests. This experiment helps us to evaluate the natural frequency of the structure, and the peak value indicates that the tested point (bottom panel) has reached the resonant frequency. As shown in this figure, the trend and peak points of the two response curves are close to each other, indicating that the structure has no structural degradation after the random vibration test, and the overall CubeSat structure conforms to the stiffness specifications of launch.

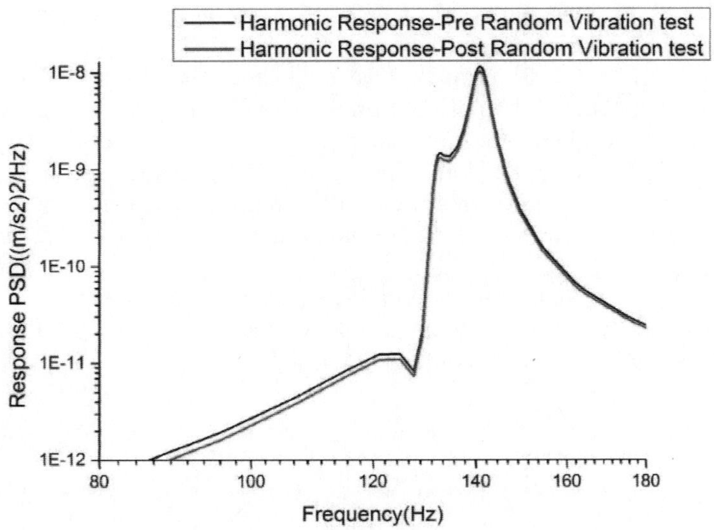


**Figure 4.** Pre/Post Random Vibration test comparison between the curves of Harmonic Response.

## 5. Conclusion

As the primary performer of today's space exploration missions, the CubeSat design considers orbit, payload, thermal balance, subsystem layout, and mission requirements. In this research, a CubeSat design for performing LEO tasks was proposed, including power budget, mass distribution, and ground testing, and the CubeSat structure for manufacturing was combined with 3D printing technology. The feasibility of the new rapid prototyping technology in CubeSat manufacturing was evaluated by simulation tests. The results show that the CubeSat can withstand the launch loads without structural damage and can meet the launch stiffness specification.

## Reference

[1] H. Sanchez, "BioSentinel - A 6U Nanosatellite for Deep Space Biological Science," in *13th Annual Summer CubeSat Developers' Workshop*, Logan, UT, 2016.

[2] W. C. Shufan Wu, Caixia Chao, "The STU-2 CubeSat Mission and In-Orbit Test Results," presented at the 30th Annual AIAA/USU Conference on Small Satellites, Logan, Utah, 2016.

[3] Y. L. B. Giulio Manzoni, Meini Zhang, "Athenoxat-1, Night Vision Experiments in LEO," presented at the 30th Annual AIAA/USU Conference on Small Satellites, Logan, Utah, 2016.

[4] J. Piattoni, G. P. Candini, G. Pezzi, F. Santoni, and F. Piergentili, "Plastic Cubesat: An innovative and low-cost way to perform applied space research and hands-on education," *Acta Astronautica,* vol. 81, pp. 419-429, 2012.

[5] R. B. Elwood Agasid, Roberto Carlino, Gregory Defouw, Andres Dono Perez, Arif Göktuğ Karacalıoğlu, Benjamin Klamm, Abraham Rademacher, James Schalkwyck, Rogan Shimmin, Julia Tilles, Sasha Weston, "Small Spacecraft Technology State of the Art," National Aeronautics and Space Administration, NASA Ames Research Center, Technical Publication NASA/TP–2015–216648/REV1, 01-12 2015.

[6] J. B. Soares, J. Finamor, F. P. Silva, L. Roldo, and L. H. Candido, "Analysis of the influence of polylactic acid (PLA) colour on FDM 3D printing temperature and part finishing," *Rapid Prototyping Journal,* vol. 24, pp. 1305-1316, 2018.

[7] M. Grasso, L. Azzouz, P. Ruiz-Hincapie, M. Zarrelli, and G. G. Ren, "Effect of temperature on the mechanical properties of 3D-printed PLA tensile specimens," *Rapid Prototyping Journal,* vol. 24, pp. 1337-1346, 2018.

[8] R. W. C. a. R. E. Wirz, "Mission Capability Assessment of CubeSats Using a Miniature Ion Thruster," *JOURNAL OF SPACECRAFT AND ROCKETS,* vol. 50, 2013.

IOP Conf. Series: Materials Science and Engineering **608** (2019) 012025 doi:10.1088/1757-899X/608/1/012025

[9] J. Wertz, Everett, D., and Puschell, J, "Space Mission Engineering: The New SMAD," *Microcosm Press, Hawthorn,* 2011.
[10] I. Nason, Puig-Suari, J., and Twiggs, R., "Development of a Family of Picosatellite Deployers Based on the CubeSat Standard," *IEEE Aerospace Conference Proceedings,* vol. 1, 2002.
[11] O. Popescu, "Power Budgets for CubeSat Radios to Support Ground Communications and Inter-Satellite Links," *Ieee Access,* vol. 5, pp. 12618-12625, 2017.
[12] H. U. Oh and T. Park, "Experimental Feasibility Study of Concentrating Photovoltaic Power System for CubeSat Applications," *Ieee Transactions on Aerospace and Electronic Systems,* vol. 51, pp. 1942-1949, Jul 2015.
[13] NASA, "NASA Space Vehicle Design Criteria (Structure)," ed, 1970.
[14] J. R. W. W.J.Larson, *Space Mission Analysis and Design, third edition*. California: SpaceTechnologyLibrary-MicrocosmPress, 2004.

# Stability Area of a Spacecraft's Partially Invariant Centre of Mass Motion Stabilization System

**Nickolay Zosimovych**

Intelligent Manufacturing Key Laboratory of Ministry of Education, Shantou University, 243, Daxue Road, Shantou, Guangdong, 515063, China

nzosimovych@stu.edu.cn

**Abstract.** In practice, a spacecraft's stabilization system, which is partially invariant to the disturbing moment, is the easiest to implement. The velocity performance limitation of the control actuator has the most influence upon stability of the system among all the nonlinearities considered in the mathematical modelling. Therefore, we shall consider a model of an invariant stabilization system with due regard to nonlinearity. For the system under consideration, it is possible to construct lines of equal values of the auto oscillation amplitude in a two-parameter plane. The study of the stability of the proposed partially invariant stabilization system revealed that it is possible to ensure sufficient stability margins in the system under consideration by choosing parameters for the stabilization controller. At the same time, it allows to provide high quality of the transition process.
***Keywords:*** *accuracy, spacecraft, stabilization, controller, nonlinearity.*

## 1. Introduction

Some publications [1-6] included synthesis of an invariant centre of mass motion stabilization system of a spacecraft and a comparative analysis of accuracy and quality of transition processes when conventional stabilization system is used. They demonstrated advantages of the proposed algorithms in terms of accuracy improvement.

An essential criterion determining practical application of the control system in addition to the control accuracy is a possibility to provide proper stability margins of the system. Consequently, the challenge is to select appropriate parameters for a stabilization controller. Therefore, a detailed study of the stability of the partially invariant algorithm proposed in [4-6] as well as selection of the proper parameters for the stabilization controller based on conducted studies, which shall provide sufficient stability margins for the system and the required quality characteristics for the transition process shall be the task to be solving herein.

The selection of stabilization controller parameters shall be more convenient and more informative if we build the system stability area in the stabilization controller parameter plane: for the angular stabilization channel and for the centre of mass motion stabilization channel. In practice, sufficient stability margins of the system shall be 1.5 – 2.0 time decrease (increase) margin of the system parameters with respect to their boundary stability conditions [7]. Therefore, ensuring of the specified stability margins is an essential criterion in the selection of parameters for automatic system in this research.

## 2. Problem formulation

Let us examine the stability of the proposed partially invariant stabilization system considering the influence of nonlinear elements of the system. As regards stability, the velocity performance limitation of the control actuator [8] has the most influence upon stability of the system among all the nonlinearities considered in the mathematical modelling [9, 10]. Therefore, we shall consider a model of an invariant stabilization system with due regard to nonlinearity. A block diagram corresponding to a model of the system allowing for nonlinearity is showing in Figure 1.

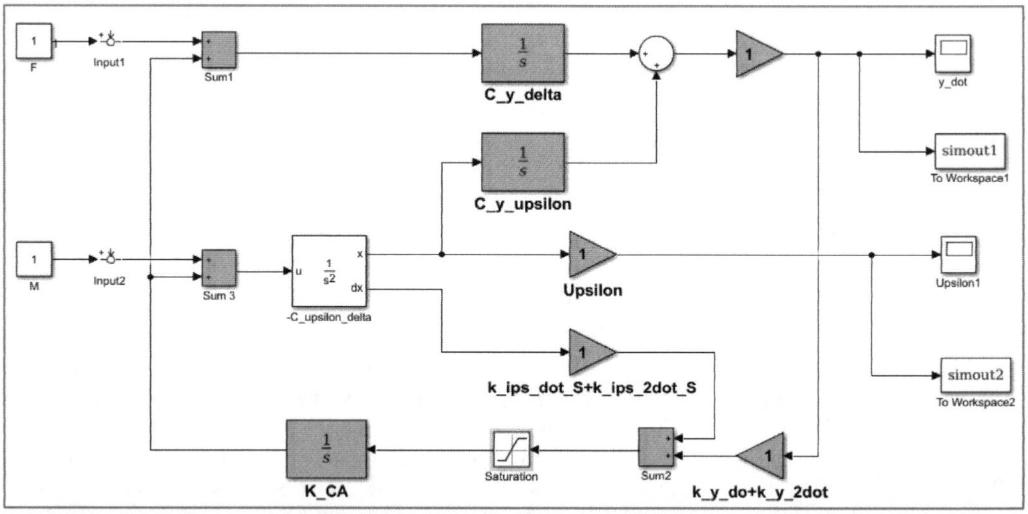


**Figure 1.** Block diagram of an invariant centre of mass motion stabilization system considering nonlinear velocity performance of the control actuator [6].

In order to study the nonlinear system we shall apply the method of harmonious balance [7]. For this purpose, we shall open the system at input of the linear element (Figure 2).

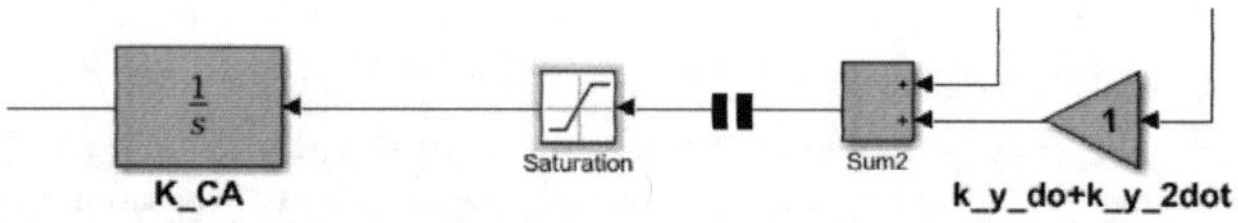


**Figure 2.** Break element at input of the nonlinear element.

We shall define the transfer functions for the linear and nonlinear parts of the system. According to the block diagram, the transfer function of the linear part of the system is as follows:

$$W_L(s)=\frac{K_{OD}}{s^4}\left[s^3\left(C_{y\delta}k_{\ddot{y}}-C_{\vartheta\delta}k_{\ddot{\vartheta}}\right)+s^2\left(C_{y\delta}k_{\dot{y}}-C_{\vartheta\delta}k_{\dot{\vartheta}}\right)-sC_{\vartheta\delta}C_{y\delta}k_{\ddot{y}}-C_{\vartheta\delta}C_{y\delta}k_{\dot{y}}\right]. \quad (1)$$

For nonlinearity with a saturation zone, the expression for the coefficient of harmonic linearization is [9]:

$$\begin{cases} q=\dfrac{2}{\pi}\left(\arcsin\dfrac{I_N}{i_{\max}}+\dfrac{I_N}{i_{\max}}\left(1-\dfrac{I_N^2}{i_{\max}^2}\right)^{1/2}\right), when: i_{\max}>I_N; \\ q=1, when: i_{\max}\le I_N. \end{cases} \quad (2)$$

where $i_{\max}$ – the amplitude of the input is signal for the control actuator; $I_N$ – is velocity performance saturation current for the control actuator.

According to the above formulas, the dependence of the coefficient of harmonic linearization $q$ from the amplitude of the input signal $i_{\max}$ is showing in Figure 3.

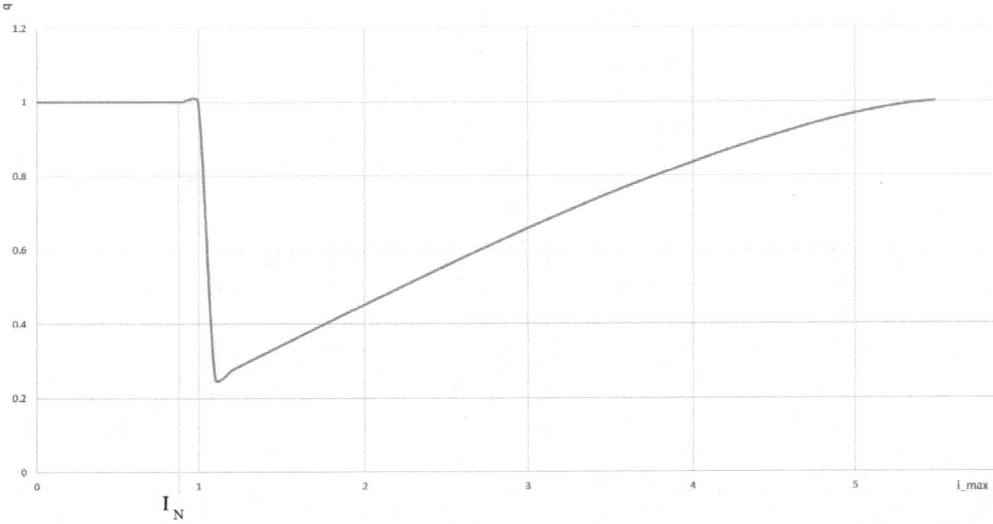


**Figure 3.** Dependence of the coefficient of harmonic linearization $q$ from the signal amplitude $i_{\max}$.

## 3. Stability area of a spacecraft's partially invariant centre of mass motion stabilization system with due regard to velocity performance saturation zone of the control actuator

To analyse periodic solutions we shall use the graphical method [11]. For this purpose, we shall build a frequency hodograph for the transfer function of the linear part of the system. By substituting Laplace operator $s$ for $j\omega$, in the expression $W_L(s)$ we obtain:

$$W_L(j\omega)=\frac{K_{OD}}{\omega^4}\left[j\omega^3\left(-C_{y\delta}k_{\ddot{y}}+C_{\vartheta\delta}k_{\ddot{\vartheta}}\right)+\omega^2\left(-C_{y\delta}k_{\dot{y}}+C_{\vartheta\delta}k_{\dot{\vartheta}}\right)-j\omega C_{\vartheta\delta}C_{y\delta}k_{\ddot{y}}-C_{\vartheta\delta}C_{y\delta}k_{\dot{y}}\right]. \tag{3}$$

Let us introduce the following notations:

$$\alpha_1=\left(C_{\vartheta\delta}k_{\ddot{\vartheta}}-C_{y\vartheta}k_{\ddot{y}}\right);\ \alpha_2=\left(C_{\vartheta\delta}k_{\dot{\vartheta}}-C_{y\vartheta}k_{\dot{y}}\right);\ \alpha_3=C_{y\delta}C_{\vartheta\delta}k_{\ddot{y}};\ \alpha_4=C_{y\vartheta}C_{\vartheta\delta}k_{\dot{y}}. \tag{4}$$

Given the above notations, the expression (1) will be as follows:

$$W_L(j\omega)=\frac{K_{OD}}{\omega^4}\left[j\omega^3\alpha_1+\omega^2\alpha_2-j\omega\alpha_3-\alpha_4\right]. \tag{5}$$

We shall separate the real and imaginary parts from the expression (5):

$$\mathrm{Re}(W_L)=\frac{K_{OD}}{\omega^4}\left(\omega^2\alpha_2-\alpha_4\right) \tag{6}$$

$$\mathrm{Im}(W_L)=j\frac{K_{OD}}{\omega^4}\left(\omega^3\alpha_1-\omega\alpha_3\right). \tag{7}$$

The characteristic equation for the linear part of the open-loop system has four zero roots, so when the frequency changes between -0 and +0 the hodograph of the transfer function $W_L(j\omega)$ will move along a circle with infinitely large radius with $\omega_1=+180^0$ till $\omega_2=-180^0$.

If $\omega=0$ the expressions (6) and (7) will take on values $-\infty: \mathrm{Re}(W_L)_{\omega=0}=-\infty; \mathrm{Im}(W_L)_{\omega=0}=-\infty$. In this case, the phase shift of frequency hodograph will be:

$$tg\varphi_{\omega=0}=\frac{\mathrm{Im}(W_L)_{\omega=0}}{\mathrm{Re}(W_L)_{\omega=0}}=\left.\frac{\omega\left(\omega^2\alpha_1-\alpha_3\right)}{\omega^2\alpha_2-\alpha_4}\right|_{\omega=0}=0; \varphi_{\omega=0}=-180^0. \tag{8}$$

If $\omega=+\infty$ the expressions (6) and (7) will take on values $0: \mathrm{Re}(W_L)_{\omega=+\infty}=0; \mathrm{Im}(W_L)_{\omega=+\infty}=0$.

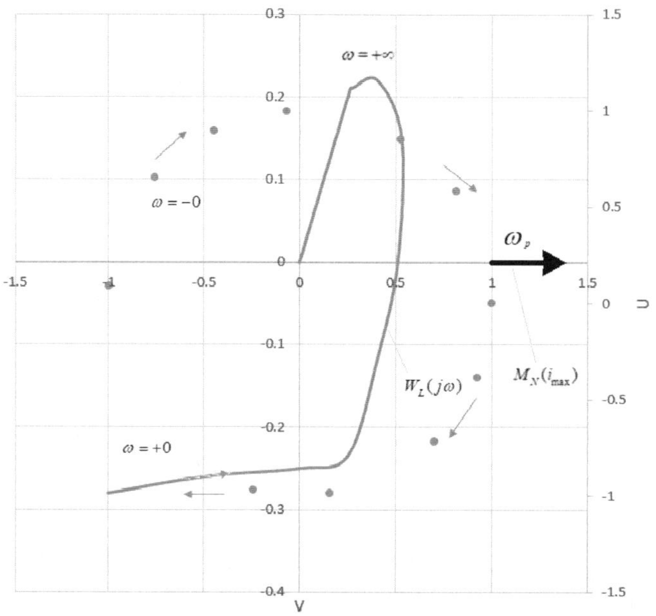


**Figure 4.** The amplitude-frequency characteristics of the partially invariant spacecraft centre of mass stabilization system.

Thus, if frequency $\omega$ changes from 0 to $+\infty$, the transfer function hodograph $W_L(j\omega)$ beginning in the third quadrant with the phase shift $\varphi = -180^0$, while moving through the 4th and 1st quadrants reaches the origin of coordinates. In this case, since closure of the system was made through +1 and not through -1, it is necessary that the branch of the phase and gain performance corresponding to the positive frequencies, together with a fragment of the circle beginning with a positive real semi axis doesn't cover the point $(+1;0j)$ to provide stability of the closed-loop system [12].

The existence of periodic solutions in the system can be determined from the condition $W_L(j\omega)W_N(i_{\max}) = 1$, where $W_N(i_{\max})$ a transfer function of the nonlinear element is, or otherwise:

$$W_L(j\omega) = \frac{1}{q(i_{\max})}. \tag{9}$$

According to (6) and (7), condition (8) shall be split into two conditions:

$$\operatorname{Im}(W_L) = 0 \tag{10}$$

$$\operatorname{Re}(W_L) = \frac{1}{q(i_{\max})}. \tag{11}$$

By substituting the expression for the imaginary part of the transfer function of the linear part of the system into the condition (8), we get:

$$\frac{K_{OD}}{\omega^4}\left(\omega^3\alpha_1 - \omega\alpha_3\right) = 0, \tag{12}$$

where from $\omega^2\alpha_1 - \alpha_3 = 0$, or

$$\omega = \left(\frac{\alpha_3}{\alpha_1}\right)^{1/2}. \tag{13}$$

The determined frequency is the frequency of a periodic solution for the given nonlinear system. Further, we will denote it as $\omega_p$. The self-oscillation amplitude in the system under consideration $i_{\max\ p}$ shall be found from the expression (13):

$$\frac{K_{OD}}{\omega_p^4}\left(\omega_p^2\alpha_2-\alpha_4\right)=\frac{1}{q(i_{\max\,p})},\quad q(i_{\max\,p})=\frac{\omega_p^4}{K_{OD}\left(\omega_p^2\alpha_2-\alpha_4\right)}. \tag{14}$$

By substituting the expression (2) for $q(i_{\max\,p})$ into the last formula, we finally obtain:

$$\frac{2}{\pi}\arcsin\left(\frac{I_N}{i_{\max\,p}}\right)+\frac{I_N}{i_{\max\,p}}\left(1-\frac{I_N^2}{i_{\max\,p}^2}\right)^{1/2}=\frac{\omega_p^4}{K_{OD}\left(\omega_p^2\alpha_2-\alpha_4\right)}. \tag{15}$$

The expression (15) is transcendental relative $i_{\max\,p}$, therefore, it can be solved only numerically. The value $i_{\max\,p}$ can be approximately determined from the diagram $q(i_{\max})$ (Figure 4).

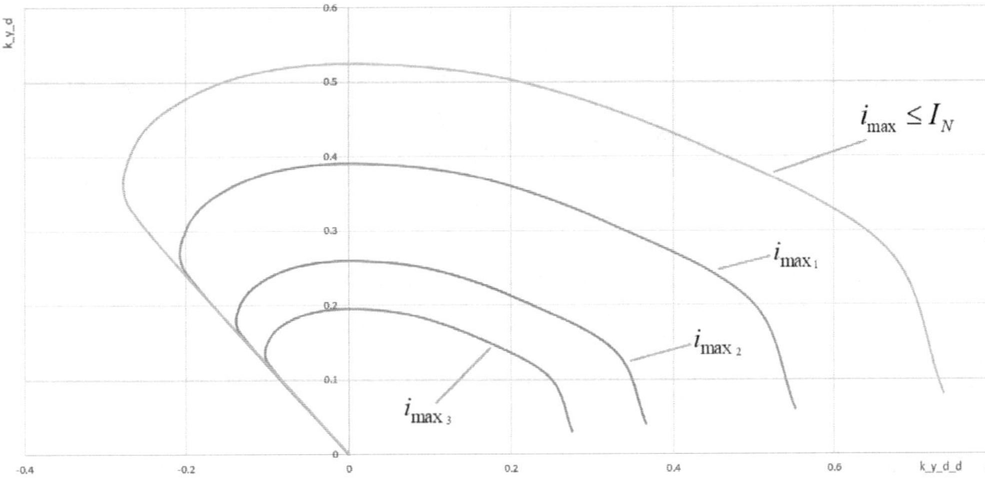


**Figure 5.** Stability area limits of a partially invariant stabilization system in the parameter plane $k_{\dot{y}}, k_{\ddot{y}}$ with due regard to velocity performance saturation zone of the control actuator.

Now we shall consider stability of periodic solution. To this end, we shall build a hodograph of the function $M_N=\frac{1}{q(i_{\max})}$ in the plane $UV$ (Figure 5). According to (2), when $0\le i_{\max}\le I_N, M_N=1$, if $i_{\max}>I_N, M_N$ monotonically increases to $+\infty$. Since hodograph $M_N$ crosses the amplitude-frequency characteristic of the linear part of system "from the outside to the inside" [13], then the periodic solution of the system under study $\left(i_{\max\,p},\omega_p\right)$ will be unstable. Therefore, when $i_{\max}<i_{\max\,p}$ the system solution will converge to 0, that is, the system will be stable, and if $i_{\max}>i_{\max\,p}$ we will observe a divergent transient process, that is, the system will be unstable.

Thus, the stability condition for the system in general shall be determined by the initial conditions of the transition process, at which the amplitude $i_{\max}$ at the input of the nonlinear element is less than the value $i_{\max\,p}$:

$$i_{\max}<i_{\max\,p}. \tag{16}$$

For the system under consideration, it is possible to construct lines of equal values of the auto oscillation amplitude in a two-parameter plane. Such parameters will be the gain of the angular velocity and angular acceleration stabilization controller $k_{\dot{\vartheta}}, k_{\ddot{\vartheta}}$. The following can be obtained from (15) and (16):

$$\omega_p = \left(\frac{\alpha_3}{\alpha_1}\right)^{1/2} = \left(\frac{C_{\vartheta\delta} C_{y\delta} k_{\ddot{y}}}{C_{\vartheta\delta} k_{\dot{y}} - C_{y\delta} k_{\ddot{y}}}\right)^{1/2} = \left(\frac{C_{\vartheta\delta}}{\frac{C_{\vartheta\delta}}{C_{y\delta}} - 1}\right)^{1/2}. \quad (17)$$

Let us set $c_1 = \left(\frac{C_{\vartheta\delta}}{C_{y\delta} k_{\ddot{y}}}\right)^{1/2}$, then $\omega_p = \left(\frac{C_{\vartheta\delta}}{c_1 k_{\ddot{\vartheta}} - 1}\right)^{1/2}$. From $\frac{K_{OD}}{\omega_p^4}\left(\omega_p^2 \alpha_2 - \omega \alpha_4\right) = \frac{1}{q(i_{\max\, p})}$, the expression (11):

$$\frac{K_{OD}}{\omega_p^4}\left(\omega_p^2 \alpha_2 - \omega \alpha_4\right) = \frac{1}{q(i_{\max\, p})}, \quad (18)$$

from which

$$\alpha_2 = \left[\frac{\omega_p^4}{K_{OD} q(i_{\max\, p})} + \alpha_4\right] \frac{1}{\omega_p^2}. \quad (19)$$

By expanding the expression for $\alpha_2$, we get after transformation:

$$k_{\dot{\vartheta}} = \frac{\frac{1}{\omega_p^2}\left[\frac{\omega_p^4}{K_{OD} q(i_{\max\, p})} + \alpha_4\right] + C_{y\vartheta} k_{\dot{y}}}{C_{\vartheta\delta}}. \quad (20)$$

After substituting in the expression for $\omega_p$:

$$k_{\dot{\vartheta}} = \frac{\frac{C_1 k_{\ddot{\vartheta}} - 1}{C_{\vartheta\delta}}\left[\frac{C_{\vartheta\delta}^2}{K_{OD} q(i_{\max\, p})(C_1 k_{\ddot{\vartheta}} - 1)^2} + \alpha_4\right] + C_{y\vartheta} k_{\dot{y}}}{C_{\vartheta\delta}}. \quad (21)$$

After transformation, we get:

$$k_{\dot{\vartheta}} = \frac{1}{K_{OD} q(i_{\max\, p})(C_1 k_{\ddot{\vartheta}} - 1)} + \alpha_4 \frac{C_1 k_{\ddot{\vartheta}} - 1}{C_{\vartheta\delta}^2} + \frac{C_{y\vartheta}}{C_{\vartheta\delta}} k_{\ddot{\vartheta}}. \quad (22)$$

By expanding the expressions for $\alpha_4$ and $c_1$, we finally have:

$$k_{\dot{\vartheta}} = \frac{C_{y\vartheta} k_{\ddot{\vartheta}}}{K_{OD} q(i_{\max\, p})(C_{\vartheta\delta} k_{\ddot{\vartheta}} - C_{y\vartheta} k_{\ddot{\vartheta}})} + \frac{k_{\dot{y}}}{k_{\ddot{y}}} k_{\ddot{\vartheta}}. \quad (23)$$

It should be noting that the right-hand side of expression (23) is different from the right side of the condition (10), which determines the stability limit for the linear systems only by the presence of harmonic linearization coefficient in the denominator of the first term. Since $q(i_{\max}) \le 1$, then the presence of nonlinearity $I_N < i_{\max\,1} < i_{\max\,2} < i_{\max\,3}$ will lead to narrowing of the stability area according to the parameters $k_{\dot{\vartheta}}, k_{\ddot{\vartheta}}$.

Accordingly, the expression for the stability limits according to the parameters $k_{\dot{y}}, k_{\ddot{y}}$ will be similar to the right side of inequality (16) taking into account the coefficient of harmonic linearization $q(i_{\max})$:

$$k_{\dot{y}} = -\frac{C_{y\vartheta}k_{\ddot{y}}^2}{K_{OD}q(i_{\max\ p})(C_{\vartheta\delta}k_{\ddot{\vartheta}} - C_{y\vartheta}k_{\ddot{y}})k_{\ddot{\vartheta}}} + \frac{k_{\dot{\vartheta}}}{k_{\ddot{\vartheta}}}k_{\ddot{y}}. \quad (24)$$

A general view of the stability area in the parameter plane $k_{\dot{y}}, k_{\ddot{y}}$ for different amplitude values at the input of the nonlinear element is shown in Figure 5.

As can be seen from the expressions (13)-(14) and Figure 5, stability of the invariant stabilization system considering nonlinearity and having a saturation zone, will directly depend on the value of the input signal in the nonlinear element, and ultimately on the disturbances acting on the spacecraft [14]. Therefore, in order to use such a stabilization system it is necessary to choose in practice such parameter values of the control action under which the system will be stable under possible perturbations. On the other hand, we will need to determine for the system under study conditions for its absolute stability regardless of perturbation.

According to [10] the absolute stability area shall be determined as an area of system parameters, which obviously escapes periodic solutions, and in which any transition process fades under any initial conditions to the value $x = 0$.

To determine an absolute stability area we can apply any linear stability criterion to the harmonically linearized system.

Let us apply the Hurwitz criterion to the system. According to the block diagram (Figure 1) and the expression for the transfer function of the linear part of the system (1), a characteristic equation for a harmonically linearized system will be:

$$s^4 + K_{OD} = \left[s^3\left(C_{\vartheta\delta}k_{\ddot{y}} - C_{y\vartheta}k_{\ddot{\vartheta}}\right) + s^2\left(C_{\vartheta\delta}k_{\dot{\vartheta}} - C_{y\vartheta}k_{\dot{y}}\right) + sC_{\vartheta\delta}C_{y\vartheta}k_{\ddot{y}} + +C_{\vartheta\delta}C_{y\vartheta}k_{\dot{y}}\right]q(i_{\max}) = 0. \quad (25)$$

Let us introduce the following notations:

$$\beta_1 = \left(C_{\vartheta\delta}k_{\ddot{\vartheta}} - C_{y\vartheta}k_{\ddot{y}}\right)K_{OD};\ \beta_2 = \left(C_{\vartheta\delta}k_{\dot{\vartheta}} - C_{y\vartheta}k_{\dot{y}}\right)K_{OD};\ \beta_3 = C_{y\vartheta}C_{\vartheta\delta}k_{\ddot{y}}K_{OD};\ \beta_4 = C_{y\vartheta}C_{\vartheta\delta}k_{\dot{y}}K_{OD}. \quad (26)$$

Then the characteristic equation for the linearized system can be written as:

$$s^4 + s^3\beta_1 q + s^2\beta_2 q + s\beta_3 q + \beta_4 q = 0. \quad (27)$$

The Hurwitz conditions for the characteristic equation of the fourth order [8]:

$$q\beta_1 > 0; q\beta_2 > 0; q\beta_3 > 0; q\beta_4 > 0; \beta_3 q\left(\beta_1\beta_2 q^2 - \beta_3 q\right) - \beta_4\beta_1^2 q^3 > 0. \quad (28)$$

To ensure absolute stability of the system it is necessary that the above conditions be satisfied for any admissible values of the coefficient of harmonic linearization $q$. For the system under study $0 \le q \le 1$. Obviously, with the boundary condition $q = 0$ of the Hurwitz conditions for this system shall be not performed. Therefore, this system escapes absolute stability with any parameters. This is also confirmed by the graphic method of periodic solutions investigation. Let us refer to Figure 4. To ensure stability of a linear system it is necessary that the frequency hodograph for the transfer function of the linear part of the system $W_L(j\omega)$ should not cover point $(+1;0j)$. When frequency changes from 0 to $+\infty$. On the other hand, the hodograph of the function $\frac{1}{q(i_{\max})}$ when $q$ changes from 0 to 1 goes through a real positive semi axis from $+\infty$ to +1. According to the above studies, the point of intersection of the frequency hodograph $W_L(j\omega)$ and the hodograph of the function $\frac{1}{q(i_{\max})}$ is a point of unstable periodic solutions and provided the conditions for the stability of the nonlinear system fulfilled (when the hodograph does not cover point $(+1;0j)$, there will be always an area of unstable periodic solutions.

## 4. Conclusion

Based upon research results we can conclude the following:

1) It is impossible to implement a centre of mass stabilization system, which is invariant regarding both the disturbing force and disturbing moment.
2) In practice, a stabilization system, which is partially invariant to the disturbing moment, is the easiest to implement. In order to comply with the invariance conditions, there must be a positive control actuator feedback with a gain equal to the object's angular deflection gain of in the angular stabilization channel. Stability of the system shall be ensuring by introduction of an additional second derivative action from the object's deflection angle into the action as well as by introduction of an equivalent delay loop in the feedback of control actuator in order to compensate for the dynamic delay of the stabilization controller.
3) It is also possible to synthesize a stabilization system, which shall be partially invariant fewer than two disturbances simultaneously. Open feedback of the control actuator and exclusion of control according to object's deflection angle and of the spacecraft centre of mass drift coordinate from the angular stabilization channel are the invariance conditions in this case. Such a stabilization system has obvious advantages over a system, which is invariant under disturbing moment M, and therefore it is more suitable for practical implementation.
4) A partially invariant under two disturbances stabilization system provides a significant increase (several times) in the accuracy of the centre of mass tangential stabilization velocities as compared to known stabilization systems.
5) The tangential velocity transition process in a partially invariant stabilization system has a significantly shorter (several times) decay time as compared to known stabilization systems.
6) Employment of additional self-regulation elements in a partially invariant stabilization system reveals the significant advantages of such a system in terms of greater accuracy when compared to known stabilization systems.
7) The study of the stability of the proposed partially invariant stabilization system revealed that it is possible to ensure sufficient stability margins in the system under consideration by choosing parameters for the stabilization controller. At the same time, it allows to provide high quality of the transition process.
8) The system under consideration has an unstable limit cycle due to "saturation zone" nonlinearity, which cannot be excluding by system parameter adjustment. It means the system is missing the absolute stability area.

## References

[1] Zosimovych N, Modeling of Spacecraft Centre Mass Motion Stabilization System 2017 *Int. Refereed Journ. of Eng. and Science (IRJES)* Vol. 6, Issue 4 **34** 41

[2] Zosimovych N, Increasing the Accuracy of the Center of Mass Stabilization of Space Probe 2017 *"Новината за Напреднали Наука - 2017", XIII Int. Scientific and Pract. Conf.,* May, 15-22, Techn. Science, Vol. 10 София, «Бял ГРАД-БГ» ООД, **31** 37

[3] Zosimovych N, Improving the Spacecraft Center of Mass Stabilization Accuracy 2017 *IOSR Journ. of Eng. (IOSRJEN)* Vol. 7, Issue 6 **7** 14

[4] Zosimovych N, Increasing the Accuracy of the Center of Mass Stabilization of Space Probe with Partially Invariant System 2017 *Science and Education, a New Dimension. Natural and Techn. Science*, Vol. 14, Issue 132 **105** 08

[5] Zosimovych N, Raj A N J. Synthesis of Stabilization Algorithms in the System Controlling Rotations of the Operating Device 2017 *«Prospects of World Science -2017», XIII Int. Scientific and Pract. Conf.,* Jul. 30 – Aug. 7 , Vol. 4, Sheffield Science and Education ltd, Yorkshire, England, S1 4LR **14** 20

[6] Zosimovych N, Invariant Stabilization Algorithms in a Control System with Rotating Operating Device 2017 *American Journ. of Eng. Research (AJER)*, Vol. 6, No. 10 **297** 11
[7] Sineglazov V M 2003 *Theory of automatic control*. In 2 Vol.: Manual for students of all specialities, Kyiv, NAU
[8] Malyshev V V, Krasilshikov M N, Bobronnikov V T, Dishel V D 1996 *Aerospace vehicle control*, Moscow, MAI
[9] Andreev V D 1967 *Theory of Inertial Navigation (Aided Systems)*, Izdatelstvo ′Nauka′, Moscow, (Available from the U.S. Department of Commerce, Springfield, Va. 22151), **409**
[10] Dorf R C, Robert Bishop R H 1973 *Modern Control Systems*, Pearson, 12th edition, **807**
[11] Popov E P 1973 *Prikladnaya teoriya processov upravleniya v nelineynyh sistemah*, Moscow: Mashinostroeniye
[12] Greensite A L 1970 *Elements of Modern Control Theory*, New York, Spartan Books
[13] Furman T T 1981 *Approximate Methods in Engineering Design*, Vol. 155: Math. in Science and Eng., Edited by Richard Bellman
[14] Zosimovych N, Xiaodong N, Simeone A, Peng X 2018 SmallSats Rational Design 6th Int. *Conf. and Exhibition on Mech. & Aerospace Eng.*, Nov, 07-08, Atlanta, USA. *Journ. of Aeronautics & Aerospace Eng.*, Nov., Vol. 7, Conf. Series LLC Ltd: 47 Churchfield Road, W3 6AY, London, UK

# Study on the Effects of Atmospheric Parameters on Space Vehicle Design

**Guo Peng-fei[1*], Fei Wang-hua[1] and Xiao Cun-ying[2]**

[1]China Academy of Launch Vehicle Technology, Beijing 100076, China;
[2]National Space Science Center, Beijing 100190, China

[*]Email: gpfmail@126.com

**Abstract.** For the space vehicle flying in the airspace and near-space, the structure safety and attitude control ability are related to the surrounding atmosphere. The distribution characteristics of atmospheric parameters, including temperature, pressure, density and wind field should be studied based on the data statistics. In this article, the spatiotemporal distribution characteristics and measurement methods of atmospheric parameters are summarized, and the effects of atmospheric parameters on the engineering design of ballistic trajectory, propulsion system, aero-thermal heating, and attitude control system are provided. These results have considerable significance in the space vehicle general design.
**Keywords:** Atmospheric Parameters; Space vehicle; Atmospheric measurements

## 1. Introduction

Driven by the strong demands for military operations and civil space transportation, research and development of new type spacecraft has been conducted extensively in the space field world-widely [1]. Recently, many new aerospace vehicles, such as reusable rockets, high-speed cruise vehicles of the United States and European space reentry vehicles, have been developed and carried out flight tests successively. More and more researchers and institutions have already begun studies relevant to space and near-space from the earth's surface to orbital space.

Traditional launch vehicles, missile weapons and space reentry spacecraft always pass through atmosphere for a short time (usually at the speed of above Mach 10). However, the new type of aerospace vehicles fly for a long time (usually at the speed of Mach 3-6) and perform specific maneuvering flight tasks in the 0-30km airspace and 30-80km near-space, where the temperature, pressure, density, wind field and other characteristic parameters of earth atmosphere change sharply along with time and space. Especially, after NASA (National Aeronautics and Space Administration) found that the variability of atmospheric density using the real flight telemetry data of the Space Shuttle, the research on the characteristics of earth atmosphere has attracted extensive attention [2]. Therefore, it is necessary to pay attention to the distribution characteristics and variability range of atmospheric parameters in the design of space vehicle to ensure sufficient control ability and safety margin for the flight process of an aerospace vehicle in the near-space and airspace.

## 2. Atmospheric parameters and measurement methods

The atmosphere studied in this paper refers to the Earth's atmosphere from the launch site ground to a height of 80km, where the new type spacecrafts usually fly. Among them, atmosphere under 15km is

IOP Conf. Series: Materials Science and Engineering **608** (2019) 012027 doi:10.1088/1757-899X/608/1/012027

the troposphere, and contains 75% of the total mass of the Earth's atmosphere. The convective movement in the vertical direction is intense and the parameters such as density and temperature change rapidly with height; atmosphere between 15km and 50km is the stratosphere, where the motion of the air is dominated by the meridional wind component and temperature is fairly well-distributed; atmosphere between 50km and 80km belongs to the mesosphere, where the air is thin, vertical movement of the air is intense and temperature decreases rapidly with increasing altitude; atmosphere above 80km is extremely thin, and is mainly composed of ionized and neutral particles. For aerospace vehicles performing space reentry, high-speed cruising and maneuvering missions, the atmospheric parameters, including temperature, pressure, density, wind field, and other characteristics, have important impact on many aspects of the vehicle including the trajectory design, power system performance and attitude control capability [3].

Since the 1960s, some international organizations and countries have formulated and published a series of atmospheric parameter models, such as CIRA [4], GRAM [5], NASA TM-100697 [6], USSA-1976 [7], and China-GJB366 [8] as shown in Table 1. These atmospheric parameters models focus on different geographical area and altitude range and show the atmospheric variability in month, season or solar activity in the form of graph or program. Among these, the USSA-1976 model of the US standard atmosphere is most widely used. In this model, atmospheric temperature, pressure, meridional wind and zonal wind in different seasons are provided for five latitudes (15,30,45,60,75) in the northern hemisphere. As shown in Figure 1, the atmospheric pressure and density decreases with increasing altitude, but the atmospheric temperature decreases from the ground surface to 10km, and then increases from 10km to 50km, after that, decreases with increasing altitude. In USSA-1976 atmosphere model, the atmospheric density is determined by ideal gas equation using its temperature and pressure according to latitude, longitude and height. Namely, $\rho = P/RT$ where R is the specific gas constant of air and R=287.05287J/(K*kg).

**Table 1.** Partial typical atmospheric model

| Atmospheric Model | Geographical Area | Altitude Range | Variables | Data Form |
|---|---|---|---|---|
| CIRA [4] | Global | 130-2000km | Season<br>Solar Activity | Graph<br>Program |
| GRAM [5] | Global | 0-2500km | Month | Program |
| NASA TM-100697 [6] | Global | 0-120km | Month | Graph |
| USSA -1976 [7] | Mid-latitude region | -5-1000km | Season<br>Solar Cycle | Graph |
| China-GJB 366 [8] | China | 0-80km | Season | Graph |

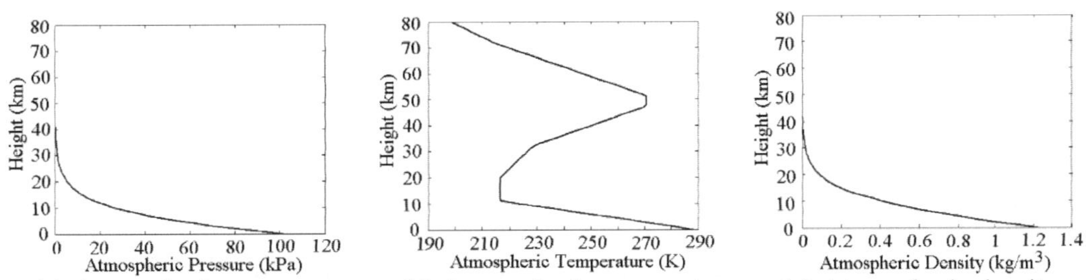


(a) Atmospheric pressure (b) Atmospheric temperature (c) Atmospheric density

**Figure 1.** The USSA-1976 atmospheric model [7]

The measurement methods for obtaining atmospheric characteristic data mainly include meteorological balloons, sounding rockets, falling spheres, lidars, meteorological radars and meteorological satellites. The meteorological balloon is a latex balloon carrying radio detector to an altitude of 30km, which transmits the observed temperature, humidity, pressure and wind field data along the way to the ground observatory. The sounding rockets and falling spheres are the field detection tools above 30km, which can detect the atmospheric structure, composition and other characteristic parameters. Near-space Rayleigh scattering lidar mainly uses the principle that the scattering intensity of atmospheric molecules is proportional to the atmospheric density. By measuring and analyzing the intensity of the scattered signal, the atmospheric density, pressure and temperature at altitudes of 30-70 km can be obtained. The meteor radar uses the reflection signal of the meteor trails on the ultra-high frequency radio waves to detect the atmospheric horizontal wind field and meteor flux in the height range of 70-110km. Meteorological satellites operate at low Earth orbit (LEO) and geostationary orbit (GEO) and obtain different locations and altitude atmospheric parameters by electronic and electromagnetic detection.

In addition, some aerospace vehicles carry the Flush Air Data System to measure and calculate more accurate flight states and atmospheric parameters in real time such as angle of attack, sideslip angle, Mach number, dynamic pressure and so on. For example, the GRAM-1999 model of the United States uses the atmospheric measurements from 32 actual flights of Space Shuttle as the reference.

## 3. The effects of atmospheric parameters on design of aerospace vehicles

The changes of the atmosphere are complex, because it is influenced by many factors such as longitude, latitude, altitude, season, day and night, irregularity of gravity waves, solar activity, cosmic rays and geomagnetism [8]. Figure 2 shows the wind speed data of 72-80km in the Yinchuan Airport in West China measured by the all-sky meteor radar (belonging to the China National Space Science Center). It can be seen the wind speed varies with day, night and altitude.

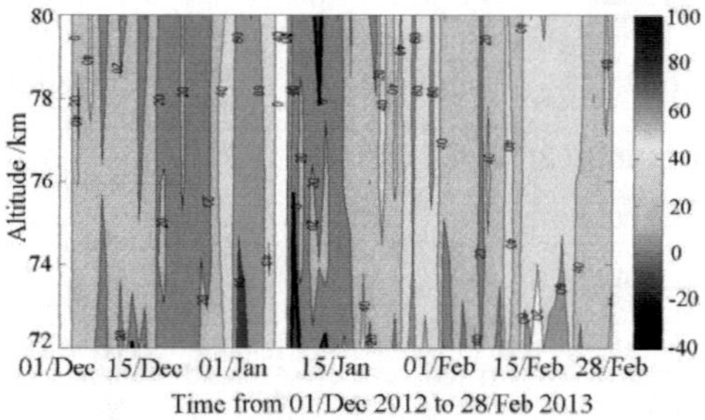


**Figure 2.** Wind speed data of 72-80km height in the Yinchuan Airport in West China

For aerospace vehicles which mainly carry out missions in 0-80km altitude airspace and near-space, it is necessary to pay more attention to the influence of characteristic parameters including atmospheric temperature, pressure, density and wind field on the design of the aircraft system, and determination of the deviation range of the main parameters so as to ensure that the aircraft has sufficient attitude control ability and flight safety.

### *3.1. Impacts on flight trajectory design*

For a long-time cruising aerospace vehicle in high altitude atmosphere, the effects of atmospheric temperature, density, pressure and wind field and their deviations should be considered in the flight trajectory design.

The variations of atmospheric pressure and temperature change the engine thrust and further affect the flight overload and trajectory; the deviation of atmospheric density change the aerodynamic drag value of the aircraft, which affect the flight speed and the track inclination, and reduce the accuracy of the drop point; the change of atmospheric wind field cause the change of flying attack angle and sideslip angle, which affect the flight attitude and direction; changes in atmospheric temperature cause changes in the flight Mach number, leading to the difference of the interpolation results of aerodynamic forces and aerodynamic moments coefficients and affecting flight overload; the emergence of the gust cause a sharp change in the angle of attack and the side slip angle, which shift the flight path.

During the actual flight of the spacecraft, there is a certain degree of difference between standard atmospheric model and actual the atmospheric parameters values including temperature, density, pressure and wind field. However, the deviation of observations from the standard atmosphere mode can be obtained by the probability distributions of the parameters in different seasons and altitudes. Then the envelope range of flight trajectory can be obtained by deviation trajectory design and trajectory model correction considering the effect of various factors, and the accuracy of flight and landing point prediction can be improved. In order to ensure the smooth execution of key flight actions, the effects of atmospheric parameters on the state parameters of key design points, including the angle of attack, sideslip angle, Mach number and dynamic pressure of aircraft, need special attention and evaluation, especially for those critical trajectory points of abrupt movements, such as space reentry, stage separation, fairing abandoning, engine restart, and maneuvering steering.

### *3.2. Impacts on power system design*

For traditional solid or liquid rockets, the thrust needs to be corrected for high altitude conditions (especially vacuum) in the design of the power system, because the oxidant and fuel used are self-contained and the thrust performance is mainly affected by the jet state under high atmospheric pressure environment.

However, for an aerospace vehicle with a new type of inspiratory power system, which only carries fuel and needs oxygen combustion in the atmosphere to provide thrust, the characteristic parameters such as pressure, temperature, density and wind field of the atmosphere have a great impact on its intake effect, fuel supply calculation, judgment of ignition status and stable working boundary. Therefore, the atmospheric parameter model needs to be pre-bound or solved in real time by the atmospheric measurement system.

Figure 3 shows the deviations of the atmospheric temperature (compared with the USSA-1976 standard atmospheric model, the same as Figures 4 and 6 based on the data of TIMED/SABER satellite [9] for 13 years from January 2002 to January 2015 in the Yinchuan Airport in West China. In Figure 3, the same lines (2 lines) with the same legend show the maximum and minimum parameter values in the same season, that is similar to Figures 4 and 6. For high dynamic suction aerospace vehicle with high-speed and low-altitude, the fuel control needs to pay special attention to the atmospheric temperature and its deviation range at different altitudes to improve the cruise distance.

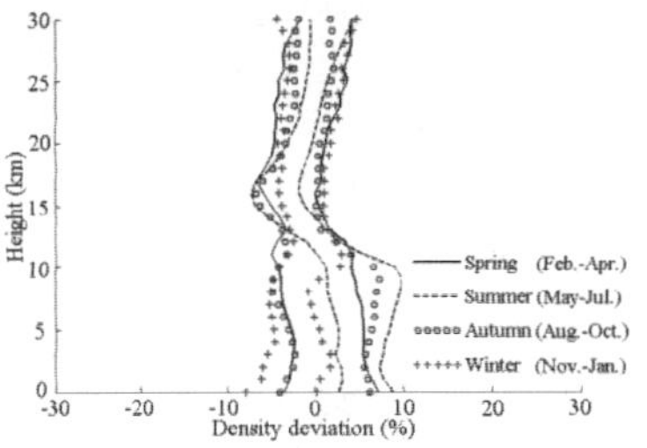


**Figure 3.** Deviation distributions of the four seasons atmospheric temperature in the Yinchuan Airport in West China

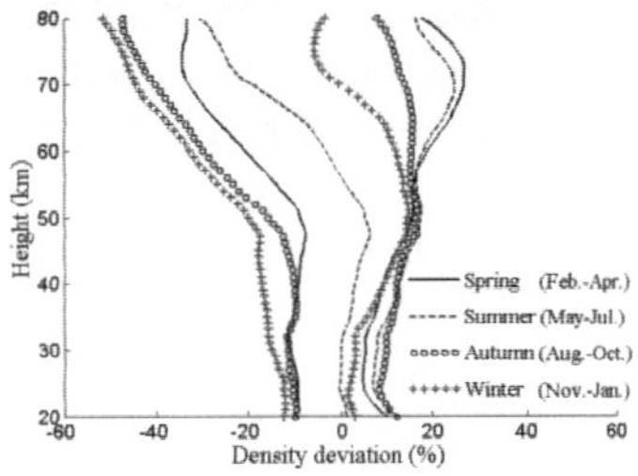


**Figure 4.** Deviation distributions of the four seasons atmospheric density data in the Yinchuan Airport in West China

### *3.3. Impacts on pneumatic thermal environment design*

When the aerospace vehicle performs high-speed cruising or space re-entry at a height of 0-80km, its head stagnation point and wing leading edge position need to withstand up to MW/m$^2$-level pneumatic heating, and the maximum temperature can reach above 1500°C. Therefore, the aircraft's nose, windward side, leeward side, tail end, wing and moving control parts must be protected with high temperature resistant materials or covered with thermal protection system. The maximum heat flow

IOP Conf. Series: Materials Science and Engineering **608** (2019) 012027 doi:10.1088/1757-899X/608/1/012027

and maximum heat load conditions need to be evaluated according to the flight trajectory in the design of pneumatic thermal environment. The atmospheric density has the most direct effect on pneumatic heat. Figure 4 shows the distribution of the deviations of the atmospheric density calculated from the atmospheric density data of TIMED/SABER satellite in the Yinchuan Airport in West China. It can be seen from Figure 4 that the summer atmospheric density is obviously higher than other seasons, which is unfavorable for aircraft reentry and long-time cruise flight. This is consistent with the US Space Shuttle's ban that does not allow summer reentry and return from the Arctic. Moreover, during the US Space Shuttle flight, there have been many dramatic changes in aerodynamic drag caused by fluctuations in atmospheric density, resulting in frequent fluctuations in flight angle of attack [10-11].

### *3.4. Impacts on attitude control system design*

The attitude control system design of aerospace vehicle needs to reasonably control the engine swing angle and the movable rudder angle to ensure the stability and controllability of the aircraft attitude during the whole flight process. The irregular distribution of the atmosphere in space and time and the effect of gust [12] cause the overall sudden change of aircraft's aerodynamic drag and rudder surface control force, which affect the important flight state parameters such as angle of attack, sideslip angle, Mach number and dynamic pressure, even cause the aircraft to become unstable and the mission failure.

When the attitude control system carries out attitude balance and six-degree-of-freedom flight simulation, it is necessary to evaluate the control capability of the active rudder and engine swing. The distribution of wind field with height has an important influence on its anti-interference ability. Figure 5 shows the distribution of 95% probability of the maximum wind speed with altitude based on the four seasons wind field data of TIMED/SABER satellite in the Yinchuan Airport in West China, which was calculated by the integrated vector wind profile method [13]. It is demonstrated by Figure 5 that within the height range of 0-30km, the maximum wind speed gradually increases to the maximum value from the ground to the height of about 11km, then the maximum wind speed decreases gradually until the height of about 18km, and then increases with the height again. Among them, the winter wind field is more severe than other seasons, so the attitude control design of the aircraft is the most severe in winter. This is the main reason why the spacecraft attitude control system usually adopts the winter wind field as the design condition.

### *3.5. Impacts on design of structural and institutional system*

In aerodynamic calculation, the influence of dynamic pressure on aerodynamic system design is avoided by using aerodynamic coefficients, so the influence of atmospheric parameters on aerodynamic system design is small. However, in the design of the structure and mechanism system, the aerodynamic coefficient needs to be converted into the load condition of the structure, and the influence of the atmospheric parameters cannot be ignored.

The airframe structure and active control components need to bear larger aerodynamic pressure in high-speed flight, while the cabin pressure is closer to the atmospheric static pressure at its altitude [14]. In the design of the aircraft structure and mechanism system, it is necessary to consider the load changes caused by atmospheric fluctuations in the flight state, to ensure the structural strength and safety, and the activity control mechanism has sufficient actuation capability. Therefore, in addition to obtaining the external aerodynamic pressure through wind tunnel test and numerical simulation, it is also crucial for the structure and mechanism system design to accurately grasp the atmospheric environment of the space vehicle flight area. Figure 6 shows the deviation of the atmospheric pressure calculated from TIMED/SABER satellite data for four seasons in the Yinchuan Airport in West China. The results show that the atmospheric pressure in summer is higher than that in winter. In addition, considering the influence of day and night changes and geographical location, the structure and mechanism system design of space vehicle should retain enough safety margin to ensure flight safety.

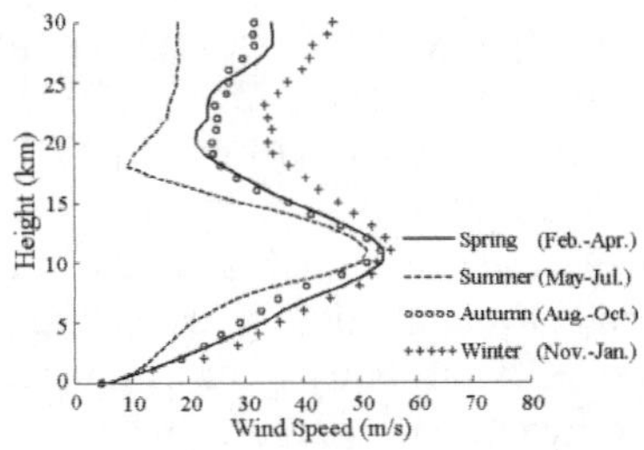


**Figure 5.** Maximum wind speed distribution in the Yinchuan Airport in West China

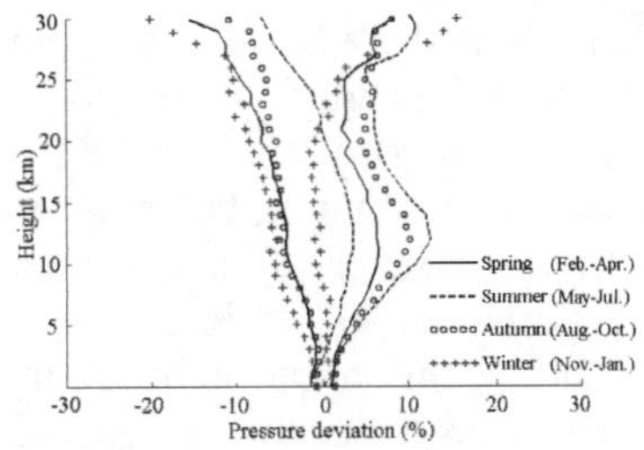


**Figure 6.** Deviation distributions of the four seasons atmospheric pressure data in the Yinchuan Airport in West China

### *3.6. Impacts on the Design of Thermal Control System*
The thermal control system of aerospace vehicle is designed to protect the temperature of important single aircraft and equipment which are sensitive to temperature in the cabin by using active and passive temperature control method. In order to reduce the spacecraft structure's load caused by the pressure difference between the interior and exterior during cruise flight at high speed and high dynamic pressure, the passive flushing and exhausting system is often used to enable the surrounding atmosphere to enter and exit the spacecraft cabin structure [14]. The atmospheric temperature gradually decreases with increasing altitude under 10km. For example, the atmospheric temperature at the height of 10km is reduced by about 60°C compared to the ground as shown in Figure 1. During the return of the aircraft to the ground, the cold air gradually enters the aircraft cabin under pressure, which results in the temperature drop in the cabin. In the design of the thermal control system, it is necessary to evaluate the effects of atmospheric temperature, density and intake and exhaust volume at different altitudes on the cabin temperature, especially on the temperature sensitive equipment.

### *3.7. Impacts on measurement system design*
The measurement antennas and windows of the flight measurement system are usually arranged on the side and back of the aerospace vehicle to avoid the influence of intense aerodynamic heating. However, the drastic changes of atmospheric temperature, density, pressure and other parameters lead to frost phenomenon in the measurement module. It is necessary to reduce the influence of atmospheric moisture by blowing, heating or dehumidifying. And for high-precision optical measurement and imaging equipment, the temperature gradient caused by the combination of atmospheric temperature and aerodynamic heating changes the curvature of the optical lens, which affects the imaging effect. It is necessary to design its temperature control system separately or to correct the deviation of the imaging results. In addition, the deviation of atmospheric temperature and pressure affects the atmospheric refraction factor, and the propagation speed of electromagnetic wave in the atmosphere will change, leading to the deviation of radar detection distance, angle and height.

## 4. Conclusions
Compared with traditional spacecraft represented by launch vehicles and missile weapons, and low-altitude aircraft represented by passenger and military fighters, aerospace vehicles should be capable of long-term cruising and maneuvering at a high speed at altitudes of 0-80km. The temperature, density, pressure and distribution of the surround atmosphere in space and time have a great influence on the design of various systems of aerospace vehicles, which is one of the important causes of aerospace accidents.

This article dedicates to the study on the effects of atmosphere parameters on the engineering design of ballistic trajectory, propulsion system, aero-thermal heating, and attitude control system of the new type spacecrafts, which usually fly in the 0-30km airspace and 30-80km near-space. Based on the temperature, pressure and density data analysis from TIMED/SABER satellite, and wind speed from the all-sky meteor radar, the distribution characteristics of atmospheric parameters at different times are provided. These results have significance in the space vehicle general design.

For the safety of the structure and the success of the flight mission, the distribution and deviation of atmospheric parameters should be considered in the design of each system on the one hand. On the other hand, it is necessary to obtain the atmospheric parameter data of launch site, landing field and mission execution area by various ways such as radio, electromagnetic, laser, sounding rocket, satellite, etc. and to enrich and improve the atmospheric model continuously by using the telemetry data in flight test so as to accurately determine the deviation range of each parameter.

**References**

[1] Wang X, Yan J, Yu Y F. Research on hypersonic aerospace vehicle (in Chinese). Measurement & Control Technology, 2006(7): 83-86

[2] Hale N, Jr Lamotte N, Garner T. Operational experience with hypersonic entry of the space shuttle. In: The Proceedings of AIAA/AAAF 11th International Space Planes and Hypersonic Systems and Technologies Conference. Orleans: AIAA, 2002. AIAA-2002-5259

[3] Tian J H, Fan Q F. The importance of study on upper atmosphere and its forecast in space flight (in Chinese). Spacecraft Environment Engineering, 1999,1: 39-46

[4] COSPAR, COSPAR International Reference Atmosphere 1972, Akademie-Verlag, 1972

[5] Justus C, Jeffries III W, Yung S P, et al. The NASA/MSFC Global reference atmospheric model-1995 version (GRAM-95). 1995, NASA TM-4715

[6] Fleming E L, Chandra S, Shoeberl M R, Barnett J J. Monthly mean global climatology of temperature, wind, geopotential height and pressure for 0-120 km. NASA TM- 100697, New York, USA. USA: NASA, 1988

[7] National Oceanic and Atmospheric Administration, National Aeronautics and Space Administration, United States Air Force. U.S. Standard Atmosphere, 1976. Technical Report, 1976. NASA-TM-X-74335

[8] Huang S Y, Wen Y. Reasearch for atmospheric density and wind field from 20km to 85km (in Chinese). Missiles and Space Vehicles, 2017,3: 17-21

[9] Remsberg E, Marshall B, Garcia-Comas M, et al. Assessment of the quality of the Version 1.07 temperature-vesus-presure profiles of the middle atmosphere from TIMED/SABER. J Geophys Res, 2008, 113:1641-1653.

[10] Hale N, Lamotte J N, Garner T. Operational experience with hypersonic entry of the space shuttle. In: the Proceedings of AIAA/AAAF 11th International Space Planes and Hypersonic Systems and Technologies Conference. Orleans: AIAA, 2002. AIAA-2002- 5259

[11] Chen M K, Du T, Hu X, et al. Effect of atmospheric parameter oscillation at high altitude in the northern hemisphere for near-space hypersonic flight aerothermodynamic prediction (in Chinese). Chin Sci Bull, 2017, 62: 1402–1409.

[12] Liu F H, Ma X P. Response for 2D airfoil due to gust with cubic nonlinearities (in Chinese). Computer Simulation, 2012,11: 122-125

[13] Zhao R L, Chen Z G, Fu W X. Wind shear and rocket design (in Chinese). Journal of Astronautics, 1998,2: 105-108

[14] Guo P F, Liu J P, Chen L. Internal pressure prediction of hypersonic vehicle (in Chinese). Journal of Astronautics, 2016,1: 55-60

IOP Conf. Series: Materials Science and Engineering **608** (2019) 012028 doi:10.1088/1757-899X/608/1/012028

# Research on Sinking Movement of Liquid Propellant in Space Vehicle

**Guo Peng-fei[1*], Xie Hai-ou[1], Zhou Bing-hong[2] and Zheng Xiao-ping[3]**

[1]China Academy of Launch Vehicle Technology, Beijing 100076, China;
[2]National Space Science Center, Beijing 100190, China;
[3]School of Aerospace, Tsinghua University, Beijing 100084, China

[*]Email: gpfmai@126.com

**Abstract.** Considering the requirement of rocket engine restart on the liquid propellant location, this paper dedicates to the simulation research on sinking movement of liquid propellant in space vehicle. The extrusion, thrust and structural types of liquid propellant management for space vehicle are summarized. The numerical simulation and drop tower test are compared to study how the liquid propellant moves in the gravity environment, and the relationship between the sinking time and flight acceleration for a new-type space vehicle are provided. These results have considerable significance for the liquid propellant management and acceleration design of space vehicle.

## 1. Introduction

With the rapid development of aerospace technology and space military demand, various new space vehicles are being developed to seize the space interests and the initiative of future wars. New space vehicles represented by the United States X-37B and X-51 will be able to perform complex missions in outer space, earth orbit space, sub-orbital space and airspace in the near future, including orbit transfer, detection, reconnaissance, strike and so on [1-2].

Complex flight trajectory requires that the power system of the new space vehicle can restart the engine many times according to the requirements of mission and control. However, in the microgravity environment after engine shutdown, the liquid propellant will present complex spatial motion in the tank. Once the propellant delivery port is exposed to the pressurized gas in the tank when the engine restarts, it will cause the propellant entrained air, which will affect the engine start, and even cause the engine structure damage and explosion [3-4]. Therefore, the management of liquid propellant before restart of power system is of great significance to enhance the safety and maneuverability of spacecraft, which is one of the key technologies for the development of new spacecraft.

In order to ensure the liquid propellant can cover the conveyor port at the tank's bottom when the engine starts, many sinking management types of liquid propellant for spacecraft are applied, such as the extruded type, the forward push type and the tank structure scheme [5]. Extruded propellant management uses high pressure gas to extrude liquid propellant into engine by designing metal or non-metal film in tank. It is often used in cruise missiles for short flight. However, due to the compatibility and fatigue between film and propellant, this method is not suitable for low temperature propellants such as liquid hydrogen and liquid oxygen. The film extrusion propellant supply tank is shown in Figure 1. Forward propellant management uses the overload generated by the thrust of small solid

IOP Conf. Series: Materials Science and Engineering **608** (2019) 012028 doi:10.1088/1757-899X/608/1/012028

rocket or attitude control engine to sink liquid propellant into the conveyor port. It is mainly applicable to the sinking management of medium and large liquid rockets, such as Saturn V in USA (as shown in Figure 2), Ariane-5 in Europe, H-2A in Japan and Long March-3 in China. Tank structured propellant management uses mechanical conveyors, current storage devices or compartments to ensure liquid propellant sinking. The design of piston, collector and other devices is common in liquid rocket and missile in 1950s (as shown in Figure 3), while the design of multilayer compartment combined with one-way valve is adopted in American space shuttle and X-34 aircraft. However, the structure and executing mechanism of tank-structured propellant management are complex and heavy.

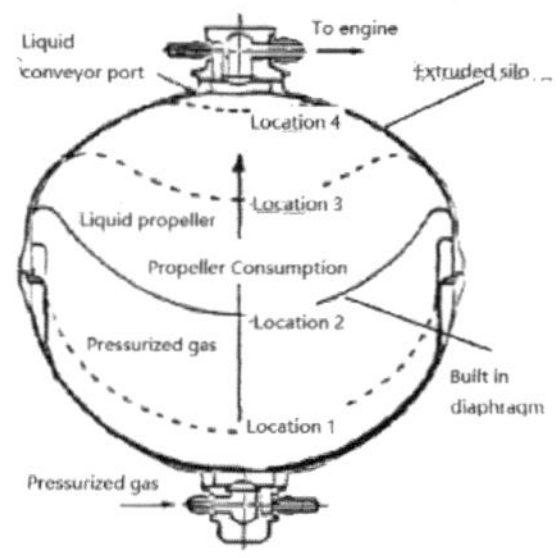


**Figure 1.** Film extrusion propellant supply tank

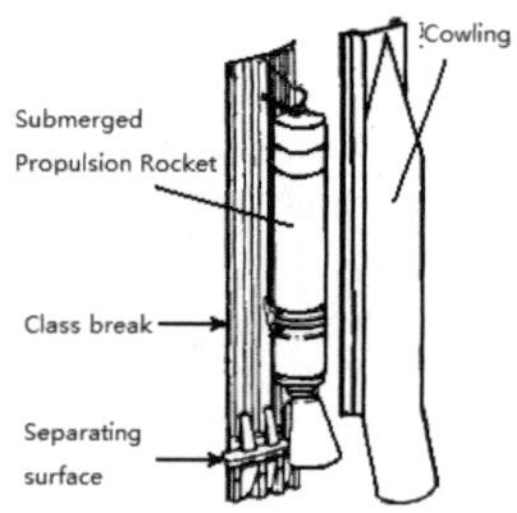


**Figure 2.** S-IVB propellant sunken rocket of Saturn V

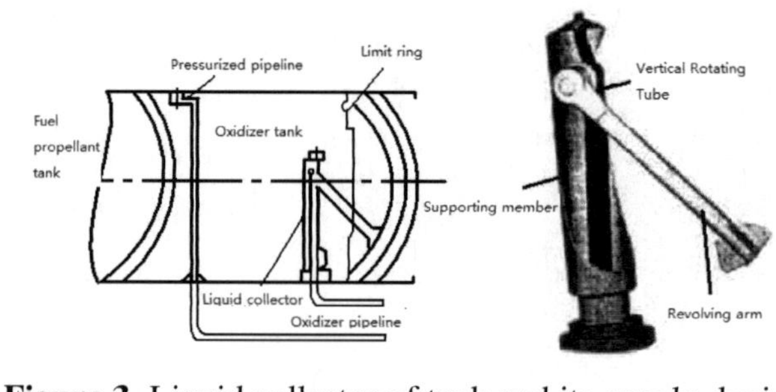


**Figure 3.** Liquid collector of tank and its supply device

With the well development and successful application of liquid hydrogen and liquid oxygen cryogenic propellant technology in the field of space transportation and weapon equipment, and the increasing size of space vehicles, the design of forward liquid propellant sinking management has become an important factor to improve the maneuverability of new space vehicles. The key is to design reasonable boost overload and fuel consumption to ensure that liquid propellant sink to the bottom and cover the conveyor port in a certain time after overcome the surface tension and sloshing effects. Larger thrust will increase fuel consumption and reduce payload, while smailler thrust will not meet the sinking needs. Therefore, the analysis of the sinking process is very important to ensure the smooth restart of the spacecraft engine, which can provide important reference for the selection and design of bottom sinking boost.

In this paper, the Volume of Fluid [6-8] (VOF) method is used to simulate the motion of liquid propellant under overload conditions for the scaled model of American Centaurus Liquid Hydrogen

Tank. The reliability and accuracy of the numerical simulation are verified by the drop tower test results. Then, according to the flight mission characteristics of a space vehicle, the relationship curves of settling time of liquid oxygen propellant under different overload conditions are obtained, which provides an important reference and basis for the selection and design of bottom sinking boost.

## 2. Liquid motion simulation under overload condition

The volume of fluid (VOF) method is a Euler numerical method for free surface tracking, in which the free surface variation is described by tracking the volume fraction of the mesh element occupied by the fluid volume. When the ratio of fluid volume to mesh volume is 1, the mesh is filled with fluid, when the ratio is 0, the mesh does not contain fluid, and when the ratio is between 0 and 1, the mesh contains free surface.

When it is assumed that the fluid is incompressible viscous flow and its flow pattern is laminar flow, then the fluid continuity equation, momentum equation and volume function conservation equation of fluid motion process can be expressed respectively as follows:

$$\begin{cases} \nabla \cdot \vec{V} = 0 \\ \rho(\dfrac{\partial \vec{V}}{\partial t} + \vec{V} \cdot \nabla \vec{V}) = -\nabla p + \mu \Delta \vec{V} + \rho \vec{a} + \vec{F}_\sigma \\ \dfrac{\partial f}{\partial t} + \vec{V} \cdot \nabla f = 0 \end{cases} \quad (1)$$

where, $\vec{V}$ is the velocity, $\rho$ is density, $p$ is pressure, $\vec{a}$ is overload, $\mu$ is kinematic viscous coefficient, $f = f(x, y, z, t)$ is volume function and $\vec{F}_\sigma$ is equivalent volume force form of surface tension of fluid. The surface tension of the fluid is calculated using the Continuous Surface Force (CSF):

$$\vec{F}_\sigma = \sigma k \nabla f \quad (2)$$

Where, $\sigma$ is the surface tension coefficient of the fluid and $k$ is the interfacial curvature.

In addition, when it is assumed that the boundary condition of fluid motion is that the velocity of fluid motion satisfies the non-slip condition on the wall of the tank and that the fluid and the wall of the tank are completely infiltrated, the following equations hold:

$$\begin{cases} \vec{V} = 0 \\ \theta_c = 0 \end{cases} \quad (3)$$

In order to analyze the liquid propellant moving process of the American Centaurus Liquid Hydrogen Scaled Tank [9-11] under overload conditions, the 3D modeling and simulation analysis were carried out by using the VOF method of FLUENT software. The scaled tank has a cylindrical section in the middle and two revolving ellipsoids with 11cm long axis, 1.38 long and short axis ratio at the top and bottom. The tank height is 21.8 cm and the radius is 5.5 cm. The established 3D mesh distribution is shown in Figure 4.

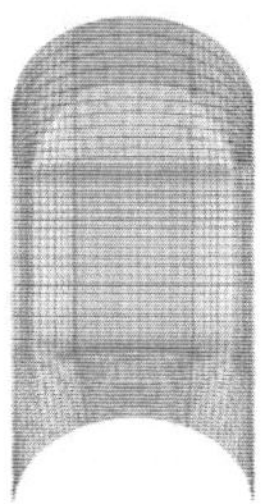

**Figure 4.** The Scale Model and Its Mesh Distribution of American Centaurus Liquid Hydrogen Tank

The liquid volume of American Centaurus Liquid Hydrogen Scaled Tank is 70%, and the initial distribution of liquids is shown in Figure 5. In this state, the liquid is all located at a front bottom away from the bottom transfer port of the tank. The liquid propellant sinking is considered under the overload conditions, the simulation analysis is carried out, and the moving process is shown in Figure 5.

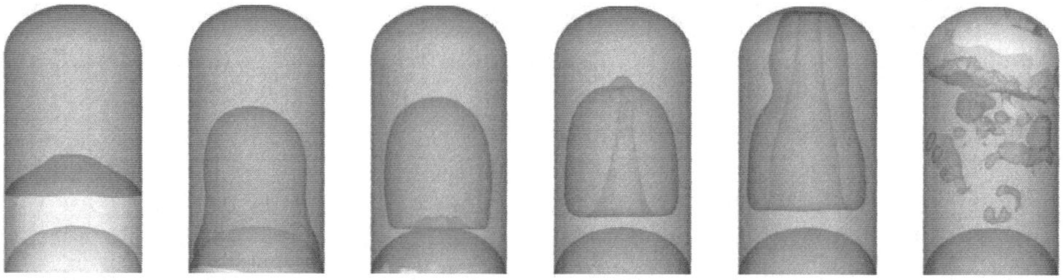

a) 0s b) 0.51s c) 0.61s d) 0.76s e) 1.11s f) 2.82s

**Figure 5.** The liquid moving process of American Centaurus Liquid Hydrogen Scaled Tank

It can be seen from Figure 5 that the liquid propellant sinking movement under overload mainly includes 6 typical process states:

1) T1=0s, initial state of liquid propellant;
2) T2=0.51s, the liquid flows back from the front bottom along the wall surface of the tank and contacts the back bottom;
3) T3=0.61s, the liquid flows from the back bottom edge to the center and squeezes each other to form upward liquid flow;
4) T4=0.76s, the liquid flow moves upward to reach the gas-liquid interface;
5) T5=1.11s, the liquid flow continues to move upward to the top bottom of the tank;
6) T6=2.82s, the liquid flow falls back, the front bottom is occupied by the gas, and then the liquid sloshing at the back bottom until it is stable.

The comparison between 3D simulation analysis results by VOF method and drop tower overload test of American Centaurus Liquid Hydrogen Scaled Tank is shown in Table 1. It can be seen from Table 1 that the maximum error is 11%, which can meet the requirement of the design of propellant sinking management. The reliability and accuracy of the numerical simulation of liquid motion in spacecraft tank using VOF method are verified.

**Table 1.** Comparison of simulation and test results of liquid moving process

| | T1(s) | T2(s) | T3(s) | T4(s) | T5(s) | T6(s) |
|---|---|---|---|---|---|---|
| Numerical simulation | 0 | 0.51 | 0.61 | 0.76 | 1.11 | 2.82 |
| Drop tower test | 0 | 0.46 | 0.58 | 0.71 | 1.04 | 2.68 |
| Error ratio | - | 11 % | 5 % | 7 % | 7 % | 5 % |

## 3. Motion Analysis of Liquid Sinking for Space Vehicle

Compared with traditional launch vehicles and missile weapons, new space vehicles perform more complex maneuvering flight and steering adjustment. The mission profile of a new type of spacecraft is shown in Figure 6. The boost stage is separated when the high thrust booster rocket reaches a certain height, and the aircraft leaps and turns in the air by using the liquid oxygen kerosene engine, then returns to the primary launch site to complete the whole flight task.

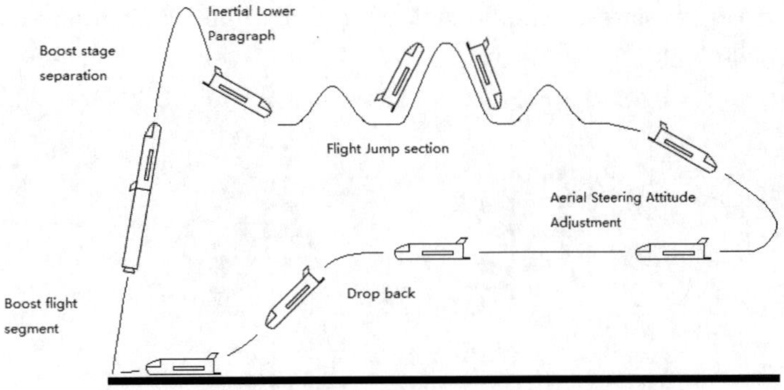


**Figure 6.** Mission Profile of a New Spacecraft

In the flight mission, the aircraft needs to undergo the process of drastic changes in flight overload, which includes boost stage separation, inertial landing, jump flight, air steering and attitude adjustment, descent and return field, etc. Especially in the flight stage when the propellant margin is small and the engine shuts down to restart, due to the inertia of aerodynamic drag and attitude adjustment, the liquid propellant will produce intense motion in the tank, exposing the conveyor port to the pressurized gas. Therefore, in order to ensure the normal restart of the liquid oxygen kerosene engine, it is necessary to analyze and predict the liquid movement process in the propellant tank, and obtain the relationship between overload and sinking time, which provides important reference for the control and power system design.

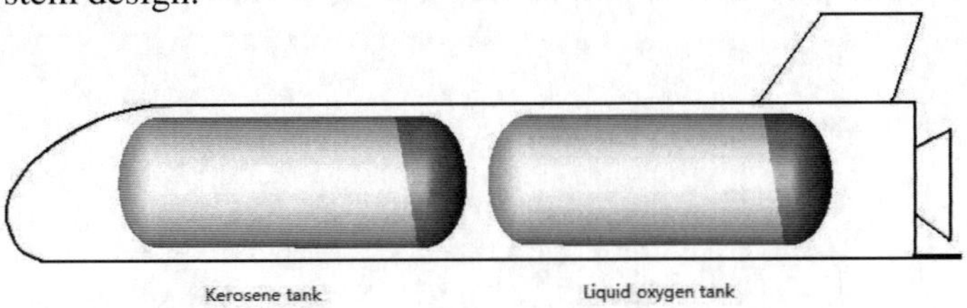


**Figure 7.** Propellant and Tank of a Space Vehicle

For the sinking motion of propellant during the jump flight and the attitude adjustment in the liquid oxygen tank of the spacecraft in Figure 7, the 3D fluid model based on VOF method is used to simulate and analyze the sinking motion. The length and radius of the straight section of the tank are 7 m and 1.44 m. and the depth of the front and back bottom of the tank is 0.4 m. During the process of aircraft steering and attitude adjustment, liquid oxygen flows along the side wall of the tank to the front bottom due to centrifugal force and overload. The distribution of liquid oxygen propellant in the tank is shown in Figure 8 at the initial moment after turning.

(a) Side view (b)Top view

**Figure 8.** Distribution of liquid oxygen in the tank of a spacecraft after steering

With the liquid oxygen propellant distribution at the initial time after the spacecraft's steering and attitude adjustment as the initial condition, if the propellant sinking control is not carried out, that is, there is no flight overload to cause the liquid sinking, the distribution of liquid oxygen in the tank in the next 20 seconds is shown in Figure 9.

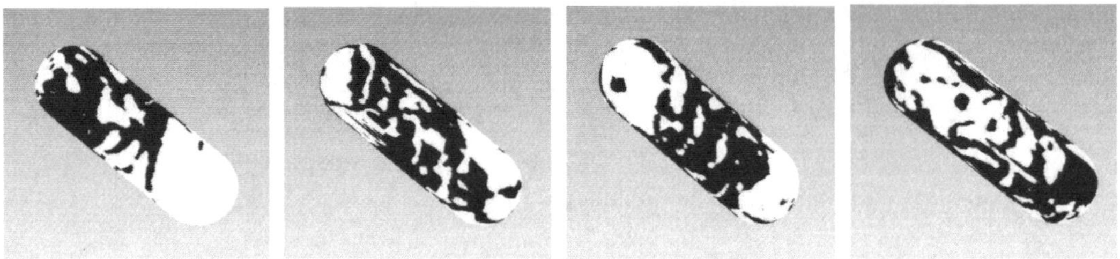

(a) 5s (b) 10s (c) 15s (d) 20s

**Figure 9**. Distribution of liquid oxygen propellant in tank in 20s

It can be seen from Figure 9 that when the liquid oxygen propellant is free to move, its distribution in the tank will become confusing. The propellant delivery port on the back and bottom of the tank is always exposed to the pressurized gas. At this time, the liquid oxygen kerosene engine ignition may lead to the engine abnormal work or even explosion.

When an axial 1g overload is applied to the liquid oxygen propellant distribution state shown in Figure 8, the distribution state change process is as shown in Figure 10. The liquid oxygen first flows along the wall surface to the rear bottom, and then rushes from the other side wall surface due to inertia. The process is repeated until the kinetic energy is consumed and sloshing to stop, the liquid propellant covers the transfer port, and bubbles in liquid are gradually precipitated.

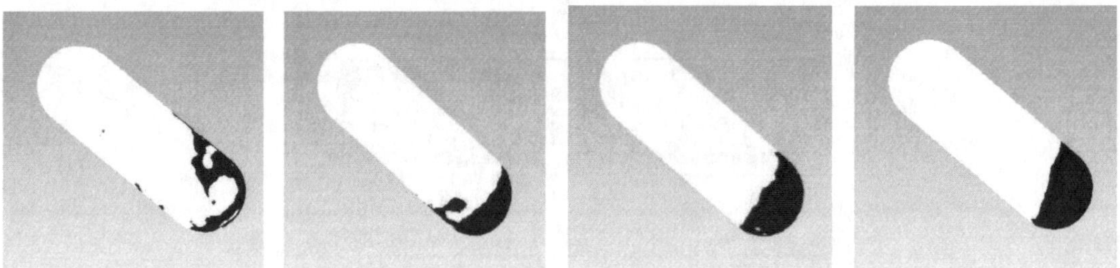

(a) 1.5s (b) 3s (c) 4.5s (d) 6s

**Figure 10.** Distribution state change process of liquid oxygen propellant after 1g overload

When the sinking time is defined as the time span from the initial time to the time when the liquid oxygen propellant completely covers the rear bottom ports without air entrainment and the sloshing is stable. The relationship between sinking time and sinking overload can be obtained by applying different sinking overload to the distribution state of liquid oxygen propellant shown in Figure 8, and the result is shown by the solid line in Figure 11. In addition, since the consumption of the sinking booster fuel is proportional to the sinking overload and the sinking time, the equivalent consumption curve of the sinking booster fuel can be obtained as shown by the dashed line in Figure 11.

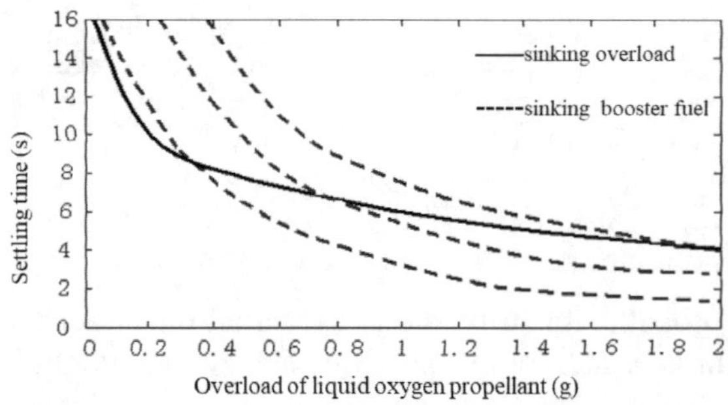


**Figure 11.** The relationship between settling time and overload of liquid oxygen propellant

IOP Conf. Series: Materials Science and Engineering **608** (2019) 012028 doi:10.1088/1757-899X/608/1/012028

It can be seen from Figure 11 that a larger sinking overload can cause the liquid oxygen propellant to sink faster, but consume more sinking boosting fuel. Smaller sinking overload is more economical, but the sinking time is longer. Therefore, the sinking overload needs to be determined according to the flight mission and maneuverability design requirements of the aircraft. The above results can provide important reference for the overall capability, attitude control and ballistic design of the new space vehicle.

## 4. Conclusions

Liquid propellant sinking movement in the tank of space vehicle is studied in this paper. The VOF method based liquid motion simulation under overload condition is developed. The reliability and accuracy of the VOF based method to simulate the sinking process under overload is verified by comparing simulation results with the results of the drop tower overload test of American Centaurus Liquid Hydrogen Scaled Tank. The sinking time of the propellant under different overload conditions is obtained according to the initial distribution of the liquid oxygen propellant of a spacecraft after air steering and attitude adjustment. The results indicate that smaller sinking overload has more economical sinking boost propellant consumption, but requires longer sinking time. Therefore, in the design of the engine restart after the spacecraft undergoes severe overload changes and attitude adjustment, it is necessary to comprehensively consider the influence of sinking time, fuel consumption, ballistics and other factors to determine the sinking management scheme of the liquid propellant. The research results and methods on the sinking movement of liquid propellant in this paper can be widely used in the design of launch vehicle, missile weapon and new space vehicle using liquid propellant.

## References

[1] Wang X, Yan J and Yu Y F 2006 Research on hypersonic aerospace vehicle, *Meas. & Control Tech.*, **7**: 83-86.

[2] Bettinger R A 2011 Spacecraft demand tasking and skip entry responsive maneuver, *Air Force Inst of Tech Wright-Patterson AFB OH Graduate School of Eng. and Management.*

[3] Ying G L 1980 Control problem of liquid propellant in large launch vehicle, *Struct. & Environ. Eng.*, **1**: 64-89.

[4] Liu C H 1997 Microgravity researches relating to engine restart of spacecraft. *Struct. & Environ. Eng.*, **1**:45-51.

[5] Liao S Y 2005 Liquid propellant management for the hypersoar mobile vehicle in the atmospheric layer. *Aerospace Shanghai*, **2**:34-38.

[6] Saghi H, Ketabdari M J, Booshi S 2012 Generation of linear and nonlinear waves in numerical wave tank using clustering technique-volume of fluid method. *Appl. Math. and Mech.*, **33** 9: 1179-1190.

[7] Hu Q, Li Y, Liang J Q, Liu J T and Song T 2017 Liquid sloshing performance in vane type tank under microgravity. *Aerospace Control*, **5**:19-24.

[8] Deng X Y, Tang Y H, Huang B and Zhang F Z 2009 Numerical simulation of propellant reorientation. *Missiles and Space Veh.*, **1**:11-14.

[9] Jack A S, William J M, and Raymond FL 1973 Low-gravity reorientation in a scale-model centaur liquid-hydrogen tank. *NASA TN D-7166*, 1973.

[10] Chu G M 2007 Propellant Management of cryogenic upper stage during coast. *Missiles and Space Veh.*, **1**:27-31.

[11] Liu Z, Wang L X, Lin H, Lin R, Zhou Y Q 2018 Thrust time-series optimization for reorientation of propellant *Missiles and Space Veh.*, **1**:106-110.

# Research on the Cabin's Local Orientation Cues of Space Station Oriented to Navigation Enhancement

**Zhu Lei[1], Bai Xianglin[2] and Zhang Yanfei[2*]**

[1]Harbin Institute of Technology.150001Harbin.China
[2]Harbin Institute of Technology Robot Group Co.Ltd.150016Harbin.China

[*]Email: zhangyf@hitrobotgroup.com

**Abstract.** The large and complex interior space of multi-cabin space station makes the astronauts easily happened to navigational disorder during the travel. With the need to enhance the navigation ability of the astronauts, based on a method study on space station intra-vehicular navigation simulation including both VR visual simulation and physiological HDBR simulation, multiple different sets of sample scenes are designed for the axial orientation cues of non-node cabin and the local orientation cues of node cabin. Through the experimental data analysis, the influence law of the cabin's local orientation cues of the space station on the navigation task performance and related physiological effects in the cabin are obtained.

## 1. Introduction

The multi-cabin space station has a complex three-dimensional axial space structure, and the astronauts are prone to navigation problems in the cabins, the astronauts mainly use visual cues to conduct space navigation [1], one of the visual environmental factors affecting intra-vehicular navigation is the whole station's interior orientation cues. In this article, the whole station's interior orientation cues refers to the orientation information which is provided by the various elements in the environment, especially the elements that are composed and implied in the space station interior, one of which is the cabin's local orientation cues, mainly refers to local visual vertical cues and axial orientation cues in the cabin.

In 2001, NASA planned the interior colors of American cabin on the International Space Station in the <International Space Station Interior Color Scheme>, clearly pointing out the main purpose was to provide visual orientation cues [2]; In 2010,NASA proposed that establishing a stable cabin local visual vertical orientational clue could shorten the adaptation time of spatial orientationin [3]; in addition, the stable cabin's local visual vertical orientation cues has been used as the default condition in the intra-vehicular navigation related research [4-7].

It can be seen that, NASA has formed a relatively complete solution through relevant documents and practical applications for the non-node cabin's local visual vertical orientation cues. However, for the non-node cabin's axial orientation cues and the node cabin's local orientation cues, the current approach adopted by NASA is relatively single and lacks relevant basis. Therefore, based on a system of space station intra-vehicular navigation simulation including both VR visual simulation and physiological HDBR simulation, this paper studies the influence law of the non-node cabin's axial orientation cues and the node cabin's local orientation cues on the intra-vehicular navigation task performance and associated physiological effects, thus to provide an effective reference for enhancing the intra-vehicular navigation capability of astronauts.

## 2. The space station intra-vehicular navigation simulation system

In related studies, the VR simulation method is used to simulate the astronauts' visual navigation in multi-cabin space station's interior [4-7], and the physiological HDBR simulation is a mature method for simulating physiological effects of weightlessness [8-9], ttherefore the VR visual simulation and physiological HDBR simulation are combined as the space station intra-vehicular navigation simulation system, as shown in Figure 1. This system is composed of the graphics workstation computer 1, the wide-field head-mounted display 2, the three-dimensional mouse 3, and the angle adjustable experimental bed 4.

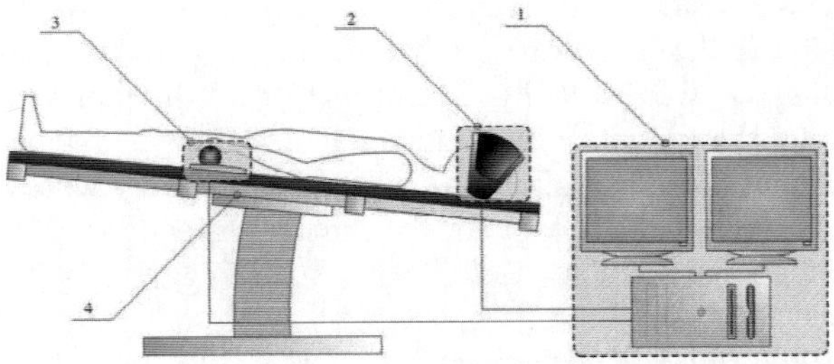


**Figure 1.** Schematic diagram of the space stationintra-vehicular navigation simulation system

The simulation system's scenario of the virtual space station refers to the structure of the international space station and is slightly modified, as shown in Figure 2. The virtual space station consists of two cubic node cabins (Node1, Node2) which are 2×2×2m$^3$ in size and seven rectangular non-node cabins which are 2×2×6m$^3$in size. The interior surface of the cabin is mapped with pictures of the international space station interior, each cabin is distinguished by different visual signs and given unique name.

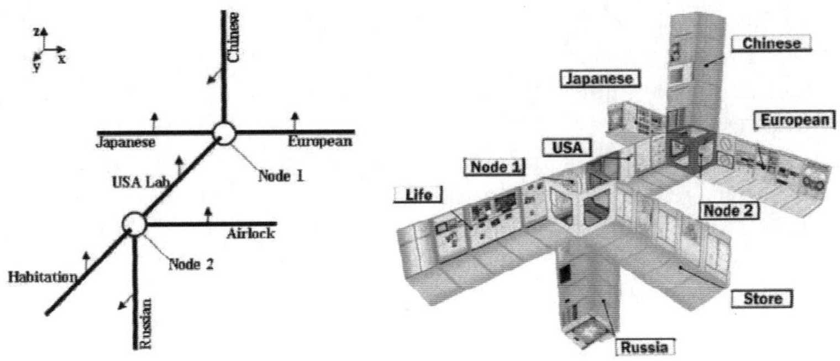


**Figure 2.** Virtual space station scenario

The layout and visual signs of the space station cabin remain unchanged but the internal visual design is different. The whole orientation cues in the interior decoration of each cabin are arranged in the gap position of the wall cabin.

## 3. The experiment

### *3.1. The experimental subjects*

A total of 60 (18-28 years old) subjects participated, all without visual or auditory impairment. The subjects were asked to participate in the mental rotation test and visual rotation ability test before the experiment. All the subjects were divided into 6 groups, with 10 people in each group. The distribution of gender and test scores in each group of subjects was balanced.

### *3.2. The experimental process*

*3.2.1. The preparation stage of experiment.* The subjects were placed on a supine position with the - 6° head-down posture and weared a wide-view helmet-mounted display, kept this pose for a period of time; then, the simulation program of the virtual multi-cabin space station scene in the graphics workstation was started. The subjects operated the three-dimensional mouse to realize the roaming in the multi-cabin space station and to check the equipment availability.

*3.2.2. The navigation training stage of experiment.* In the training part, the subjects firstly followed the simulated astronauts walking along nine preset path in the space station, they went through three rectangular cabins and two node cabins that required the observation orientation of the subjects be consistent with the simulated astronaut. The subjects secondly passed through each cabin of the virtual space station, they must used the keyboard or handle to switch the orientation to complete the corresponding task. The subjects were asked to select one of all the paths on the screen and to go to the cabin on their own under the guidance of the simulated astronaut. After reaching the destination, the subjects needed to return to the starting point, if it was wrong, they were asked to try it again. The subjects visited each cabin three times (except for the American laboratory in the middle of the space station—all paths would pass through it).

*3.2.3. The navigation testing stage of experiment.* In the testing part, the subjects were asked to perform orientational and path-finding task, they were asked to indicate the orientation of their destination firstly, then to move as fast as possible to complete the path-finding task, and then to indicate the location of the starting point. There were two parts in this testing part, and each part had 9 tests. In each test, the wrong type r indicating the orientation, the response time indicating the destination and the operation time of completing the evacuation task were recorded. In the orientational task, if the orientational was correct, r=r0; or else r=r1.

In addition, the subjects were asked to dictate oral the space station cabin layout after the completion of all tasks, the staff drew a virtual space station layout map according to the description of the subject and confirmed with the subjects whether it matched the description, the error type e in this layout was recorded. If it was exactly true, e=e0; if the cabin sequence (structure) described by the subjects was correct, but the cabin interior orientation was wrong, e=e1; if the cabin sequence (structure) was wrong, e=e2.

In the process of simulated navigation and autonomous navigation, the head of the subjects received EEG continuous real-time monitoring on the head. Before and after the experiment and related nodes, they also received the head resting FMRI scan to obtain the objective information of the physiological state of the brain. Finally, the staff also asked and recorded the symptoms of motion sickness such as dizziness during the experiment.

### *3.3. The design of experimental sample scenes*

*3.3.1. Sample scene design of the non-nodal cabin's axial orientation cues.* Three sets of sample scenes were designed, and each group showed different setting strategy for the non-node cabin's axial orientation cues, as shown in Figure 3. The C1 scheme reflected the absence of any non-node cabin's axial orientation cues setting strategy, the C2 scheme reflected the node cabin's landmark setting strategy, and C3 reflected the whole orientation setting strategy. Before the experiment began, the subjects were informed of the orientation meaning represented by every setting strategy of the non-node cabin's axial orientation cues.

*3.3.2. Sample scene design of the node cabin's orientation cues.* Three sets of sample scenes were designed, and each group showed different setting strategy for node cabin's orientation cues. The D1 scheme reflected the absence of any node cabin's orientation cue setting strategy, the D2 scheme reflected the node cabin's vertical orientation cues setting strategy, and D3 reflected the node cabin's axial orientation cues setting strategy. Before the experiment began, the subjects were informed of the orientation meaning represented by every setting strategy of the node cabin's axial orientation cues.

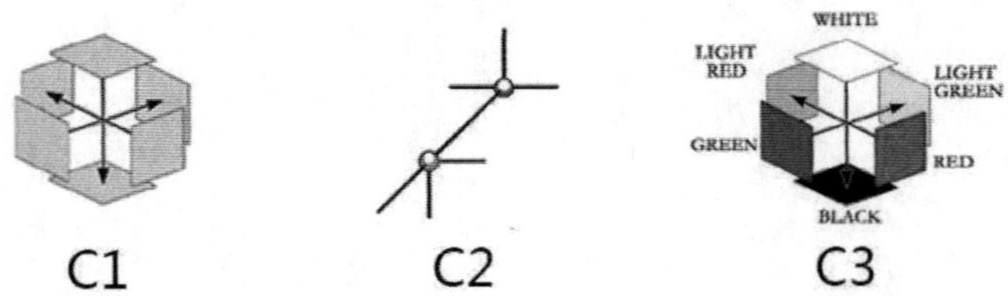


**Figure 3.** A sample scheme showing different non-node cabin's axial orientational cues setting strategies

*3.4. The experimental results and analysis*

*3.4.1. The experimental results of the sample scenes which reflect the non-nodal cabin's axial orientation cues.* Through the statistics of the obtained experimental data, we can obtain the comparison chart of the accuracy rate of orientational task, the average time of path-finding task,the accuracy rate of psychological map task and the proportion distribution of symptoms of motion sickness for different sample scenes which reflect the non-nodal cabin's axial orientation cues.

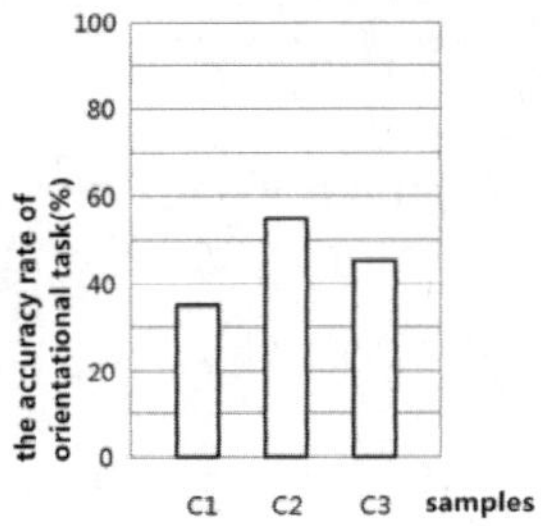


**Figure 4.** Comparison of the accuracy rate of orientational task for different samples

In Figure 4, the accuracy rate of orientational task corresponding to C1 (no setting) was significantly lower than others ($P<0.05$); the accuracy rate of orientational task corresponding to C2 (cabin landmark setting) was significantly higher than C3 (whole orientation setting) ($P<0.05$).

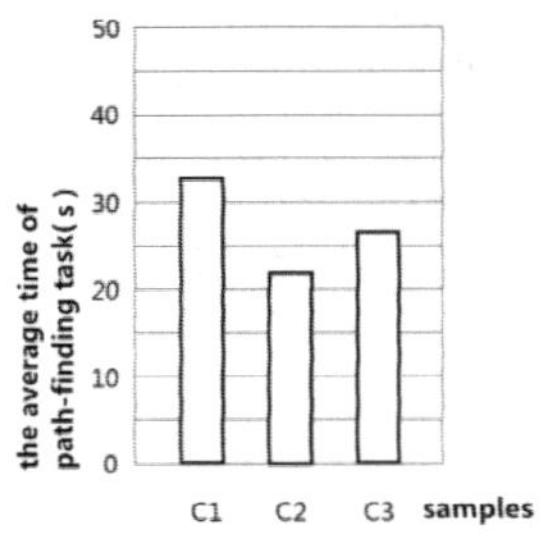


**Figure 5.** Comparison of the average time of path-finding task for different samples

In Figure 5, the average time of path-finding task corresponding to C1 (no setting) was significantly higher than others ($P<0.05$); the average time of path-finding task corresponding to C2 (cabin landmark setting) was significantly lower than C3 (whole orientation setting).

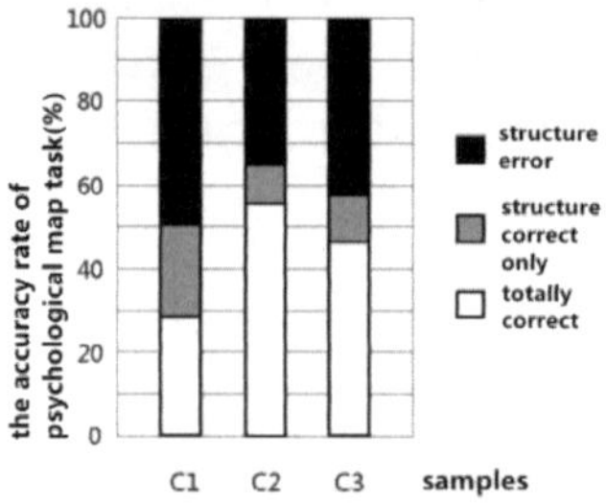


**Figure 6.** Comparison of the accuracy rate of psychological map task for different samples

In Figure 6, the accuracy rate of psychological map task corresponding to C1 (no setting) was significantly lower than others (P<0.05); the accuracy rate of psychological map task corresponding to C2 (cabin landmark setting) was significantly higher than C3 (whole orientation setting) (P<0.05).

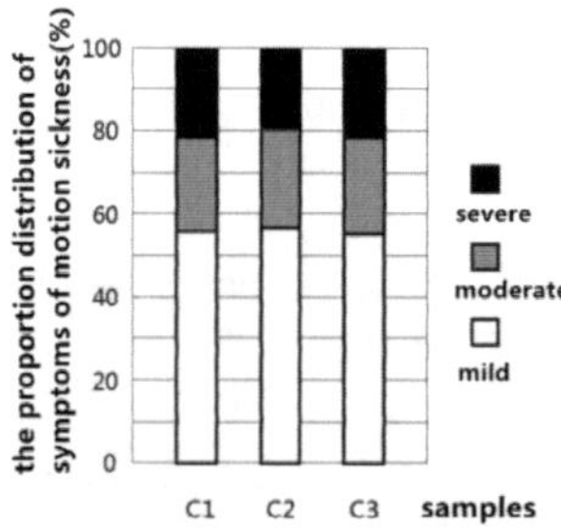


**Figure 7.** Comparison of the proportion distribution of symptoms of motion sickness for different samples

In Figure 7, there was no significant difference in the proportion of mild, moderate and severe motion sickness symptoms in all samples.

Conclusion and analysis: compared with the situation of no local cabin's orientation cues, the setting of the non-node cabin's axial orientation cues can improve the navigation task performance to some extent, but the symptoms of motion diseases such as dizziness have not been significantly improved; the setting strategy of the node cabin's landmark can improve the navigation task performance more effectively than the whole orientation cues.

*3.4.2. The experimental results of the sample scenes which reflect the node cabin's orientation cues.* Through the statistics of the obtained experimental data, we can obtain the comparison chart of the accuracy rate of orientational task, the average time of the path-finding task, the accuracy rate of psychological map task and the proportion distribution of symptoms of motion sickness for different sample scenes which reflect the node cabin's orientation cues.

In Figure 8, there was no significant difference in the accuracy rate of orientational task in all samples.

In Figure 9, there was no significant difference in the average time of path-finding task in all samples.

In Figure 10, there was no significant difference in the accuracy rate of psychological map task in all samples.

In Figure 11, there was no significant difference in the proportion of symptoms of mild, moderate and severe motor diseases in all samples.

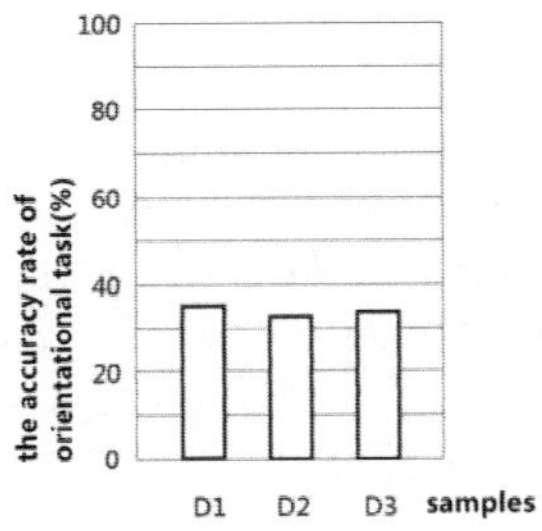


**Figure 8.** Comparison of the accuracy rate of orientational task for different samples

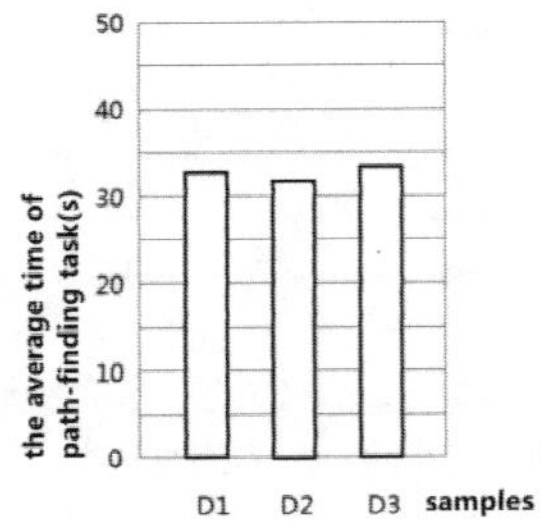


**Figure 9.** Comparison of the average time of path-finding task for different samples

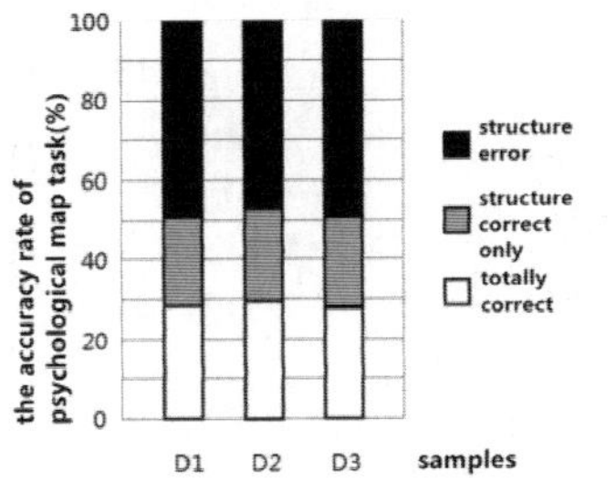


**Figure 10.** Comparison of the accuracy rate of psychological map task for different samples

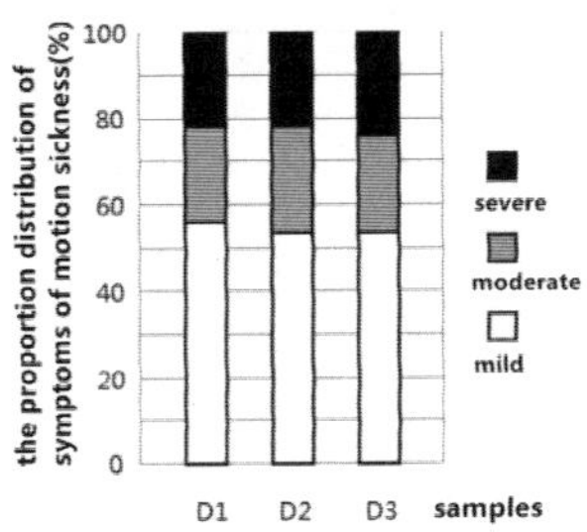


**Figure 11.** Comparison of the proportion distribution of symptoms of motion sickness for different samples

Conclusion and analysis: compared with the case of no cabin's local orientation cues, the setting of the node cabin's vertical orientation cues or axial orientation cues can not significantly improve navigation task performance or symptoms of motion sickness.

## 4. Conclusion

With the need to enhance the navigation ability of the astronauts, based on a method study on space station intra-vehicular navigation simulation including both VR visual simulation and physiological HDBR simulation, multiple different sets of sample scenes are designed for the axial orientation cues of non-node cabin and the local orientation cues of node cabin. Through the experimental data analysis, the influence law of the cabin's local orientation cues of the space station on the navigation task performance and related physiological effects in the cabin are obtained. The research could provide reference for the space station interior design engineering in the orientation of the cabin visual navigation capability.

## Acknowledgements

Project 51305099 supported by National Natural Science Foundation of China.

## References

[1] Hirofumi Aoki, Oman CM, Alan Natapoff. Virtual-reality Based 3D Navigation Training for Emergency echamber from Spacecraft. Aviation, Space, and Environmental Medicine.**78**(8):774 ~783(2007)
[2] NASA,*International Space Station Interior Color Scheme* (2001)
[3] NASA,*Human Integration Design Handbook*(2010)
[4] Manuel Vidal, Mark Lipshits, Joseph McIntyre, Alain Berthoz. Gravity and spatial orientation in virtual 3d-mazes. Journal of Vestibular Research.**13**: 273~286(2003)
[5] M. Vidal, M. Amorim, A. Berthoz. Navigating in A virtual three-dimensional maze: how do egocentric And allocentric reference frames interact?.Cognitive Brain Research,**19**: 244~258(2004)
[6] Hirofumi Aoki, Ryuzo Ohno, TakaoYamaguchi. The Effect of The Configuration and The Interior Design of a Virtual type less Space Station on Human Spatial Orientation. Conference. **56**: 1005-1016(2005)
[7] Wouldiam B. Lathrop, Mary K. Kaiser. Acquisition Spatial Knowledge While Traveling Simple and Complex Paths with Immersive and Nonimmersive Interfaces. Presence.**14**(3): 249~263(2005)
[8] Yan Xiaoxia. Long-term ground simulation of weightlessness - bed rest and its physiological research. Advances in biochemistry and biophysics.**17**(3): 181~184(1990)
[9] Wang Linjie, Zhang Dan. Studies on motion diseases in space before, during and after spaceflight. Aerospace medicine and medical engineering.**16**(5): 382~386(2003)

# Research Progress on Requirements Integrated Preprocessing and Mission Planning for Earth Observation Satellites

**Shu-Jian Chen, Zhi Li and Min Hu***

Space Engineering University, Beijing 101416, PR China

*Email: jlhm09@163.com

**Abstract.** The scale of Earth observation satellite system continues to expand and its application is becoming more and more extensive, which puts forward higher requirements for satellite mission planning system. This paper summarizes and analyzes some key issues in the mission planning of earth observation satellites. Firstly, the development status and the trend of earth observation satellites are introduced. Secondly, we summarize and analyze complex requirements of observation decomposition method, the priority of requirements determination method, the model construction and solving methods of single-satellite scheduling, multi-satellite integrated scheduling and multi-satellite dynamic scheduling in the process of requirements Integrated Preprocessing and mission planning. Finally, we discuss the problems that still need to be solved and the direction of research.

## 1. Introduction

Earth observation satellites include optical imaging satellites, SAR (Synthetic Aperture Radar) imaging satellites, missile early warning satellites, electronic reconnaissance satellites, ocean surveillance satellites, meteorological satellites and other types [1]. Earth observation satellites use the sensors they carried to obtain images and signals of specific areas according to the user requirements, and send the information back to the ground for analysis and judgment [2]. Earth observation satellite is currently the satellite with the largest number of launches and the widest application range in the world. Its unique space location advantage plays an important role in national defense construction and social development.

In recent years, the boom in the development of civil and commercial earth observation satellites has led to the continuous expansion of the scale of space earth observation systems and their wider application. It has also given rise to many problems such as complicated and diverse user requirements and difficulties in the unified scheduling of satellite. How to use operational research knowledge and computer technology to coordinate and dispatch satellite and ground station resources has become an urgent problem in space engineering. Therefore, relevant researchers in various countries have been committed to this aspect of research.

## 2. Development status of earth observation satellites

Since the United States successfully launched the world's first earth observation satellite Discovery 1 on February 28, 1959, the earth observation satellite has undergone nearly 60 years of development. The capability of space earth observation system has been steadily improved, its scale has been continuously expanded, and its application has become more and more extensive. As of September 31, 2018, the world has successfully launched 8,126 spacecraft, of which Earth observation satellites are the most numerous satellites, with 2,865, accounting for more than 35%. In terms of imaging type, the number of optical imaging satellites is the largest, accounting for more than 60% [3-6]. In addition, there are satellites carrying infrared, ultraviolet, microwave and other sensors, which can integrate the acquired information. These satellites carrying one or more sensors tend to form constellations for observation more and more, so as to enhance the capability of quick revisit, such as Flock constellation, Lemur constellation and SkySat constellation. The development of global civil space has accelerated this trend. Satellites with different sensors, different orbits and different administrative departments have entered space in large quantities, which has brought great challenges to the management of satellite resources and to meet the requirements of users. This paper is based on this background to introduce the research status of earth observation satellite requirements integrated preprocessing planning and mission planning.

## 3. Research on requirements integrated preprocessing

The tasks of satellite observation can be classified from different perspectives.How to accurately understand the user requirements and make a priority ranking is a key step before the mission planning. This chapter will focus on summarizing and analysing the integrated preprocessing of the user observation requirements.

### *3.1. Complex requirements decomposition*

The requirement that satellites cannot complete at a time is called the complex requirement. The complex observation requirement mainly includes: regional target observation, periodic observation, multi-sensor cooperative observation, multi-satellite cooperative observation, moving target searching and tracking, etc. This paper analyses the following three complex observation modes and introduces the research progress on them.

*3.1.1. Regional target observation.* Regional target refers to the target that satellites cannot completely cover by one-ff observation due to the limitation of remote sensing width. Now, there are mainly two methods to decompose regional targets: one is Rivett C's proposal to decompose regional targets into multiple point targets for independent observation [7], but this method may lead to more overlap of observation bands, thus reducing observation efficiency. The other is to realize the decomposition of regional targets by splicing observation strips [8]. This method is also divided into static decomposition method and dynamic decomposition method. Static decomposition refers to decomposing regional targets into parallel strips according to a fixed strip width in advance, while dynamic decomposition method selects the width and direction of strips according to the satellites that transit through the targets. Static decomposition method is suitable for single-satellite observation, while dynamic decomposition method is suitable for multi-satellite joint observation [9-10].

*3.1.2. Multi - satellite cooperative observation.* The efficiency of satellites cooperative observation with different sensors is higher than that of large and complex single satellite observation, and the robustness of multiple satellites is stronger, but the integrated mission planning of multiple satellites is far more complex than that of single satellite. Different researchers design different rules and algorithms for different emphases to solve the problem of multi-satellite integrated mission planning. Literature [11] proposed a multi-satellite cooperative observation strategy based on target priority ranking and space-time constraints for satellite observation in natural disaster areas. Literature [12] established a constraint satisfaction model for mission planning of cooperative observation of high and low orbit satellites. Literature [13] greatly improved the recognition capability and positional accuracy of multi-satellite cooperative observation.

*3.1.3. Moving target searching and tracking.* The searching and tracking of moving targets generally need some prior information. If the region of the moving target is known, the LEO imaging satellite constellation can be used to search for the specified region, and a motion prediction model of the target needs to be established. Berry P E [14-15] established a target motion prediction model based on Gaussian distribution using Bayesian estimation technology. If the target has electromagnetic signals, it is more reliable to use electronic reconnaissance satellites and optical imaging satellites to search for moving targets in cooperation [16].At present, there is still no general and effective method for searching and tracking moving targets with camouflage, no electromagnetic signals and agile movement, which is a research focus in the field of military reconnaissance satellites.

### *3.2. Prioritization of Requirements*

At present, determining priority of user requirements is still mainly through relevant experts or commanders. However, with the increasing number of requirements, the efficiency of relying on people to determine the priority of these needs is becoming lower. How to use computers to automatically determine an appropriate observation priority according to certain rules is of great importance to subsequent mission planning. On the basis of analysing the engineering requirements of TT&C resources, literature [17] studied nine basic principles of resource priority and proposed a method to solve the problem of priority based on improved TOPSIS method (Technique for Order Preference by Similarity to an Ideal Solution). Literature [18] used the same method to analyse the priority of earth observation satellite resources. In the research on the quantification of requirements value, literature [19] analysed the STT (Strategy To Task) framework of RAND Company and the application of STT in ISR(Intelligence, Surveillance and Reconnaissance), and made some improvement on STT framework in the calculation of requirements value. Literature [20] proposed a concept of time margin to reflect the degree of satisfaction of meeting tasks' urgency, and established a mission planning model based on priority and time margin. We concluded that there are many researches on priority evaluation of requirements or tasks, but it is mainly similar to expert systems that establish certain rules to actively assign parameter values to some influencing factors, with lacking evaluation of results.

## 4. Research on Mission Planning Model

Mission planning of Earth observation satellites is a complex combinatorial optimization problem including multi-task and multi-constraint. Different mission planning models and corresponding solving methods need to be established according to different types of earth observation satellites and tasks including conventional and emergency tasks or static and dynamic task. From the current research situation, satellite mission planning modeling can be divided into single-satellite mission planning and multi-satellite mission planning modeling. The established models are shown in Figure 1.

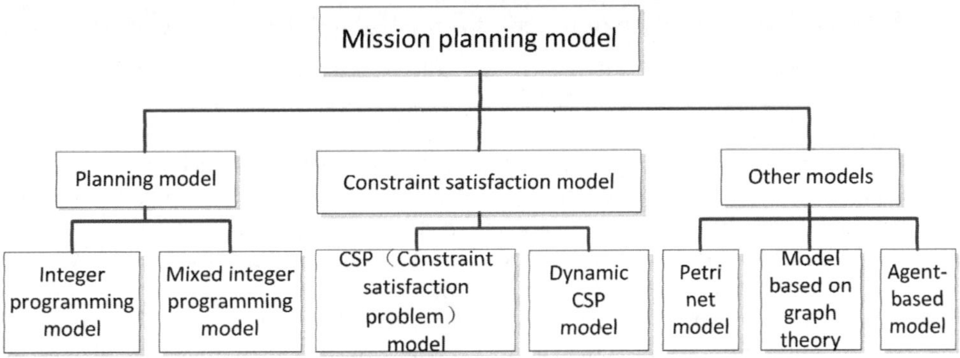


**Figure 1.** Mission planning model.

### *4.1. Modeling of Single Satellite Mission Planning*

Different earth observation satellites have different characteristics in their mission planning. When SAR satellites perform imaging missions to the earth, they are not affected by meteorological factors such as cloudiness and illumination, can switch working wave positions for observation, and the power consumption of equipment in standby state is very small, which is quite different from visible satellites in terms of working mode and sensor constraints. Therefore, there are differences in the mission planning models and algorithms between visible light satellites and SAR satellites.

For SAR imaging satellites, De Florio S [21] proposed two mission planning methods, sequential insertion method and intermediate insertion method. Bianchessi N and Righini G [22] have studied the mission planning of COSMO-SkyMed SAR satellite, developed a construction algorithm with foresight and retrospective ability, and achieved the goal of shooting and transmitting 1800 images every day. For optical imaging satellites: Bensana E [23], Vasquez M [24], and Gabrel V [25] have all studied mission planning of French SPOT5 satellite. [25] established an integer programming model, but did not consider the constraints of data transmission. [26] established a constraint satisfaction model similar to the 0-1 knapsack problem, which takes into account the constraints stored on the satellite but also does not consider the data transmission constraints. [27] established an integer programming model based on graph theory and mathematical programming knowledge, but neither on-board storage constraints nor data transmission constraints were considered.

### *4.2. Modeling of Multi - Satellite Integrated Mission Planning*

The planning of multi-satellite joint observation mission is very complicated. One of the most important factors is that it contains many constraints, and the actual constraints for specific problems are not the same. This paper summarizes the main constraints of mission planning for earth observation satellites, as shown in Figure 2.

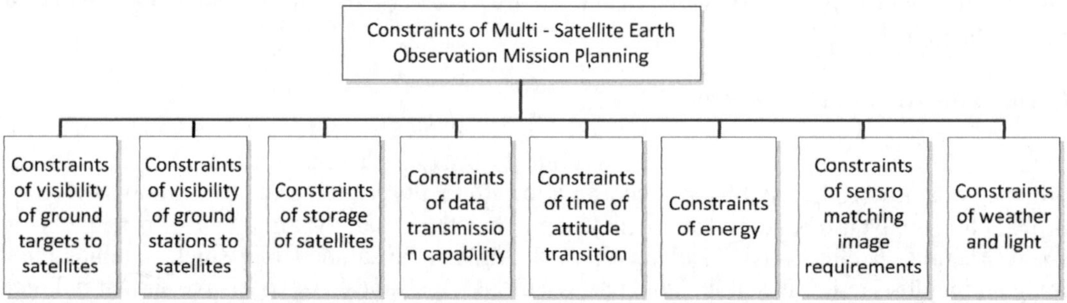


**Figure 2.** Constraints of Multi - Satellite Earth Observation Mission Planning.

For the complex constraints, the researchers have made targeted studies respectively. Literature [26] considered various data compression ratios of data storage, and on this basis establishes a time-ordered acyclic directed constraint model. Literature [27] divided the multi-satellite mission planning into two phases: imaging to the ground and data return, and establishes a model based on phase optimization. Literature [28] proposed a closed-loop control model for satellite imaging, which takes into account uncertainties such as targets, imaging sensors and environment. Frank J [29] considered the constraints of storage capacity, data download speed and task priority in his model, but the disadvantage is that the required resources need to be declared before planning.

In multi-satellite mission planning, mission composition is a key link, which is an important strategy to reduce mission conflicts and improve the efficiency of satellite observation. The composition of multiple atomic task must meet certain conditions. If we assuming the composite task $COM_{il}$ includes atomic-task set as $SO_{il}$, Start-time set of $SO_{il}$ as $SWS_{il}$, end-time set of $SO_{il}$ as $SWE_{il}$, and observation angle set as $SA_{il}$. The time window of atomic-task is $[we_{ijkv}, ws_{ijkv}]$ and the angle of it is $a_{ijkv}$. $Span_i$ and $\Delta a_i$ represent the maximum single boot time and viewing angle of satellite $S_i$. Then the conditions that the atomic task in $SO_{il}$ can be composed are:

$$\max_{we_{ijkv} \in SWE_{il}} we_{ijkv} - \min_{ws_{ijkv} \in SWS_{il}} ws_{ijkv} \leq Span_i \tag{1}$$

$$\max_{a_{ijkv} \in SA_{il}} a_{ijkv} - \min_{a_{ijkv} \in SA_{il}} a_{ijkv} \leq \Delta a_i \tag{2}$$

Many researchers have studied the methods of satellite observation missions composition. Liu X L [30] eveloped a dynamic planning algorithm to find the best task composition plan for each satellite. Literature [31] proposed a dynamic merging (DM) strategy and designs a candidate merging task set establishment (CMTSE) algorithm. Literature [32] proposed a repair technology to overcome the problem of reduction of task imaging opportunity caused by composite.

In addition to task composition, revenue calculation of multi-satellite task planning is also one of the keys. At present, the research on revenue function of task planning model is still mainly focused on task completion based on the sum of task priorities $P_j$, as shown in Equation (3).

$$C = \sum_{j=1}^{N_j} x_j \times P_j \quad , \quad x_j = \begin{cases} 1, T_j & \textit{is arranged to observe} \\ 0, T_j & \textit{is not arranged to observe} \end{cases} \tag{3}$$

However, as the satellite mission planning model becomes more and more complex, it is not comprehensive to only consider the completion of the mission, and some other optimization criteria are also proposed in some literatures. Literature [33] put forward the optimization objectives of maximizing the observation benefit, minimizing the number of sidesteps and minimizing the total sidesteps angle. Literature [34] put forward a comprehensive evaluation model considering the conflict of ground receiving resources.

### *4.3. Modeling of Satellite Dynamic Mission Planning*

In 2002, Pemberton of Veridian Company of France conducted the first research on satellite dynamic mission planning and discussed the general characteristics of this problem. Literature [36] summarizes five characteristics of dynamic scheduling of earth observation satellite missions on the basis of it, as shown in Table 1.

**Table 1.** Dynamic scheduling characteristics of earth observation satellites.

| | |
|---|---|
| Insertion of new tasks | Users urgently need to observe some targets and request temporary insertion of some tasks. |
| Cancellation of scheduled tasks | Some changes in user's requirements result in cancellation of corresponding tasks. |
| Changes in tasks attributes | It mainly refers to the change of priority of certain tasks. |
| Changes in external environment | The sudden change of weather causes the task to be unable to be completed or delayed. |
| Changes in satellites state | The satellite encounters a failure during its operation, which prevents it from completing its observation tasks on time. |

When disturbances occur, the imaging mission may undergo the following changes:

- The mission is completed as required, but the satellite resources for completing the mission change.
- The completion time of the mission changes and the satellite resources for completing the mission remain unchanged.
- The time for completion of the mission changes, and the satellite resources for completing the mission change.
- The task is deleted.

Although there are various disturbance factors in the implementation of imaging satellite scheduling schemes, they can be essentially attributed to a kind of dynamic scheduling problem oriented to newly inserted tasks. Compared with conventional scheduling, dynamic scheduling has higher requirements, mainly reflected in the following three aspects:

- The robustness of the scheduling scheme is as strong as possible.
- The difference between the new plan and the original plan is as small as possible.
- The new scheme is generated as fast as possible.

Robustness and profitability of dynamic scheduling are often in conflict. Schemes with good profitability may have poor robustness, resulting in difficulties in adjusting in interference. Schemes with good robustness may have poor profitability, resulting in waste of satellite resources.

Aiming to the dynamic scheduling problem of agile imaging satellites, literature [35] summarized three kinds of uncertain factors and proposed a proactive scheduling strategy. Literature [36] proposed a rolling dynamic task planning strategy, which aims to decompose the complex dynamic planning problem into a plurality of simple static planning subproblems. Literature [37] established a dynamic constraint satisfaction model with two-level optimization objectives, in which the first-level index is still the sum of maximizing the task priority and the second-level index is the minimum disturbance.

## 5. Research on Mission Planning Algorithm

The algorithms for solving imaging satellite mission planning problems are mainly divided into two categories: deterministic algorithms and uncertain algorithms. Deterministic algorithm is also called exact algorithm, which searches the solution space completely. Uncertainty algorithm includes intelligent optimization algorithm and heuristic algorithm, as shown in Figure 3. The classical intelligent optimization algorithms include genetic algorithm, tabu search algorithm, simulated annealing algorithm and the improvement methods of these methods. Heuristic algorithm is an algorithm based on intuitive or empirical construction.

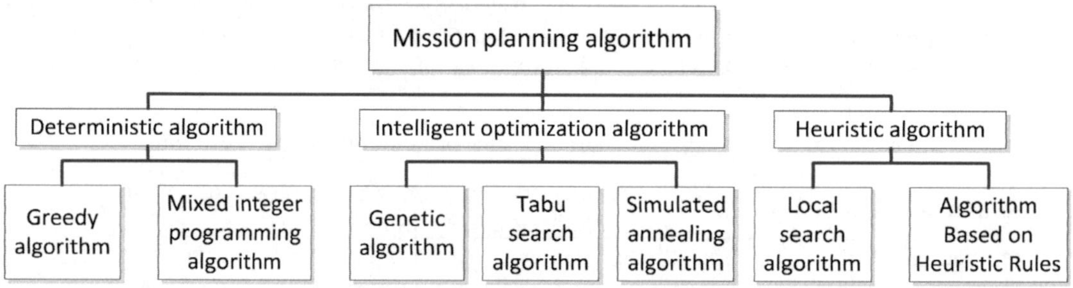


**Figure 3.** Mission planning algorithm.

Lemaitre M [38] studied and compared greedy algorithm, constrained programming algorithm, dynamic programming algorithm and local search algorithm for the mission planning system of Pleiades (Agile Satellite). Bensana E [39-40] studied and compared the computational performance of different algorithms under different scale problems. Li H J [41] designed a heuristic search algorithm of the graph. A large number of studies show that the deterministic algorithm can obtain the optimal solution for small-scale problems and the algorithm is stable. However, for large-scale combinatorial optimization problems such as multi-satellite task planning, deterministic algorithms are prone to fall into local optimal solutions and take too long to solve. Therefore, most of the current research focuses on uncertain algorithms. At the same time, some researchers proposed hyper-heuristic algorithms [18] For the multi-satellite cooperative observation problem.

## 6. Conclusions

In the research of requirements integrated preprocessing, the input of existing requirements is mostly accurate input, which requires users have relevant space knowledge. However, if the mission planning system has the ability to deal with natural language knowledge, it will greatly improve the intelligence level of the system, which is conducive to the promotion of military operations and commercial operations.In the research of mission planning, the current research is still mostly focused on batch planning tasks. In the future, user requirements are likely to be proposed at any time, which requires high timeliness of observation information so offline scheduling can no longer meet user requirements. Therefore, the on-line scheduling of satellites is the trend of development, which requires researchers to make breakthroughs in the field of satellite autonomous mission planning.

Moreover, the integration of mission planning and multi-source observation data is also an important trend. Using the fusion of multi-source data to realize automatic target recognition, automatic labelling and selecting imaging mode, and the most effective information for users is directly provided to reduce the time for users to process and analyze by themselves. Nowadays, the human society is changing from the information era to the intelligent era, and the overall planning of earth observation satellites will inevitably develop towards intelligence and refinement.

## Reference

[1] Zhang B and Zou H X 2014 The model of imaging satellite task scheduling and research progress of algorithm *J.Computer Engineering and Applications.* **50(S1)** pp 116-120
[2] Jiang W, Hao H C and Li Y J 2013 Review of task scheduling research for the earth observing satellites *J.Systems Engineering and Electronics* **35(9)** pp 1878-85
[3] Yuan M H and Liu T 2018 Recent development of space-based earth observation system and application *J.Space International* **4** pp 8-15
[4] Fu Y 2018 Space launch report of Q1 2018 *J.Space International* **6** pp 52-58
[5] Fu Y 2018 Space launch report of Q2 2018 *J.Space International* **9** pp 62-58
[6] Fu Y 2018 Space launch report of Q3 2018 *J.Space International* **12** pp 58-63
[7] Rivett C and Pontecorvo C 2004 Improving satellite suiveillance through optimal assignment of assets *Australian Government Department of Defense: DSTO-TR-1488*
[8] He R J, Li J F, Yao F and Xing L N 2011 Earth Observing Satellite Mission Planning (Beijing:Science Press) pp 49-63
[9] Li J F, Yao F, Bai B C and Ran Q M 2008 Multi - satellite cooperative earth observation mission planning for regional targets *J.Science of Surveying and Mapping* **S3** pp 56-58
[10] Wang H F and Li X Z 2012 Cooperate method of planning area target for earth observation satellite *J.Computer Simulation* **29(4)** pp 101-103
[11] He H X, Wu B, Li R and Han X Z 2018 Research on satellite cooperative observation strategy for natural disaster emergency response *J.Spacecraft Recovery & Remote Sensing* **6** pp 91-101
[12] Pang Z H 2013 *Research on Collaborative Mission Scheduling Methods of High-orbit and Low-orbit Earth Observation Satellites* (Harbin:Harbin Institute of Technology)
[13] Chen J Y, Zhang C and Li Y B 2018 Multi-star cooperative task planning based on hyper-heuristic algorithm *J.Journal of CAEIT* **3** pp 254-259
[14] Berry P E and Fogg D A B 2003 GAMBIT: Gauss-Markov and Bayesian inference technique for infor-mation uncertainty and decision making in surveillance simulations *Australia: DSTO Information Sciences Laboratory*
[15] Berry P E, Pontecorvo C and Fogg D A B 2003 Optimal search location and tracking of sur-face mariti-me targets by a constellation of surveillance satellites *Australia: DSTO Information Sciences Laboratory*
[16] Li J J, Guo H and Zhu J H 2011 Planning method for multi-satellite cooperating search for a moving target on the sea *Proc. of the International Workshop on Multi-platform/Multi-sensor Remote Sensing and Mapping* pp 1-6.
[17] Ran C X, Wang H L and Xiong G Y 2009 Priority design method analysis for TT&C resource *J.Chinese Space Science and Technology* **3** pp 78-83
[18] Wang H L, Ma M H and Ran C X 2010 Priority design method analysis for earth observation satellite *J.Computer Simulation* **27(11)** pp 41-43
[19] Tan Q, Qiu D S, Li Z M and Zhao Y 2012 An improved STT method for quantifying the value of satellite reconnaissance needs *J.Fire Control & Command Control* **37(10)** pp 59-62

[20] Guo C, Xiong W and Liu C X 2016 Mission planning of satellite emergency observations based on priority and time margin degree *J.Telecommunication Engineering* **56(7)** pp 744-749
[21] De Florio S, Zehetbauer T and Ne T 2005 optimal operations planning for SAR satellite constellations in low earth orbit *Conf. on Reducing the Costs of Spacecraft Ground Systems & Operations*
[22] Bianchessi N and Righini G 2008 Planning and scheduling algorithms for the COSMO-SkyMed constellation *J. Aerospace Science & Technology* **12(7)** pp 535-544
[23] Bensana E, Verfaillie G, Agnèse J C, et al. 1996 Exact &inexact methods for daily management of earth observation satellite *Proc. of the Fourth International Symposium on Space Mission Operations and Ground Data Systems: European Space Agency* pp 507-514
[24] Vasquez M and Hao J K 2001 A "logic-constrained" knapsack formulation and a tabu algorithm for the daily photograph scheduling of an earth observation satellite *J.Computational Optimization & Applications* **20(2)** pp 137-157
[25] Gabrel V and Murat C 2003 *Mathematical programming for earth observation satellite mission planning* Operations Research in Space and Air(US:Spinger) pp 103-122
[26] Wang Z Y, Wang Y Q, Wang J, Li J and Wu Q Y 2012 New multi-satellite scheduling method *J.Chinese Space Science and Technology* **1** pp 8-14
[27] Sun K, Chen C, Chen Y W and He R J 2012 Modeling and simulation on imaging satellites and ground stations joint scheduling problem *J.Control Engineering of China* **19(4)** pp 695-698
[28] Ci Y Z, Tan Y J, He R J and Li J F 2008 Mission planning for joint search by multi-satellite *J.Journal of Astronautics* **29(2)** pp 653-658
[29] Frank J, Ari J, Morris R, et al. 2001 Planning and scheduling for fleets of earth observing satellites *J.Proceedings of Sixth Int.symp.on Artificial Intelligence Robotics Automation & Space* p307
[30] Liu X L, Baocun B, Chen Y W, et al. 2014 Multi satellites scheduling algorithm based on task merging mechanism *J.Applied Mathematics and Computation* **230** pp 687-700
[31] Qiu D S,Wang J J, Wu C B and Liu G 2013 Emergency scheduling method of earth observation satellites based on task merging *J.Systems Engineering and Electronics* **35(7)** pp 1430-37
[32] Wang J J, Zhu X M, Wu C B and Qiu D S 2013 Multi-satellite dynamic scheduling method for emergencies *J.Acta Aeronautica et Astronautica Sinica* **34(5)** pp 1151-64
[33] Wang M C, Dai G M and Song Z M 2011 *Research on Mission Planning and Scheduling Algorithms for Imaging Satellites* (Beijing: Science Press ) pp 29-30
[34] Wang W D, Zhou Z X, Li X, Zhan M and Wang H Q 2011 Optimization method to resolve the confliction of ground station receiving resources for multi-satellites data *J.Systems Engineering and Electronics* **33(6)** pp 1299-1304
[35] Li Z L, Li X J and Sun W 2017 Task scheduling model and algorithm for agile satellite considering imaging quality *J.Journal of Astronautics* **38(6)** pp 590-597
[36] Liu X L, Yang B, Gao C H and He C D 2017 Research on rolling dynamic mission scheduling technique for remote sensing satellites *J.Radio Engineering* **47(9)** pp 68-72
[37] Zhu J H, Huang W, Li J J and Wang D 2011 Mission planning for electronic reconnaissance satellites oriented the insertion of the new missions *J.Fire Control & Command Control* **36(7)** pp 174-177
[38] LemaîTre M, Verfaillie G, Jouhaud F, et al. 2002 Selecting and scheduling observations of agile satellites *J.Aerospace Science & Technology* **6(5)** pp 367-381
[39] Bensana E, Lemaitre M andVerfaillie G 1999 Earth observation satellite management *J. Constraints* **4(3)** pp 293-299
[40] Bensana E, Verfaillie G, Michelonedery C, et al. 1999 Dealing with uncertainty when managing an earth observation satellite *Proc. of Isairas* pp 205-207
[41] Li H J 2014 The Analyses of imaging satellites ability on emergency mission Proc. of 2014 *International Conference on Appliced Science,Engineering and Technology* **10**

IOP Conf. Series: Materials Science and Engineering **608** (2019) 012031 doi:10.1088/1757-899X/608/1/012031

# One Real-time Micro-thrust Measurement System Based on Barkhausen Effect

**Xinru Du*, Jianjun Wu, Yang Ou, Jian Li and Biqi Wu**

College of Aerospace Science and Engineering, National University of Defense Technology, 410073, Changsha, China

*Corresponding author:1289253951@qq.com

**Abstract.** The thrust of the electric thruster for micro-nano satellites is very small, so it is very difficult to accurately measure the real-time variation of micro-thrust. In order to solve this problem, a real-time micro-thrust measurement system based on Barkhausen effect is proposed. Buckhausen effect is that under the action of external stress, the change of internal stress or strain of ferromagnetic material will cause the change of permeability, reluctance and flux of ferromagnetic material, which will result in Barkhausen noise under the external alternating magnetic field. Therefore, the change of stress or strain in ferromagnetic materials can be obtained by measuring Barkhausen noise. The micro-thrust measurement system based on Barkhausen effect proposed exactly in this paper transforms the change of thrust into the change of internal stress of the magnetoelastic element [1], which can realize real-time and high-precision measurement of instantaneous micro-thrust with the advantages of simple structure, easy operation and high reliability.

## 1. Introduction

As a product of the development of high and new technology and space exploration missions in the fields of microelectronics, micro-machinery, new materials and computers, micro-nano satellites have the advantages of light weight, small volume, short development period, high power density, flexible launch, low cost, fast networking and strong survivability, which play an important role in civilian and military aspects. Whether it is a single flight or a formation flight, it is necessary to control the orbit and attitude of the satellite accurately, which requires that the thrust generated by the propulsion device of the satellite is small, wide and controllable, usually within the order of μN~mN [2].

The thrust produced by the electric thruster is generally small, and its order of magnitude is between nN and N, making electric thrusters the proper propulsion devices of micro-nano satellites. For example, the thrust produced by pulsed plasma thruster is generally of μN order, while that of laser plasma micro-thruster based on laser ablation of solid working fluid is of nN magnitude. Although the micro-thrust of mN or N is easy to measure, it becomes very difficult to measure the micro-thrust of the order of μN~nN as the thrust decreases further. Moreover, in order to control the satellite more accurately, it is necessary to monitor the real-time change of the thrust.

At present, the micro-thrust measurement systems of electric thrusters mainly involve the measurement system based on the single pendulum, the electromagnetic balance based on the inverted pendulum, the measurement system based on the torsion pendulum and the measurement system based on the piezoelectric sensor. The measuring precision of the first two kinds of force measuring systems is low. Although the measuring system based on torsion pendulum has high precision and sensitivity,

IOP Conf. Series: Materials Science and Engineering **608** (2019) 012031 doi:10.1088/1757-899X/608/1/012031

and the lower limit of thrust measurement is low, twisting wires are easy to be affected by vibration and noise, and can only measure the average thrust. The measurement system based on the piezoelectric sensor has high sensitivity and is able to measure the change of instantaneous thrusts accurately. However, it is vulnerable to electromagnetic interference, which results in inaccurate measurement results or the measurement process not being carried out normally.

In summary, the existing measuring methods can not measure the tiny transient thrust generated in the working process of the electric thruster accurately and real-timely. To solve this problem, a transient micro-thrust measurement system based on Barkhausen stress detection method is proposed.

## 2. Measurement principal

Magnetoelasticity is a phenomenon in which the magnetic properties of ferromagnetic materials change with the action of mechanical stress or strain. Ferromagnetic materials are magnetoelastic elements.

Barkhausen effect was first discovered by German physicist Bardhausen in 1919. When he observed the hysteresis loop of the ferromagnetic material, he found that the curve was not smooth. After amplification, he saw the variation of step jitter [3] as shown in Figure 1. When placing the receiving coil on the surface of the sample, the voltage pulse and noise are generated. This phenomenon is called Barkhausen effect, and the corresponding magnetic noise is called Barkhausen noise, or MBN.

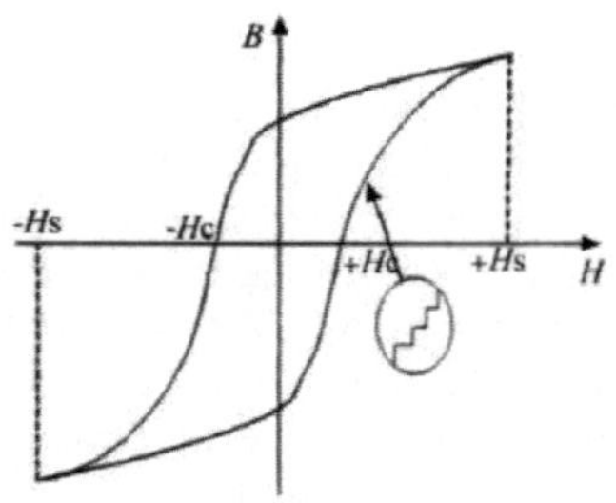


**Figure 1.** Barkhausen noise generation diagram

Barkhausen stress detection method is a nondestructive testing method based on magnetoelastic principle and Barkhausen effect, which is usually used to detect the stress state of ferromagnetic materials widely used in engineering. When ferromagnetic materials are subjected to external stress, the change of internal stress or strain will cause the change of permeability, reluctance and flux of ferromagnetic materials. Barkhausen noise induced by the external alternating magnetic field will reflect this change. Therefore, the change of stress or strain in ferromagnetic materials can be obtained by measuring Barkhausen noise. In this paper, the method is applied to micro-thrust measurement of electric thruster, which transforms the change of thrust into the change of internal stress of the magnetoelastic element. This method not only has the advantages of simple structure, convenient operation and high reliability, but also has high detection sensitivity and accuracy. It can measure the change of stress in real time, which means that it can accurately measure the tiny transient thrust of electric thrusters.

## 3. System composition

As shown in figure 2, the system consists of two parts: the stress detection system and the thruster system.

### *3.1. The stress detection system*

The Barkhausen stress detection system based on digital signal processing (DSP) [4,5] includes an ARM human-computer interaction module, a DSP system, a power amplifier circuit, a signal

IOP Conf. Series: Materials Science and Engineering **608** (2019) 012031 doi:10.1088/1757-899X/608/1/012031

generation circuit, a signal conditioning circuit, a signal acquisition circuit, a magnetoelastic element, a sensor and two shields. Among them, the ARM human-computer interaction module is the key of human-computer interaction control, which realizes the display of stress and MBN signal eigenvalues. DSP is responsible for digital signal processing, controlling the generation of excitation and signal acquisition, and realizing the interactive communication with ARM. The power amplifier circuit can amplify the excitation signal and improve the load capacity. The signal generation circuit is responsible for generating the excitation signal and providing the external alternating magnetic field needed to generate the Barkhausen signal. The signal conditioning circuit filters and amplifies the received signal. The analog signal is sampled by the signal acquisition circuit, and then it is converted into a digital signal that can be processed by DSP. The sensor detects the MBN signal. When the exciting coil passes the alternating excitation signal, the alternating magnetic field is produced in the yoke, and the magnetoelastic element is magnetized locally. The receiving coil can sense the MBN signal generated by the magnetoelastic element due to domain reversal or domain wall movement. Shields Shield the electromagnetic interference making the measurement results more accurate.

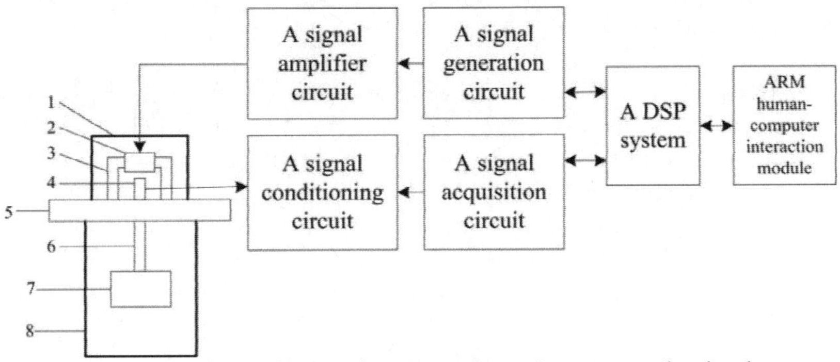


1-the first shield box, 2-excitation coil, 3-yoke, 4-receiver, 5-magnetoelastic element, 6-rigid plate, 7-thruster, 8-the second shield box

**Figure 2.** The micro-thrust transient measurement system

*3.1.1. The ARM man-machine interaction module.* The ARM embedded hardware consists of the ARM microprocessor, the liquid crystal display (LCD), the flash memory, the synchronous dynamic random access memory (SDRAM), the universal asynchronous receiving / transmitting device (UART) interface, the universal serial bus (USB) peripheral interface, the SD card interface, the real-time clock (RTC), the external clock and the power supply battery. ARM embedded system is the key to realize man-machine interaction. The user can operate and control the whole system through LCD. At the same time, the ARM embedded system can store the detected stress data, display the stress curve, export data stored in U disk and drive some related peripherals. Flash memory is used to store programs and testing data. SDRAM provides guarantee for system operation. UART interface realizes communication with DSP system. USB peripheral interface can insert U disk and export testing data. SD card interface can be connected with SD card, realizing expansion of the system storage capacity. RTC provides clock signal for the whole system. The power supply battery provides the guarantee for the stable operation of the whole system.

*3.1.2. The DSP digital signal processing system.* The DSP mainly realizes the generation of the excitation signal and the processing of the acquisition signal. DSP has powerful data processing ability and can process a large number of data quickly. Thus, the design goal of real-time thrust detection can be realized.

*3.1.3. The signal producing circuit.* The essence of ferromagnetic materials generating MBN signal is domain reversal and domain wall movement. Therefore, the premise of MBN signal generation is the

IOP Conf. Series: Materials Science and Engineering **608** (2019) 012031 doi:10.1088/1757-899X/608/1/012031

external alternating magnetic field and alternating force, while the signal generation circuit can provide the excitation signal to produce alternating magnetic field. EMIFA expand the signal generation circuit, that is, digital-to-analog converter (DAC).In order to complete the conversion and control quickly, the conversion speed of DAC must meet the index requirements. In order to ensure the correct data processing, DAC must have enough conversion accuracy to reduce the error, so the conversion speed and accuracy are important parameters to measure DAC. In general, the more DAC bits, the higher its resolution and conversion accuracy. Most DACs used in industrial automatic control systems are of 10 or 12 bits, and the conversion accuracy can reach 0.5% to 0.1%.

*3.1.4. The signal acquisition circuit.* The signal acquisition circuit is a circuit that converts the analog signal which has been filtered and amplified by the signal conditioning circuit to the digital signal. EMIFA expand signal acquisition circuit, that is, analog-to-digital converter (ADC). In order to ensure the accuracy of the system processing results and realize the real-time control and detection of the rapidly changing signals, ADC must have sufficient conversion accuracy and high conversion speed, and it should satisfy the requirements of the input voltage range, the output digital coding, the working temperature range and the voltage stability, etc.

*3.1.5. The power amplifier circuit.* The function of the power amplifier circuit is to amplify the excitation signal generated by the DSP, and then to send the amplified excitation signal to the excitation coil, which magnetizes the magnetoelastic element to produce Barkhausen signal. In general, the load capacity of the signal generation circuit is poor and the impedance of the excitation coil is large, so it is necessary to pass the excitation signal generated by the signal generation circuit to the power amplifier circuit firstly and then pass the amplified signal to the excitation coil. The power amplifier circuit is essentially to provide enough power signals for the power equipment. The requirements for the power amplifier circuit are as follows: output power maximization, high conversion efficiency $\eta$ and low nonlinear distortion.

*3.1.6. The signal conditioning circuit.* MBN is a weak signal of mV level, and its bandwidth is 1 kHz ~2 MHz. The frequency bands of MBN signals produced by general materials are usually 1~500 kHz, so it is necessary to use appropriate signal conditioning circuit to amplify and filter MBN signals.

*3.1.7. The magnetoelastic element.* The ferromagnetic material of Q235 ordinary carbon steel is chosen as the magnetoelastic element. Q235 has moderate carbon content, and its strength, plasticity and welding property are relatively good, so Q235 has been widely used in engineering, especially in automobile, shipbuilding, building and aerospace, etc. The yield limit of Q235 is 235MPa, and the tensile limit is 400MPa. The composition and content of Q235 are as follows: carbon 0.14%~0.19%, silicon 0.12%~0.33%, manganese 0.3%~0.65%, chromium 0.3%, nickel 0.3%, beryllium < 0.045%, sulfur < 0.05% and the rest is iron. Q235 can adapt to the change of external environment and obtain good mechanical properties.

The size of the magnetoelastic element, that is, length, width and thickness, should be reasonably selected according to the measuring thrust value and the sensitivity of the sensor.

*3.1.8. The sensor.* As the core component of the whole Barkhausen stress detection system, the sensor's ability to excite and receive the MBN signal directly affects the energy and signal-to-noise ratio of the original MBN signal. The necessary condition for the generation of the MBN signal is the jump of the magnetic domain. In practice, the method of electromagnetic induction is often used to excite, that is, the alternating current in the coil generates the dynamic alternating magnetic field. The coil is wound on the soft magnetic material with high permeability, the magnetoelastic element with which form a closed excitation loop to magnetize the magnetoelastic element.

As shown in figure 3, the sensor consists of an excitation section and a receiver. The excitation part is composed of the excitation coil winding the U-shaped yoke and the receiver consists of the receiving coil winding the cylindrical iron core. In detection, the receiver is placed in the middle of the U-shaped yoke and one end of it is in contact with the surface of the magnetoelastic element, and a sinusoidal alternating current of a certain frequency is passed through the excitation coil to produce an alternating magnetic field, and the magnetoelastic element is magnetized by the U-shaped yoke, generating the MBN signal, then the signal is received by the receiver. The frequency range of the Barkhausen signal of most ferromagnetic materials is 1~250 kHz. The resonance frequency of the excitation part needs to be much higher than the highest frequency of the MBN signal, which is generally above 10 times, so the material chosen for the yoke is Mn-Zn ferrite.

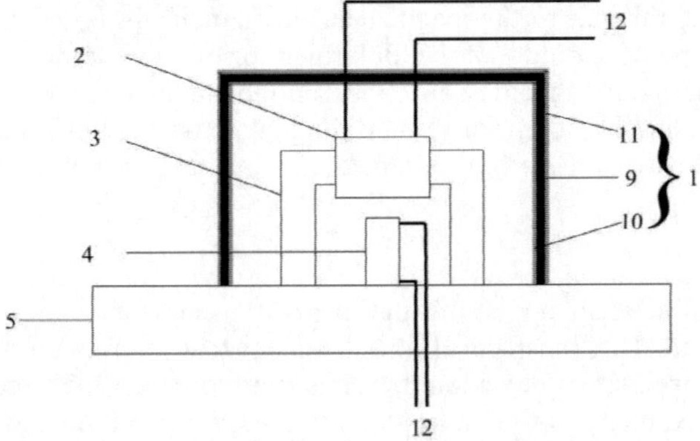


1-the first shield box, 2-excitation coil, 3-yoke, 4-receiver, 5-magnetoelastic element, 9-low carbon steel shell, 10-silicon steel sheet, 11-double side shielded copper tape, 12-twisted pair wire

**Figure 3.** The sensor

The function of the receiver is to convert the inductive magnetic signal into the voltage signal, which plays a decisive role in the detected MBN voltage signal. When the frequency of magnetic core is close to the central frequency of Barkhausen signal, the inductive signal is relatively good, so Mn-Zn ferrite is chosen to be the core still. Mn-Zn ferrite is an ideal magnetic core with high permeability, low loss and low cost.

In order to make the detection equipment more dexterous and portable, the size of the sensor should be as small as possible. It is necessary to determine the diameter and the number of turns of the excitation coil and the receiving coil by certain tests to match the size and sensitivity of the thrust to be measured. The excitation frequency of the sensor is 5~500 Hz, and the excitation intensity is 0.1~2.0 T.

*3.1.9. The first shield box.* All kinds of electromagnetic devices in the measuring environment will cause electromagnetic interference to the sensor of the stress detection system, and the plasma produced by the electric thruster itself in the working process will also bring electromagnetic interference to the sensor. It is necessary to reduce the influence of stray electromagnetic field on the sensor by magnetic shielding technology. The first shield box adopts the annealed low carbon steel shell as the first shield box. Silicon steel sheet, which is the soft magnetic material of high magnetic conductivity, is pasted inside, and double-sided shielded copper tapes are wound on the external surface. The sensor is placed in the shield box.

In order to shield the external interference during the signal transmission, twisted-pair wires are used as the sensor signal wires to reduce the capacitive and inductive coupling, and to improve the common mode suppression ability.

### *3.2. The thruster system*

The main function of thruster system is to generate thrust and transfer the thrust to the magnetoelastic element to keep it in a stress state. At the same time, it is necessary to ensure that the thruster will not cause electromagnetic interference to the magnetoelastic element. As shown in figure 2, the thruster system includes a thruster body, a rigid plate and a second shield box.

The second shield box is the same as the first shield box, and the annealed low carbon steel shell is also used as the box, and the silicon steel sheet of high magnetic conductivity is attached inside, and the external windings are made of double-sided shielded copper tapes. However, the size of the second shield box should be large enough, otherwise, the plasma ejection of the thruster will be blocked and the size of the thrust will be affected, so the normal thrust value of the thruster can't be obtained.

The role of the rigid plate is to connect the thruster with the magnetoelastic element and transfer the thrust produced by the thruster to the magnetoelastic element. In order to ensure the accuracy of measurement, the rigid plate should not be deformed or nearly deformed under thrust impact. Changing the shape (width and thickness) of the rigid plate and the cross-section moment, and properly arranging the position of the thruster and the magnetoelastic element on the rigid plate can change the thrust magnification, therefore, this method can measure the micro-thrust in the field of electric propulsion.

## 4. Summary

In this paper, Barkhausen effect is applied to the micro transient thrust measurement of the electric thruster. The change of thrust is transformed into the change of internal stress of the magnetoelastic element, then the Barkhausen stress detection device is used to detect the stress and obtain the thrust. This method has the advantages of simple structure, easy operation and high reliability. Most importantly, it solves the problem that the traditional thrust measurement method is difficult to measure the transient micro-thrust. It can realize real-time and high precision measurement of instantaneous micro-thrust.

## Acknowledgments

The authors would like to thank National Natural Science Foundation of China for the financial assistance provided under the grant number 11772354 for this work.

## Reference

[1] Ou Yang, Wu Jianjun, Zhang Yu, et al. One micro-thrust measurement method based on the acoustoelastic theory [J]. *MATEC Web of Conferences*, 2018: 1010.

[2] Tan Sheng, Wu Jianjun, Zhang Yu, et al. Research progress of laser supported space micro propulsion technology [J]. *Propulsive Technology*, 2018 **39 (11)**: 2415-2428.

[3] Yin Hechi, Chen Ligong, Zhang Guangye, et al. Application of magnetic Barkhausen residual stress measurement technique in heat treatment process evaluation [J]. *Heat Treatment*, 2008 **23 (2)**: 17-20.

[4] Wang Ping, Ren Zifeng, Tian Guiyun, et al. Barkhausen noise stress detection system based on ARM platform [J]. *Nondestructive testing*, 2012N **34 (12)**: 12-14.

[5] Zhu Xiaoxue. Development of Barkhausen portable stress detector based on DSP [D]. Nanjing: Nanjing University of Aeronautics and Astronautics, 2013.

# Review on the Development of Satellite Formation and Collision Avoidance Technology

**Sun Hongqiang[1], Zhang Zhanyue[2] and Hu Yiwen[1]**

[1] Space Engineering University, Graduate School, Huairou, 101416, China
[2] Space Engineering University, Huairou, 101416, China

Email: sjtushq@163.com

**Abstract.** Formation satellites have a very broad application prospect in many aspects due to the characteristics of unlimited baselines, multi-satellites cooperative working, flexible control and strong survivability, such as distributed radar, electronic reconnaissance, three-dimensional imaging, regional navigation and positioning. In the aspect of avoiding space debris collision, the maneuvering avoidance for a single satellite is often adopted, but there are few studies on the coordinated avoidance of formation satellites. This paper introduces the concept and research status of formation satellites, and lists the common formation satellites in the world, then introduces collision warning technologies and collision geometry relation determination of formation satellites, including BOX regional warning method and collision probability warning method. In addition, the strategies of satellite evasive maneuvering are analysed, and various maneuver strategies on different purposes are introduced. Finally, analysing and comparing the research status on debris collision avoidance of formation satellites, a method to keep the working efficiency of formation satellites while maneuvering is put forward.

## 1.Introduction

With the development of space science, more and more spacecrafts are flying into the space, the space environment has become more and more complex. As of October 2017, the US space surveillance network has catalogued about 23,000 space targets larger than 10cm, and 18,747 public catalogued objects in orbit, of which 14,133 are space debris [1][2]. All kinds of signs show that the space field will face congestion, collision and other safety problems in the future. For the spacecraft safety in the space, collision warning is becoming more and more important. To obtain the early warning information, complete the evading within the early warning time, and ensure the safety, are the primary guarantee for the spacecraft to perform the mission in space.

At the same time, with the level of space technology continues to improve, the concept of "distributed spacecraft"comes into being——it separates some functions into several small spacecrafts. Its advantages are significant: not limited by the physical baseline, multiple satellites work together to broaden the application field, stronger flexibility, and higher survivability. For such formation satellites, collision avoidance is much more complicated. It is necessary to adjust the attitude and orbit of one or even more satellites in the formation by means of propulsion devices, at the same time, it is also necessary to keep the working status and working efficiency of the formation as much as possible. In this context, it is of great practical significance and necessity to study the control maneuver process of formation satellites under the condition of early warning.

IOP Conf. Series: Materials Science and Engineering **608** (2019) 012032 doi:10.1088/1757-899X/608/1/012032

## 2. Research status in the world

### *2.1. The conception and research status of formation satellites*

*2.1.1. The conception of formation satellites.* The concept of spacecraft formation was proposed in the 1990s with the rapid development of small satellite technology, which is an important innovation in the field of aerospace technology. The formation of each satellite forms a virtual spacecraft by information interacting and collaborative working, in order to achieve the function that a traditional single large satellite cannot achieve in the space mission. Therefore, this concept quickly became a hot spot of academic research in the field of aerospace as soon as it was proposed.

Compared with the traditional satellites, the spacecraft formation has been improved significantly in function and performance. By placing system loads on different member spacecrafts, the spacecraft formation can achieve some special configurations that are difficult for traditional satellites to achieve, such as providing long baselines and large apertures, expanding the field of space applications. Member spacecrafts can use different modes to work independently or cooperatively. For example, in the earth observation mission, several member spacecrafts observe adjacent areas respectively, and then the observation data are spliced and arranged, which improves the efficiency of space applications. Members spacecrafts can be grouped into another configuration, or access to different loads of spacecraft, the satellite system's flexible switching could enhance the flexibility of space applications. When some members of the spacecraft failure, the rest of the spacecrafts can be reconfigured to achieve the continuation of the system function, or just launch another spacecraft access to the system, enhancing maintainability of spaceflight in orbit.

*2.1.2. research status of formation satellites.* Germany's next generation earth observation plan, TanDEM-X mission (shown as Figure 1), aims to generate high-precision 3D Digital Elevation Models (DEM) with positioning accuracy of 12m and relative height accuracy of 2m. It was launched on 21 June 2010. Due to the TanDEM-X scheme's dual technical advantages of no time loss and InSAR height detection, the expected DEM quality will reach the HRTI-3 standard with a relative elevation accuracy which is better than 2m, making it become the most accurate DEM data in the world.

**Figure 1.** TanDEM-X satellite

The Gravity Recovery And Climate Experiment (GRACE) mission, a twin-satellite formation flight project between the German aerospace center and NASA, was proposed in 1997 and put into orbit in 2002.The system consists of two satellites in a serial formation with a distance of 170-270km, shown as Figure 2. The relative distance between the two satellites is measured by a K-band microwave measuring device with a precision of micron and a relative speed of less than 1 m/s. The project brings the high-precision global gravity field observation experiment into a new stage and provides a powerful tool for monitoring global environmental changes.

IOP Conf. Series: Materials Science and Engineering **608** (2019) 012032 doi:10.1088/1757-899X/608/1/012032

**Figure 2.** GRACE satellite configuration

The "white cloud" satellite system is an ocean surveillance satellite system developed by the US in the 1970s, which is used to track and locate radar and communication signals of ships and moving targets at sea. Analyzing of the early "white cloud" satellites information and satellites observations, the "white clouds" satellite constellation is a group of three satellites, the distance between the three satellites are 30-110km, using time difference of arrival techniques to realize positioning. There is no data storage on the satellites. The detection area is limited to the corresponding area of the ground station's visibility [3]. Figure 3 shows the scenario that the "white cloud" passes over the moon.

**Figure 3.** The "white cloud" passes over the moon

### *2.2. Research status of collision early warning technology*

*2.2.1. Typical collision warning technology.* Due to the errors of spacecraft and space debris in the process of orbit observation, determination and prediction, the occurrence of collision is uncertain. Currently, in collision risk assessment, the main criteria are minimum approach distance (the corresponding determination method is Box region method) and probability (the corresponding determination method is the collision probability method).

Box region method is to define a space area around the spacecraft (the main target). The space area is centered on the spacecraft, and if other space objects enter into this area, an early warning will occur. The space area defined by the US space shuttle collision warning system is a cuboid with the space shuttle as the center, track direction of ±25km, track surface normal direction and vertical track direction of ±5km. When the debris enters this region, the system will give an early warning and provide more detailed orbit prediction data. Figure 4 shows the space region defined in the Box region method. Box region method has a very obvious drawbacks, it will give a dangerous warning as long as a debris enters into the box region, which could cause waste of space resources.

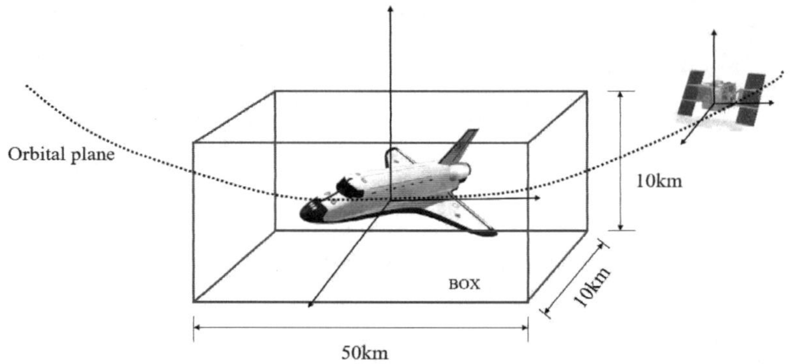


**Figure 4.** BOX region method

Collision probability is a more accurate, effective and objective collision criterion widely used in the world. In addition to taking the two values of Miss Distance (MD) and Time of Closet Approach (TCA) as input data, the calculation of collision probability also involves the motion state of two objects and the information of orbit error covariance at the close of time. This method determines the collision probability according to the error covariance information at TCA. When probability is larger than the given threshold value, the warning system will send out an early warning signal. For the international space station and the space shuttle, NASA divides the probability threshold of evasive maneuvers into Yellow Threshold PY ($10^{-5}$) and Red Threshold PR ($10^{-4}$). When probability was larger than PR, evasive maneuver was performed. When the probability is larger than the PY, the maneuver can be carried out without causing harm to the target of the task. The collision probability method can make the spacecraft to avoid unnecessary maneuvering. Therefore, the space debris collision warning method based on collision probability has become the main method in this field around the world.

In addition, Foster and Estes proposed a numerical method for solving the double integral of polar coordinates according to a certain step size. Patera proposed a method to calculate the collision probability using an equivalent numerical integration model, which can be applied to objects with arbitrary irregular shapes. Chan chooses the representative error and actual approximate parameters to realize the reduction of dimension of double integral, thus obtaining one-dimensional PDF (presented as Rician distribution), and deriving the approximate analytic expression in infinite series. In addition, National University of Defense Science and Technology, Liu Jing's team from Space Center of Chinese Academy of Sciences,Space Engineering University and other universities and research institutes researched the calculation method of collision probability from different perspectives.

*2.2.2. Collision relative geometry determination and collision location determination.* In the space environment, it is possible to have multiple satellites in formation warning at the same time. According to the calculation results of the relative distance, collision position, collision direction and relative collision velocity between the formation spacecraft and dangerous debris, the geometric relationship between large debris and spacecraft can be determined. Take the 3-satellites positioning system as an example, there are two main types of possible collision:

(1) Single satellite collision warning.

As Figure 5 shows below, only one satellite in the 3-satellites formation is under collision warning. At this time, the maneuver of this satellite can be considered while keeping the working efficiency of 3-satellites positioning as much as possible.

**Figure 5.** Single satellite collision warning

(2) Multiple satellite collision warning.

As shown in the figure 6 below, a number of satellites in the 3-satellites formation are under collision warning. At this time, the maneuver of these satellites can be considered while keeping the working efficiency of three satellites positioning as much as possible.

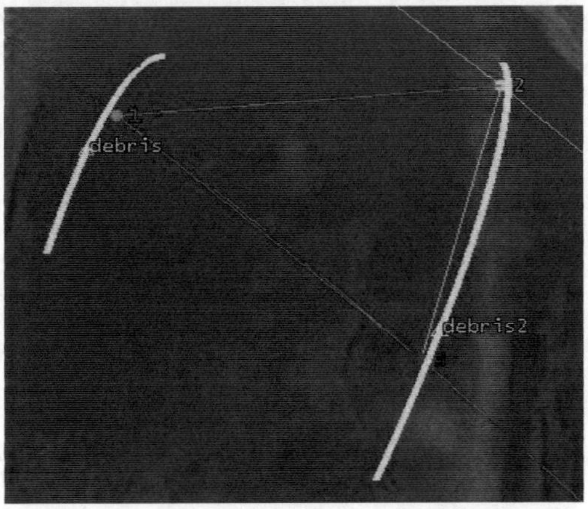


**Figure 6.** Three satellites collision warning

In the orbital coordinate system, Wang Yuefeng's team [4] (from National Defense University) reached the following conclusions based on the collision relationship:

1.Head-on collision:

Before the collision happen, the relative distance in X and Z directions is almost 0. At this time, if the relative distance is Y >0, a head-on collision will occur.

2. Rear-end collision:

before collision happen, the relative distance in X and Z directions is almost 0. If the relative distance Y is less than 0, a rear-end collision will occur.

3. Other coplanar collision:

The relative distance in the Z direction is almost zero before the collision happen:

If the relative distances are X>0 and Y>0, the large debris will collide with the aircraft in the front from upward side.

If the relative distance X<0 and Y>0, then the large debris collided with the aircraft in the front from bottom side (These two collisions would be dangerous enough to cause the craft to disintegrate and potentially explode violently).

If the relative distance is X>0 and Y<0, the large debris will collide with the aircraft in the rear from upward side.

If the relative distance X<0 and Y<0, the large debris will collide with the aircraft in the rear from bottom side (These two collisions are less harmful, but it is still possible to make the spacecraft disintegrated or flight mission interrupted).

In addition, Guo Rong from this team proposed a local collision location determination method: when large debris approaches the space vehicle, the relative motion of two space objects can be regarded as a linear motion, and all perturbation effects can be ignored. In any two adjacent moments, the orbital coordinate system can be "stationary" and the linear equation of the relative distance of the corresponding two moments can be obtained.

The intersection point between the linear equation and the aircraft configuration is the initial collision point. Then, the relative position of the next moment and the previous relative position are taken to form the linear equation, and the intersecting point coordinates with the aircraft configuration can also be obtained. The iterative calculation can be terminated within the required range until the error is acceptable. Through verification, the calculation results of this method can control the precision of the collision point in meters.

### *2.3. Research status of orbital maneuvering*

Short-term avoidance strategy (height separated method):

To forecast the intersection of incident, in N orbital circles before the incident happen, providing a single direction speed increment $\Delta V$ along the track, as a result, the orbital height of the objects is separated by a certain distance. This evasive strategy is suitable for the events with short prediction time, and the events have a close geometric relationship. As shown in the Figure 7, A is the orbital transition point, B is the collision point, and C is the position of satellites after the maneuvering at the moment of collision

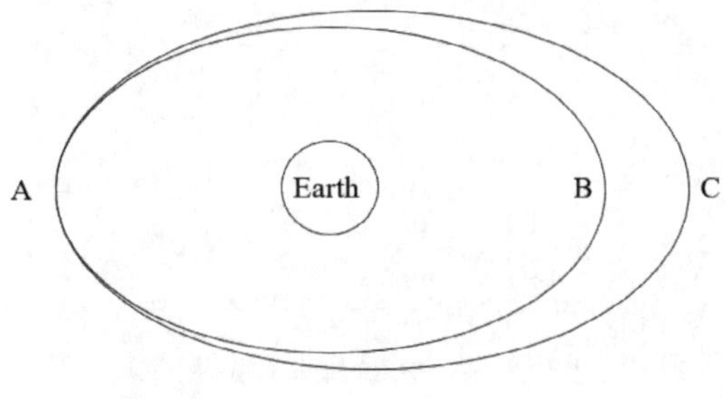


**Figure 7.** Height separated method

This was used by the European Space Agency (ESA) to resolve a high-stakes rendezvous between its ERS-1 satellite and Russia's Kosmos-614 satellite successfully. According to the forecast, this event would take place on June 25, 1997 13:24:46 UTC, ESA control the orbit to provide satellite ERS-1 of 1 m/s (reverse) orbital velocity increment before incident happen, through this maneuver, making ERS-1 satellite orbital altitude is 4 km lower than Kosmos-614 satellite's altitude, thus greatly reduce the collision probability.

Mid-term avoidance strategy (trace separation method):

If the focus is on saving fuel, or if the forecast of a rendezvous event is sufficiently early (at least a day or more in advance), a medium-term avoidance strategy can be adopted.Before the incident, in the direction along the track, giving one or several small velocity increment $\Delta V$ motor to avoid collision, the result is to increase the distance between satellite and debris along the track direction. Since the separation distance along the track direction between the satellite and debris is proportional to the length of the glide arc between the target body's maneuver and the rendezvous time, the increment of velocity needed is approximately 0.06/n(m/s) for each 1km of separation distance along the track direction (where n is the number of loops of the glide track after maneuver). If the given rendezvous warning time is early enough, the trace separation method should be used, and it is best suited for non-head-on rendezvous events. The Figure 8 shows the situation of the satellite returning to the original orbit after the time of collision. If the satellite returns to the original orbit before the time of collision, the satellite will also pass the collision point B, but it will miss a certain time (or distance) at the time of collision.

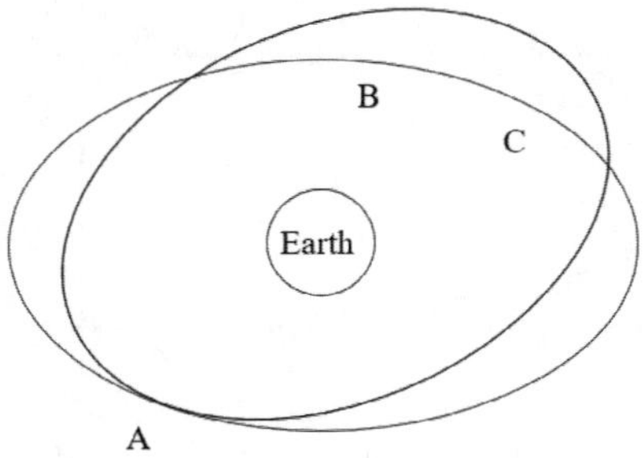


**Figure 8.** Trace separation method

French space agency (Centre National d'Etudes Spatiales, CNES) has used this method to resolve a high-risk rendezvous between its SPOT-2 satellite and debris from the explosion of ARIANE-1 v-16. According to the forecast, this event would occur on July 24, 1997 02:27 UTC. Before the rendezvous happen, debris are flying in front of the SPOT-2 satellite, and the relative velocity is about 4.67 km/s, the CNES provides a speed increment ΔV on the along track direction, made the distance increasing to more than 2 km in the direction along the trails, greatly reduced the collision probability.

In addition, Zhang Yasheng [5] took the maximum invisible time as an indicator to analyze the constellation coverage performance when one or more satellites were missing. Considering the fuel consumption and time the maneuver need, a satellite phase adjustment method is proposed, which is shown as Figure 9. Then based on the worst combination of satellites, using the satellite phase adjustment method, proposed a method to minimizing the transfer time basing on rapid reconfiguration of uniform phase constellations.

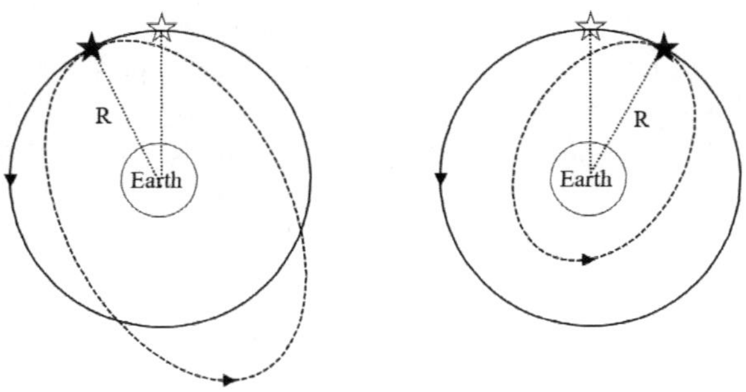


**Figure 9.** Satellite phase adjustment method

Zhang Dongqing [6] of Harbin Institute of Technology proposed a short-term avoidance strategy based on gradient descent method in his paper. The collision probability density function is obtained at the collision point. the satellite position after maneuver can be shifted in the direction of the negative gradient of the collision probability density function.

In the literature, Feng Hao [7] from the department of Beijing Space Vehicle proposed a maneuver strategy based on closed-loop trajectory. The drift of the normal satellite ground trajectory and the adjustment process are as Figure 10: When the trajectory of the ground is $\Delta L/2$ located at the eastern boundary (positive to the east), adjust the half-length axis to $a = a_0 + \Delta a$ ($a_0$ is the nominal value of the half-length axis). The trajectory of the ground drifts westward. Then, due to the influence of atmospheric resistance, the half-length axis gradually decreases. At $a = a_0$, the trajectory just drifts to the western boundary $-\Delta L/2$. Thereafter, the semi-major axis continues to decrease, but the trajectory drifts to the east instead. When $a = a_0 - \Delta a$, the trajectory drifts to the eastern boundary, then the half-length axis needs to be adjusted again, otherwise the trajectory will continue to drift eastward beyond the boundary allowed. The adjustment is $2\Delta a$. Therefore, the ground trajectory constitutes a closed loop, that is, the trajectory maintaining control loop, as shown in the figure below.

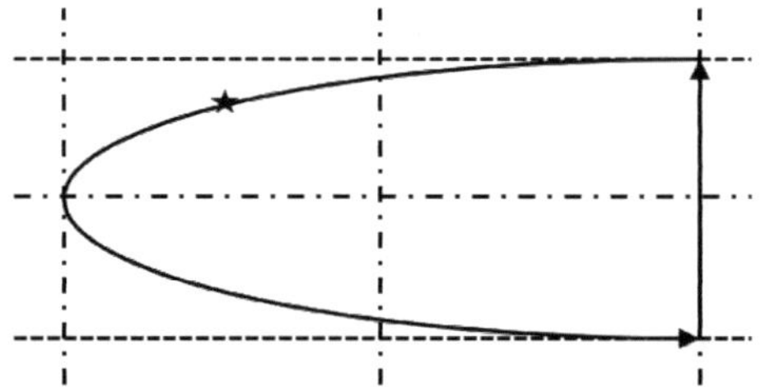


**Figure 10.** Trajectory maintaining control loop

## 3. Comparison of research status

Based on the comprehensive research status, the following conclusions can be drawn:

(1) Satellite formation technology is put into use worldwidely. In the space target evasive maneuver, the method widely used in the world is to give the spacecraft an instantaneous pulse speed, so that it deviates from the original orbit, so as to reduce the risk of collision. However, when the orbit changes, the working state of the vast majority of satellites will change, and there are few studies on how to keep the working efficiency of formation satellites to the greatest extent when maneuvering.

(2) In terms of collision warning technology, BOX area method and collision probability method are mainly used for collision warning, among which collision probability method can avoid many unnecessary maneuvering evasions of spacecraft and save maneuvering resources. Therefore, collision probability method is a commonly used collision warning method around the world. At the same time, the accuracy of the collision point can be controlled in meters by the local collision position determination method.

(3) In the aspect of spacecraft maneuvering, the methods most commonly used are height separated method and trace separation method. However, such avoidance methods fail to take the working efficiency of satellites during the avoidance period into consideration. Therefore, with the goal of keeping the working efficiency during the avoidance maneuver, multiple factors such as positioning effect, energy consumption and avoidance time can be taken into account comprehensively to build a function optimization model, so as to formulate the most appropriate orbital maneuver mode.

## Reference

[1] Fletcher K. *Space Operations* [M] European Space Agency, 2017:2-2
[2] NASA Orbital Debris Program Office. *Orbital Debris Quarterly News*, Volume 21, Issue 4[J].2017.11
[3] Li H.S. *Estimation of TDOA And Research on Three-Satellite Location Algorithms* [D] Harbin: Harbin Institute of Technology, 2010
[4] Guo R. *Study on Space Debris Collision Early Warning and Maneuver Strategy to Low Earth Orbit Spacecraft* [D]. University of national defence science and technology, 2005.
[5] Zhang Y.S, Zhang Y.L. *Research on the Fast Reconfiguration Method for the Performance Restore Type Constellation* [J]. Journal of equipment command and technology institute, 2005, 16(4).
[6] Zhang D.Q. *Study on the Collision Risk Assessment and Security Control of Satellite Formation Flying* [D]. Harbin: Harbin Institute of Technology 2015.
[7] Feng H, Zhao Q, Gao S. *A Practical Collision Avoidance Maneuver Method of Space Debris for Remote Sensing Satellite* [C] China aerospace security conference. 2015.

IOP Conf. Series: Materials Science and Engineering **608** (2019) 012033 doi:10.1088/1757-899X/608/1/012033

# Research on Optimization of Working Mode of Long-range Early Warning Radar Guided by Space-based Information

**ZHOU Tong[1], ZHANG Yasheng[2], ZHOU Haijun[2] and HAN Jingbin[1]**

[1]Graduate School, Space Engineering University, Beijing 101416, China;
[2]Space Engineering University, Beijing 101416, China;

Email: 465325519@qq.com

**Abstract.** The long-range early warning radar is the key equipment in the early defense process of ballistic missile. For the problem of low working efficiency of long-range early warning radar without guidance information, this paper studies the working process of long-range early warning radar with space-based guidance information, and proposed a new long-range warning radar working mode. Firstly, the coordinate system transformation and guidance information extraction of space-based guidance information are given. Then the concept of coherent accumulation of radar signals is introduced and a new radar operation mode is proposed. Finally, the radar signal coherent accumulation process is established. A mathematical model of radar search distance and intercept probability. The typical scene is built and simulated. The results show that the proposed radar working mode will effectively reduce the search airspace of the long-range early warning radar and improve the radar detection range.

## 1. Introduction

In the process of long-range ballistic missile defense, early warning satellites and long-range early warning radars are the key equipment for early warning of ballistic missiles. Because of its high position, the early warning satellite will first detect the tail flame signal of the ascending missile and find the missile target[1]. The long-range early warning radar has a long detection range and can capture the missile target earlier because of its own power aperture[2]. In the early warning process of missiles, long-range early warning radars usually search and capture incoming missiles under the guidance of early warning satellites[3-4].

The general long-range early warning radar can use multiple working modes in combat readiness, and use different beam scanning methods for target detection, such as search for warning lines, search walls and small window search windows, but no matter which working mode, if no Guided by the guidance of information, the radar will increase the detection range to ensure the target's falling probability. This way of working will seriously waste radar resources and reduce the working effect of the radar[5-6].

Aiming at the problem of low efficiency of long-range early warning radar without guidance information, this paper will study the working process of long-range early warning radar with space-based guidance information, and propose a new long-range early warning radar working mode.Firstly, the paper studies the conversion process of the warning information generated by the early warning satellite to the guidance informationthe needed by the long-range early warning radar, and extracts the content of the guidance information needed for the radar work. The concept of coherent accumulation of radar signals is introduced, and analysis how the cumulative amount of radar signal coherent increases the radar's working effect. On this basis, the space-based guidance information content is combined with the radar working process, and a new long-range early warning radar working mode is proposed, and the radar search interception process is modeled. Finally, through the scene simulation, the working effect of the long-range early warning radar working mode proposed in this paper is quantitatively analyzed.

## 2. Coordinate system transformation and extraction of space-based guidance information

The early warning information generated by the early warning satellite is usually expressed in the geocentric inertial coordinate system. If the space-based information is to be used to guide the radar work, the warning information needs to be converted to the radar pointing coordinate system. The transformation from the geocentric inertial coordinate system to the radar pointing coordinate system involves the geocentric inertial coordinate system, the geocentric fixed coordinate system, the radar rectangular coordinate system and the radar pointing coordinate system. The definition and conversion relationship of these coordinate systems can be referred to the literature [7]. The conversion matrix of the geocentric inertial coordinate system to the geocentric fixed coordinate system, the geocentric fixed coordinate system to the radar Cartesian coordinate system and the radar Cartesian coordinate system to the radar pointing coordinate system can be expressed as $\boldsymbol{M}_g, \boldsymbol{M}_p$ and $\boldsymbol{M}_s$. Assume that the covariance matrix in the geocentric inertial coordinate system is $\mathrm{Cov}(C_{ei})$. According to the above derivation, the missile motion state covariance matrix $\mathrm{Cov}(C_s)$ in the radar pointing coordinate system can be obtained. The conversion relationship between the two can be expressed as:

$$\mathrm{Cov}(C_s) = \boldsymbol{J}_t \mathrm{Cov}(C_{ei}) \boldsymbol{J}_t^{\mathrm{T}} \tag{1}$$

Where $\boldsymbol{J}_t$ is the total transformation matrix during the entire conversion process:

$$\boldsymbol{J}_t = \boldsymbol{M}_s \boldsymbol{M}_p \boldsymbol{M}_g \tag{2}$$

In the covariance matrix $\mathrm{Cov}(C_s)$ under the radar pointing coordinate system, the three elements on the diagonal are $\delta_{sx}^2$, $\delta_{sy}^2$ and $\delta_{sz}^2$, respectively representing the variances in the X, Y and Z directions. At present, most scholars express the radar search radius as $r = \sqrt{\delta_{sx}^2 + \delta_{sy}^2 + \delta_{sz}^2}$ in the research process. At the same time, according to the principle of small probability events in probability science, in order to ensure the radar capture of the target, the radius of the search area of the radar is expanded by three times. This method of division is relatively rough and lacks scientific theoretical derivation. In this paper, the uniform sphere model proposed in the literature [8] is used to determine the radar search radius. The maximum eigenvalue $\lambda_{V,1}$ of the matrix is calculated from the covariance matrix $\mathrm{Cov}(C_s)$ in the radar pointing coordinate system, and the search radius can be expressed as $\sqrt{\kappa \lambda_{V,1}}$. The resulting search radius is lower than the traditional method, which is conducive to saving radar resources and improving radar work efficiency.

For the long-range early warning radar, in addition to the size of the search area, its work is still inseparable from the pitch angle and azimuth information of the search area. When the coordinate system is converted, the azimuth angle $\varphi$ and the pitch angle $\theta$ of the missile in the radar Cartesian coordinate system are The angle information required for the radar to set the screen. At this point, the space-based guidance information for guiding the long-range early warning radar work is extracted.

## 3.Radar signal coherent accumulation principle

The classic single-pulse radar equation can be expressed as:

IOP Conf. Series: Materials Science and Engineering **608** (2019) 012033 doi:10.1088/1757-899X/608/1/012033

$$R_{\max} = \left[ \frac{P_t G_t^2 \lambda^2 \sigma}{(4\pi)^3 k T_0 B_n F_n L (S/N)_{\min}} \right]^{1/4} \tag{3}$$

$P_t$ is the radar peak radiated power, $G_t$ is the transmit antenna gain, $\lambda$ is the radar wavelength, $\sigma$ is the target radar cross-sectional area (RCS), and $k$ is the Boltzmann constant 1.38×10-23 (J/K), $T_0$ For noise temperature, $B_n$ is the receiver noise bandwidth , $F_n$ is the system noise figure, $L$ is the total loss factor of the radar system, $(S/N)_{\min}$ is the minimum detectable signal-to-noise ratio, that is, the ratio of signal to noise, in the case where the guaranteed target is detected, and is usually 13 dB or more.

It is noted that the minimum detectable signal-to-noise ratio $(S/N)_{\min}$ in the radar equation has an important influence on the maximum detection range of the radar. In order to more clearly see the influence of the minimum detectable signal-to-noise ratio on the maximum detection range of the radar, the equation (3) is rewritten as energy form:

$$R_{\max} = \left[ \frac{P_t G_t^2 \lambda^2 \sigma}{(4\pi)^3 k T_0 B_n F_n L (S/N)_{\min}} \right]^{1/4} \overset{E_t = P_t \tau}{=} \left[ \frac{E_t G_t^2 \lambda^2 \sigma}{(4\pi)^3 k T_0 F_n L (S/N)_{\min}} \right]^{1/4} \tag{4}$$

It can be seen that the radar detection range has nothing to do with the radar bandwidth.

In the process of radar target detection, the coherent accumulation process of the radar received signal will also have a greater impact on the radar detection range. For stationary targets, the signal phase remains unchanged, and the signal energy can be cumulatively enhanced. For slow moving targets, the signal phase changes slowly and the cumulative efficiency is also high. For space fast moving targets, the signal phase changes rapidly, and there is also a cross-cell migration. The signal can also be accumulated by first performing signal compensation.

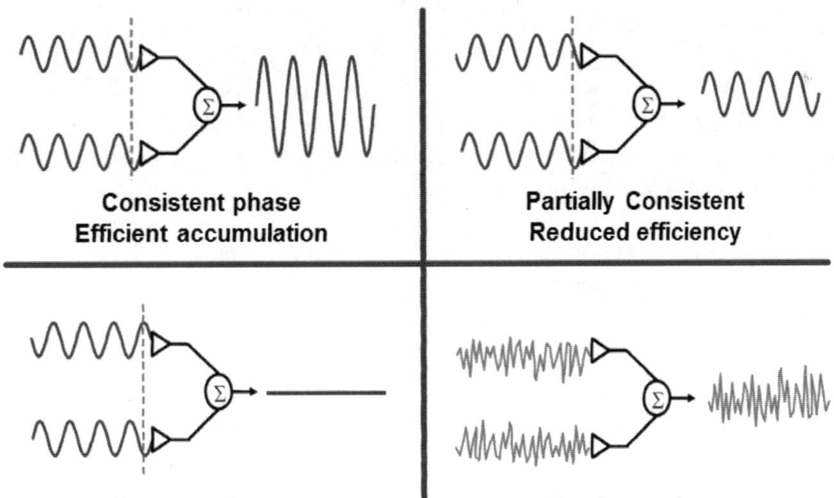


**Figure 1.** Coherent cumulative effect of different phase signals

Figure 1 shows the coherent accumulation effect of the signal in the case of four phase differences. When the two signals are in phase, the best coherent accumulation effect can be obtained, and the signal energy after superposition is significantly increased. If the two signals are opposite in phase, the two signal energies will cancel each other, and the superimposed signal energy is zero; The signal phase is partially consistent, and the coherent accumulation effect will be between the above two cases, and the signal energy is partially enhanced; if the superimposed signal is random noise, the signal energy is inefficiently enhanced.

Taking the two signals in phase as an example, if there are $M$ equal amplitude pulse coherent accumulations, the signal-to-noise ratio power of the accumulated radar will be increased to $M$ times, the radar signal power will increase by $M^2$, and the corresponding noise power will increase by $M$ times. After superposition, the signal-to-noise ratio power of the signal will eventually increase by $M$ times. The effect of the coherent cumulative result on the radar signal can be expressed by the following formula:

$$(S/N)'_{\min} = (S/N)_{\min} / M \tag{5}$$

Bringing equation (5) into the energy equation of the radar's maximum detection distance, we can more clearly see the improvement of the effect of the coherent accumulation process on the radar. In the actual working process of the radar, when the transmitted signal is reflected back to the receiver by the target, the phase of the multiple echo pulses of the radar is the same or the phase change is known, and the phase accumulation can form an efficient accumulation of energy. When $M$ equal-amplitude pulses accumulate, the signal-to-noise ratio is increased by a factor of $M$; if it is non-coherent, the signal-to-noise ratio is increased by a factor of $\sqrt{M}$. That is, if the phases of multiple consecutive echo signals are different and the variation law is unknown, the signal-to-noise ratio of the phase-accumulated pulses is increased by at least $\sqrt{M}$ time, that is:

$$\left[\frac{\sqrt{M}P_tG_t^2\lambda^2\sigma}{(4\pi)^3kT_0B_nF_nL(S/N)_{\min}}\right]^{1/4} \le R_{\max} \le \left[\frac{MP_tG_t^2\lambda^2\sigma}{(4\pi)^3kT_0B_nF_nL(S/N)_{\min}}\right]^{1/4} \tag{6}$$

It can be seen that the cumulative number $M$ of coherent has an important influence on the working effect of the radar.

## 4.Radar search work mode optimization

If the long-range early warning radar does not have the guidance information provided by the early warning center during the daily combat readiness, a 120° search screen is usually established at the far end when the combat readiness is on duty. The principle of this setting is that all the targets in this direction will pass through the search screen and be captured by the radar, as shown in Figure 2 ($\theta_0$ is the azimuth range of the radar search screen).

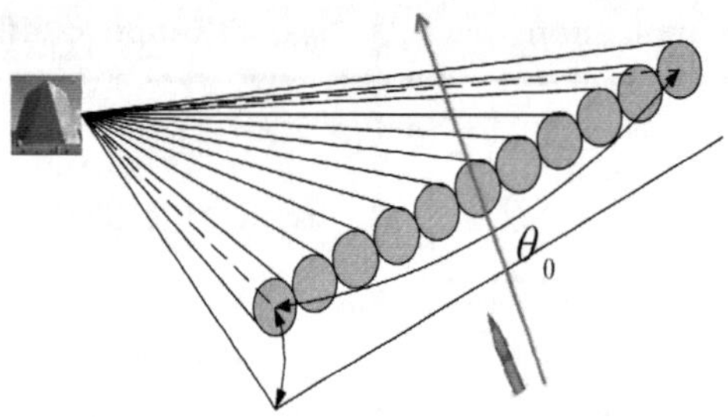


**Figure 2.** Long-range early warning radar beam diagram without guidance information

If there is guidance information, if the radar continues to maintain the target search in the large search screen in Figure 2, it will not only waste radar resources, but also reduce the radar performance. In response to this problem, this paper combines space-based guidance information with radar work, and proposes a radar search mode based on space-based guidance information, as shown in Figure 3 ($\theta_1$ is the azimuth range of the radar search screen).

IOP Conf. Series: Materials Science and Engineering **608** (2019) 012033 doi:10.1088/1757-899X/608/1/012033

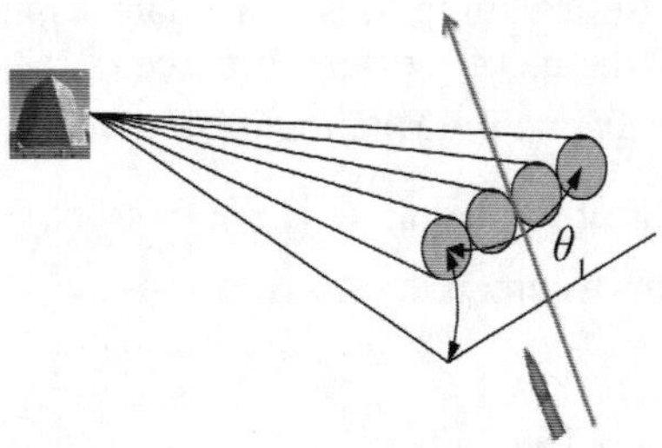


**Figure 3.** Long-range early warning radar beam diagram with guidance information

In this radar mode of operation, the radar will no longer maintain a large search screen. Instead, the radar search area will be determined in real time based on the content of the space-based guidance information. The reduction of the radar search range will effectively improve the radar work and enhance the radar work effect. The radar search interception process in both modes of operation is modeled below in conjunction with the coherent cumulative quantity.

## 5.Radar search interception process modeling

In this section, the detection range and discovery probability model are established according to the two radar search screen settings in the presence or absence of space-based guidance information, and the expressions of radar detection distance and discovery probability under two setting modes are given.

1) Under the condition of no guidance information, for the scanning operation mode of the early warning radar, the search time of the radar to complete the search area is $T_{s1}$, and the time during which the missile crosses the radar beam range is $t_p$. To ensure that the target is not missed, $T_{s1} < t_p$ should be satisfied. Assume that the horizontal search airspace azimuth range is $\mathrm{A}_p$, the beam width is *a*, and for the beam at the distance $R_0$, the beam span is $R_0 \cdot a$, assuming that the target's velocity in the vertical direction of the beam is $v_p$, and the time of the target crossing the search screen is:

$$t_p = \frac{R_0 \cdot a}{v_p} \tag{7}$$

The dwell time of each position beam can be calculated by $PRT \cdot M_p = T_{s1} / (\mathrm{A}_p / a)$. Combined with $T_{s1} = t_p = R_0 \cdot a / v_p$, the number of pulses of the radar at one position can be obtained as follows:

$$M_p = \frac{R_0 \cdot a^2}{PRT \cdot \mathrm{A}_p \cdot v_p} \tag{8}$$

Where $R_0$ is the single-pulse search distance of the radar without guidance; *PRT* is the single beam dwell time; $\mathrm{A}_p$ is the radar azimuth range, and the radar uses a large fan search screen under no-boot conditions. The search screen azimuth range is usually set to 120°. $v_p$ is the speed perpendicular to the radar beam front when the missile is screened.

2) Under the condition of guiding information, the radar does not need to have a large fan-shaped screen for wide-range warning. The range of the search screen can be narrowed according to the

guidance of the space-based guiding information to save radar resources and improve the radar working effect. Analogy without the guidance of the search process, the radar to complete the search time for the search area when there is guidance information is $T_{s2}$, the time of the missile crossing the radar beam range is $t_w$, in order to ensure that the target is not missed, should meet $T_{s2} < t_w$, in this case, the time the target traverses the search screen is:

$$t_w = \frac{R \cdot a}{v_w} \tag{9}$$

The dwell time of each position beam can also be calculated by $PRT \cdot M_w = T_{s2} / (A_w / a)$, and the number of pulses of the radar at one position can be obtained as follows:

$$M_w = \frac{R \cdot a^2}{PRT \cdot A_w \cdot v_w} \tag{10}$$

*PRT* and $v_w$ have the same meaning under no-boot condition, but $R$ and $A_w$ have different meanings from no-boot. At this time, $R$ and $A_w$ are no longer a fixed value, rather, they are variable related to the guidance information, which is determined by the size of the error radius in the guidance information. As shown in equation (11):

$$A_w = 2 \cdot \arctan(r / R) \tag{11}$$

Where $r$ is the radar search radius determined under the radar pointing coordinate system. Therefore, equation (10) can be rewritten as:

$$M_w = \frac{R \cdot a^2}{2 \cdot PRT \cdot \arctan(r / R) \cdot v_w} \tag{12}$$

Missile defense is a dynamic process. The distance between the target and the radar changes at the same time. At the same time, the error radius of the guidance information also changes at any time. Therefore, the cumulative number of coherent parameters is not a fixed value, that is, the effective detection distance of the radar also changes with time. Timely capture of missile targets requires real-time update of warning guidance information. This paper uses the following methods to pre-push to ensure timely detection of missile targets:

① Pre-push from time *t*, it is assumed that the maximum effective detection distance of the radar is equal to the distance between the target and the radar, that is, *R*=*D*, where *D* can be obtained by the position of the missile under the radar pointing coordinate system. Under this condition, according to the radar beam distance *R* and the search radius in the radar guidance information, the cumulative number *M* of coherent parameters of the radar at the current time can be obtained by using equation (12).

② Using the coherent cumulative number *M* at the current time, the range *Rr* of the radar after the coherent accumulation can be obtained according to equation (6). Compare *Rr* with *D*. If *Rr*<*D*, that is, the radar effective range is smaller than the current radar and target distance, return ①; if *Rr*>*D*, the target enters the radar power range, and the radar captures the target.

The single intercept probability of the radar to the target can be expressed as the product of the target's probability of discovery and the probability of falling into:

$$P_j = P_l \cdot P_f \tag{13}$$

Where $P_l$ is the probability of falling, and $P_f$ is the probability of discovery, where the probability of falling $P_l$ is about 1. In the actual acquisition process, the radar beam will detect multiple times on the target, and the probability of accumulation through multiple interception is:

$$P = 1-(1-P_j)^M = 1-(1-P_l \cdot P_f)^M \tag{14}$$

Where $P_j$ is the single intercept probability, $P$ is the final intercept probability, and $M$ is the coherent cumulative number.

The different $M_p$ and $M_w$ obtained are brought into the radar distance and the intercept probability equation, and the radar detection distance under the condition of the space-based guidance information can be compared.

**6.Simulation analysis**

In order to verify the effectiveness of the proposed method, the following simulation scenarios are established. Based on an intercontinental ballistic missile, the algorithm verification is carried out. The simulation parameters are set as follows:

1) The missile launch time is 11:57:00 on June 1, 2018, the launch point is (40°N, 127°E), and the drop position is (42°N, -120°E), the missile shutdown time is 190s and the flight time is 1981.6s.

2) The position of a certain type of long-range early warning phased array radar is (38°N, -118°E). The radar operating parameters refer to the US Army AN/TPY-2 radar. For a warhead target radar with an RCS of 0.1 $m^2$, the single-pulse detection distance is 779 km, the beam opening angle is 120°, the radar beam width is 0.5°, and the single beam dwell time is 1 ms. The radar operates in a non-coherent accumulation mode with a target single detection probability of 0.8.

3) The state average $m_0$ and the covariance $p_0$ of the missile's shutdown point detected by the high-altitude early-warning satellite are :

$m_0$ =[-3198710.02m -3701918.82m 4501927.48m 3836.31m/s -1995.50m/s 5089.47m/s],

$$p_0 = \begin{bmatrix} 250000 & 0 & 0 & 0 & 0 & 0 \\ 0 & 250000 & 0 & 0 & 0 & 0 \\ 0 & 0 & 250000 & 0 & 0 & 0 \\ 0 & 0 & 0 & 250 & 0 & 0 \\ 0 & 0 & 0 & 0 & 250 & 0 \\ 0 & 0 & 0 & 0 & 0 & 250 \end{bmatrix}$$

. The error propagation analysis was performed using the covariance analysis description method (CADET), and the result of the ballistic error propagation was used as the initial value of the space-based guidance information[9-10].

In order to compare and analyze the improvement of the working effect of the guiding information on the radar, the following simulation scenarios are respectively established:

a) Without guidance information, the radar uses a single pulse fixed search range and beam angle to detect.

b) With guidance information, the radar calculates the target error radius and the number of coherent accumulations in real time according to the guidance information, and updates the search screen opening angle and the radar action distance after the coherent accumulation.

Due to the influence of the curvature of the earth and the position of the radar, the missile is not in the radar line of sight for a period of time after the shutdown, and the scene pre-push time is calculated to be 12:17:18. That is, $t_0$=1218s, the simulation step size is 0.1s.

In the scene simulation process, the change of the radius of the radar search area is shown in Figure 4. The black solid line in the figure is the radar search area radius calculated by the traditional radar search area radius determination method, and the red dotted line is the radar search area proposed by the paper. It can be seen that using the uniform sphere model to determine the radar search radius will effectively reduce the radar search area.

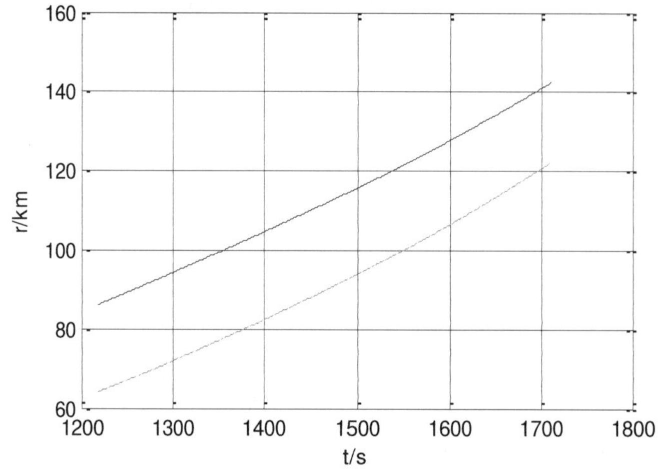


**Figure 4.** Radar search area radius change

After scene simulation, the calculation results are shown in the Table 1:

**Table 1.** Scene simulation results

| | Discovery time | Discovery distance | Coherent cumulative quantity | Interception probability |
|---|---|---|---|---|
| Scene a | 1889.3s | 779km | 87 | About 1 |
| Scene b | 1711.9s | 1869.8km | 1103 | About 1 |

As can be seen from Table 1,In scene b, the radar uses the small screen search mode under the guidance of the space-based guidance information, determines the size of the search airspace, and obtains the azimuth and elevation angle required for the radar to set the search screen. According to the calculated cumulative number of coherents, the search screen is set at the far end in advance, so that the radar detection distance is greatly increased. There is also a certain coherent accumulation in the radar in scene a, but there is no guiding information when the radar works, and the target position cannot be predicted in advance, and the search screen cannot be set at the far end; For the radar intercept probability, the cumulative number of radar coherent is higher in both cases, and the radar can obtain a good interception effect.

## 7. Conclusion

Based on the principle of coherent accumulation of radar signals, this paper proposes a new radar working mode for the problem that the traditional long-range early warning radar is not efficient. The traditional radar working mode is optimized by using the space-based guiding information, and the size of the radar search airspace and the azimuth and elevation angle of the radar search screen are determined in real time according to the content of the guiding information. The simulation results show that the working mode of the long-range early warning radar proposed in this paper will effectively increase the cumulative number of coherent radar signals and improve the radar detection range.

## Reference

[1] Zhang W C, Chen J and Gao Y. *Overview of the Development of American Infrared Early Warning Satellite System*[J]. Ordnance Automation, 2018, v.37; No.236(06): 4-8.

[2] Liu X, Jia C Y. *Overview of Sea-Based X-Band Radar in the United States*[J]. Military Abstracts, 2018(3): 45-47.

[3] Xiao J K, Du Y and Li W M, et al. *Analysis of Ballistic Missile Defense System and Its Key*

*Technology Development*[J]. Tactical Missile Technology, 2017(3): 18-22.

[4] Ding X D, He W X and Li Y X. *Searching for Airspace Optimization of Ground-Based Early Warning Radar Supported by Space-Based Information*[J]. Shanghai Aerospace, 2011, 28(1): 56-60.

[5] Chen Y Z, Guo Z L and Wu T, et al. *Accuracy Analysis of Ground-based Early Warning Radar Supported by Space-based Information*[J]. Journal of Projectiles, Missiles and Guidance, 2006, 26(2): 989-993.

[6] Zhou Y, Wang X S and Feng D J, et al. *Research on monitoring airspace of phased array radar based on ballistic prediction*[J]. Journal of Electronics & Information Technology, 2006, 28(7): 1209-1214.

[7] Yan D D, Li Z and Xu C. *Research on phased array radar search area guided by low-orbit early warning satellite*[J]. Journal of University of Electronic Science and Technology of China, 2016, 46(6):911-916.

[8] Zhou T, Zhang Y S, Zhou H J and Han J B. *Research on the Method of Generating Early Warning Information of High-orbit Early Warning Satellite*[J]. Modern Defence Technology, 2018(06): 15-23.

[9] Sun Y, Meng F K and Wu N, et al. *Ballistic Prediction Performance Analysis of Analytic Geometry and Numerical Integration Method*[J]. Journal of Information Engineering University, 2016, 17(5): 635-640.

[10] Chai H, Xing J J and Tang G J. *Ballistic Missile Fall Point Prediction Based on Covariance Analysis Descriptive Function Technology*[J]. Missile and Space Vehicle Technology, 2011(4):37-40.

IOP Conf. Series: Materials Science and Engineering **608** (2019) 012034 doi:10.1088/1757-899X/608/1/012034

# A Method of Communication Constellation Resilience Evaluation

**Han Jingbin**[1], **Zhang Ya-sheng**[2], **Tang Ya-feng**[2] **and Zhou Tong**[1]

[1]Department of Graduate Management, Space Engineering University, Beijing 101416, China;
[2] Space Engineering University, Beijing 101416, China;

Email: dngshjb@163.com

**Abstract.** The resilience of the communication constellation is an important measure of the communication mission assurance capability of the communication constellation. In order to effectively measure the strength and weakness of the communication constellation, the research on the elastic evaluation method of communication constellation is carried out. First, according to the definition of the space system elasticity of the US Air Force Space Command, the definition of the communication constellation elasticity is given. Subsequently, the covering ability and reconstruction energy index are proposed as the objective function of the restored constellation performance recovery, and the damaged constellation performance recovery model based on NSGA-II algorithm is constructed. Finally, a resilience evaluation method for communication constellation with a single communication constellation as the evaluation target is established, and the simulation calculation is carried out by taking the MUOS constellation as an example. It is concluded that the constellation is poor in terms of robustness, and because of the lack of satellite performance, the task guarantee capability is limited.

## 1. Definition of communication constellation resilience

Resilience, originated from the Latin resiliere, that is, "rebound or recoil." In a broad sense, Resilience refers to a property of the object itself. It is a property that can restore its original state after a destructive event occurs. Resilience is widely used in many disciplines and fields such as physics, sociology, ecology, and internet. [1]

In recent years, with the advent of “resilience engineering” in the field of engineering systems, the United States has proposed to strengthen the resilience of critical infrastructure. The resilience of the space system is thus proposed and regarded as an important reference standard for future space capacity building. [2]

The resilience of the space system given by the Air Force Space Command (AFSPC) is defined as: " Resiliency is the ability of a system architecture to continue providing required capabilities in the face of system failures, environmental challenges, or adversary actions. "[3-4]

According to AFSPC’s research and analysis on the resilience of space system, this paper proposes the definition of the resilience of communication constellation:

Resiliency of communication constellation is the ability of a system architecture to continue providing required communication capabilities in the face of system failures, environmental challenges, or adversary actions.

Unlike the resilience of space system proposed by the AFSPC, the resilience of the communication constellation is considered to be the internal property of the constellation itself and does not involve external effects. Therefore, the spatial support actions outside the constellation (such as reissuing satellites and mobilizing other constellation satellites) improve the performance of the communication constellation through external actions. It does not belong to the behavior of the constellation to improve the constellation performance, and does not belong to the category of communication constellation resilience.

## 2. Research on resilience recovery method of damaged communication constellation

The fast recovery capability of a compromised communication constellation refers to the ability to recover partially compromised communication performance through short-term adjustments. It can reduce the impact of adverse events on communication mission assurance, and it is an important part of communication mission assurance. It is also a key link to evaluate the resilience capabilities of communication constellations.

When multiple satellites fail, the methods for quickly recovering the damaged constellation include rapid response satellites, near space vehicles, and damaged constellation reconstruction. However, the rapid response satellite needs to have backup satellites and can only supplement the low-orbit constellation. The near space vehicles can only meet the service needs of a specific area. Moreover, both fast-sounding satellites and adjacent-space vehicles recover damaged communication capability through external support. But the purpose of this paper is to evaluate the resilience of the communication constellation itself, and its recovery is reflected in the ability to restore the damaged communication function only through the internal adjustment of the communication constellation. Therefore, the damaged constellation reconstruction is not only the main way to quickly recover the communication capability of the damaged communication constellation, but also an important means to ensure the elastic capability of the communication constellation.

### *2.1. Reconstruction indicators*

(1) Covering ability

Coverage is a measure of the extent to which a system can support a communication transmission when a user makes a request. To simplify the indicator model, the inter-satellite link problem is ignored. That is to say, when the satellite can cover the area, it can provide communication services for the area without considering whether the communication link can be established.

The location of the satellite determines the range of services it can provide. Due to mission needs, national interests, etc., users must have different levels of communication needs for different regions of the world. Based on this consideration, regional coverage based on communication requirements is proposed. The world is divided into 60 grid areas. Because the high latitudes in the map have a certain deformation, the area grid size has some differences at different latitudes. According to the degree of demand for communication satellites in the region, communication needs are divided into three levels: necessary, urgent need and general need.

Taking the United States as an example, the distribution and coverage of a total of 29 satellites in the DSCSIII, Milstar, MUOS, SDSIII, SDSIV, UFO, WGS series, etc., which were launched after 2000, are shown in figure 1[5]. Based on this, the communication demand distribution map is drawn. As shown in figure 2, necessary is red, urgent need is yellow, and general need is colorless.

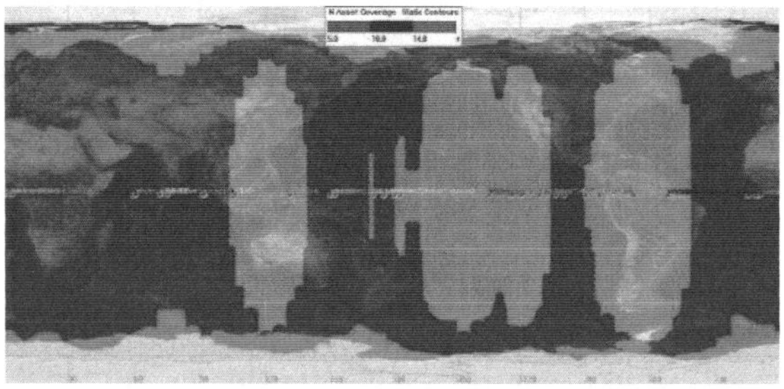

**Figure 1.** Multiple coverage of US communications satellites

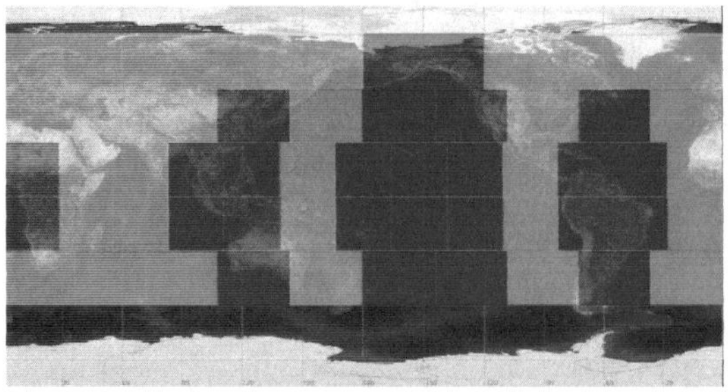

**Figure 2.** Communication demand distribution

The calculation formula for coverage is: $C = \sum_{i=1} \lambda_i c_i / \sum_{i=1} \lambda_i$ (1)

Where $\lambda_i$ is the communication requirement of the i region ($\lambda$=1, 5, 10), and $c_i$ is the average coverage of the communication constellation for the i region in one regression period.

(2) Reconstruction energy

Reconstructing energy is an indicator of the energy consumption required to measure a constellation reconstruction scheme. Since most of the communication satellites are located in the GEO orbit, the launch is difficult and costly. Every energy is invaluable without considering the on-orbit energy filling. At the same time, the residual energy of the satellite determines the remaining life of the satellite to a certain extent. Therefore, the energy consumption should be minimized when performing reconstruction optimization.

The total energy consumed for reconstruction refers to the sum of the energy required to participate in the orbital maneuver of all reconstructed satellites, which is represented here by the total speed increment $\Delta V_{all}$.

$$\Delta V_{all} = \sum_{i=1}^{n} \Delta V_i \alpha_i \tag{2}$$

Where n is the number of satellites required to reconstruct the constellation, $\Delta V_i$ is the total speed increment required for the reconstruction of the i-th satellite, and $\alpha_i$ is the energy weighting factor determined from the remaining energy of the i-th satellite. The more residual energy of the satellite, the smaller the factor. The less energy available, the larger the factor. In this way, the possibility of orbital maneuvering of a satellite with relatively few remaining energy is reduced, so that the life of the reconstructed communication constellation is longer. A detailed comparison table is shown in table 1 below:

**Table 1.** Energy weighting factor comparison table

| Residual energy ratio | Energy weighting factor |
|---|---|
| 80~100 | 1 |
| 60~80 | 2 |
| 40~60 | 3 |
| 20~40 | 4 |
| 0~20 | 5 |

### *2.2. Damaged constellation reconstruction method based on NSGA-II*

Due to the small number of communication satellites in the medium and high orbits, the configuration of the communication constellation is relatively simple. When a satellite failure event occurs, the orbital maneuver is mainly based on the mission requirements and the experience of the expert. The reconstruction scheme can achieve the mission requirement, but it is not necessarily optimal. With the rise of multi-objective optimization algorithms in recent years, the method of modeling and analysis through simulation software has gradually become the main way to solve multi-objective optimization problems [6-9].

Debu et al. proposed the NSGA-II algorithm in 2000. Most of the NSGA algorithm's flaws have been improved, making it one of the most widely used multi-objective optimization algorithms. The NSGA-II algorithm is simple and efficient, including initialization, hybridization, and mutation. Combined with the actual requirements of on-orbit reconstruction of the damaged communication constellation, the main flow of the algorithm is shown in table 2.

**Table 2.** Damaged communication constellation reconstruction method calculation process

| Step | Process |
|---|---|
| **Input:** | **Initialization parameters:** population number N; evolution algebra M; hybrid probability Cr; mutation probability α.<br>Set the objective function $\min F$ ; communication demand distribution; communication constellation current state; constraints and so on. |
| **Step1:** | **Generate initial population:** g=1<br>$P_g=\left(X_{1,g},X_{2,g},\cdots,X_{N,g}\right)$<br>**Evaluation of the initial population:**<br>$F\left(P_g\right)=\left[F\left(X_{1,g}\right),F\left(X_{2,g}\right),\cdots,F\left(X_{N_p,g}\right)\right]$ |
| **Step2:** | **Hybridization:** $for\ t=1:\left(N\times Cr\right)$<br>$X_c=\left[C_1\ C_2\right]\begin{bmatrix}X_i\\X_j\end{bmatrix}$<br>$P_c\left(t\right)=X_c$<br>*End for* |
| **Step3:** | **Mutation:** $for\ k=1:\left(N\times\alpha\right)$<br>$X_m=X_{new}$<br>$P_m\left(k\right)=X_m$<br>*End for* |
| **Step4:** | **Select:** $P=\left[P_g\ P_c\ P_m\right]$<br>Sorting population P based on Pareto's dominant idea, retaining the first N:<br>$P_s=sort(P)$<br>$P_g=P_s\left(1:N\right)$ |
| **Step5:** | **Evolution:** $if\ \ g\le M$<br>g=g+1, And return Step2<br>*End if* |
| **Output:** | Deriving the Pareto optimal solution set and determining the optimal reconstruction scheme according to the constraints |

## 3. Analysis of communication constellation resilience

### *3.1. Communication constellation resilience assessment method*

Resiliency of the communication constellation is the ability of the communication constellation architecture to continue providing required communication for mission success in the face of system failures, environmental challenges, or adversary actions. When the communication constellation encounters an adverse event, if the degree of functional degradation of the communication constellation is lower, the performance recovery is better, the recovery time is shorter, and the recovery cost is smaller, the resilience of the constellation is better.

Refer to MITRE's assessment method for the resilienec of the space system, the process of constellation performance in the event of an adverse event is defined as a function that varies over time [10-11]. And the constellation resilience is embodied as the residual communication capability under complex events where multiple satellites are at risk of failure. When a multi-satellite failure event occurs, the adverse event is divided into several sub-level scenarios according to the number and location of the possible failed satellites, and the subsequent performance changes of the sub-level scenario are analyzed and calculated to obtain performance over time in the scenario. The probability of occurrence of the sub-level scene is the Y-axis, and the performance of the time and communication constellation is changed to the X and Z axes. Construct a three-dimensional matrix of coverage loss, and the remaining volume of the matrix is a measure of the communication constellation resilience. A schematic diagram of the three-dimensional matrix is shown in figure 3.

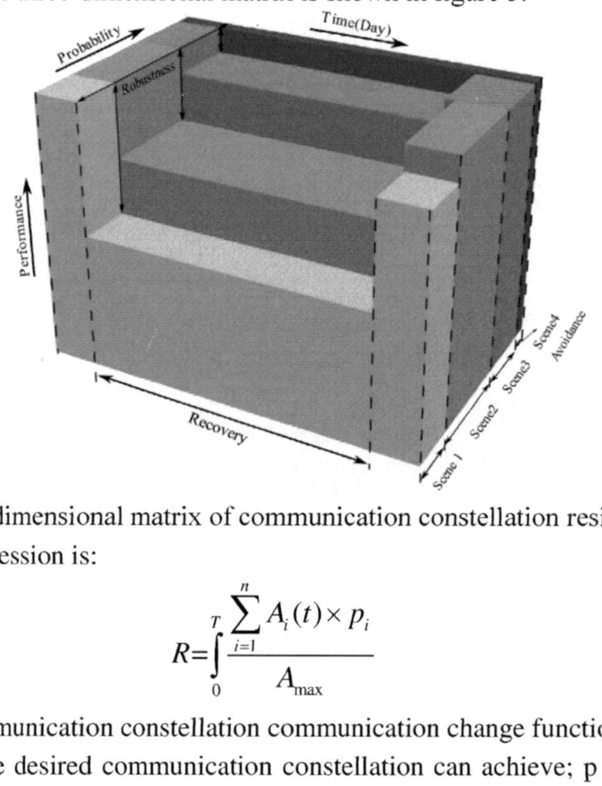


**Figure 3.** Three-dimensional matrix of communication constellation resilience evaluation

Its mathematical expression is:

$$R=\int_{0}^{T}\frac{\sum_{i=1}^{n}A_i(t)\times p_i}{A_{max}} \tag{3}$$

Where $A_i$ is the communication constellation communication change function in the i-th scene, $A_{max}$ is the capability that the desired communication constellation can achieve; p is the current sub-level scene occurrence probability, and $\sum_{i=1}^{n} p_i = 1$; t is the time; n is the total number of scenes.

### *3.2. Simulation case*

(1) Parameter setting

Referring to the current US MUOS military communications constellation construction plan, the simulation scenario is set to five communication satellites. According to the launch time of the satellites in the constellation and the planned service time, the remaining fuel level of the satellite can be roughly estimated, and the energy weighting factor can be determined. The specific parameters are shown in table 3.

**Table 3.** MUOS constellation satellite parameter

| Satellite number | Track semi-major axis(km) | Sub-satellite longitude (°) | Energy weighting factor |
|---|---|---|---|
| 1 | 42166.3 | -176.96 | 3 |
| 2 | 42166.3 | -100.1 | 2 |
| 3 | 42166.3 | -15.48 | 2 |
| 4 | 42166.3 | 74.8 | 1 |
| 5 | 42166.3 | -104.13 | 1 |

The multiple coverage of the complete constellation is shown in figure 4:

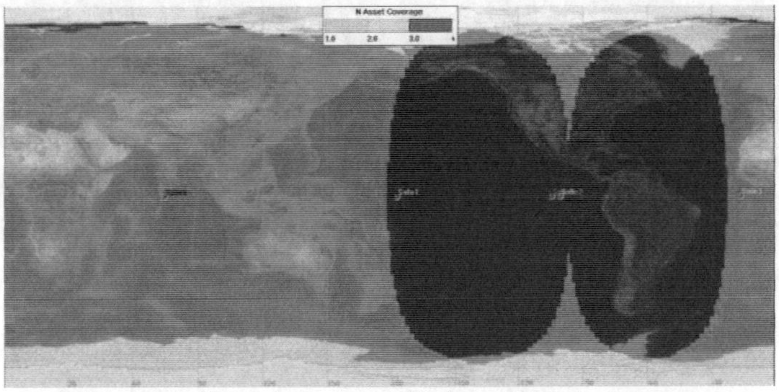

**Figure 4.** Complete constellation multiple coverage

At this time, the global coverage rate based on communication requirements is 99.73%, which ensures that all key areas are covered. It can be seen from the multiple coverage maps that the important areas can basically guarantee double coverage, and the surrounding areas of the United States can guarantee triple coverage, which provides a good guarantee for completing communication tasks.

**Constellation performance:**

The main communication performance of the MUOS constellation is shown in table 4.

**Table 4.** MUOS constellation communication performance

| Indicator | Safety protection | Coverage | Information transmission | Business | Communication |
|---|---|---|---|---|---|
| Capability | 0.8816 | 0.9973 | 1 | 0.9162 | 0.8055 |

**Mission requirements:**

- The communication constellation needs to ensure that the global coverage based on communication needs reaches more than 99%.
- The performance recovery of the damaged constellation is no more than 3 days.

**Scene settings:**

- Due to the high orbital position of the GEO satellite, the factors and possibilities for satellite failure are relatively small. Therefore, the evaluation scenario is set to have three satellites with a risk of failure.
- The start time of the failure event is 0.5 days after the start of the scene.
- After the constellation performance is restored, it must be debugged for 0.5 days to ensure the normal operation of the satellites in the constellation. When all the sub-scene debugging ends, the event is terminated.

(2) Failure event analysis and functional degradation

Considering various satellite protection methods, it is considered that there are three satellites with risk of failure, and the probability of failure of each satellite is 70%. At this time, the probability of failure of all three satellites is 34.3%, the probability of failure of only two satellites is 44.1%, the probability of failure of only one satellite is 18.9%. and the probability of no satellite failure is 2.7%.

(3) Performance recovery analysis

When the satellites 1, 3, and 4 fail, the constellation coverage is shown in figure 5. It can be seen that there is a large area without satellite coverage at this time, and the global coverage of the constellation based on communication needs is only 50.93%. At this time, it is necessary to adopt constellation orbit reconstruction to restore the global coverage of the communication constellation.

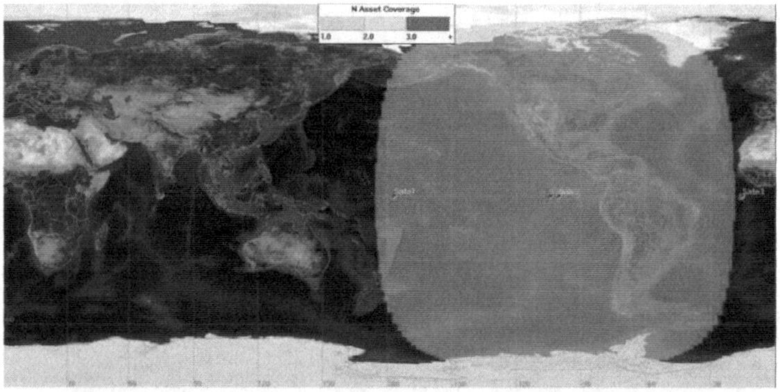

**Figure 5.** Multiple coverage when satellites 1, 3 and 4 fail

**Objective function:**

$$\min \quad F = \left(C, \Delta V_{all}\right) \tag{4}$$

- Global coverage based on communication needs;
- Reconstruct the total energy consumed.

The optimization results are shown in figure 6.

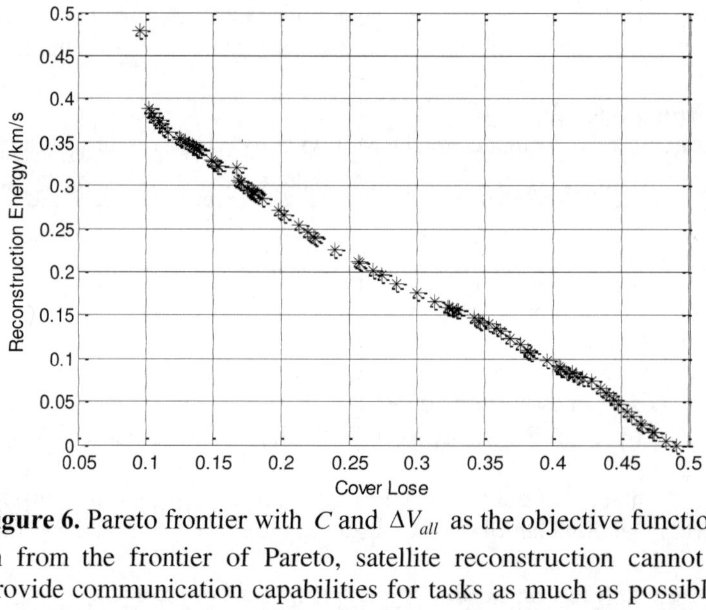


**Figure 6.** Pareto frontier with $C$ and $\Delta V_{all}$ as the objective function

As can be seen from the frontier of Pareto, satellite reconstruction cannot meet the mission requirements. To provide communication capabilities for tasks as much as possible, the solution that rebuilds to achieve greater coverage is used as the final refactoring scheme. Considering the cost-effective factor, the scheme of (0.1015, 0.3878) coordinate points is selected as the reconstruction scheme. The total energy consumption required for the reconstruction is 0.3878km/s, and the global coverage based on communication requirements after reconstruction is 89.85%. The global coverage is shown in figure 7.

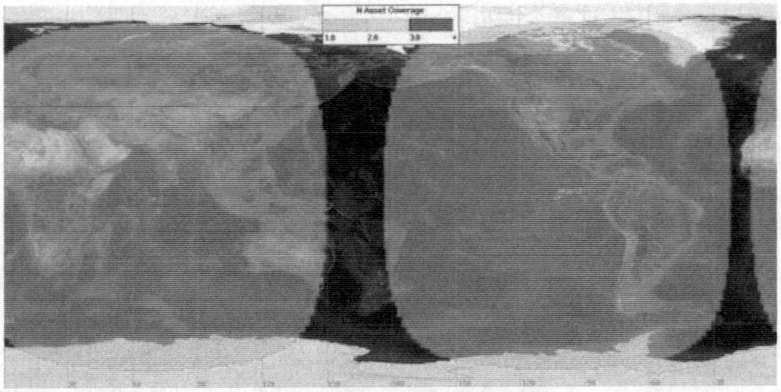

**Figure 7.** Reconstructed communication constellation global coverage

In the same way, the performance recovery process of other scenes can be obtained.

(4) Building a capacity loss matrix

The three-dimensional matrix takes the time, scene occurrence probability, and constellation performance as the X, Y, and Z axes. The specific parameters are shown in table 5 and table 6.

**Table 5.** Relationship between scene performance and time

| Time(day) | 0.5 | 2.8762 | 0.1052 | 0.0071 | 0.5 |
|---|---|---|---|---|---|
| Scene 1 | 0.8055 | 0.4114 | 0.4114 | 0.4114 | 0.7257 |
| Scene 2 | 0.8055 | 0.5793 | 0.5793 | 0.8 | 0.8 |
| Scene 3 | 0.8055 | 0.7277 | 0.7999 | 0.7999 | 0.7999 |
| Scene 4 | 0.8055 | 0.8055 | 0.8055 | 0.8055 | 0.8055 |

**Table 6.** Probability of scene occurrence

| Scenes | Scene 1 | Scene 2 | Scene 3 | Scene 4 |
|---|---|---|---|---|
| Probability | 0.343 | 0.441 | 0.189 | 0.027 |

The three-dimensional evaluation matrix of the communication constellation resilience can be obtained as shown in figure 8.

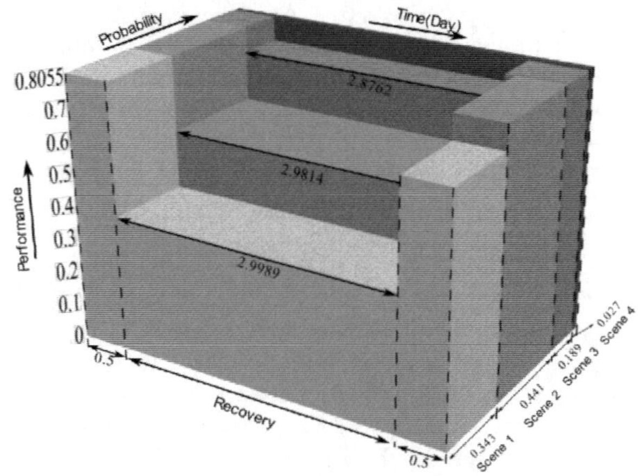


**Figure 8.** MUOS constellation three-dimensional evaluation matrix

At this time, the resilience characteristics of the communication constellation are as shown in table 7 below:

**Table 7.** MUOS constellation communication performance

| Indicator | Avoidance | Robustness | Recovery | Resilience |
|---|---|---|---|---|
| Performance | 0.027 | 0.6901 | 0.9347 | 0.6133 |

As can be seen from the table 7, due to the relatively small number of satellites, the MUOS communication constellation is inferior in terms of robustness. And because the task satisfaction of the complete communication constellation is only 0.8055, the resilience of the constellation in the face of adverse events is not very satisfactory.

## 4. Conclusions

For the single communication constellation, according to the definition and understanding of the space system resilience of the US Air Force Space Command, the definition of the communication constellation elasticity is proposed to provide an evaluation method for the communication constellation mission guarantee capability. The in-orbit reconstruction model of the damaged constellation based on NSGA-II algorithm is constructed and used as the main way of fast recovery performance after constellation damage, which is the main component of the constellation resilience recovery capability. The communication constellation resilience evaluation method is proposed, and the remaining communication capability in the process of encountering adverse events in the communication constellation is taken as a measure of resilience. Taking the MUOS constellation as an example, the simulation calculation is carried out, and its resilience ability is only 0.6133. The mission guarantee capability under adverse conditions needs to be strengthened. In addition, the evaluation of the performance of the communication constellation is relatively simple, and a more detailed study can be conducted in subsequent studies.

## References

[1] Tang Y F, Xu Y L and Li J L 2017 A Summary of Space Architecture Resilience Assessment Methods *Journal of Equipment Academy* 03 pp 74-80

[2] Wieland A and Marcus Wallenburg C. 2013 The influence of relational competencies on supply chain resilience: a relational view *International Journal of Physical Distribution & Logistics Management* 43(4) pp 300-320.

[3] Edlund and Gregory 2013 Space Systems Resiliency: Definition, Measurement, and Application Retrieve d September 23, 2017 from http://www.ndia.org/Divisions/Divisions/Space/Documents/NDIA%20Resilie ncy%20Update_NGAS13-1829_091613.pdf.

[4] McCarthy I P, Collard M and Johnson M 2017 Adaptive organizational resilience: an evolutionary perspective *Current Opinion in Environmental Sustainability* 28 pp 33-40

[5] UCS Satellite Database Retrieved September 20, 2018 from https://www.ucsusa.org/nuclear-weapons/space-weapons/satellite-database?_ga=2.33431677.29022504.1537235795-307381898.1537235795#.W6OepIWf7at

[6] Sheng W, Liu K Y and Liu Y 2015 Optimal Placement and Sizing of Distributed Generation via an Improved Nondominated Sorting Genetic Algorithm II *IEEE Transactions on Power Delivery* 30(2) pp 569-578.

[7] Lu Y X, Zhao C F, Wu X F and Han X X. 2018 Multi-objective optimization method research based on improved NSGA-Ⅱ *Application Research of Computers* 35(6) pp 1733-1737.

[8] Zhao S, Zhang Y S, Dai H Y and Qi Y 2018 Failure Performance and Reconstruction Method of Satellite Navigation System *Aerospace Control and Application* 44(2) pp 49-55.

[9] Zhao S, Zhang Y S and Dai H Y. 2018 Orbit reconstruction configuration of navigation constellation based on differential evolution algorithm *Chinese Space Science and Technology* 38(4) pp 27-35.

[10] McLeod G, Nacouzi G, Dreyer P and et al. 2016 Enhancing Space Resilience Through Non-Materiel Means (California: Rand Corporation)

[11] Dreyer P, Langeland K S, Manheim D and et al. 2016 RAPAPORT (Resilience Assessment Process and Portfolio Option Reporting Tool) (California: Rand Corporation)

IOP Conf. Series: Materials Science and Engineering **608** (2019) 012035 doi:10.1088/1757-899X/608/1/012035

# Assembly Method for Satellite Propulsion System based on HoloLens

**Wei Zhang[1], Jie Zhang[1], Huajun Chen[1], Bin Zhang[1], Feng Liu[1] and Jun Zhao[1]**

[1] Beijing Institute of Spacecraft Environment Engineering, Beijing Engineering Research Center of the Intelligent Assembly Technology and Equipment for Aerospace Product, Beijing, China

E-mail: zhangwei308308@163.com

**Abstract.** As the satellite design changes from the traditional 2D mode to the 3D digital mode, the satellite assembly also enters the 3D assembly mode. An effective and accurate propulsion system assembly information transmission method from designers to operators is not formed, in the process of complex propulsion assembly, operators need to frequently confirm status back and forth between satellite and computer stations, and accidents with incorrect polarity in component installation often occur. In order to improve efficiency and quality of satellite propulsion system installation, a system based on HoloLens is designed. The successful application of this system provides a new solution for satellite propulsion system installation and marks a new step in the digital assembly capacity.

## 1. Introduction

The satellite propulsion system is a complete set of devices that use the principle of reaction to provide thrust to the satellite and is an important part of the satellite. The propulsion system has a complicated layout on the satellite, and the components involved are numerous. The large devices are mainly propellant tanks and pressure tanks; the small devices include engine, thrusters, pressure transducers, latch valves, filters, fill drain valves and etc. the exclusive parts for integration include pipe brackets, pipe clamps and thermal insulation pads; each type of components are subdivided into several specifications. The propulsion system assembly mainly includes the installation of devices and exclusive parts for integration on the satellite, the fixing of the propulsion system piping and the thermal control implementation of the propulsion system.

Most satellites belong to one-piece customized production, the propulsion system layouts of different satellite have large differences, the final assembly space is small, the operation conditions are complex, the various layouts of small devices and exclusive parts for integration are scattered and coded, and therefore, the assembly of the satellite propulsion system is heavily dependent on manual work. In order to complete the installation, the worker needs to obtain the required information from the drawing or 3D model before installation and confirm installation polarity and fastener mounting specifications after installation in the satellite cabin. The disadvantage of this type of method is that the relevant design information is manually matched to the real physical environment, and the identification of the exclusive parts for integration which are densely distributed is inefficient and error-prone. Work efficiency is reduced significantly because workers have to bear a lot of additional cognitive burden to confirm the correctness of assembly.

The propulsion system assembly is complicated and error-prone, in recent years, virtual assembly technology is widely used[1-5]. In order to improve efficiency and quality of satellite propulsion system assembly, a propulsion system assembly system based on AR technology is designed. This method automatically matches the propulsion system installation information with the real physical environment, and guides the workers to complete the assembly. In the field of complex mechanical assembly, enterprises have researched the assembly guidance method based on augmented reality, but cannot meet the needs of satellite single-piece production and unable to systematically solve the problem of automatic conversion of propulsion system assembly 3D models to visual process information.

## 2. Propulsion System Assembly in AIT Process of Satellite

Propulsion system assembly is the first stage in AIT process of satellite, and the process of propulsion system installation in AIT process of satellite is showed in Figure 1, devices with polarity installation requirements needs to be thermally implemented before installation. In addition to installing devices, there are a large number of exclusive parts for integration that need to be installed, the exclusive parts for integration include pipe brackets, pipe clamps and thermal insulation pads; each type of components are subdivided into several specifications. After the installation of devices and exclusive parts for integration is pipe welding and screwing, then start the pipe thermal control implementation.

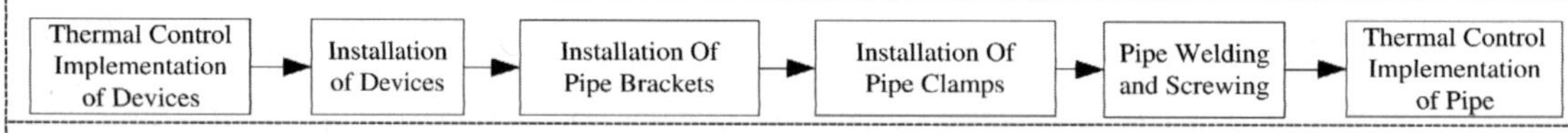


**Figure 1.** Process of Propulsion System Assembly

The diagram of pipe assembly on pipe bracket is showed in Figure 2, there are many types of pipe brackets, pipe clamps and thermal insulation pads, and each component has a large number and is distributed to various locations on the satellite.

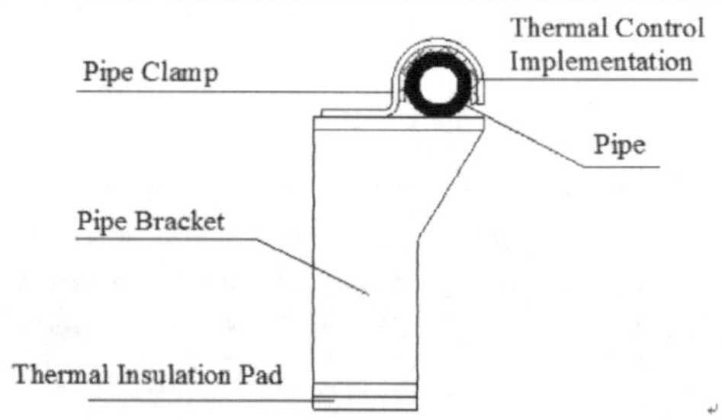


**Figure 2.** Diagram of Pipe Assembly on Pipe Bracket

As the satellite propulsion system design changes from the traditional 2D mode to the 3D digital mode, the satellite propulsion system assembly also enters the 3D assembly mode. Because the 3D design input is inconvenient to use, and an effective and accurate propulsion system assembly information transmission method from designers to operators is not formed, accidents with incorrect polarity in component installation often occur. The current 3D model based on desktop computer has poor human-computer interaction, In the process of complex propulsion system assembly, operators need to frequently confirm status back and forth between satellite and computer stations, and therefore, new technology is needed to improve propulsion system assembly efficiency.

## 3. Microsoft HoloLens

Compared with Virtual Reality technology, Augmented Reality technology can not only see virtual scenes, but also superimposes virtual scenes in real scenes by calculating operator position and perspective in real time. The Microsoft HoloLens[6~7] was released as a development edition in 2016

IOP Conf. Series: Materials Science and Engineering **608** (2019) 012035 doi:10.1088/1757-899X/608/1/012035

and is now available as a consumer version. It is the ideal choice for investigating an AR assembly application on a commercially available device. The HoloLens features four Intel Atom x5- Z8100 1.04 GHz Intel Airmont Logical Processors, a HPU/GPU Holographic Processing Unit, 64 GB Flash, 2 GB RAM and 2-3 hours of active battery life that allows standalone operation of this device. All of this processing power is used to run 2 HD 16:9 light engines that project light through holographic lenses leading to a total resolution of 2.3 million light points. High resolution spatially located 3D content is generated by this system. The HoloLens also consists of an Inertial Measurement Unit (IMU), 4 environment-processing cameras, an RGB camera, and 1 depth camera to map its surroundings and allow interaction between the real and virtual world while tracking the device's position.

## 4. The Process of Application of Augmented Reality in Satellite Propulsion System Assembly

A system based on AR technology is designed to improve efficiency and quality of satellite propulsion system installation, the flow chart of the method is showed in Figure 3. Compared with the traditional method, this method automatically matches the propulsion system information with the real physical environment, and guides the workers to complete the propulsion assembly.

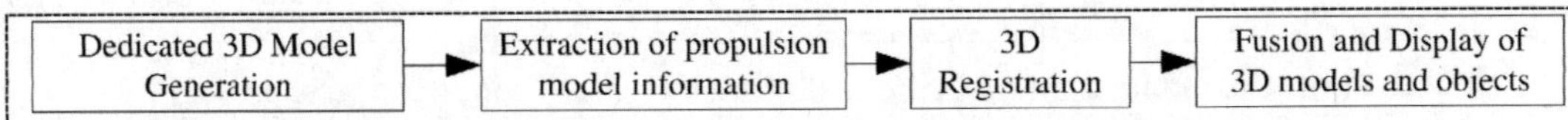


**Figure 3.** Process of Application of Augmented Reality in Propulsion System Assembly

### *4.1. Dedicated 3D Model Generation*

As the satellite design changes from the traditional 2D mode to the 3D digital mode, the 3D digital mode of a satellite contains much information, such as structural, cable and instruments. It is necessary to simplify the data by creating a dedicated 3D model which removed unused component parts. propulsion system dedicated model tree which is shown in Figure 4 need to be reconstructed to facilitate subsequent display in HoloLens. The new model tree is divided into 4 nodes, the deck structure, device related, pipe support related and pipe. Pipe support related contains pipe bracket and pile clamp, so the operator can selects the object to display by type. The 3D propulsion system model is large, and the required computer performance is relatively high, so the 3D design model which is need to be lightweight is not suitable for application in mobile terminals. In order to improve the subsequent display rendering effect, the unnecessary model information should be deleted first, and then compress the 3D model by format conversion. The lightweight 3D model has the advantages of small file size, easy operation and smooth display.

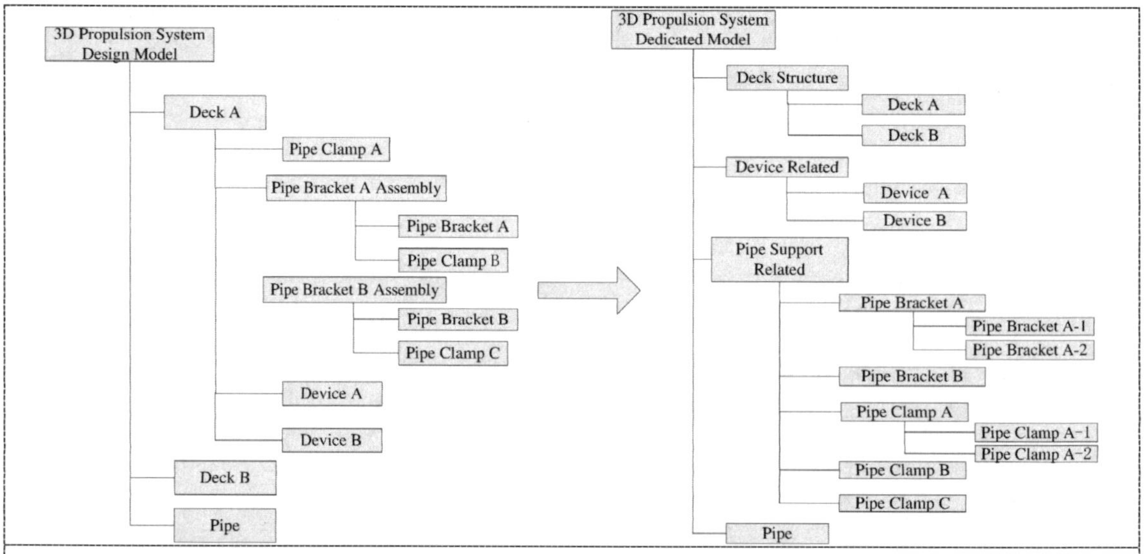


**Figure 4.** Generation of Dedicated Model Tree of Propulsion System

### *4.2. Extraction of Propulsion Model Information*

While installing device and pipe support component, the operator needs to know the code, the position, the fastener and the related exclusive parts for integration, so it is necessary to extract propulsion model information in design 3D model through secondary development of software. The propulsion model information is showed in Table 1, and they will be added to the corresponding position of the lightweight and dedicated 3D model.

**Table 1.** The propulsion model information.

| Type | Device | Pipe Support |
|---|---|---|
| Code | T104a | 23-02 |
| Name | Pressure transducer A | Single pipe clamp |
| ModelName | T104a-3.asm | danka.prt |
| Fastener | 4*M3×12 | 1*M4×12 |
| Thermal Control Part | 1*23-012 | / |
| Other exclusive parts | 1*23-013 | / |
| Position(X,Y,Z) (mm) | 732,189,208 | 341,445,883 |

### *4.3. 3D Registration*

The 3D registration technology mainly realizes the correct fusion of virtual propulsion components and real assembly scenes through the conversion between coordinate systems. Accurate 3D registration of virtual and real scenes is a key technical problem to be solved in the real-time fusion, to accurately register the virtual assembly object in the corresponding position in the real environment, the conversion relationship between the virtual space and the real space coordinate system must be established in the virtual fusion assembly system. Registration error reaches a certain value will result in erroneous results. In this system based on AR technology, 2D codes which can be recognized very quickly are used in 3D registration, and it is pasted on satellite deck structure.

### *4.4. Fusion and Display of 3D Models and Objects*

The system based on AR technology was developed using Unity3D, Vuforia and the Microsoft HoloLens. Unity3D is an excellent development tool for the Microsoft HoloLens. To achieve higher precision, the embedded Vuforia system was used to perform marker based tracking. After 3D registration, the HoloLens can properly track position as the user moves around in the assembly area using its IMU and environment processing cameras. An assembly application with a user friendly UI has been built to improve the user experience.

## 5. The Typical Case of Application in Satellite Propulsion System Assembly

The assembly method for satellite propulsion system based on HoloLens is tested during the AIT process of a remote sensing satellite in the paper. Dedicated and lightweight 3D propulsion system model of the service module is showed in Figure 5. As is shown in Figure 6, a 2D QR code is printed on paper and pasted on the satellite deck structure where is easy to determine the location. The operator must first wear Microsoft HoloLens before propulsion system assembly and use the gesture to enter the corresponding module of the remote sensing satellite, secondly, scan 2D QR code in the satellite deck structure to complete 3D registration, and then select device or pipe support to be installed from the UI interface. The process of this system is very simple, and the assembly information is clear at a glance, thereby greatly improving the visibility of the 3D model and guiding principle. Figure 7 and Figure 8 shows a typical case of application in a satellite propulsion system assembly, device assembly with polarity installation requirements is shown in Figure 9 and Figure 10, and assembly of exclusive parts for integration is shown in Figure 11 and Figure 12. With the help of the assembly system based on AR technology, the time of installing device and all component of a remote sensing satellite propulsion system can be reduced from 2 days to 1 day, and the satellite propulsion system assembly accuracy can reach up to 100%.

**Figure 5.** Dedicated and Lightweight 3D Propulsion System Assembly Model of a Satellite

**Figure 6.** The 2D QR Code Pasted on the Satellite Deck

IOP Conf. Series: Materials Science and Engineering **608** (2019) 012035 doi:10.1088/1757-899X/608/1/012035

**Figure 7.** The Typical Case of Application.

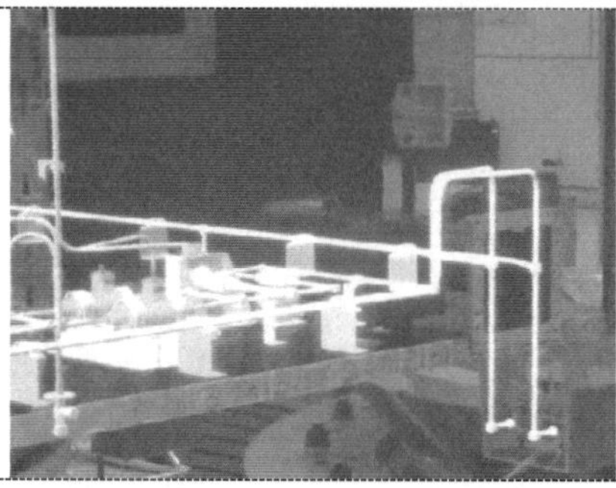

**Figure 8.** Fusion and Display of 3D Satellite Propulsion System.

**Figure 9.** Device Assembly with Polarity Installation Requirements.

**Figure 10.** Fusion and Display of Device with Polarity

**Figure 11.** Assembly of Exclusive Parts for Integration.

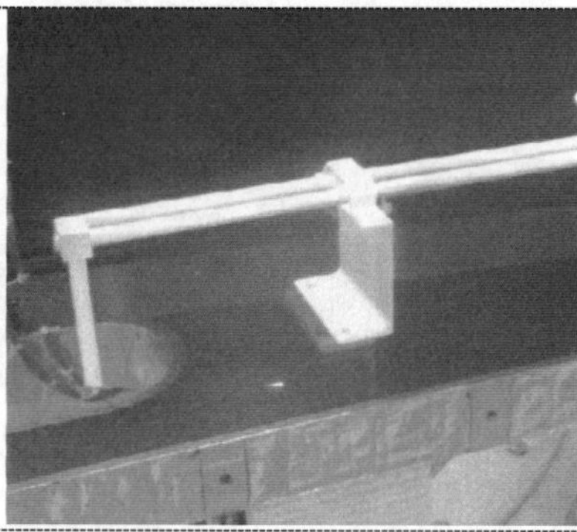

**Figure 12.** Fusion and Display of 3D Models and Objects.

## 6. Conclusion

This paper reviews the traditional propulsion system installation in AIT process of satellite. As the satellite design changes from the traditional 2D mode to the 3D digital mode, the satellite assembly also enters the 3D assembly mode. In order to improve efficiency and quality of satellite propulsion system installation, a system based on HoloLens is designed. The process of this system is very simple, and the assembly information is clear at a glance, especially for device with polarity installation requirements, thereby greatly improving the visibility of the 3D model and guiding principle.

IOP Conf. Series: Materials Science and Engineering **608** (2019) 012035 doi:10.1088/1757-899X/608/1/012035

**References**

[1] ZHANG Qiuyue,AN Luling. Application of Virtual Reality and Augment Reality in Aircraft Assembly, Aeronautical Manufacturing Technology,2017,530(11)
[2] LI Shiqi, PENG Tao, WANG Junfeng, and XU Chi, Mixed Reality-based Interactive Technology for Aircraft Cabin Assembly, CHINESE JOURNAL OF MECHANICAL ENGINEERING, 2009, 22(3)
[3] YIN Xuyue,FAN Xiumin,WANG Lei,QIU Kai,LIU Rui,Augmented Reality Training System for Aerospace Product Assembly Process Guidance and Its Application,Aeronautical Manufacturing Technology,2018,61(1)
[4] Gao Xiang,An Hui,Chen Wei,Pan Zhigeng,A Survey on Mobile Augmented Reality Visualization,Journal of Computer-Aided Design & Computer Graphics,2018,30(1)
[5] YANG Qing,WU Xiaobin,LU Ali,ZHANG Jiande,TU Lizhong,LIN Zhong,Development of the intelligent assembly assistant system for complex product based on augmented reality,MACHINE DESIGN AND MANUFACTURING ENGINEERING,2017,46(11)
[6] G Evans,et al. Evaluating the Microsoft HoloLens through an augmented reality assembly application, Proc.of SPIE Vol.10197
[7] https://www.microsoft.com/zh-cn/hololens

# In-orbit Failure Analysis and Verification for Thermistors outside Satellites

**Bin Zhang[1,2,*], Dongmei Wang[1,2], Yufeng Fan[3], Li Liu[1,2] and Peirong Zhao[1,2]**

[1]Beijing Institute of Spacecraft Environment Engineering,Beijing, China, 100094

[2] Assembly Processing Technology Center for Aerospace Product of CASC. Beijing, China, 100094

[3] Beijing Institute of Spacecraft System Engineering, Beijing, China, 100094

*E-mail: zhangbin_bisee@126.com

**Abstract:** Satellite thermistor is used to measure the in-orbit temperature to support thermal control so that the on-board equipment can work in the appropriate temperature to get higer reliability and more life-time. In recent years, there have been several in-orbit failure events of thermistors outside the satellite, which have brought troubles to the thermal control of the whole satellite. In this paper, failure modes of in-orbit thermal resistors outside cabin are analyzed by fault tree, and ground simulation experiment is carried out by high-low temperature impact test. Fault cause is given according to test result and analysis.
**Keywords:** Thermistor, failure analysis, in-orbit, satellite

## 1. Introduction

As we know, all the satellites work in the dark cold space environment.If thermal control measures are not taken, when the satellite is facing the sun, the surface temperature will rise sharply under the sun's illumination. At the same time, the temperature on the side against the sun will be very low. The instruments on the satellite can not endure the wide range of cold and hot alternating environment. This may harm even damage the instrument inside the satellite[1]. Therefore, thermal control design and manufacturing should be carried out in the stage of satellite development in order to keep the in-orbit temperature in proper range to get good working and storing performance for on-board instruments, so as to help us improve the overall performance and prolong the life-time of the whole satellite. Accurate acquisition of on-board temperature is the first step in thermal control[2,3].

Thermal resistor, which is short for thermistor, is used to measure the temperature of satellite during the in-orbit work in order to get the working temperature of the instruments or panels.

But in recent years, some thermistors outside the satellite lost their functions during the in-orbit work. It brought troubles for thermal control of the satellites. So it is very necessary to analysis the failures of the satellite thermistors to find the reasons to help us to improve the design and assembly of the thermistor circuit.

## 2. Composition of the thermistor circuit

Thermistor circuit is composed by thermistor and conductive wires, as shown in Figure 1. The thermistor pin is Dumet wire. The conductive wire is copper wire. The Dumet wire and the conductive wire are welt together before installed on the satellite.

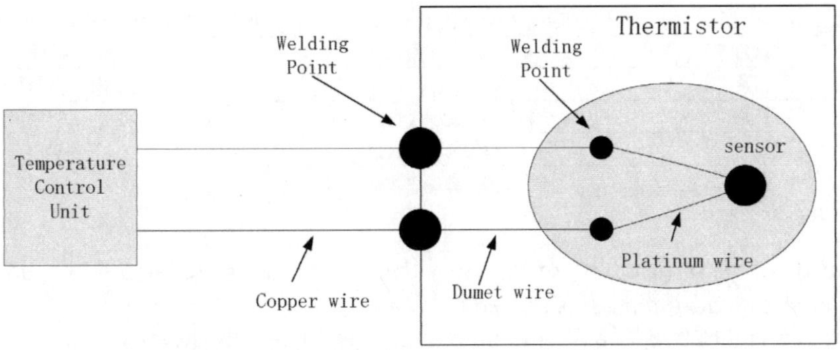


**Figure 1.** Schematic diagram of thermistor circuit

For insulation considerations, the Dumte wire of thermistor is covered by single heat-shrinkable sleeve before welt, and the welding point between Dumet wire and Copper wire is covered by double layers of heat-shrinkabllye sleeve, as shown in Figure2. After the thermistor is fixed on the surface of the satellite, it is covered by silicone rubber glue. One case of the installation state of thermistor outside the satellite is shown by Figure 3.

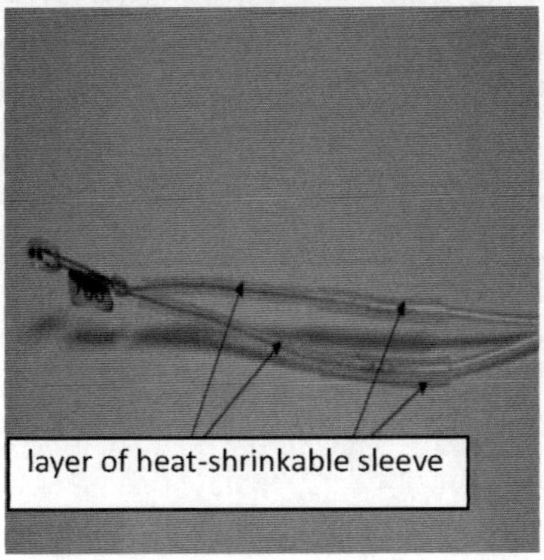


**Figure 2.** Insulation sleeves for thermistor circuit.

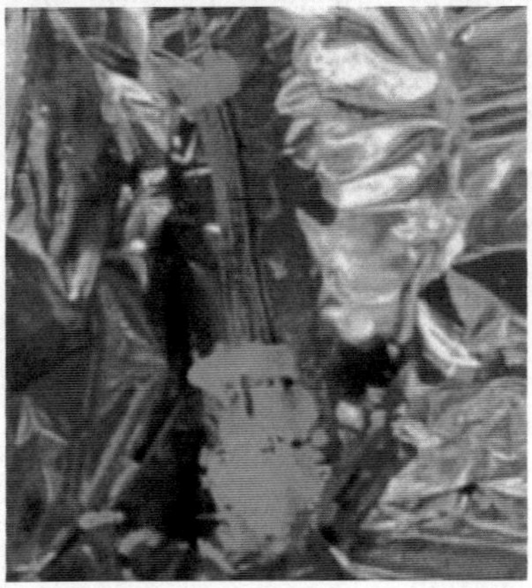

**Figure 3.** Installation state of thermistor outside the satellite.

### 3. Features of in-orbit failure

In recent years, 15 cases of in-orbit failure of out-cabin thermistors from several satellites are counted. The statistical results are shown in Table 1.

**Table 1**. In-orbit failure statistic of thermistors outside satellite

| No. | Working time (month) | Installation state | In-orbit temperature range(°C) |
|---|---|---|---|
| 1 | T0[a]+0 | Uncovered | -140~+110 |
| 2 | T0+2 | Uncovered | -100~+100 |
| 3 | T0+2 | Uncovered | -140~110 |
| 4 | T0+6 | Uncovered | -45~ +13 |
| 5 | T0+8 | Uncovered | -80~+95 |
| 7 | T0+8 | Uncovered | -130~+120 |
| 8 | T0+8 | Uncovered | -40~-+10 |
| 9 | T0+11 | Uncovered | -140~+110 |
| 10 | T0+12 | Uncovered | -150~50 |

| No. | Working time (month) | Installation state | In-orbit temperature range(°C) |
|---|---|---|---|
| 11 | T0+12 | Uncovered | -150~+100 |
| 12 | T0+18 | Uncovered | -140~+110 |
| 13 | T0+19 | Uncovered | -80~++95 |
| 14 | T0+19 | Uncovered | -80~++95 |
| 15 | T0+28 | Uncovered | -140~+110 |

[a]*T0 means the launching time*

Though the analysis of the in-orbit failure statistics, the features are summarized as follow:

- Signal jump-normal-jump before failed

1 case is invalid right after launch, and there is no signal for it. The others are working until the signal jumps and at last there is no signal reserved.

- Less working hours

Compared with 6 -10 years life time of the whole satellites, most of the failure occurred within 1 year after launch.

- Wide and alternated working temperature rang

Most of the failure thermistors have been working in the alternating temperature changing from the temperature below -80°C to the temperature above +90°C. In some worse cases, the working temperature is alternated from -150°C to +100°C, or from -140°C to +110°C.

According to the in-orbit temperature measuring data statistics, most of the working temperature of thermitstors inside the satellite is in the range of -60°C to +60°C, and the thermistors can be working stably in that situation.

## 4. Cause analysis

In view of in-orbit failure problem mentioned above, a failure mode fault tree is established for failure reason analysis[4-6], as shown in Figure 4.

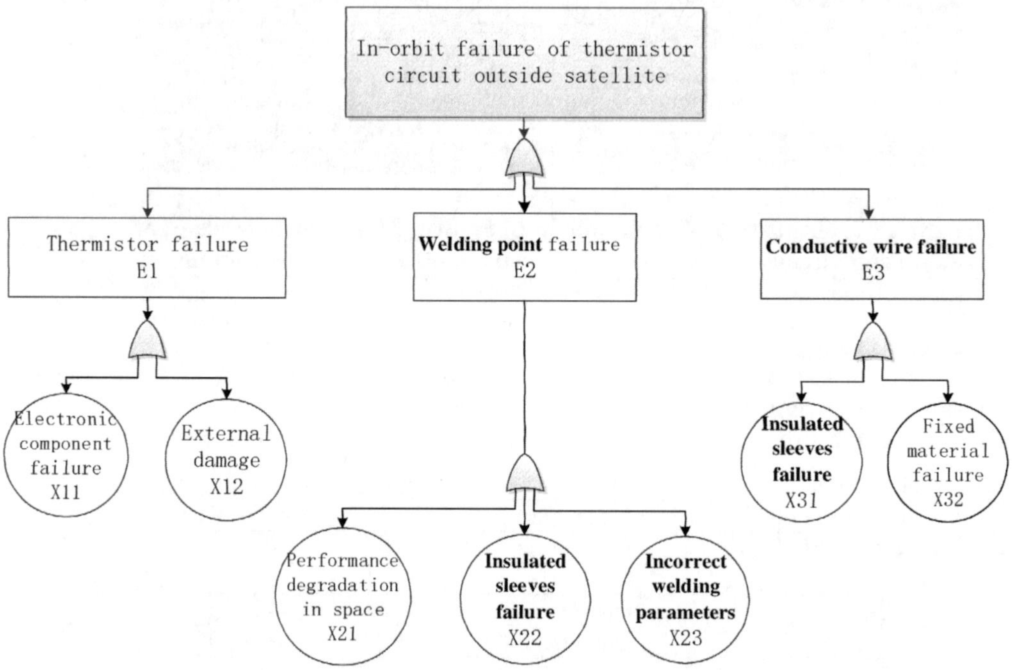


**Figure 4.** Fault tree of thermistor in-orbit failure

The specific analysis is as follows:

1) Reason 11 : Thermistor failure by electronic component failed

Thermistors used outside the cabin are processed and manufactured according to relevant standards, and have been tested for electrical performance, mechanics and adaptability of in-orbit environment. They are selected according to the in-orbit working range of the satellite, and their performance can satisfy various requirements of the satellite on the ground testing. But the satellite will experience a bad and complex space environment during its in orbit work. The thermistor exposed outside the cabin without the protection of Multi-layer Insulation materials, which is used to control the temperature in suitable range in order to protect the instruments and the temperature sensitive components, may cause performance degradation or failure. The influence degree has not been tested and verified. Therefor, reason 11 cannot be exclude.

2) Reason 12: Thermistor failure by external damage

The technical design and process requirements for the final assembly and distribution of thermistor is implemented. The conformance of the operation and implementation state has been examined by professional inspection, electric test, vibration test and thermal test. The failure caused by external force can be completely avoided by means of standardized operation and strengthened management. Reason 12 can be excluded.

3) Reason 21: Welding failure by performance degradation and failure caused by space environment

For weld exposed to space environment, it is analyzed that their performance is subject to extreme cold-hot alternating environment, which will cause degradation and failure of welding performance. Especially under the environment of in-orbit high-low temperature cycle, certain internal stress is generated both in Dumet wire and Copper wires because of the different expansion coefficient of different materials (thermal expansion coefficient of thermistor dumet wire is about $10^{-6}$, and copper wire is$10^{-5}$). The stress could further aggravate the displacement of each other, causes welding alloy layer lattice moved, reduces the adhesion strength of the alloy layer and substrate, eventually makes the welding joint cracking. That caused the failure. Reason21 can not be excluded.

4) Reason 22: Welding failure caused by insulated sleeves failure

If the heat-shrinkable sleeves coated on the welding points and the wires closed to them cannot withstand cold-hot alternating environment, it expands when heated and shrinks when cooled. Frequent thermal expansion and contraction produces stress acting on the welding points , which leads to the cracking of the welding points. That caused the failure. Reason 22can not be excluded.

5) Reason 23: Welding failure caused by incorrect welding parameters

Because of the neglect of the cold-hot alternating environment outside the satellite, there is no special validation welding experiment for thermistors outside the cabin The welding parameters are the same as those inside the satellite. This makes the welding points cannot withstand the space environment of alternating heat and cold outside cabin, resulting in cracking and failure of welding point. That caused the failure. Reason 23 can not be excluded.

6) Reason 31: Wire failure caused by insulated sleeves failure

Just like the reason 22, frequent thermal expansion and contraction of the insulated sleeves which cover the wiers and welding points produces stress acting on the Copper wires, which leads to the borked of the wires. That caused the failure. Reason 31 can not be excluded.

7) Reason 32: Wire failure caused by fixed material failure

If the wire fixing materials cannot bear the cold-hot alternating environments, the fixed materials fail and cannot play a fixed role on the wire. It results swings of the wire in-orbit work. Multiple bending of the wire under stress will break the wire, resulting in failure of the thermistor circuit. Reason 32 can not be excluded.

## 5. Verification

In order to verificate which is the real reason among the courses that analyzed above, a simulated experiment was taken. In the experiment, two kinds of thermistors, six of each, were employed. They were installed on an aluminium board with GD414 silicone rubber the same installation method as the thermistors installed on satellite.

The temperature impact tests were carried out and the temperature was cycling as+20°C-(-196)°C-(+125)°C-(+20)°C, which is wider than the in-orbit failed themistors' working temperature range. Electrical performance tests and appearance inspections were done before impact test, after 50 period cycling, after 500 period cycling, and after 1500 period cycling to see if there is something wrong during the temperature impact tests. The result is shown in Table2.

**Table2.** Resistance value at room temperature during the experiment

| | Resistance value at room temperature(kΩ) | | | |
|---|---|---|---|---|
| | before impact test | after 50 period | after 500 period | after 1500 period |
| A1 | 27.86 | 29.08 | 23.59 | open |
| A2 | 26.04 | 27.12 | 22.10 | 26.85 |
| A3 | 24.73 | 25.77 | 21.02 | 25.38 |
| A4 | 26.31 | 27.43 | 22.30 | 27.21 |
| A5 | 27.11 | 28.28 | 22.98 | open |
| A6 | 26.96 | 28.07 | 22.92 | 27.63 |
| B1 | 30.57 | 31.42 | 28.23 | 31.24 |
| B2 | 28.14 | 28.85 | 25.94 | 28.68 |
| B3 | 24.89 | 25.57 | 23.05 | 25.52 |
| B4 | 26.58 | 27.33 | 25.51 | open |
| B5 | 24.87 | 25.56 | open | open |
| B6 | 25.89 | 26.61 | 24.01 | open |

After 50 times temperature impact test, all the 12 themistor circuits can measure the room temperature, and there is no significantly deformation for each themistor. After 500 times of temperature impact test, one test circuit of thermistor appeared to be open. After 1500 times of temperature impact test, five test circuits of thermistor appeared to be open. There is no obvious deformation on the head of the thermistor. It means reason 1 can be excluded.One case was shown by Figure 5. But there is obvious deformation of both the Dumet wire and the Copper wire, especially the copper wire where is close to the welding point. One case was shown by Figure 6. Figure 6 also shows there is deformation of insulated sleeves during the temperature experiment. That means reason 22 and reason 31 can not be excluded. In the experiment, there is no fixing material failed, so that the reason 32 can be excluded.

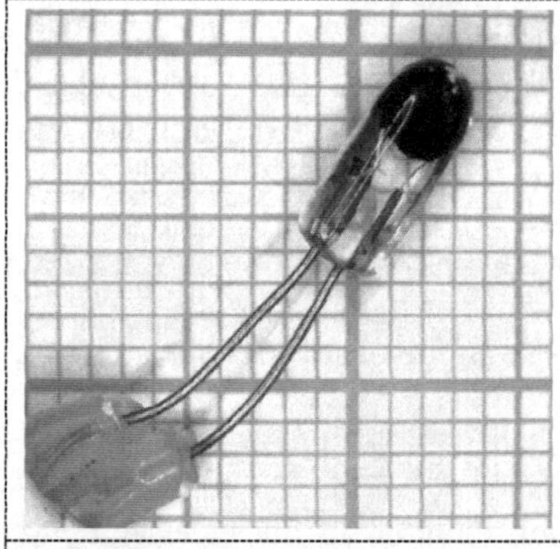

**Figure 5.** Photo of thermistor of one failed case after 1500 times temperature impact test

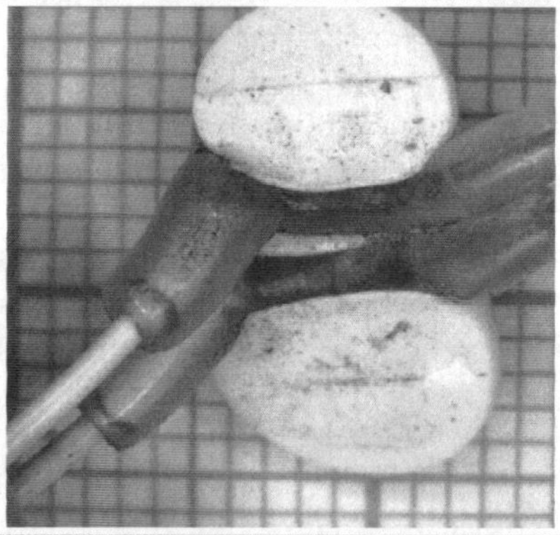

**Figure 6.** Photo of welding point and wire after 1500 times temperature impact test

Using test pins, the thermistor resistance values of all 5 open thermistor circuits were measured correctly. It means that the components are normal. The reason 1 can be excluded.

For the further inspection, X-ray detection was employed. In the X-ray film, fractures can be seen clearly between the Copper wires and welding points in all 5 open thermistor circuit cases, and there is no damage on the welding points and the thermistors, as shown by Figure7 and Figure8. That means the reason21, reason 22, and reason 23 can be excluded. The reason 31 is the course of the in-orbit failure of thermistor outside the satellite.

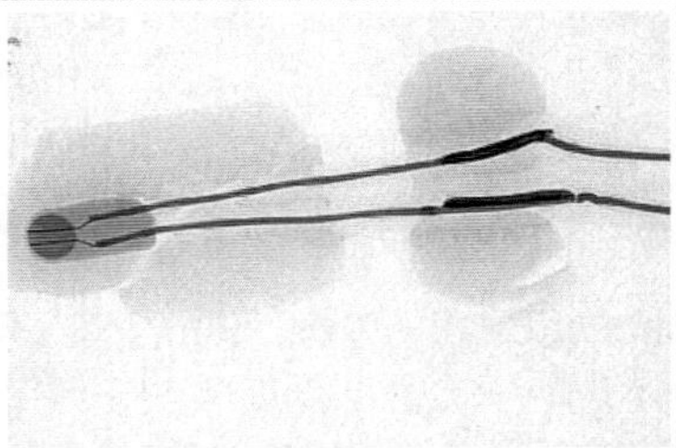

**Figure 7.** X-ray film of one failed case after 1500 times temperature impact test

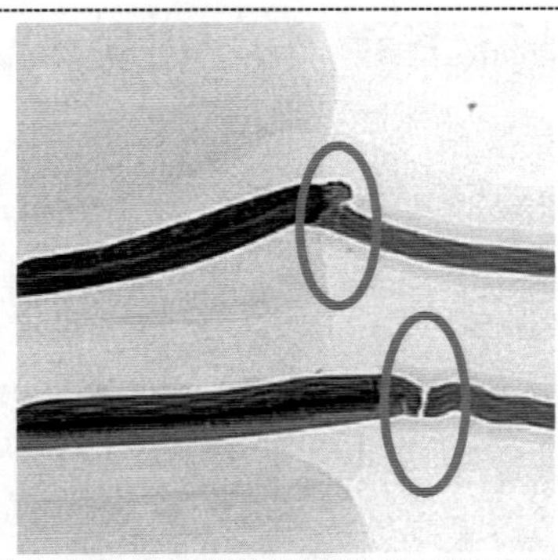

**Figure 8.** Fractures of one failed case after 1500 times temperature impact test by X-ray film

## 6. Conclusion

In this paper, the failure causes of in-orbit thermistors are analyzed. Through fault tree analysis, 7 possible causes of in-orbit failure of thermistor outside satellites are analyzed. Through cold - hot temperature impact test, the causes are further analyzed and verified, and with the help of X-ray detection, the causes of the problems are finally determined. It was caused by the fractures between the Copper wires and Welding points due to the expansion and contraction stress of insulated heat-shrinkable sleeves in the cold-hot alternating environment in space.

However, due to the limitations of the samples and the inadequate fidelity of the in-orbit environment model, there may be other reasons that cannot be exposed, and further analysis and verification can be carried out in the follow-up.

## Reference

[1] Xun J, Hong W, and Jialin S,The Application Analysis of Thermistor in Spacecraft, *Chinese Space Science and Technology*.200424(6),V4.

[2] Lili W, Jianhua W, Weiqun Y, Application of NTC thermistor in high accurate temperature Measurement, J*ournal of Transducer Technology*.2004 23(12), TP223.

[3] Jifeng W, Xiaodong X, Min X, Yaping L, Accuracy Analysis of On-board Telemetry Acquirement, *Spacecraft Engineering*. 2007 16(5), V557.

[4] PLACCA L, KOUTA R. Fault tree analysis for PEM fuel cell degradation process modeling[J]. Int. J. Hydrogen Energ.2011, 36:393-405.

[5] CHENG S, LI Z, MANG H, et al. Application of fault tree approach for technical assessment of small-sized biogas systems in Nepal[J].

[6] ABDUL RF VARUTTAMASENI A. KINTNER-MEYER M. et al.Application of fault tree analysis for customer reliability assessment of a distribution power system[J]. Reliab. Eng. Syst. Safe, 2013, 111:76-85

IOP Conf. Series: Materials Science and Engineering **608** (2019) 012037 doi:10.1088/1757-899X/608/1/012037

# Research on Three-Dimensional Digital Assembly Technology of Communication Satellite Cable

**Zhibin Liu[1,2*], Xiaohui Song[1,2], Pu Yin[3] and Xuewei Zhou[1]**

[1]Beijing Institute of Spacecraft Environment Engineering, Beijing 100094, China
[2]Beijing Engineering Research Center of the Intelligent Assembly Technology and Equipment for Aerospace Product, Beijing 100094, china
[3]Communications Satellite Division, Beijing 100094, China

*Corresponding author's e-mail: raindown@126.com

**Abstract.** With the development of information technology and the increasing task of communication satellite products, higher requirements are put forward for the design and manufacture of cables. The traditional cable design is based on the two-dimensional deployment layout. The design efficiency is low and the error of cable length is large. It is difficult to check the interference in the cable channel. In view of the new technology status of a telecommunication satellite, a comprehensive study has been carried out from cable design, production and manufacture to field laying of assembly. The design and production modes have been analysed and expounded, and the field laying of assembly has been summarized, which not only improves the efficiency, but also strengthens the quality and reliability. The three-dimensional digitization of subsequent satellites has been carried out in an all-round way. Design, manufacture and assembly technology provide technical basis and experience for reference.

## 1. Introduction

Digital technology is based on virtual reality and simulation technology. It unifies the modelling of product design, production, manufacturing and assembly process, realizes product design, processing, manufacturing and assembly on computer, simulates and simulates the whole life cycle, so as to optimize product design quality and manufacturing process, optimize production management and resource planning, and effectively reduce product research. The risks and costs of the system. [1][2]

Cables play a very important role in the manufacture of communication satellites. All instructions are transmitted through cables. In order to ensure the accuracy and smoothness of signal transmission on the satellite, ensure the normal and stable operation of the satellite in space, ensure the quality of the whole satellite, and avoid any abnormal quality. [3]

Therefore, the digital prototype design of cable can be realized in product design stage by using computer three-dimensional modelling technology, focusing on the rationality of cable layout and wiring, processing and assembly technology, so as to ensure the smooth progress of the actual production process.

The power requirement of a satellite is 7200W, which not only has a large number of cables, but also has a very complex trend. In order to ensure the high quality and high efficiency of the whole life cycle work, from the beginning of the design stage, three-dimensional tools have been used to design the cable network, and a new mode has been applied to design, produce, manufacture and assembly work.

## 2. 3-D Digital Design

### *2.1. Traditional mode*

AUTOCAD is used to design cable alignment. As shown in Figure 1, this model only contains the information of cable alignment and possible binding points. It cannot grasp the information of cable length, nor can it accurately and stereoscopically describe the rationality and correctness of cable alignment. [4]

### *2.2. Digital mode*

In CATIA software, every cable is modelled entirely, as shown in Figure 2. In this way, the BOM tables with key information such as cable orientation, binding point, length, cable code, plug start and end code can be obtained in advance. This information can be transferred and transmitted in the form of Excel table, as shown in Table 1. The whole process realizes the online collaborative work mode between designers and designers, designers and manufacturing technicians, designers and assembly process. The technicians are involved in the process of cable three-dimensional digital design in advance. According to the connection relationship of cable, the cable diameter is calculated and assigned, so that the three-dimensional design result of cable is similar to the actual cable diameter. In the process of digital design, the rationality of the main channel and layout of cable design can be checked, and the design efficiency can be improved.

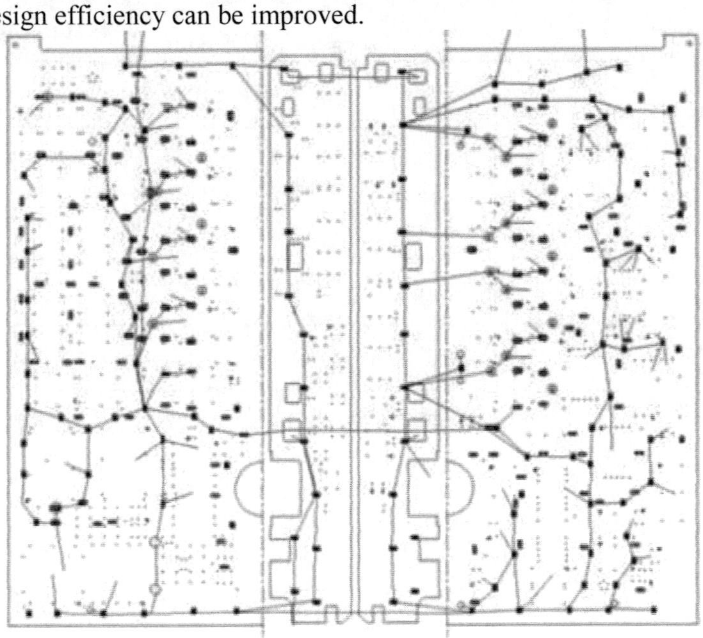

**Figure 1.** Traditional cable design mode

IOP Conf. Series: Materials Science and Engineering **608** (2019) 012037 doi:10.1088/1757-899X/608/1/012037

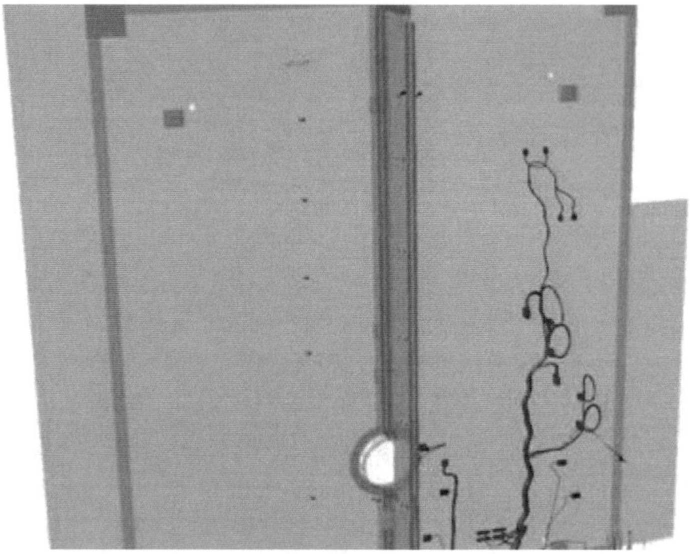

**Figure 2.** Digital Cable Wiring Effect

**Table 1.** BOM Table of Cable Information.

| No | Cable number | Name | Line type | Length (mm) | Beginning end | terminal | Remarks |
|---|---|---|---|---|---|---|---|
| 1 | XXW110 | X2Z(T). 30-> XPN1-1.4 | 1124-24 | 7253 | X2Z (T) | XPN1-1 | Twisted pair |
| 2 | XXW110 | X2Z(T). 38-> XPN1-1.2 | 1124-24 | 7253 | X2Z (T) | XPN1-1 | Twisted pair |
| 3 | XXW110 | X2Z(T). 37-> XPN1-2.2 | 1124-24 | 7255 | X2Z (T) | XPN1-2 | Twisted pair |
| 4 | XXW110 | X2Z(T). 29-> XPN1-2.4 | 1124-24 | 7255 | X2Z (T) | XPN1-2 | |
| 5 | XXW110 | X2Z (T). 36 -> XPN2-1.2 | 1124-24 | 5628 | X2Z (T) | XPN2-1 | Twisted pair |
| 6 | XXW110 | X2Z(T). 28-> XPN2-1.4 | 1124-24 | 5628 | X2Z (T) | XPN2-1 | |
| 7 | XXW110 | X2Z(T). 35-> XPN2-2.2 | 1124-24 | 5629 | X2Z (T) | XPN2-2 | Twisted pair |
| 8 | XXW110 | X2Z(T). 27-> XPN2-2.4 | 1124-24 | 5629 | X2Z (T) | XPN2-2 | |

## 3. 3-D Digital Manufacturing

### *3.1. Traditional mode*

According to the cable direction, the accurate cable length can be obtained through the actual simulation of wood mould processing. It depends on the operator's experience, and takes a long time and period, and the accuracy needs to be improved. [5]

### *3.2. Digital mode*

According to the BOM table of cable information given in the digital model, the cable is manufactured and processed according to the pin layout projected by the digital model. The pin layout is shown in Figure 3.

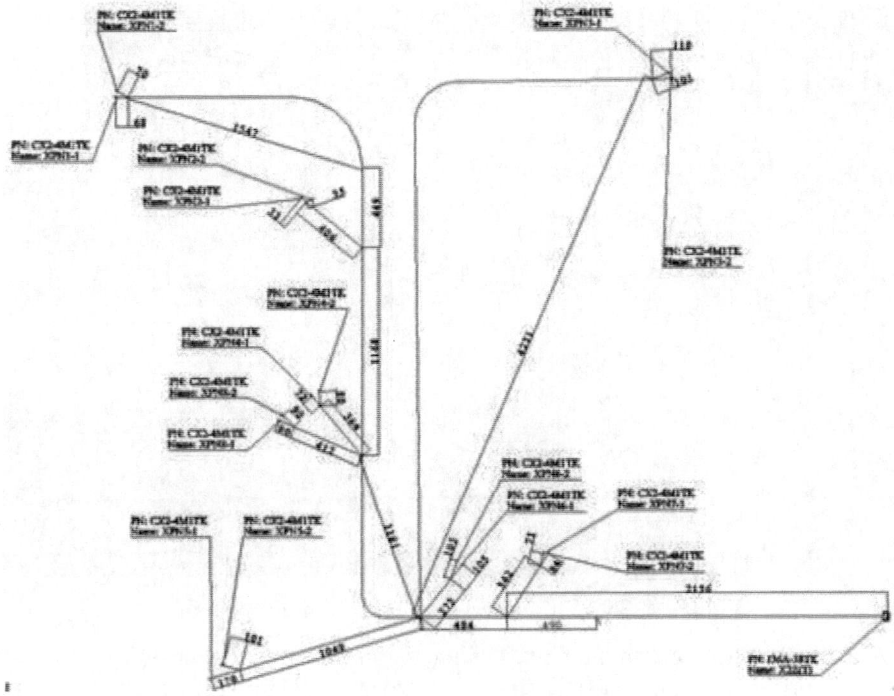

**Figure 3.** Digital nail layout.

## 4. 3-D Digital Technology

### *4.1. Traditional mode*

Previously, in the assembly of communication satellites, the traditional assembly process card was still used as the main form of process documents. Assembly process planning process cards usually have only text content. Process documents are compiled by CAPP software. Kanban Production System (MES) is used to view process documents and two-dimensional drawings on the assembly site. For complex assembly operations, it is difficult to accurately transfer information.

### *4.2. Digital mode*

Through the construction of a three-dimensional structured process planning system, the assembly process simulation and assembly implementation based on digital model are completed on some satellites, and a new mode of collaborative design, process design, assembly simulation and verification and field execution based on unified data source is preliminarily realized. As shown in Figure 4.

IOP Conf. Series: Materials Science and Engineering **608** (2019) 012037 doi:10.1088/1757-899X/608/1/012037

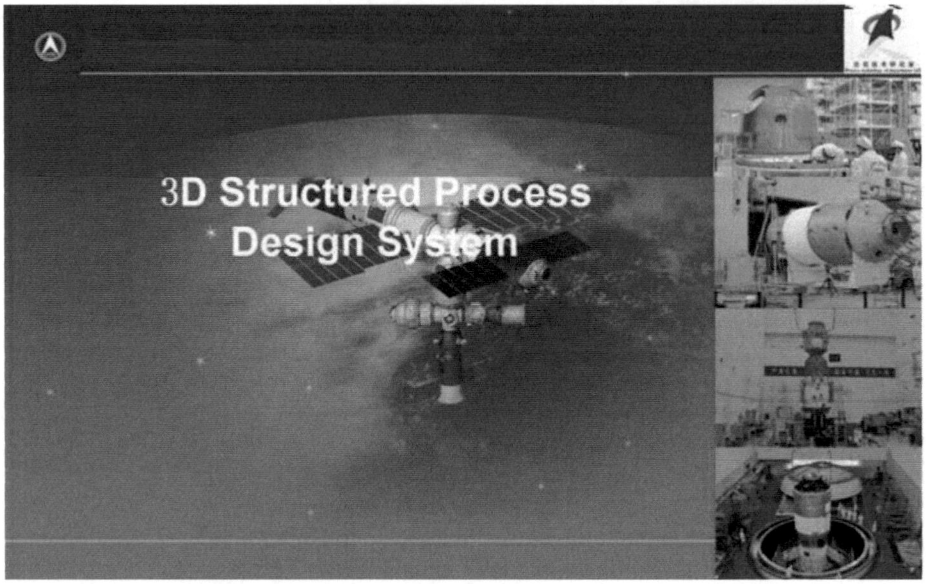


**Figure 4.** 3-D Structured Process Planning System

In the digital mode, through combing the design requirements and forming a three-dimensional process digital prototype, as a unified data source of process design. It links the preceding with the following, takes the three-dimensional design model and file as input, is the source of information for process design and material preparation, and is also the basic framework for connecting satellite package. At the same time, the three-dimensional process digital prototype includes product structure tree, three-dimensional model and product attributes, in which product attributes refer to generic attributes such as name, code number, specification and special attributes that different types of products need to contain.

The organization of three-dimensional process documents refers to the structure of current two-dimensional process documents, which includes five parts: cover, process description, process route, process catalogue and process card. The structure of three-dimensional temporary process documents is the same.

## 5. 3-D Digital Assembly

### *5.1. Cable marking*

In view of the change of cable design, manufacture and process mode, digital tools and means are adopted to carry out assembly work.

Traditional model: Cables need to be laid on wooden moulds for the following identification work:

1) Mark the direction of the cable on the main beam;

2) Make detailed labelling for each branch, branch point and reference points (span cable bridge, hatch hole, etc.).

3) The exit cables should indicate the location and length of the exit cabin.

Digital mode: In the process of cable production and manufacturing, the direction of cable is marked, and the work is carried out synchronously in advance, without occupying the main line time.

### *5.2. Site Laying Work*

Traditional model: According to the logo, the two-dimensional orientation is not intuitive and takes a long time to lay the cables according to the cable orientation diagram and documents.

Digital mode: Field configuration workstation and large display screen, relying on three-dimensional process documents, combined with product lightweight model, model on-site display, by switching different views, you can see the corresponding cable model, and see the laying details, greatly simplify the operation, improve the laying efficiency, as shown in Figure 5:

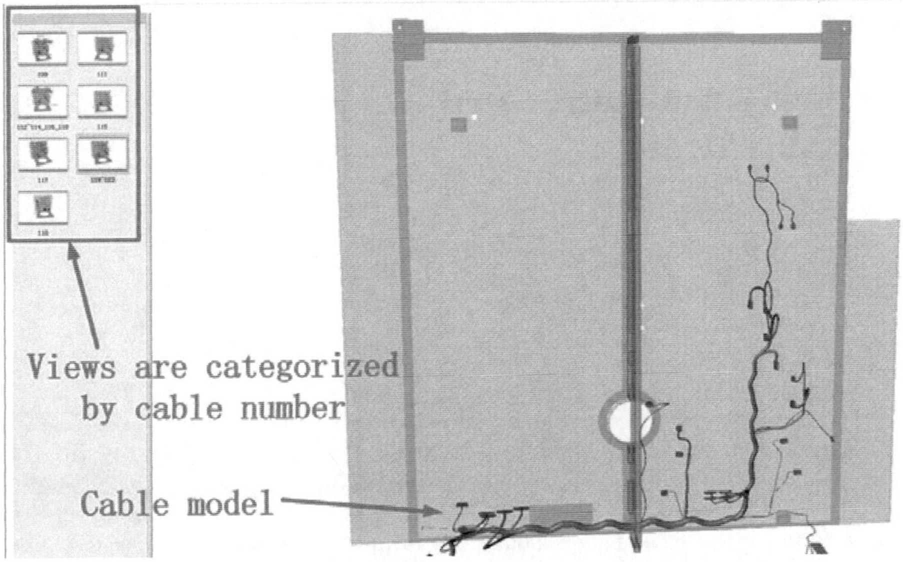


**Figure 5.** Schematic diagram of field lightweight model.

## 6. Concluding remarks

The three-dimensional digital design, manufacture and assembly of telecommunication satellite cables have realized real-time communication, improved design efficiency, laying speed and implementation quality, avoided the need for secondary processing of input files and error accumulation, and directly went to the factory using three-dimensional model of cables. It is more intuitive, the length of cable branches is more accurate, the weight and occupied space of cables on the device are reduced, and ultimately through. After displaying the three-dimensional model on the assembly site, the cable laying is more accurate and the efficiency is greatly improved.

Through the analysis of engineering requirements and the summary of practical results, the three-dimensional digital cable technology can greatly improve the efficiency and quality of cable design, manufacture and assembly, and meet the increasing demand of communication satellite production.

## References

[1] Xu Fuxiang, Lin Huabao and Hou Deep 2003. Introduction to Satellite Engineering (Beijing: China Astronautics Press)
[2] Yi Wangmin, Li Qi, Liu Hongyang and Sun Gang 2008. Study on electrostatic protection during satellite assembly J. Spacecraft environmental engineering. 25 (4) pp 387-390
[3] Xu Fuxiang 2002. Satellite Engineering (Beijing: China Aerospace Press)
[4] Zhu Guibai and Ma Shijun 2002. Introduction to Aerospace Technology (Beijing: Aviation Industry Press)
[5] Wang Shouzun and Liu Xia 2005. Research on Web-based Virtual Assembly Visualization Technology J. Journal of Naval Engineering University, 4 pp 120-124

# Solving Velocity Ambiguity for Pulse Doppler Radar Space Target Measurement

**Zhihua Yuan[1], Defeng Chen[1,*], Shan Xue[1] and Tuo Fu[1]**

[1]School of Information and Electronic, Beijing Institute of Technology, Beijing 100081, China

[*]Corresponding author: defengchenbit@gmail.com

**Abstract.** In order to solve the problem of severe velocity ambiguity for pulse Doppler radar high-speed space orbital targets measurement, a new velocity ambiguity resolution method based on the short arc orbit determination technique is presented in this paper. By means of a modified orbit determination model, this method can adopt the original radar measurements with the range-Doppler coupling term directly. It uses the numerical integration method to solve the differential equation of the state transition matrix and corrects the range-Doppler coupling term simultaneously, and then the optimal reference orbit is obtained by using the iterative least squares method. Finally, the velocity ambiguity is determined with the accurate reference velocity which can be calculated from the above reference orbit. The experimental results indicate that the proposed method can estimate the reference velocity accurately and satisfies the performance requirement of velocity ambiguity resolution.

## 1 Introduction

The range and velocity ambiguities are frequently encountered problems for pulse Doppler radar measurement. Because of the periodicity of transmitting pulses, the waveforms with low pulse repetition frequency (PRF) are ambiguous in velocity [1].

Current common methods for solving the velocity ambiguity in pulse Doppler radar detection are divided into two types. One is to calculate the unambiguous velocity by multiple measurements with multiple PRFs [2]-[5], among which the most representative is the Chinese Remainder Theorem [6]. There are some algorithms based on the Chinese Remainder Theorem, such as one-dimensional set algorithm [7], group algorithm [8] and the method of resolving ambiguity based on the theory of compressed perception [9]; the other is to use the reference velocity which can be obtained by the range differentiation after getting the accurate range measurement [10]. As long as the error between the reference velocity and the actual velocity does not exceed one ambiguity unit, the unambiguous velocity can be obtained correctly. Range-difference algorithm and invariant embedding algorithm[11] are typical algorithms in this type.

The above two conventional methods solve the velocity ambiguity problem from the radar waveform characteristics and the mathematical points, respectively. In the space target observation scenario, the orbits of non-maneuvering or weakly maneuvering space targets have significant feature. Therefore, from the perspective of the observed target, it is also possible to obtain a reference velocity by constructing an optimized trajectory from the original observation data. Based on this idea, a velocity ambiguity resolution method based on short arc orbit determination model is proposed in this paper and its effectiveness is verified.

The rest of the paper is organized as follows. The model and algorithm for solving the velocity ambiguity are introduced in the second section, detailed algorithm steps are also given at the end of this section. A performance comparison and analysis of the short arc orbit determination method and the conventional range-difference method to solve the velocity ambiguity are presented in the third section. The last section concludes the paper.

## 2 Model and algorithm

*2.1 Doppler shift and velocity ambiguity*

There often exist errors in the guided trajectory derived from the orbit motion model with inaccurate state vector or orbital elements, and those errors will lead to a Doppler deviation when the echo signal accumulates. We assume $\lambda$ is the carrier wavelength of the transmitting signal, the velocity deviation can be calculated according to the relation between velocity and Doppler frequency, and then the actual velocity of the target can be obtained by modifying the velocity deviation to the guided velocity:

$$\Delta V = \lambda \Delta f_d / 2 \tag{1}$$

$$v_m = \hat{v}_l + \Delta V \tag{2}$$

where $\Delta f_d$ and $\Delta V$ are the Doppler deviation and the velocity deviation respectively, $\hat{v}_l$ is the guided velocity and $v_m$ is the actual velocity of the target.

Due to the periodicity of radar transmitting signal, the Doppler ambiguity will occur when the Doppler deviation is bigger than half of the PRF, which is $|\Delta f_d| \geq PRF/2$, so the actual velocity cannot be got by (2) directly. At this point, the actual velocity of the target should be

$$v_m = \hat{v}_l + \Delta V - N v_{PRF} \tag{3}$$

where $v_{PRF}$ is the velocity corresponding to a PRF, which is $v_{PRF} = \lambda PRF/2$, and the integer $N$ is a fold factor which is usually determined by the reference velocity $v_{ref}$ in engineering:

$$N = round\left(\frac{\hat{v}_l + \Delta V - v_{ref}}{v_{PRF}}\right) \tag{4}$$

(4) requires an accuracy of the reference velocity less than $v_N$, and $v_N$ is called the Nyquist velocity[12], which is $|v_m - v_{ref}| \leq v_N = v_{PRF}/2 = \lambda PRF/4 = cPRF/(4f_c)$, where $c$ is the speed of light and $f_c$ is the carrier frequency. It can be seen that the improvement of carrier frequency or the decrease of PRF will require higher reference velocity accuracy.

The range-Doppler coupling of the chirp signal is another problem should be considered when using the short arc orbit determination method to solve the velocity ambiguity [13]. For a typical negative chirp signal, the relation between delay $t_{delay}$ and Doppler shift $f_d$ is

$$t_{delay} = -\frac{f_d \cdot T_p}{B} \tag{5}$$

where $T_p$ is the pulse width. The additional range deviation caused by $t_{delay}$ is

$$\Delta R = t_{delay} \cdot \frac{c}{2} = -\frac{f_d \cdot T_p}{B} \cdot \frac{\lambda \cdot f_c}{2} = -\frac{f_c \cdot T_p}{B} \cdot V \tag{6}$$

where $B$ is the bandwidth of the transmitting signal, $V$ is the unknown radial velocity of target. The range after the correction of range-Doppler coupling is

$$R_t = R + \Delta R = R - \frac{f_c \cdot T_p}{B} \cdot V \tag{7}$$

where $R$ is the radial range before the correction of range-Doppler coupling. Obviously, we should use the $R_t$ in the short arc orbit determination model, but $R_t$ needs to be solved by the unknown $V$ .Therefore, in order to obtain a more accurate reference velocity, it is necessary to update $R_t$ simultaneously during short arc orbit determination.

*2.2 Orbit and measurement model*

We use $\mathbf{r}, \mathbf{v}$ to indicate the position and velocity vectors of the target in the Earth Centered Inertial (ECI) Coordinate System respectively; $\mathbf{r}_S, \mathbf{v}_S$ to indicate the position and velocity vectors of the target in the Topocentric Horizon (ENZ) coordinate system respectively. The positional relationship between the ECI coordinate system and the Earth Centered Earth-Fixed (ECEF) Coordinate System is the rotation with respect to the hour angle, and the relationship between the ECEF coordinate system and the ENZ coordinate system is the translation of the origin and the rotation of the axis. Then we have

$$\mathbf{r}_S = \mathbf{EU}\left(\mathbf{r} - \mathbf{r}_{Site\text{-}ECI}\right) \tag{8}$$

$$\mathbf{v}_S = \mathbf{EU}(\mathbf{v} - \mathbf{v}_{Site\text{-}ECI}) + \mathbf{E}\frac{\mathrm{d}\mathbf{U}}{\mathrm{d}t}(\mathbf{r} - \mathbf{r}_{Site\text{-}ECI}) \tag{9}$$

$$\mathbf{E} = \mathbf{R_x}(-\pi/2 + lat) \cdot \mathbf{R_z}(-\pi/2 - lon) \tag{10}$$

$$\mathbf{U} = \mathbf{R_z}(\theta_t) \tag{11}$$

where $\mathbf{E}$ is the rotation matrix of ECEF coordinate system to ENZ coordinate system, $\mathbf{U}$ is the rotation matrix of ECI coordinate system to ECEF coordinate system, $\mathbf{r}_{Site\text{-}ECI}$ is the coordinates of the radar station in the ECI coordinate system, $\mathbf{v}_{Site\text{-}ECI}$ is the velocity vector of the radar in the ECI coordinate system, $lat, lon$ is the latitude and longitude of the radar station respectively, $\theta_t$ is the Greenwich Hour Angle corresponding to the measurement time, $\mathbf{R_x}(\theta)$ represents the rotation matrix corresponding to the YZ plane around the X axis with rotation angle $\theta$, $\mathbf{R_z}(\theta)$ has the same meaning as $\mathbf{R_x}(\theta)$, but represents the XY plane rotating around the Z axis [14].

We suppose the measurement vector at the time $t$ is $\mathbf{z}(t)=(R \quad A \quad E)^{\mathrm{T}}$, where $A$ and $E$ are the azimuth and elevation respectively, the relative position and velocity vectors in the ENZ coordinate system corresponding to $\mathbf{z}(t)$ are $\boldsymbol{\rho} = (\rho_E \quad \rho_N \quad \rho_Z)$ and $\dot{\boldsymbol{\rho}} = \left(\dot{\rho}_E \quad \dot{\rho}_N \quad \dot{\rho}_z\right)$ respectively. According to the conversion between the ENZ coordinate system and the polar coordinate system, there is

$$\begin{cases} R_t = R - \dfrac{f_c \cdot T_p}{B} \cdot V = \sqrt{\rho_E^2 + \rho_N^2 + \rho_Z^2} - \dfrac{f_c \cdot T_p}{B} \cdot \dfrac{\rho_E \dot{\rho}_E + \rho_N \dot{\rho}_N + \rho_Z \dot{\rho}_Z}{\sqrt{\rho_E^2 + \rho_N^2 + \rho_Z^2}} \\ A = \begin{cases} \arccos \dfrac{\rho_N}{\sqrt{\rho_E^2 + \rho_N^2}}, \rho_E \geq 0 \\ 2\pi - \arccos \dfrac{\rho_N}{\sqrt{\rho_E^2 + \rho_N^2}}, \rho_E < 0 \end{cases} \\ E = \arctan \dfrac{\rho_Z}{\sqrt{\rho_E^2 + \rho_N^2}} \end{cases} \tag{12}$$

We suppose the state vector at the time $t$ in the ECI coordinate system is $\mathbf{x}(t) = (\mathbf{r} \quad \mathbf{v})$, where $\mathbf{r} = (rx \quad ry \quad rz)$ and $\mathbf{v} = (vx \quad vy \quad vz)$, then the partial derivative of $\mathbf{z}(t)$ to $\mathbf{x}(t)$ can be expressed as

$$\left(\frac{\partial \mathbf{z}(t)}{\partial \mathbf{x}(t)}\right)_{3\times 6} = \begin{pmatrix} \dfrac{\partial R_t}{\partial \mathbf{r}} & \dfrac{\partial R_t}{\partial \mathbf{v}} \\ \dfrac{\partial A}{\partial \mathbf{r}} & \dfrac{\partial A}{\partial \mathbf{v}} \\ \dfrac{\partial E}{\partial \mathbf{r}} & \dfrac{\partial E}{\partial \mathbf{v}} \end{pmatrix} = \begin{pmatrix} \dfrac{\partial R_t}{\partial \boldsymbol{\rho}} \mathbf{EU} & \mathbf{O}_{1\times 3} \\ \dfrac{\partial A}{\partial \boldsymbol{\rho}} \mathbf{EU} & \mathbf{O}_{1\times 3} \\ \dfrac{\partial E}{\partial \boldsymbol{\rho}} \mathbf{EU} & \mathbf{O}_{1\times 3} \end{pmatrix} \tag{13}$$

the $\mathbf{O}_{1\times3}$ in the (13) represents a zero matrix of 1×3 order. According to (8),(9) the $\boldsymbol{\rho}$ and $\dot{\boldsymbol{\rho}}$ can be obtained by the state vector $\mathbf{x}(t)$, and then $\frac{\partial \mathbf{z}(t)}{\partial \mathbf{x}(t)}$ can be solved by the (12),(13).

Assuming that $\mathbf{x}(t_0)$ represents the state vector at time $t_0$, the relation between $\mathbf{x}(t)$ and $\mathbf{x}(t_0)$ can be represented by the following state transition matrix

$$\boldsymbol{\varphi}(t,t_0)=\left(\frac{\partial \mathbf{x}(t)}{\partial \mathbf{x}(t_0)}\right)_{6\times6} \tag{14}$$

In order to adapt to the actual radar observation requirements, only the influence of gravity field on the earth is considered in this paper. The effects of precession, nutation and the earth pole shift are ignored. Since the partial differential equation corresponding to the state transition matrix considering the perturbation term is very complicated, it is often impossible to obtain an analytical solution and must be solved by numerical integration. The variable order variable step method proposed by Shampine & Gordon 1975 [15] is employed in this paper. This method can obtain a numerical solution with satisfactory precision because the order and step size can be changed during the integration process.

Combining the above partial derivatives, the relationship between $\mathbf{z}(t)$ and the initial state vector $\mathbf{x}(t_0)$ is obtained:

$$\left(\frac{\partial \mathbf{z}(t)}{\partial \mathbf{x}(t_0)}\right)_{3\times6}=\left(\frac{\partial \mathbf{z}(t)}{\partial \mathbf{x}(t)}\right)\cdot\left(\frac{\partial \mathbf{x}(t)}{\partial \mathbf{x}(t_0)}\right)=\left(\frac{\partial \mathbf{z}(t)}{\partial \mathbf{x}(t)}\right)\cdot\boldsymbol{\varphi}(t,t_0) \tag{15}$$

*2.3 Linearization and least squares estimation*

We denote an n-dimensional vector of measurements at times $t_1,...,t_n$ is

$$\mathbf{Z}=(\mathbf{z}_1 \ \cdots \ \mathbf{z}_n)^T \tag{16}$$

The measurement vector $\mathbf{z}_i(t_i)$ are described by

$$\mathbf{z}_i(t_i)=\mathbf{h}_i(t_i,\mathbf{x}(t_0))+\varepsilon_i \tag{17}$$

where $i=1,2,..,n$, $\mathbf{h}_i(t_i,\mathbf{x}(t_0))$ denotes the *i*th value of the measurement vector as a function of time $t_i$ and the state $\mathbf{x}(t_0)$ ; the quantities $\varepsilon_i$ account for the difference between actual and modelled observations due to measurement errors. The (17) can be simplified to

$$\mathbf{Z}=\mathbf{h}_{\mathrm{N}}(\mathbf{x}_0)+\boldsymbol{\varepsilon} \tag{18}$$

The (18) is a standard vector least squares model. According to the definition of least squares, for the measurements vector $\mathbf{Z}$, we need to find the state vector $\mathbf{x}_0^{lsq}$ to minimize the loss function

$$\mathbf{J}(\mathbf{x}_0)=(\mathbf{Z}-\mathbf{h}_{\mathrm{N}}(\mathbf{x}_0))^{\mathrm{T}}(\mathbf{Z}-\mathbf{h}_{\mathrm{N}}(\mathbf{x}_0)) \tag{19}$$

The practical solution of (19) is complicated due to the fact that $\mathbf{h}_{\mathrm{N}}(\mathbf{x}_0)$ is a highly nonlinear function of the unknown vector $\mathbf{x}_0$. So it is necessary to linearize the problem, the Herrick-Gibbs initial orbit method [16] is employed to obtain an approximation value $\mathbf{x}_0^{ref}$ of the initial state firstly, and then all quantities are linearized around $\mathbf{x}_0^{ref}$. The residual vector is approximately given by

$$\begin{aligned}\mathbf{Z}-\mathbf{h}_{\mathrm{N}}(\mathbf{x}_0)&\approx\mathbf{Z}-\mathbf{h}_{\mathrm{N}}(\mathbf{x}_0^{ref})-\frac{\partial \mathbf{h}_{\mathrm{N}}(\mathbf{x}_0)}{\partial \mathbf{x}_0}(\mathbf{x}_0-\mathbf{x}_0^{ref})\\&=\Delta\mathbf{Z}-\mathbf{H}\Delta\mathbf{x}_0\end{aligned} \tag{20}$$

where $\Delta\mathbf{Z}$ denotes the difference between the actual measurements and the predicted measurements derived from the reference orbit, $\Delta\mathbf{x}_0$ denotes the difference between the initial state and the initial reference state, and $\mathbf{H}$ is the Jacobian matrix denotes the partial derivatives of the predicted measurements vector derived from the reference orbit model with respect to the state vector $\mathbf{x_0}$ at time $t_0$ .Combined with (15), we have

$$\mathbf{H}=\frac{\partial \mathbf{h}_N}{\partial \mathbf{x_0}}\bigg|_{\mathbf{x_0}=\mathbf{x_0}^{ref}} = \left.\begin{pmatrix} \frac{\partial \mathbf{h_1}}{\partial \mathbf{x_1}} \cdot \frac{\partial \mathbf{x_1}}{\partial \mathbf{x_0}} \\ \vdots \\ \frac{\partial \mathbf{h_n}}{\partial \mathbf{x_n}} \cdot \frac{\partial \mathbf{x_n}}{\partial \mathbf{x_0}} \end{pmatrix}_{3n\times 6}\right|_{\mathbf{x_0}=\mathbf{x_0}^{ref}} \quad (21)$$

where $\mathbf{x_1}\cdots\mathbf{x_n}$ denote the state vectors corresponding to the vectors of predicted measurements $\mathbf{h_1}\cdots\mathbf{h_n}$ . So the partial derivative matrix $\mathbf{H}$ corresponding to n-dimensional vector of measurements should be 3n×6 dimensions.

In this way, the orbit determination problem is reduced to the linear least squares problem of finding $\Delta\mathbf{x_0}^{lsq}$ . Consider the weighted matrix $\mathbf{W}$ , we have

$$\Delta\mathbf{x_0}^{lsq} = (\mathbf{H}^{\mathrm{T}}\mathbf{W}\mathbf{H})^{-1}\mathbf{H}^{\mathrm{T}}\mathbf{W}\Delta\mathbf{z} \quad (22)$$

$$\mathbf{W} = \mathrm{diag}(\boldsymbol{\sigma}_0^{-2},,\ldots\boldsymbol{\sigma}_{n-1}^{-2}) \quad (23)$$

$$\boldsymbol{\sigma}_i=\mathrm{diag}\left(\sigma_{\mathrm{R}i} \quad \sigma_{\mathrm{A}i} \quad \sigma_{\mathrm{E}i}\right), i=1,2...n. \quad (24)$$

where $\boldsymbol{\sigma}_i$ denotes the diagonal matrix consisting of the random noise measured at the range, azimuth and elevation of the point $i$ . Because of the approximation of $\mathbf{x_0}^{ref}$ , the value of $\mathbf{x_0}^{lsq}=\mathbf{x_0}^{ref}+\Delta\mathbf{x_0}^{lsq}$ determined so far is not yet the exact solution of the orbit determination problem. In practical engineering applications, we can let $\mathbf{x_0}^{ref}=\mathbf{x_0}^{lsq}$ and repeat the same procedure. After several iterations, the reference orbit can basically meet the accuracy requirements.

### *2.4 The steps of algorithm*

Step 1: Obtaining the approximate initial reference state $\mathbf{x_0}^{ref}$ using the Herrick-Gibbs method based on the measurements vector $\mathbf{Z}$ .

Step 2: Obtaining $\mathbf{x}_1,\ldots,\mathbf{x_n}$ by propagating $\mathbf{x_0}^{ref}$ to times $t_1,\ldots,t_n$ , and then obtaining the vectors $\mathbf{h}_1,\ldots,\mathbf{h_n}$ .

Step 3: Solving the partial derivative matrix of $\mathbf{h}_1,\ldots,\mathbf{h_n}$ with respect to $\mathbf{x}_1,\ldots,\mathbf{x_n}$ by (13), solving the state transition matrix by the numerical integration method, and then obtaining the Jacobian matrix $\mathbf{H}$ .

Step 4: Solving the $\Delta\mathbf{x_0}^{lsq}$ by the least squares method.

Step 5: Judging that the $\Delta\mathbf{x_0}^{lsq}$ whether or not satisfies the requirement of solving the velocity ambiguity, if satisfied, turns to step 6 otherwise let $\mathbf{x_0}^{ref}=\mathbf{x_0}^{lsq}$ and repeats step 2 to step 5.

Step 6: Solving the reference velocity $v_{ref}$ by coordinate transformation, and then solving the velocity ambiguity.

## 3 The experimental results

### *3.1 The simulation*

In order to verify the effectiveness of the short arc orbit determination method proposed above, we choose to compare it with the range-difference method. The principle of the range-difference method is to obtain the reference velocity of the intermediate point by the range-difference between the front and back points. Assume that the radial range and radial velocity of the target at three consecutive times $t_1,t_2,t_3$ are $\{r_i,v_i\},i=1,2,3$ , since the original measurements include the range-Doppler coupling term, the reference velocity at time $t_2$ obtained by the range-difference method is

$$v_{ref} = \frac{(r_3 - \frac{f_c \cdot T_p}{B} \cdot v_3) - (r_1 - \frac{f_c \cdot T_p}{B} \cdot v_1)}{t_3 - t_1} \tag{25}$$

In this simulation, we use KOMPSAT 5 satellite (NORAD ID: 39227) as the target, and the working parameters of the radar are shown in Table 1. The theoretical guidance is extrapolated by the SGP4 (Simplified General Perturbations 4) model using the TLE (Two Line Element) of KOMPSAT 5, which is considered a precise orbit in this simulation. The measurements are constructed by adding Gaussian white noise and other terms to the theoretical guidance. Finally, the range measurement includes a theoretical range term, a noise term with a standard deviation of 5 m, and the range-Doppler coupling term. The azimuth and elevation measurements contain respective theoretical term and noise term with a standard deviation of 0.14 mrad. The velocity measurement includes the theoretical velocity term, the noise term with a standard deviation of 0.01 m/s, and the ambiguous velocity term $Nv_{PRF}, (-5 \le N \le 5)$.

**Table 1.** Working parameters of radar

| | Carrier frequency | Signal bandwidth | Duration | Pulse width | Pulse repetition time |
|---|---|---|---|---|---|
| Waveform 1 | 8.5 GHz | 10 MHz | 66.7 s | 1 ms | 4 ms |
| Waveform 2 | 8.5 GHz | 10 MHz | 66.7 s | 2 ms | 7 ms |
| Waveform 3 | 8.5 GHz | 10 MHz | 66.7 s | 4 ms | 14 ms |

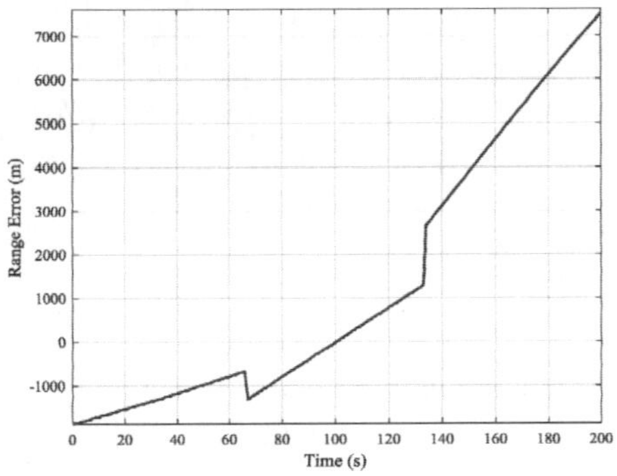


(a)

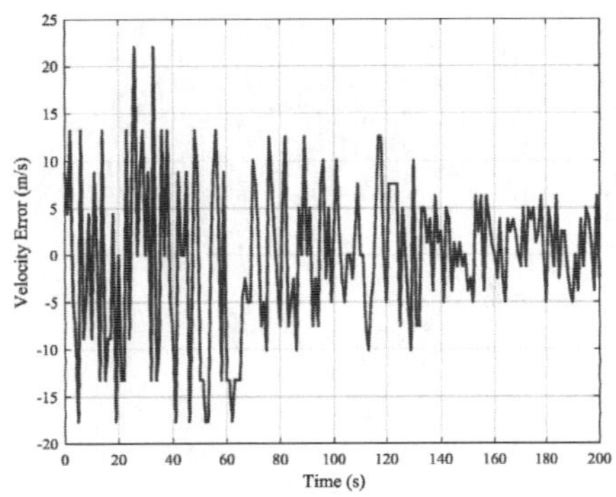


(b)

**Figure 1.** The errors before ambiguity resolution.

Figure 1(a) and Figure 1(b) describe the errors of the measurements and theoretical values of the range and velocity before velocity ambiguity resolution respectively. It can be seen from Figure 1 that the error of range measurement reaches several kilometers and the velocity measurement has an error of more than ten meters per second. In addition, we can see the range error in Figure 1(a) has a jump at the waveform switching point, which is due to the existence of the range-Doppler coupling term, and the velocity error in Figure 1(b) shows the velocity measurement is obviously ambiguous.

We use the range-difference method and the short arc orbit determination method to solve the reference velocity respectively, and then obtain the unambiguous velocity by the reference velocity, and finally, the more accurate range measurement can be obtained. Figure 2(a) and Figure 2(b) show the range error obtained by the range-difference method and the short arc orbit determination method respectively. Figure 2(c) and Figure 2(d) show the velocity error obtained by the range-difference method and the short arc orbit determination method respectively. We can see from Figure 2(a) and Figure 2(c) that the range error obtained by the range-difference method does not exceed 70 m and the velocity error does not exceed 23 m/s. According to the formula shown in the previous section $v_N = cPRF/(4f_c)$, the Nyquist velocities corresponding to waveform 1, 2, and 3 are 2.2 m/s, 1.26 m/s, and 0.63m/s, respectively. Therefore, although the range-difference method can improve the range measurement accuracy, it cannot solve the velocity ambiguity problem. It can be seen from Figure 2(b) and Figure 2(d) that the range error is no more than 18 m and the velocity error is no more than 0.04 m/s, thence the velocity ambiguity had been completely solved by the short arc orbit determination method.

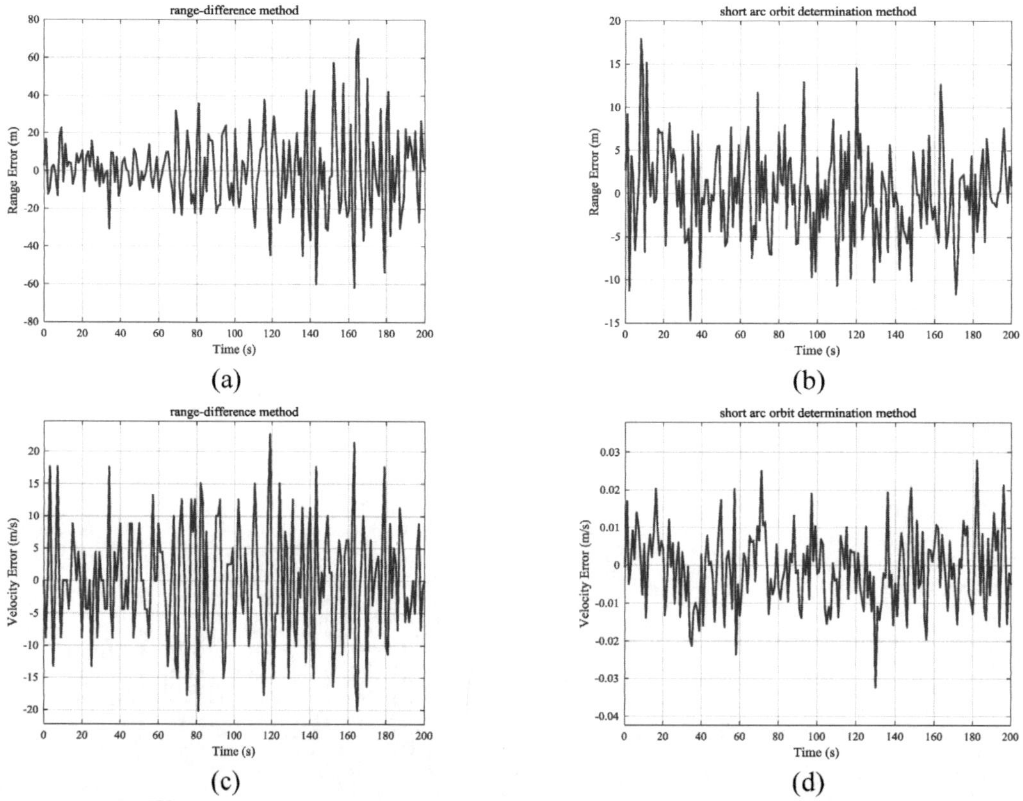


**Figure 2.** The measurements and errors after ambiguity resolution.

*3.2 The observation data*

We used the observation data of SENTINEL 1B satellite (NORAD ID: 41456) from a radar station as the measurements. The observation start time is 18:53:22 on July 21, 2018 (Beijing time), and the total data length is 200 seconds. The actual orbit is generated by the public precise ephemeris of SENTINEL 1B. Table 2 shows the waveforms used in the observation and the Nyquist velocity corresponding to the waveforms.

**Table 2.** The waveforms and the corresponding Nyquist velocities

| Carrier frequency | Pulse width | Pulse repetition time | Nyquist velocity |
|---|---|---|---|
| 8.5 GHz | 4 ms | 14 ms | 0.63 m/s |
| | 3.7 ms | 13 ms | 0.68 m/s |
| | 3.5 ms | 12 ms | 0.73 m/s |

Figure 3 and Figure 4 are similar to Figure 1 and Figure 2 respectively. Except that the measurements are obtained by observation rather than simulation. From Figure 3 we can see that the range error has a jump at the waveform switching point, and the velocity error shows the velocity measurement is obviously ambiguous.

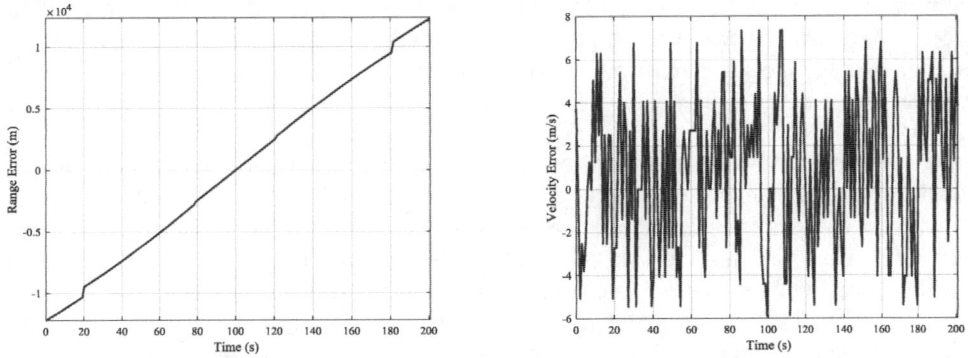


(a) (b)

**Figure 3.** The measurements and errors before ambiguity resolution.

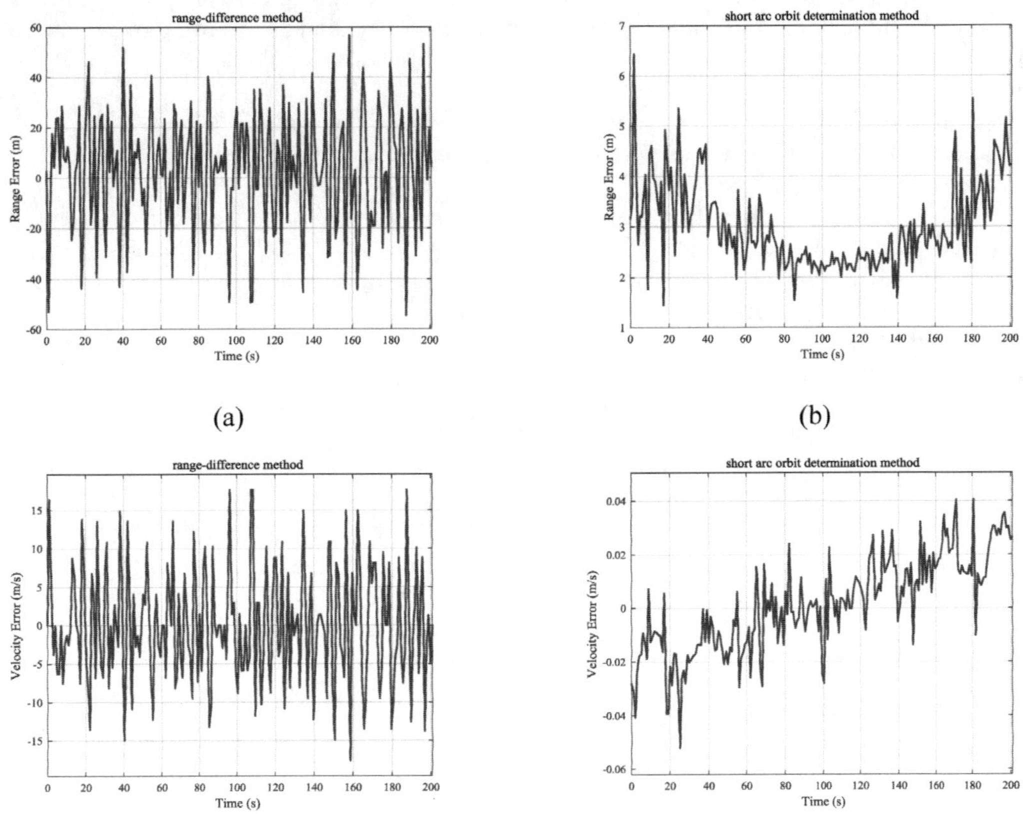


(c) (d)

**Figure 4.** The measurements and errors after ambiguity resolution

Figure 4 shows the range error and velocity error obtained by the range-difference method and the short arc orbit determination method respectively. It can be seen from the Figure 4(a) and Figure 4(b) that the range error obtained by the range-difference method does not exceed 57 m, and the range error obtained by the short arc orbit determination method does not exceed 7 m. From the Figure 4(c) and

Figure 4(d), we can see that the velocity error obtained by the range-difference method does not exceed 18 m/s, and the velocity error obtained by the short arc orbit method can reach 0.06 m/s. It can be seen from the Table 2 that the smallest Nyquist velocity is 0.63 m/s of the waveform with a period 14 millisecond. We can see that the range-difference method cannot solve the velocity ambiguity problem at this time and the short arc orbit determination method can be well solved.

## 4 Conclusion

In this paper, the application of short arc orbit determination technique to the velocity ambiguity problem of the ground-based pulse Doppler radar space target detection is studied. By analysing the relationship between radar measurements and target state vector, as well as considering the range-Doppler coupling effect, an improved short arc orbit determination model is established, and then the iterative least squares method is used to estimate a more accurate reference velocity, resulting in the unambiguous velocity. The experimental results show that the short arc orbit determination method can be applied directly to the original radar observation data with range-Doppler coupling term, especially when observing the space targets with a high carrier frequency and low PRF radar, and its performance of velocity ambiguity resolution is better than the conventional range-difference method. Generally, the proposed method can improve the velocity measurement accuracy from several meters per second to a few centimeters per second. Using the estimated unambiguous velocity to eliminate the range-Doppler coupling term, the range measurement accuracy can also be increased by several times.

## Reference

[1] M. A. Richards, *Fundamentals of radar signal processing*, New York: McGraw-Hill (2005).
[2] X. Zhu, Z. Cui, W. Cui, and S.Wu, IET, *Range and velocity ambiguity resolution based on screening method*, 1-3 (2009).
[3] L. Zedel, A. E. Hay, IEEE JOE, *Resolving Velocity Ambiguity in Multi frequency, Pulse-to-Pulse Coherent Doppler Sonar*, **35**, 847-851 (2010).
[4] T. Ma, H. Chen, and D. Peng, Modern Electronics Technique, *A fusion solution of velocity ambiguity of moving target in dual band SAR/GMTI*, **38**, 14, 32-35 (2015).
[5] M. Xia, W. Su, H. Gu, and J. Lu, Journal of Electronics & Information Technology, *Research on Velocity Ambiguity Resolution for Multiple PRF Radar*, **39**, 12, 1009-5896 (2017).
[6] Y. Song, H. An, Fire Control Radar Technology, *A New Algorithm for Range and Velocity Ambiguity Resolution at Medium PRF*, **38**, 1, 15-18 (2009).
[7] G. Trunk, S. Brockett, Radar Conference, *Range and velocity ambiguity resolution* (1993).
[8] T. Zeng, T. Long, AES, *A New Algorithm for PR Radar Ambiguity Resolution*, **28**, 12, 99-101 (2000).
[9] Y. Zhang, Z. Deng, J. Shi, L. Ye, M. Fu, and C. Zhao, JECE, *Doppler Ambiguity Resolution Based on Random Sparse Probing Pulses*, **2015**, 1-8 (2015).
[10] X. Liu, K. Xie, X. Wu. Modern Electronics Technique, *A new method of eliminating high-speed target's velocity ambiguity in LFMCW radar*, **35**, 9-10 (2012).
[11] F. Meng, L. Xie, and R. Li, IET, *Research on invariant imbedding method of velocity ambiguity resolution*, 1-4 (2009).
[12] T. Sakamoto, A. Matsuoka, and H. Yomo, IEEE SJ, *Estimation of Doppler Velocities from Sub-Nyquist Ultra-Wideband Radar Measurements*, **16**, 23, 8557-8565 (2016).
[13] H. Hu, Modem Radar, *Range Error Analysis of Range-Doppler Coupling Effect to LFM Radar*, **33**, 12, 47-50 (2011).
[14] O. Montenbruck, E. Gill, Springer Verlag, *Satellite Orbits: Models, Methods, Applications* (2012).
[15] L. Fausett, *Applied Numerical Analysis Using MATLAB* (1999).
[16] D A. Vallado, Nitrogen in the Marine Environment, *Fundamentals of Astrodynamics and Applications* (2013).

IOP Conf. Series: Materials Science and Engineering **608** (2019) 012039 doi:10.1088/1757-899X/608/1/012039

# Preliminary Discussion on the Spacecraft Electrostatic Discharge and Standards

**R J Jia, B Li and Z C Shen**

Beijing Institute of Space Environment Engineering, Beijing 100094, China

aiwotest@126.com

**Abstract.** The antistatic design of spacecraft is an important measure to improve the life and reliability of spacecraft in orbit. The standard and specification of spacecraft electrostatic discharge is an important guarantee for improving the reliability and long life of spacecraft. Analyzes the mechanism of electrostatic discharge effect of spacecraft, introduces the current status of electrostatic discharge test standards and specifications at home and abroad, and looks forward to the development direction of China's spacecraft electrostatic discharge standard, in order to improve the standards and specifications of spacecraft electrostatic discharge.

## 1. Introduction

Space-charged particle radiation is one of the spatial environmental factors that have a significant impact on spacecraft. It produces three important spatial effects, including spacecraft electrostatic discharge effects, single-event upset and blocking effects, total dose effects. Among these three effects, the electrostatic discharge effect on the spacecraft is particularly prominent, which will cause failure or damage of materials, components, components and single machines, and threaten the safe operation of the spacecraft in orbit. Therefore, in order to ensure the reliability of the spacecraft in orbit operation, in the process of design, manufacture, test, flight and return of the spacecraft, it is necessary to fully consider the influence of the electrostatic discharge effect on the spacecraft to improve the environmental resistance of the spacecraft and receptive and on-orbit stability.

Such as NASA and ESA, These foreign space agencies discovered the threat of electrostatic discharge to spacecraft very early. The study of electrostatic discharge effects has been carried out, which has been obtained from a large number of flight tests, various environmental parameters and the influence of surface charging on the instrument on the star, Since the 1960s and 1970s. And then research on antistatic design and reinforcement technology. At the same time, these space agencies have also established relatively complete standards and specifications for spacecraft electrostatic discharge, providing support for designers to design and test live space protection. Although China's research on the electrostatic discharge effect of spacecraft is relatively late, based on the successful experience of foreign countries, it has also explored a series of effective methods for antistatic, and gradually developed relevant enterprise standards and norms, but Electrostatic discharge test standards and specifications at the national or industry level and related space environmental standards and norms are still lacking. The mechanism of electrostatic discharge effect of spacecraft is preliminarily analyzed and the current status of electrostatic discharge test standards and specifications at home and abroad is summarized in this paper. Based on this, the development recommendations of the standards and specifications for electrostatic discharge test of spacecraft are given in China.

IOP Conf. Series: Materials Science and Engineering **608** (2019) 012039 doi:10.1088/1757-899X/608/1/012039

## 2. The nature of the electrostatic discharge effect of spacecraft

The electrostatic discharge effect of spacecraft, also known as the charge-discharge effect, refers to the electrostatic charge accumulation and venting process that occurs when the satellite interacts with the environment such as space plasma and high-energy electrons. The essence of the spacecraft is the mutual interaction between space-charged particles and spacecraft materials effect.

### *2.1. Surface charge and discharge effect*

The surface charge and discharge effect is caused by the interaction between the surface of the spacecraft and the space environment. It is the process of charge accumulation and release in the surface material of the satellite. The spacecraft is orbiting in a low-energy plasma environment that interacts with the surface material of the spacecraft to accumulate charge on the surface of the spacecraft to cause surface charging. The surface potential difference will occur between the surfaces of the spacecraft, the surface and the deep layer, the surface and the satellite ground, due to the dielectric properties and geometrical shapes of the surface material. When this potential difference reaches the discharge threshold, the spacecraft will be triggered, Electrostatic discharge occurs [1], causing electromagnetic interference, surface contamination, etc. If there is an exposed power component such as a solar panel at the location where the electrostatic discharge occurs, a secondary arc may occur and cause a greater risk. In severe cases, even Threats to spacecraft safety caused various on-orbit anomalies[1].

### *2.2. Internal charging effect*

The internal charging effect is another satellite charging effect caused by the radiation of spatial high energy particles that is different from the surface charging effect. It refers to the process of establishing an electric field through high-energy charged particles passing through the surface of the satellite (the energy of these high-energy electrons is mainly in the range of 0.1-7 MeV, which has strong penetrating power), which is transmitted and deposited inside the dielectric material of the satellite component to establish an electric field. When the electric field generated by the deep charge of the medium exceeds the breakdown threshold of the dielectric material, a discharge occurs, and the electromagnetic pulse generated by the discharge may interfere with or even destroy the normal operation of the in-star electronic system, especially the poorly shielded cable, printed circuit board and the thermal protection layer is particularly vulnerable to damage. In severe cases, the entire star will fail [2]. The internal charging effect mainly occurs in the middle and high orbits, many spacecraft on-orbit faults have been classified as internal charging effects in recent year [2].

## 3. Effect of electrostatic discharge on spacecraft reliability

The electrostatic discharge effect of spacecraft is closely related to the disturbance of space environment. Different track heights will have different electrostatic discharge effects. The space environment is complex and unique. It includes parameters such as charged particles, plasma, electric field, solar radiation, space debris. Each space environmental parameter will have a fatal impact on the spacecraft, resulting in reduced product performance, shortened working life, and static electricity in the orbit. The failure and damage caused by the discharge effect has become one of the main causes of the failure and damage of the spacecraft in orbit.

### *3.1. Different track charge and discharge effects*

In the earth orbit space, there is a great difference in spatial particle radiation, space plasma and high-energy electrons change with height and latitude.

The surface charge and discharge effect is related to the temperature of the plasma. The LEO orbital plasma has a high density and a low temperature. The surface charge and discharge problem of the spacecraft is not serious. It is only necessary to consider the surface charge and discharge problems of a high voltage system exposed to the outside of the spacecraft and larger than 150V. The spatial plasma of the GEO orbits, MEO orbits and polar regions have high temperature and low density, is often accompanied by geomagnetic sub-explosion activities, therefore, spacecraft operating in these two regions will face serious surface charging problems.

The internal charging effect is related to the energy and total fluence of high energy electrons. High-energy electrons have two distribution peaks of the inner radiation band and the outer radiation band. Spacecraft operating in MEO orbits face the worst-case orbital environment at the center of the outer radiating zone and therefore face the most severe internal charging effects. The GEO orbit is close to the outer edge of the outer radiating zone. When the space environment is disturbed, the high-energy electron flux will change by an order of magnitude, and the electrified electric effect in the orbit can not be ignored. The LEO orbiting spacecraft does not have an internal charging effect.

### *3.2. The effect of discharge on the reliability of spacecraft*

In 1996, the Marshall Space Flight Center (MSFC) published a statistical data on the failure and anomaly of the spacecraft system caused by the space environment. Among the 114 spacecraft faults and anomalies, the charge and discharge abnormalities caused by plasma and high-energy electrons were 41. There are 39 faults caused by particle radiation effects, 12 faults caused by thermal environment, and 11 faults caused by micrometeor/space debris. These four environmental incentives account for about 90% of the total faults. The main influencing factors of anomalies. In particular, the electrostatic discharge effect caused the spacecraft to account for 36% of the total number of failures, which is the biggest killer of spacecraft anomalies. Therefore, for the electrostatic discharge effect and the particle radiation effect, the importance of the reliability work of aerospace products is unquestionable.

The most important effect of the spacecraft's electrostatic discharge effect is caused by electrostatic discharge (ESD). The main hazards of spacecraft surface charge and discharge effects include: the discharge current causes the power supply and distribution system to burn out and short circuit, destroying the satellite energy system; the electrostatic discharge will break through the components and destroy the satellite electronic system; the electromagnetic pulse generated by the discharge will cause the circuit The device flips; the electrostatic discharge will break the surface material and affect the material performance; the satellite charging will cause the structure potential to drift, affecting the measurement system; the surface of the satellite material will also enhance the surface pollution, affecting the performance of the sensor, window glass, lens and so on. The internal charging effect is different from the surface charging. Internal charging and discharging often occur inside the in-star electronic system, which is very close to sensitive devices, such as printed circuit boards. The surface charge occurs on the surface of the satellite, and the discharge pulse it causes must be coupled to the internal sensitive component, but will decay due to the presence of the coupling factor. Compared with the surface charging, the internal charge and discharge pulse is directly coupled into the electronic system with almost no attenuation, so the internal charging is more harmful to the in-star electronic system. The occurrence of internal charging is often accompanied by a large space radiating environmental disturbance event, and the probability of occurrence is relatively small, but once it occurs, the impact on the satellite will be fatal.

### *3.3. Antistatic Strengthening Technology*

In order to resist the impact of harsh environment on spacecraft materials, components, subsystems, etc. during on-orbit operation, antistatic reinforcement is required. The criterion for surface discharge effect reinforcement design is to control the grounding of the surface material and make the surface potential of the aerospace product within a safe range. Its guiding principle is to select materials with good electrical conductivity and ensure that the surface material is well grounded. The starting point of the foundation. Therefore, its design focus includes three points: surface material control, grounding requirements and capacitance requirements. The criterion for the internal charging effect reinforcement design is to reduce the discharge occurrence as much as possible by carrying out the deep charge and discharge protection design; however, when the discharge is unavoidable, the damage caused by the discharge should be reduced as much as possible. Therefore, the goal of the internal

IOP Conf. Series: Materials Science and Engineering **608** (2019) 012039 doi:10.1088/1757-899X/608/1/012039

charging effect reinforcement design is to reduce the deposition current by shielding measures, limit the electric field between the insulating material or the isolated conductor and the local ground, and select the material with high conductivity as much as possible.

## 4. Status of electrostatic discharge standards for spacecraft at home and abroad

With the development of spacecraft technology, long life and high reliability have become the basic capabilities that future spacecraft need. International standards organizations and aerospace powers have developed a series of international standards, national standards and industry standards, which provide a good description of the threat and protection design of electrostatic discharge to guide spacecraft design and ground test. These standards are mainly formulated around the antistatic reinforcement of spacecraft, which can solve the problems of failure or damage of materials, components, components and single machines caused by electrostatic discharge. Aerospace practices over the past half century have shown that spacecraft electrostatic discharge standards (or specifications) have played an important role in the design and operation of spacecraft.

### *4.1. Foreign spacecraft electrostatic discharge standard*

NASA has released the Spacecraft Charge Effect Evaluation and Control Design Guide (NASA TP-2361) since 1984. This document has been the authoritative document for GEO's spaceborne analysis and protection design. It is used in GEO spacecraft live analysis and protection design. Play an important role[3]. In 2007, NASA released the Design Standard for Low Earth Orbiting Spacecraft (NASA-STD-4005), which provides a design standard for high-voltage space energy systems (>55 volts) that must be used in low-Earth plasma environments, followed by The “Low Earth Orbiting Spacecraft Design Manual” (NASA-HDBK-4006) was released, which provides a detailed design for the charging of low-Earth orbit spacecraft and provides design guidance for NASA-STD-4005[4]. In 2011, NASA released the NASA-HDBK-4002A “Guide to the Protection of Spacecraft's Charge Effect”, which became an important basis for the current US spacecraft's electrification protection design[5]. The Space Engineering - Spacecraft Charged (ECSS-E-ST-20-06C) developed by European ESA is the European standard for the design, protection and testing of spacecraft charging effects. ESA has also released space electrostatic discharge sensitivity test. Standard methods such as the method and the environmental induction effect of electrostatic behavior of aerospace systems, how to carry out spatial electrostatic discharge sensitivity test and space electrostatic effects are described. Japan JAXA has developed the "Design Standard for Charge and Discharge" (JERG-2-211A), which is based on ECSS-E-ST-20-06C and specifies the surface charge simulation analysis and protection design[6]. In addition, the ISO standard "Space System - Solar Panel - Spacecraft Electrostatically Induced Electrostatic Discharge Test Method" (ISO-11221) drafted by Japan, the surface charge calculation is described in the appendix, and the surface is given. Software example of charging effect calculation. The relevant content of these standards represents the advanced level of current international charge and discharge tests, and can be better used in the design and development of spacecraft.

### *4.2. Domestic spacecraft electrostatic discharge standard*

Domestic and spacecraft static-related standards are mostly for the electronic component level and electronic products to develop electrostatic protection requirements and electrostatic discharge related factors test methods, these standards mainly focus on electrostatic protection of electronic components and stand-alone products, as well as some static electricity The protection standard is a universal electrostatic protection standard, which mainly includes: GB/T 1410-2006 "Test method for volume resistivity and surface resistivity of solid insulating materials", GB/T 17626.2-2006 "Electromagnetic compatibility test and measurement technology electrostatic discharge resistance Disturbance test", GB/T 32304-2015 "Aerospace electronic products electrostatic protection requirements", QJ 1693-89 "electronic components anti-static requirements" and so on.

The standards related to spacecraft charging and discharging are mainly GB/T 15463-2008 "Static safety terminology", GB/T 32452-2015 "Spacecraft space environment terminology", QJ 20409-2016 "Spacecraft surface material charging and discharging characteristics parameter test method" and QJ 20422.1-2016 "Spacecraft component environmental test method Part 1: Surface charge and discharge test". There are also some enterprise-level and below standards, such as the yard standard of the Spacecraft Five Institutes, "Satellite Surface Discharge Effect Test Method", "High-altitude Spacecraft Surface Charge and Discharge Protection Design Guide", "Spacecraft Component Environmental Test Method Part 8" : Magnetic layer substorm environment surface charge and discharge test" and The company standard of Beijing Institute of Spacecraft Environment Engineering, "Spacecraft solar cell array electrostatic discharge test method", "Spacecraft surface charging effect simulation analysis requirements", "Spacecraft surface charging effect simulation analysis method" and so on.

Since the surface charge and discharge problem of spacecraft is one of the main reasons for the anomaly of the spacecraft in orbit, it is not uncommon for the spacecraft to experience electrostatic discharge in the early stage. The research on the electrostatic charging and discharging effects of spacecraft was carried out abroad earlier, and relevant standards and norms were formed according to the research results to guide the antistatic design of spacecraft. China started relatively late. The domestic research on the electrostatic discharge effect of spacecraft began in the mid-1980s. The first unit to carry out satellite charging and discharging experiments was Beijing Institute of Spacecraft Environment Engineering, National Space Science Center, CAS, Lanzhou Institute of Physics.

They have also achieved many results in the mechanism of electrostatic discharge effects, simulation analysis and ground test [2]. In particular, Beijing Institute of Spacecraft Environment Engineering is the first to break through the three-dimensional simulation method of charged medium in arbitrary configuration under complex boundary conditions, and developed the internal simulation analysis software ATICS (Assessment Tool of Internal Charging for Satellite) in China. In the process of external cooperation of satellites such as NIGCOMSAT-1 and PAKSAT-1R, the supervisory party entrusted the simulation analysis and risk assessment of the charging effect in the spacecraft, which was well received by the supervisory party. Although China has made certain achievements in the study of electrostatic discharge effects, it is still lacking in standards and norms, and there are no relevant top-level standards and norms. In the research, China usually draws on foreign standards and norms, but the adaptability of key indicators needs to be studied.

## 5. Future prospects

The failure and damage caused by the spacecraft's on-orbit charging and discharging effect has become one of the main causes. At present, there is no unified standard specification for spacecraft electrostatic discharge protection design. In the spacecraft design, ground simulation test evaluation, on-orbit fault analysis and on-orbit prediction and early warning process, each unit carries out its work according to its own understanding, lacks a unified reference standard specification to constrain, and aerospace material space environment adaptability test method. The relevant standards are also relatively scarce.

In addition to the research and work on the mechanism, test and evaluation technology of the spacecraft's electrostatic discharge effect, the standards and specifications related to the electrostatic discharge of spacecraft cannot be delayed. First, improve the national standards and norms related to the electrostatic discharge source of spacecraft, and formulate relevant national standards for solar cosmic rays, galaxy cosmic rays, and space plasma on the basis of existing environmental standards and norms related to electrostatic discharge of spacecraft. Standards for solar electromagnetic radiation, earth radiation belts, etc. The second is to learn from NASA, ESA and some of Japan's protective design standards and impact assessment standards to improve the surface charge and discharge effects and internal charging effects standards and specifications. Establish relevant standard specifications and general standards and specifications for surface simulation tests of spacecraft surface charge and discharge effects and internal charging effects, and fully test and verify at the design stage to ensure that space materials can adapt to them without adequate application of antistatic design. Space environment.

Beijing Institute of Spacecraft Environment Engineering has a series of space charging and discharging environmental test equipments such as large-scale geomagnetic substorm charging and discharging test equipment, 1.2m surface charging and discharging equipment, plasma charging and discharging equipment, and participated in a number of satellite model discharge verification tests. It has a group of experts and technicians who have been engaged in the analysis and research of spacecraft charging effects for a long time, and has conducted many exchanges on electrostatic discharge tests with internationally advanced units with relevant test technology (such as Japan KIT, France ONERA, etc.). Since the "Eleventh Five-Year Plan", the research on electrostatic discharge of spacecraft in Beijing Institute of Spacecraft Environment Engineering has been supported by national defense basic scientific research projects, technical basic projects, key technical research projects, national defense 973, etc., and has one set of software copyrights and corresponding patents. The results have been analyzed and evaluated many times on the internal charging effects of key components such as Nixing and Baxing, and good results have been achieved. In recent years, Beijing Institute of Spacecraft Environment Engineering has led and participated in the preparation of several GB, GJB, and QJ standards, including Spacecraft Space Environmental Terminology, Spacecraft Thermal Control Coating Test Method Part 7: Vacuum-Electronic Irradiation Test, and Aerospace. Environmental testing methods for components (Part 1: Surface charge and discharge test), etc., with the ability to write the corresponding standard specifications, I believe that in the future, we can continue to deepen theoretical, methodological and application research, and establish environmental national standards related to electrostatic discharge. Contribute to the national military standards or industry standards related to the test.

**References**

[1] NASA-HDBK-4002. Avoiding Problems Caused by Spacecraft On-Orbit Internal Charging Effects[R].1999,1-45.
[2] NASA-HDBK-4002A.Mitigating In-Space Charging Effects-A Guideline[R]. 2011,1-181F.
[3] NASA TP-2361. Design Guidelines for Assessing and Controlling Spacecraft Charging Effects[R]. 1984,1-42.
[4] NASA STD-4005. Low earth orbit spacecraft charging design standard[R].2007,1-16
[5] NASA-HDBK-4006. Low Earth Orbit Spacecraft Charging Design Handbook[R]. 2007:1-63
[6] ECSS-E-20-06 Standard on spacecraft charging: environment-induced effects on the electrostatic behaviour of space systems[S].

# Adaptive Backstepping Sliding Mode Fault Tolerant Control for Satellite Attitude under Actuator Faults

**Junhai Huo[1], Tao Meng[1,*] and Zhonghe Jin[1]**

[1]School of Aeronautics and Astronautics, Zhejiang University, Hangzhou, 310027, China

[*]Corresponding author: mengtao@zju.edu.cn

**Abstract.** Modern spacecraft missions place stringent performance requirements on the reliability and stability of spacecraft attitude control system. However, during long-term in-orbit mission, actuators may suffer faults. In this paper, we proposed an adaptive fault tolerant control law for attitude control under actuator failures and external disturbances. Firstly, the attitude dynamic model under actuator faults and attitude kinematics are described. Then, in order to solve the attitude control problem under actuator faults and external disturbances, an adaptive backstepping sliding mode fault tolerant control scheme is designed. According to Lyapunov theory, the closed-loop system is proved to be globally asymptotically stable. Finally, the numerical simulation results demonstrate that the proposed adaptive controller can achieve anticipative attitude control performance, fault information estimation and well fault-tolerant capability.

## 1. Introduction

The growing development of different space exploration missions has motivated significant demands for the reliable and autonomous control of spacecraft. Attitude control system (ACS), as one of spacecraft's subsystems, plays a valuable role in guaranteeing the success of space missions. During the past decades, far-ranging efforts have been devoted to this issue, and a large amount of modern control techniques are documented in the literature [1-5]. However, most of these existing methods focus on spacecraft with healthy actuators. During the long-term in-orbit mission, the possibility of malfunction in actuator increases due to spacecraft's challenging operating conditions. Once the anomaly do occur, there would be huge economy loss. Hence, how to guarantee the reliability and safety of the controlled plants has become a serious problem to be solved.

Generally speaking, the existing fault tolerant control (FTC) methods can be divided into two parts: passive FTC and active FTC. The passive FTC [6, 7] is based on the robust technique and implemented as a fixed controller. However, the passive FTC only tolerates a limited number of faults, and it achieves robustness at the cost of decreased nominal performance. In contrast, active FTC solves the fault control problem by adjusting and modifying the controller based on real-time fault detection, identification and controller reconfiguration so that the stability of system and well control performance can be maintained. Taking advantage of greater performance and ability to handle a wider range of failure, the active FTC has motivated more significant research during the past years, such as multiple model [8], feedback linearization [9], sliding mode control [10], adaptive control [11, 12], control allocation [13, 14], eigenstructure assignment [15], model predictive control[16, 17], etc. However, most satellite attitude fault tolerant control methods can only deal with a single type of

IOP Conf. Series: Materials Science and Engineering **608** (2019) 012040 doi:10.1088/1757-899X/608/1/012040

actuator fault. For example, [10, 11] can only deal with partial failure of actuator. Therefore, the fault tolerance of these control methods is very limited. On the other hand, spacecraft in-orbit will inevitably be affected by external disturbances such as gravity gradient moment, aerodynamic moment, etc. This problem will limit the practical application of the existing attitude fault tolerant control methods. External disturbance and actuator fault are two key problems which affect the performance of fault tolerant controller. For this reason, how to design attitude fault tolerant controller to realize reliable and stable control under the condition of multi-type failure of actuator and external disturbance has become an urgent problem to be solved.

In this paper, we concentrate on the design of an active fault tolerant control algorithm for spacecraft attitude under multiple faults and external disturbances. Based on the backstepping control algorithm, the sliding mode was introduced because of its strong robustness. The system change caused by fault can be regarded as an uncertainty, and this uncertainty can be estimated by adaptive algorithm and used for controller design. Simulation results illustrate the well performance of the adaptive backstepping sliding mode fault tolerant controller.

The rest of this paper is organized as follows. Section 2 is devoted to describe the attitude kinematics and attitude dynamics with unknown actuator faults. In Section 3, an adaptive fault tolerant controller is proposed and closed-loop system stability analysis is derived. The effectiveness and superiority of the control scheme are verified by numerical simulation in Section 4. Finally, the conclusion of this paper is shown in Section 5.

## 2 Dynamics model

### *2.1 Attitude dynamics*

The satellite attitude dynamics is

$$\boldsymbol{J}\dot{\boldsymbol{\omega}} + \boldsymbol{\omega} \times \boldsymbol{J}\boldsymbol{\omega} = \boldsymbol{u}_c + \boldsymbol{d} \tag{1}$$

where $\boldsymbol{J}$ denotes the inertia matrix of the satellite; $\boldsymbol{\omega}$ is the satellite attitude velocity; $\boldsymbol{u}_c$ is the control torque; $\boldsymbol{d}$ is the external disturbance torque.

For a healthy-actuator system, the actual output can accurately respond to the command input, $\boldsymbol{u} = \boldsymbol{u}_c$. However, actuator faults are usually inevitable. These faults include: 1) Partial loss of energy: actuator can only respond partially to the expected control command; 2) Continuous float: actuator cannot output the expected actuating power accurately, and there exists a small bias; 3) Complete loss of energy: actuator completely fails to work; 4) Locking: actuator has a fixed and uncontrolled output. In this study, loss of effectiveness fault and bias fault are considered. The faults can be modeled as

$$\boldsymbol{u} = \boldsymbol{e}\boldsymbol{u}_c + \boldsymbol{f} \tag{2}$$

where $\boldsymbol{u}_c = [u_{c1}, u_{c2}, u_{c3}]$ is the command torque input of actuators; $\boldsymbol{e} = \mathrm{diag}([e_1, e_2, e_3])$ represents the effectiveness matrix of actuator; $\boldsymbol{f} = [f_1, f_2, f_3]$ denotes the bias fault. Ultimately, the fault satellite attitude dynamics model can be written as

$$\boldsymbol{J}\dot{\boldsymbol{\omega}} + \boldsymbol{\omega} \times \boldsymbol{J}\boldsymbol{\omega} = (\boldsymbol{e}\boldsymbol{u}_c + \boldsymbol{f}) + \boldsymbol{d} \tag{3}$$

### *2.2 Attitude kinematics*

The satellite attitude kinematics employing quaternion can be written as

$$\dot{\boldsymbol{q}}_v = \frac{1}{2}(q_0\boldsymbol{I}_3 + \boldsymbol{S}(\boldsymbol{q}_v))\boldsymbol{\omega} \tag{4}$$

$$q_0 = -\frac{1}{2}\boldsymbol{q}_v^T\boldsymbol{\omega} \tag{5}$$

where $\boldsymbol{q} = [\boldsymbol{q_v}, q_0]$ is the unit quaternion; $\boldsymbol{I_3}$ is an identity matrix; $\boldsymbol{S}(\cdot)$ means the skew-symmetric matrix as follows:

$$\boldsymbol{S}(\boldsymbol{x}) = \begin{bmatrix} \boldsymbol{0} & -\boldsymbol{x_3} & \boldsymbol{x_2} \\ \boldsymbol{x_3} & \boldsymbol{0} & -\boldsymbol{x_1} \\ -\boldsymbol{x_2} & \boldsymbol{x_1} & \boldsymbol{0} \end{bmatrix}$$

Define the error quaternion $\boldsymbol{q_e} = [\boldsymbol{q_{ev}}, q_{e0}]$ as the maneuvering quaternion from the initial quaternion $\boldsymbol{q}$ to the desired quaternion $\boldsymbol{q_d} = [\boldsymbol{q_{dv}}, q_{d0}]$, then we have

$$\boldsymbol{q_e} = \boldsymbol{q_d} \otimes \boldsymbol{q} \tag{6}$$

where $\otimes$ is quaternion product. The error angular velocity $\boldsymbol{\omega_e}$ is written as

$$\boldsymbol{\omega_e} = \boldsymbol{\omega} - \boldsymbol{\omega_d} \tag{7}$$

where $\boldsymbol{\omega}$ is the initial angular velocity; $\boldsymbol{\omega_d}$ is the desired angular velocity. Substituting Eq.6 and Eq.7 into Eq.4 and Eq.5, and the error quaternion kinematics can be derived as

$$\dot{\boldsymbol{q}}_{\boldsymbol{ev}} = \frac{1}{2}(\boldsymbol{q_{e0}}\boldsymbol{I_3} + \boldsymbol{S}(\boldsymbol{q_{ev}}))\boldsymbol{\omega_e} \tag{8}$$

$$\boldsymbol{q_{e0}} = -\frac{1}{2}\boldsymbol{q_{ev}^T}\boldsymbol{\omega_e} \tag{9}$$

**3 Fault tolerant controller design**

In this section, the adaptive backstepping sliding mode fault tolerant controller is proposed to deal with the attitude control problem under the actuator faults and external disturbances.

The first-order backstepping variable is

$$\boldsymbol{z_1} = [\boldsymbol{q_{ev}}; \boldsymbol{1} - \boldsymbol{q_{e0}}] \tag{10}$$

The second-order virtual backstepping variable is

$$\boldsymbol{z_2} = \boldsymbol{\omega_e} - \boldsymbol{\sigma} \tag{11}$$

Then the time derivative of $\boldsymbol{z_1}$ is

$$\begin{aligned} \dot{\boldsymbol{z}}_1 &= \frac{1}{2}\begin{bmatrix} \boldsymbol{q_{e0}}\boldsymbol{I_3} + \boldsymbol{S}(\boldsymbol{q_{ev}}) \\ \boldsymbol{sgn}(\boldsymbol{q_{e0}})\boldsymbol{q_{ev}^T} \end{bmatrix}\boldsymbol{\omega_e} \\ &= \frac{1}{2}\boldsymbol{Q^T}(\boldsymbol{z_2} + \boldsymbol{\sigma}) \end{aligned} \tag{12}$$

where $\mathbf{Q} = \begin{bmatrix} q_{e0}\boldsymbol{I_3} + \boldsymbol{S}(\boldsymbol{q_{ev}}) \\ \mathrm{sgn}(q_{e0})\boldsymbol{q_{ev}^T} \end{bmatrix}^T$, $\boldsymbol{\sigma}$ is a virtual variable that need to be chosen.

Firstly, we define a Lyapunov function as

$$\boldsymbol{V_1} = \boldsymbol{z_1^T}\boldsymbol{z_1} \tag{13}$$

$$\dot{\boldsymbol{V}}_1 = \boldsymbol{2}\boldsymbol{z_1^T}\dot{\boldsymbol{z}}_1 = \boldsymbol{z_1^T}\boldsymbol{Q^T}(\boldsymbol{z_2} + \boldsymbol{\sigma}) \tag{14}$$

Thus, the virtual variable is chosen as $\boldsymbol{\sigma} = -\boldsymbol{k_1}\boldsymbol{Q}\boldsymbol{z_1}$, where $\boldsymbol{k_1}$ is a diagonal gain matrix. Then we have

$$\dot{\boldsymbol{V}}_1 = -\boldsymbol{z_1^T}\boldsymbol{Q^T}\boldsymbol{k_1}\boldsymbol{Q}\boldsymbol{z_1} + \boldsymbol{z_1^T}\boldsymbol{Q^T}\boldsymbol{z_2} \tag{15}$$

Secondly, differentiating Eq.11, we have

$$\dot{z}_2 = \dot{\omega}_e - \dot{\sigma} \tag{16}$$

Combining Eq.3, Eq.7 and Eq.16, we have

$$J\dot{z}_2 = J\dot{\omega}_e - J\dot{\sigma} = -\omega \times J\omega + Eu_c + F + d - J\dot{\omega}_d - J\dot{\sigma} \tag{17}$$

Then the second Lyapunov function can be defined as

$$V_2 = V_1 + \frac{1}{2} z_2^T J z_2 \tag{18}$$

Differentiating Eq.18, we get

$$\dot{V}_2 = \dot{V}_1 + z_2^T J \dot{z}_2 \tag{19}$$

Define the sliding mode surface as

$$s = z_2 \tag{20}$$

Substituting Eq.17 and Eq.20, we get

$$\dot{V}_2 = \dot{V}_1 + s^T(-\omega \times J\omega + eu_c + F + d - J\dot{\omega}_d - J\dot{\sigma}) \tag{21}$$

There exists shaking in the sliding mode control, and it can be weakened by introducing the approach law. In order to guarantee system robustness and reduce the shaking caused by external disturbances, the exponential approach law $s = -\varepsilon \mathrm{sgn}(s) - k_2 s$ is adopted in this paper.
In order to make $\dot{V}_2 \leq 0$, the controller can be designed as follows

$$u_c = (-\varepsilon sgn(s) - k_2 s + \omega \times J\omega - \hat{F} + J\dot{\omega}_d + J\dot{\sigma} - Qz_1)/\hat{E} \tag{22}$$

where $\hat{F}$ and $\hat{E}$ are the estimation of $E$ and $F$, respectively. We define the estimation error as $e_E$ and $e_F$, and thus we have

$$e_E = E - \hat{E} \tag{23}$$

$$e_F = F - \hat{F} \tag{24}$$

Substituting the proposed controller Eq.22 into Eq.21, and combining Eq.23 and Eq.24, after some simplification, we get

$$\dot{V}_2 = -z_1^T Q^T k_1 Q z_1 + s^T(-\varepsilon sgn(s) - k_2 s + e_E u_c - \hat{F} + F + d) \tag{25}$$

Define the following Lyapunov function

$$V_3 = V_2 + \frac{e_F^2}{2\gamma} + \frac{e_E^2}{2\lambda} \tag{26}$$

Differentiating Eq.26, we get

$$\dot{V}_3 = \dot{V}_2 - \frac{e_F \dot{\hat{F}}}{\gamma} - \frac{e_E \dot{\hat{E}}}{\lambda}$$
$$= -z_1^T Q^T k_1 Q z_1 + s^T(-\varepsilon sgn(s) - k_2 s + e_E u_c - \hat{F} + F + d) - \frac{e_F \dot{\hat{F}}}{\gamma} - \frac{e_E \dot{\hat{E}}}{\lambda}$$

$$= -z_1^T Q^T k_1 Q z_1 + s^T(-\varepsilon sgn(s) - k_2 s + d) + e_E\left(s^T u_c - \frac{\dot{\hat{E}}}{\lambda}\right) + e_F\left(s^T - \frac{\dot{\hat{F}}}{\gamma}\right)$$

The adaptive laws of $\hat{E}$ and $\hat{F}$ are designed as

$$\dot{\hat{E}} = \lambda s^T u_c \quad (27)$$

$$\dot{\hat{F}} = \gamma s^T \quad (28)$$

If $\varepsilon$ satisfies

$$\|d\| < \varepsilon$$

Ultimately, the derivative of Lyapunov function $V_3$ is

$$\dot{V}_3 = -z_1^T Q^T k_1 Q z_1 + s^T(-\varepsilon sgn(s) - k_2 s + d)$$
$$\leq -z_1^T Q^T k_1 Q z_1 - s^T k_2 s \leq 0$$

## 4 Simulation

In this section, numerical simulations are conducted to evaluate the performance of the proposed fault tolerant attitude controller. The objective of the simulation is to maintain the stable attitude under the condition of actuator faults and external disturbances. The moment of inertia of satellite is given as $J = \text{diag}[1.08, 1.08, 1.08]\text{kgm}^2$; the initial satellite attitude angle is [3; 3; 3]°; the initial angular velocity is [0.1; 0.1; 0.1]°/s; the desired attitude angle and angular velocity are [0; 0; 0]° and [0; 0; 0]°/s, respectively. Because of the periodicity of orbital disturbances, we randomly suppose the external disturbance torque to be

$$d = \begin{bmatrix} d_x \\ d_y \\ d_z \end{bmatrix} = \begin{bmatrix} 2e^{-5} * cos(pi * t) + 1e^{-5} \\ 1e^{-5} * cos(pi * t) + 2e^{-5} \\ 3e^{-5} * sin(pi * t) - 1e^{-5} \end{bmatrix}$$

The controller parameters are chosen as $k_1 = \text{diag}[0.005, 0.005, 0.005]$ ; $k_2 = \text{diag}[0.04, 0.04, 0.04]$ ; $\lambda = 2000$ ; $\gamma = 0.001$ ; $\varepsilon = 5 \times 10^{-5}$ ; the actuators partial loss of effectiveness matrix is chosen as $e = \text{diag}([1, 0.5, 1])$ ; the bias fault matrix is $f = [0, 0, 1 \times 10^{-5}]\text{Nm}$.

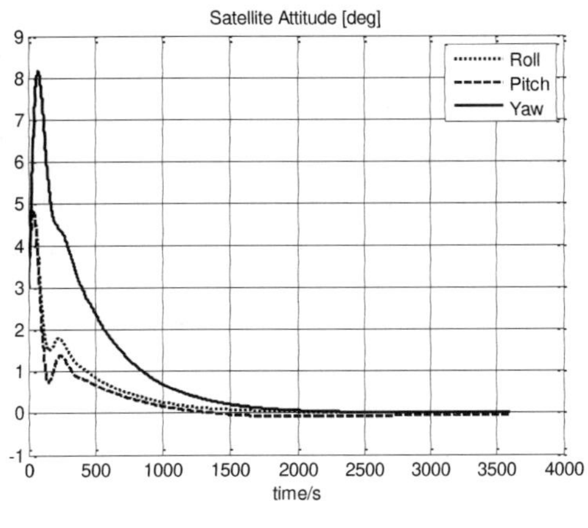


**Figure 1.** Satellite attitude

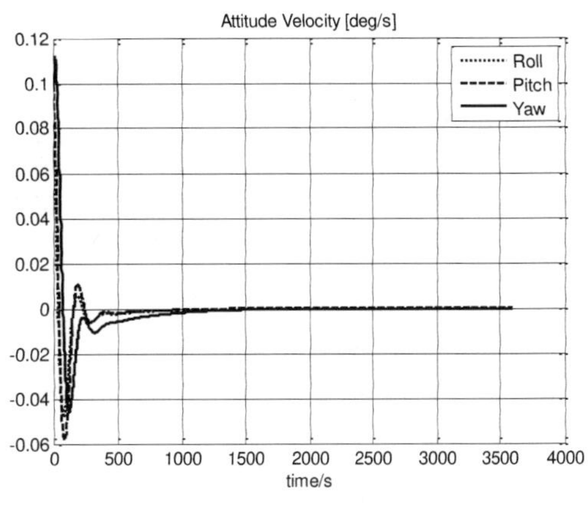


**Figure 2.** Satellite attitude velocity

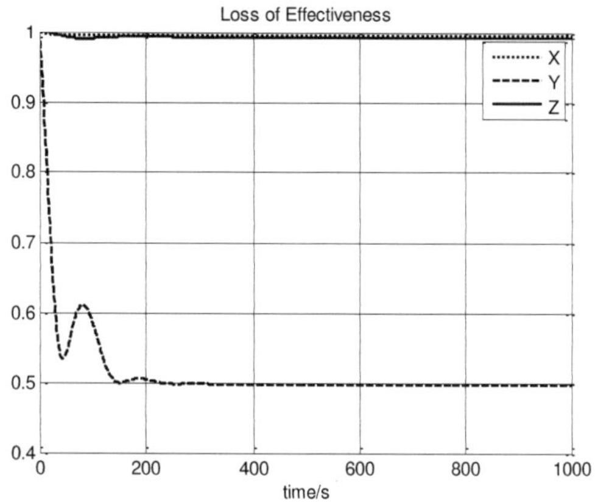


**Figure 3.** Estimation of loss of effectiveness

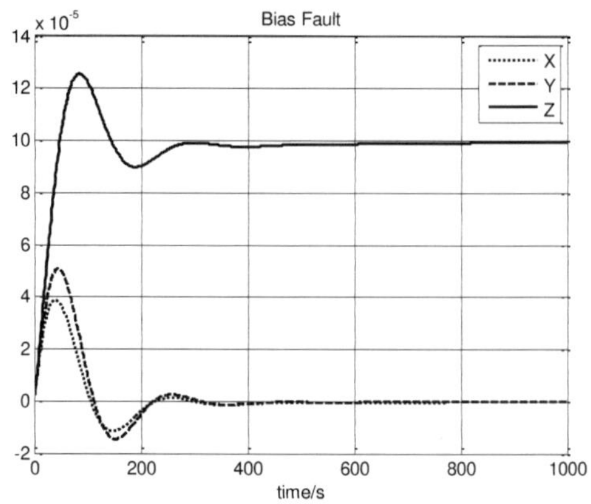


**Figure 4.** Estimation of bias fault

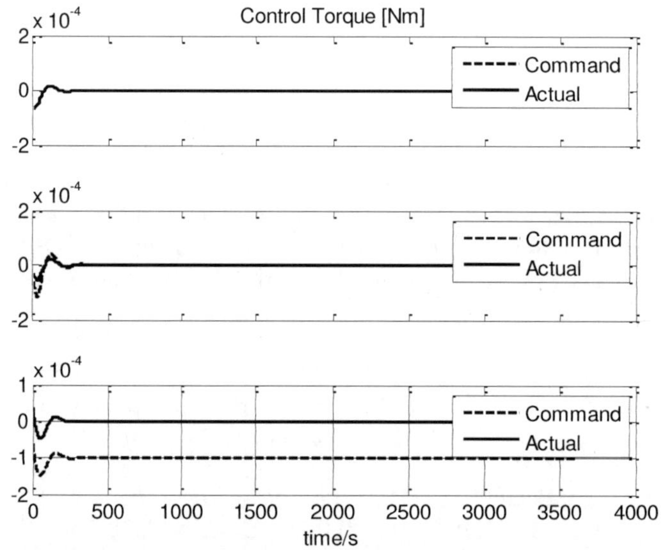


**Figure 5.** Control torque (command and actual)

Applying the controller Eq.22 and adaptive update laws Eq.27 and Eq.28, the simulation results are illustrated as Fig.1-5. The time responses of the attitude and attitude velocity are presented in Fig.1-2. The satellite attitude and attitude velocity converge to zero asymptotically under the adaptive fault tolerant controller. Evolution of the estimations of actuator faults are displayed by Fig.3-4. It can be seen that the estimated loss of effectiveness and bias fault are both bounded and converge to the true value in finite time. Fig.5 displays the comparison between the command control torque and actual control torque. The adaptive controller can adjust the actual control output according to the estimated fault to achieve better control performance.

## 5 Conclusion

In this paper, an adaptive fault tolerant controller for satellite attitude under actuator faults and external disturbances is presented. The loss of effectiveness and bias fault are considered in the actuators. The adaptive controller is designed by employing backstepping and sliding mode technique. According to the Lyapunov theory, the proposed controller ensures that the closed-loop system is globally ultimately uniformly bounded. Numerical simulation results demonstrate that the proposed controller can guarantee the attitude control system to converge with fine performance and the faults are effectively estimated by the adaptive law.

**References**

[1] Tayebi A, Unit Quaternion-Based Output Feedback for the Attitude Tracking Problem, IEEE Transactions on Automatic Control, **53**, 6, 1516-1520 (2008)
[2] Safa A, Baradarannia M, Kharrati H, et al., Robust attitude tracking control for a rigid spacecraft under input delays and actuator errors, International Journal of Control, 1-29 (2017)
[3] Jian Z, Hu Q, Xie W, Integral sliding mode-based attitude coordinated tracking for spacecraft formation with communication delays, International Journal of Systems Science, **48**, 15, 1-13 (2017)
[4] Luo W, Chu Y, Ling K, Inverse optimal adaptive control for attitude tracking of spacecraft, IEEE Transactions on Automatic Control, **50**, 11, 1639-1654 (2005)
[5] Xia Y, Zhang J, Lu K, et al., Adaptive Attitude Tracking Control for Rigid Spacecraft with Finite-Time Convergence, Automatica, **49**, 12, 3591-3599 (2013)
[6] Liang J, Dong C, Wang Q, A fault-tolerant attitude control system for a satellite based on fuzzy global sliding mode control algorithm, Proceedings of SPIE - The International Society for Optical Engineering, **7128**, 3, 35-39 (2008)
[7] Godard G, Kumar K, Fault Tolerant Reconfigurable Satellite Formations Using Adaptive Variable Structure Techniques, Journal of Guidance, Control, and Dynamics, **33**, 3, 969-984 (2010)
[8] Tudoroiu N, Khorasani K, Fault detection and diagnosis for satellite's attitude control system (ACS) using an interactive multiple model (IMM) approach, IEEE Conference on Control Applications, (2005)
[9] Calise A, Hovakimyan N, Idan M, Adaptive output feedback control of nonlinear systems using neural networks, Pergamon Press, (2001)
[10] Ran D, Sheng T, Wang Y, et al., Adaptive Non-singular Terminal Fault-tolerant Control for Rigid Spacecraft Attitude Maneuver, (2017)
[11] Hu Q, Xiao B, Zhang Y, Fault-Tolerant Attitude Control for Spacecraft Under Loss of Actuator Effectiveness, Control & Decision, **34**, 3, 263-268, (2015)
[12] Li Y, Ma Z, Tong S, Adaptive Fuzzy Fault-Tolerant Control of Non-Triangular Structure Nonlinear Systems with Error-Constraint, IEEE Transactions on Fuzzy Systems, **99** 1-1 (2017)
[13] Alwi H, Edwards C, Fault tolerant control using sliding modes with on-line control allocation, Automatica, **44**, 7, 1859-1866, (2008)
[14] Liao F, Lum K, Wang J, et al., Adaptive control allocation for non-linear systems with internal dynamics, IET CONTROL THEORY AND APPLICATIONS, **4**, 6, 909-922, (2010)
[15] Cai G, Hu C, Duan G, Eigenstructure assignment for linear parameter-varying systems with applications, Mathematical & Computer Modelling, **53**, 5-6, 861-870, (2011)
[16] Rodil S, Fuente M, Fault tolerance in the framework of support vector machines based model predictive control, Engineering Applications of Artificial Intelligence, **23**, 7, 1127-1139, (2010)
[17] Deshpande A, Patwardhan S, Narasimhan S, Intelligent state estimation for fault tolerant nonlinear predictive control, Journal of Process Control, **19**, 2, 187-204, (2009)

IOP Conf. Series: Materials Science and Engineering **608** (2019) 012041 doi:10.1088/1757-899X/608/1/012041

# Nano-Pico Satellite Actuator Fault Diagnosis Based on Adaptive Observer

**Y Fei[1], T Meng[1] and Z H Jin[1]**

[1]School of Aeronautics and Astronautics, Zhejiang University, Hangzhou, 310027, China

mengtao@zju.edu.cn

**Abstract.** This paper concerns with actuators fault diagnosis of nano-pico satellite, and a double adaptive estimation algorithm is proposed. Based on the non-linear mathematical model of reaction flywheel, a parameterized description method of flywheel fault type is established. Using the input and output information of the flywheel, a local adaptive observer is constructed and can identify the fault type of the flywheel. A global adaptive observer based on the satellite dynamics model uses the gyroscope sampling signal and the estimation of local observer as inputs, so failure moment of the other actuator can be identified. Combining two observers' identification results, the fault diagnosis of flywheel and the other actuator can be completed. In this paper, a control system composed of reaction flywheel and magnetorquer is simulated. Results show the effectiveness of proposed method for nano-pico satellite actuator fault diagnosis.

## 1. Introduction

Nano-Pico satellite denotes the small satellite with mass is ranging from a few kilograms to tens of kilograms and based on microelectronics and micromechanical technology. Due to the constraints of size, weight and power consumption, nano-pico satellite cannot improve its reliability by using the redundant backup method of traditional satellite. Once the failure occurs, the entire space mission will be directly affected.

In the on-orbit failures of spacecraft that have been disclosed abroad, the attitude control subsystem and the power subsystem have the highest failure ratio [1-2]. Attitude control subsystem contains many sensors and actuators, and integrates a variety of algorithms. Focusing on the actuator faults, fault diagnosis methods for spacecraft actuators can be divided into three categories [3-5].

The first is based on the analytical model method, using software redundancy instead of hardware redundancy to obtain residuals, and processing the residuals to achieve detection. In [6], an adaptive observer was proposed for reaction flywheels, which realized the identification of flywheel gain and constant friction moment.

Second, detection methods based on signal processing analyze the measurement signal directly, and extract the characteristics of spectrum, amplitude and variance to achieve fault detection. These methods do not need mathematical model and have strong adaptability. In [7], the wavelet transform is used to analyze the flywheel speed, and the detection of the stalling fault is realized. Vertex principal component analysis (VPCA) was applied to realize fault detection and isolation of actuators in [8-9].

The third methods are based on artificial intelligence, which do not need precise mathematical model. At present, these researches are the most popular, and indicate the direction of the development of autonomous detection in the future. Fault detection of flywheel was realized by using neural network in [10-11].

In summary, the existing research methods are mostly directed to the fault detection of single actuator, and some algorithms have heavy computing burden. In this paper, an adaptive identification algorithm is proposed. The local adaptive observer is designed by using the input and output information of the flywheel. Combined with the dynamic equation, the global observer is designed by using the output information of the gyroscope, and the local observer identification result is taken as input. This can identify the fault torque that does not contain the flywheel fault, thus achieving fault isolation and identification of the actuator. This scheme adapts to the control combination of the flywheel with any other actuator and does not require the establishment of a mathematical model of another actuator.

## 2. Attitude control system fault analysis

The Attitude Control System (ACS) is the key to ensure the stability of the satellite attitude and the realization of the satellite function. This paper takes the control system composed of the magnetorquer and the reaction flywheel group.

### 2.1. Flywheel failures

The common faults of the flywheel can be divided into three categories: abnormal torque output, which is caused by increased bearing temperature and poor lubrication, which lead to the increase of friction torque and the decrease of reaction wheel gain efficiency. The idling fault refers to that reaction wheel cannot respond to the normal control torque command due to certain factors, the rotation speed remains almost constant (slowly decreases under the action of friction), the output torque is almost zero; the stalling fault refers to the situation that the output torque of reaction wheel rapidly turns to zero after a huge reverse disturbance is generated.

Consider single reaction flywheel mathematical model:

$$\dot{w}_{\mathrm{w}} = J_{\mathrm{w}}^{-1}\left(kT - vw_{\mathrm{w}} + \operatorname{sgn}\left(w_{\mathrm{w}}\right)T_{\mathrm{fc}}\right) \tag{1}$$

Where $w_w$ denotes the angular velocity of the flywheel; $J_{\mathrm{w}}$ is the rotational inertia of each flywheel; $k$ denotes output gain; $v$ is viscous friction coefficient; $T$ denotes command input; $T_{\mathrm{fc}}$ is friction moment.

In combination with the flywheel model, its fault can be described by the change of the gain $k$ and the viscous friction coefficient $v$ . $v_{\mathrm{f}}$ is the normal viscous friction coefficient.

**Table 1.** Flywheel fault mode and its parameterized description.

| fault mode | $k$ | $v$ |
|---|---|---|
| idling | $k$=0 | $v<v_{\mathrm{f}}$ |
| stalling | $k$=0 | $v>v_{\mathrm{f}}$ |
| gain decrease | $k$<1 | - |

### 2.2. Magnetorquer failures

The magnetorquer is mainly composed of a magnetic bar and a drive control circuit. In the time domain, its faults can be divided into: sudden fault, gradual fault and intermittent fault. A sudden fault refers to sudden damage to the hardware, resulting in a large continuous deviation between the output torque and the target value; the gradual failure is due to component wear aging, and the performance slowly decreases with time; the intermittent fault is caused by component aging and poor contact, it is instantaneous sudden change[12].

## 3. FDI design based on parameter adaptation

Using the output information of the flywheel to design an adaptive observer, the parameters of the flywheel can be identified. The flywheel fault mode can be determined according to Table 1. The global adaptive observer can be designed by using the output information of the gyroscope, and the actual output torque of the flywheel is calculated by using the parameter identification result, use it as the input information of the global sensor, the global identification result only contains the fault torque of the magnetic torque device, and can estimate the fault condition of the magnetic torque device. The algorithm flowchart is shown in Fig.1:

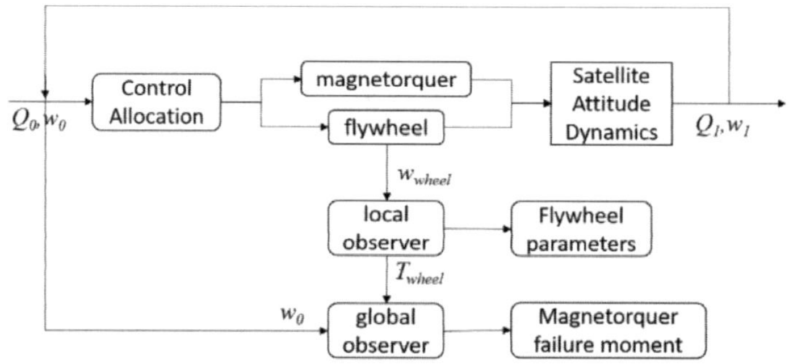


**Figure 1.** Algorithm flowchart.

### *3.1. Local parameter adaptive algorithm*

The following adaptive observer is designed according to the mathematical model of the flywheel:

$$\dot{\hat{w}}_w = J_w^{-1}\left(\hat{k}T - \hat{v}\hat{w}_w + \text{sgn}(\hat{w}_w)T_{fc}\right) + L(w_w - \hat{w}_w) \quad (2)$$

Where $\hat{w}_w$ denotes the observer state variable, and observer gain L>0.

Define the following variables:

$$\begin{aligned} &\hat{T}^i_{wheel} = J_w\left(\hat{w}^i_{w1} - \hat{w}^i_{w0}\right) \\ &\tilde{w}_w = w_w - \hat{w}_w, \tilde{k} = k - \hat{k}, \tilde{v} = v - \hat{v} \end{aligned} \quad (3)$$

Given the output gain $k$ and viscous friction coefficient $v$ are positive constant under normal circumstances, we can obtain that:

$$\dot{\tilde{k}} = -\dot{\hat{k}}, \dot{\tilde{v}} = -\dot{\hat{v}} \quad (4)$$

From (1) and (2), the error function describes as:

$$\dot{\tilde{w}}_w = J_w^{-1}\left(\tilde{k}T - v w_w + \hat{v}\hat{w}_w - L\tilde{w}_w\right) \quad (5)$$

The adaptive parameter estimation algorithms are designed as follows:

$$\dot{\hat{k}} = \tau_1 J_w^{-1} T\tilde{w}_w, \dot{\hat{v}} = -\tau_2 J_w^{-1}\hat{w}_w\tilde{w}_w \quad (6)$$

Proof: Consider the following Lyapunov function

$$V = \frac{1}{2}\tilde{w}_w^2 + \frac{1}{2\tau_1}\tilde{k}^2 + \frac{1}{2\tau_2}\tilde{v}^2 \quad (7)$$

From (5) and (6), differentiating function (7) with respect to time yields:

$$\begin{aligned} \dot{V} &= \tilde{w}_w\dot{\tilde{w}}_w + \frac{1}{\tau_1}\tilde{k}\dot{\tilde{k}} + \frac{1}{\tau_2}\tilde{v}\dot{\tilde{v}} \\ &= J_w^{-1}\left(\tilde{k}T - v w_w + \hat{v}\hat{w}_w - L\tilde{w}_w\right)\tilde{w}_w - \frac{1}{\tau_1}\tilde{k}\dot{\hat{k}} - \frac{1}{\tau_2}\tilde{v}\dot{\hat{v}} \\ &= J_w^{-1}(-v - L)\tilde{w}_w^2 - J_w^{-1}\tilde{v}\hat{w}_w\tilde{w}_w + J_w^{-1}\tilde{k}T\tilde{w}_w - \frac{1}{\tau_1}\tilde{k}\dot{\hat{k}} - \frac{1}{\tau_2}\tilde{v}\dot{\hat{v}} \\ &= -J_w^{-1}(v + L)\tilde{w}_w^2 \end{aligned} \quad (8)$$

IOP Conf. Series: Materials Science and Engineering **608** (2019) 012041 doi:10.1088/1757-899X/608/1/012041

Because $v$ and $L$ are positive constant, one can obtain that $\dot{V} \le 0$, which means that ( $\hat{w}_w, \hat{k}, \hat{v}$ ) asymptotically converges to the true values according to Lyapunov stability theory.

### *3.2. Global fault adaptive algorithm*

Consider the following satellite dynamic function:

$$\dot{\mathbf{w}} = \mathbf{J}^{-1}\left\{\mathbf{Iw}\times\mathbf{w} + \mathbf{C}_w\mathbf{h}\times\mathbf{w} + \mathbf{T}_m + \hat{\mathbf{T}}_{wheel} + \mathbf{T}_{d0} + \mathbf{f}_a\right\}. \tag{9}$$

Where the state $\mathbf{w} \in R^3$ is the satellite angular velocity; $\mathbf{J} \in R^{3\times3}$ is the inertia matrix of the satellite; $\mathbf{C}_w \in R^{3\times4}$ denotes the transformation matrix of the reaction flywheels; $\mathbf{h} \in R^4$ is the reaction flywheel momentum; $\mathbf{T}_m \in R^3$ is the control torque of magnetorquer; $\mathbf{T}_{d0} \in R^3$ is external disturbances; $\mathbf{f}_a \in R^3$ denotes fault moment; $\mathbf{T}_{wheel} \in R^3$ is the estimated output torque of the flywheels.

Define:

$$\begin{aligned} g(\mathbf{w}) &= \mathbf{J}^{-1}(\mathbf{Iw}\times\mathbf{w} + \mathbf{C}_w\mathbf{h}\times\mathbf{w}) \\ \mathbf{u} &= \mathbf{T}_m + \hat{\mathbf{T}}_{wheel} \\ \mathbf{B} &= \mathbf{J}^{-1} \\ \mathbf{T}_d &= \mathbf{J}^{-1}\mathbf{T}_{d0} \end{aligned} \tag{10}$$

Substituting (10) into (9), it can be derived that:

$$\dot{\mathbf{w}} = g(\mathbf{w}) + \mathbf{Bu} + \mathbf{Bf}_a + \mathbf{T}_d \tag{11}$$

Assumption 1: $g(\mathbf{w})$ satisfies the criterion of Lipschitz, and given a scalar $r > 0$ ,one can obtain that

$$\left\|g(\mathbf{w}_1) - g(\mathbf{w}_2)\right\| \le r\left\|\mathbf{w}_1 - \mathbf{w}_2\right\| \tag{12}$$

Design the following adaptive observer:

$$\begin{aligned} \dot{\hat{\mathbf{w}}} &= g(\hat{\mathbf{w}}) + \mathbf{Bu} + \mathbf{B}\hat{\mathbf{f}}_a + \mathbf{L}(\mathbf{y} - \hat{\mathbf{y}}) \\ \hat{\mathbf{y}} &= \mathbf{C}\hat{\mathbf{w}} \end{aligned} \tag{13}$$

Where $\hat{\mathbf{w}}$ denotes the observer state vector, and observer gain matrix $\mathbf{L}$ is positive definite.

Denote

$$\mathbf{e} = \mathbf{w} - \hat{\mathbf{w}}, \tilde{\mathbf{f}}_a = \mathbf{f}_a - \hat{\mathbf{f}}_a \tag{14}$$

Then the error differentiating function is described by

$$\dot{\mathbf{e}} = g(\mathbf{w}) - g(\hat{\mathbf{w}}) + \mathbf{B}\left(\mathbf{f}_a - \hat{\mathbf{f}}_a\right) + \mathbf{T}_d - \mathbf{LCe} \tag{15}$$

**Lemma 1** [13]**:** Given a scalar $\mu > 0$ and a symmetric positive definite matrix **Q**, the following inequality holds:

$$2\mathbf{x}^T\mathbf{y} \le \frac{1}{\mu}\mathbf{x}^T\mathbf{Qx} + \mu\mathbf{y}^T\mathbf{Q}^{-1}\mathbf{y} \qquad \mathbf{x}, \mathbf{y} \in R^n \tag{16}$$

**Lemma 2**[14]: Given a symmetric positive definite matrix **P**, the following inequality holds:

$$2\mathbf{e}^T\mathbf{P}\left[g(\mathbf{w}) - g(\hat{\mathbf{w}})\right] \le r^2\mathbf{e}^T\mathbf{PPe} + \mathbf{e}^T\mathbf{e} \tag{17}$$

The adaptive fault estimation algorithms are designed as follows:

$$\begin{aligned} \dot{\hat{\mathbf{f}}}_a &= \mathbf{GF}(\mathbf{y} - \hat{\mathbf{y}}) - \mathbf{TG}\hat{\mathbf{f}}_a \\ \mathbf{FC} &= \mathbf{B}^T\mathbf{P} \end{aligned} \tag{18}$$

Where **G** and **T** are rate matrixes.

Proof: Choose the following Lyapunov function:

$$V = \mathbf{e}^T\mathbf{Pe} + \tilde{\mathbf{f}}_a^T\mathbf{G}^{-1}\tilde{\mathbf{f}}_a^T > 0 \tag{19}$$

Differentiating function (19) with respect to time, one can obtain that:

$$\begin{aligned}\dot{V} &= 2\dot{\mathbf{e}}^{\mathrm{T}}\mathbf{Pe} + 2\dot{\tilde{\mathbf{f}}}_{\mathrm{a}}^{\mathrm{T}}\mathbf{G}^{-1}\tilde{\mathbf{f}}_{\mathrm{a}} \\ &= \mathbf{e}^{\mathrm{T}}\left[(-\mathbf{LC})^{\mathrm{T}}\mathbf{P} + \mathbf{P}(-\mathbf{LC})\right]\mathbf{e} + 2\left[g(\mathbf{w}) - g(\hat{\mathbf{w}})\right]^{\mathrm{T}}\mathbf{Pe} \\ &+ 2\mathbf{T}_{\mathrm{d}}^{\mathrm{T}}\mathbf{Pe} + 2\left[\mathbf{B}\left(\mathbf{f}_{\mathrm{a}} - \hat{\mathbf{f}}_{\mathrm{a}}\right)\right]^{T}\mathbf{Pe} + 2\dot{\tilde{\mathbf{f}}}_{\mathrm{a}}^{\mathrm{T}}\mathbf{G}^{-1}\tilde{\mathbf{f}}_{\mathrm{a}} \\ &\le \mathbf{e}^{\mathrm{T}}\left[(-\mathbf{LC})^{\mathrm{T}}\mathbf{P} + \mathbf{P}(-\mathbf{LC})\right]\mathbf{e} + r^2\mathbf{e}^{\mathrm{T}}\mathbf{PPe} + \mathbf{e}^{\mathrm{T}}\mathbf{e} + 2\mathbf{T}_{\mathrm{d}}^{\mathrm{T}}\mathbf{Pe} \\ &+ 2\left[\mathbf{B}\left(\mathbf{f}_{\mathrm{a}} - \hat{\mathbf{f}}_{\mathrm{a}}\right)\right]^{\mathrm{T}}\mathbf{Pe} + 2\tilde{\mathbf{f}}_{\mathrm{a}}^{\mathrm{T}}\mathbf{G}^{-1}\left(\dot{\mathbf{f}}_{\mathrm{a}} - \dot{\hat{\mathbf{f}}}_{\mathrm{a}}\right) \\ &\le \mathbf{e}^{\mathrm{T}}\left[(-\mathbf{LC})^{\mathrm{T}}\mathbf{P} + \mathbf{P}(-\mathbf{LC})\right]\mathbf{e} + r^2\mathbf{e}^{\mathrm{T}}\mathbf{PPe} + \mathbf{e}^{\mathrm{T}}\mathbf{e} + 2\mathbf{T}_{\mathrm{d}}^{\mathrm{T}}\mathbf{Pe} \\ &+ 2\left(\mathbf{B}\tilde{\mathbf{f}}_{\mathrm{a}}\right)^{\mathrm{T}}\mathbf{Pe} + 2\tilde{\mathbf{f}}_{\mathrm{a}}^{\mathrm{T}}\mathbf{G}^{-1}\left(\dot{\mathbf{f}}_{\mathrm{a}} - \mathbf{GFCe} + \mathbf{TG}\hat{\mathbf{f}}_{\mathrm{a}}\right) \\ &= \mathbf{e}^{\mathrm{T}}\left[(-\mathbf{LC})^{\mathrm{T}}\mathbf{P} + \mathbf{P}(-\mathbf{LC})\right]\mathbf{e} + r^2\mathbf{e}^{\mathrm{T}}\mathbf{PPe} + \mathbf{e}^{\mathrm{T}}\mathbf{e} \\ &+ 2\mathbf{T}_{\mathrm{d}}^{\mathrm{T}}\mathbf{Pe} + \left[2\tilde{\mathbf{f}}_{\mathrm{a}}^{\mathrm{T}}\mathbf{G}^{-1}\dot{\mathbf{f}}_{\mathrm{a}} + 2\tilde{\mathbf{f}}_{\mathrm{a}}^{\mathrm{T}}\mathbf{G}^{-1}\mathbf{TG}\hat{\mathbf{f}}_{\mathrm{a}}\right]\end{aligned} \tag{20}$$

From Lemma 1, one can obtain the following inequalities:

$$2\mathbf{T}_d^T\mathbf{Pe} \le \frac{1}{\mu_1}\mathbf{T}_{\mathrm{d}}^{\mathrm{T}}\mathbf{Q}_1\mathbf{T}_{\mathrm{d}}^{\mathrm{T}} + \mu_1(\mathbf{Pe})^{\mathrm{T}}\mathbf{Q}_1^{-1}(\mathbf{Pe}) \tag{21}$$

$$2\tilde{\mathbf{f}}_{\mathrm{a}}^{\mathrm{T}}\mathbf{G}^{-1}\dot{\mathbf{f}}_{\mathrm{a}} \le \frac{1}{\mu_2}\tilde{\mathbf{f}}_{\mathrm{a}}^{\mathrm{T}}\mathbf{Q}_2\tilde{\mathbf{f}}_{\mathrm{a}}^{\mathrm{T}} + \mu_2\left(\mathbf{G}^{-1}\dot{\mathbf{f}}_{\mathrm{a}}\right)^{\mathrm{T}}\mathbf{Q}_2^{-1}\left(\mathbf{G}^{-1}\dot{\mathbf{f}}_{\mathrm{a}}\right) \tag{22}$$

$$2\tilde{\mathbf{f}}_{\mathrm{a}}^{\mathrm{T}}\mathbf{G}^{-1}\mathbf{TG}\hat{\mathbf{f}}_{\mathrm{a}} \le -2\tilde{\mathbf{f}}_{\mathrm{a}}^{\mathrm{T}}\mathbf{G}^{-1}\mathbf{TG}\tilde{\mathbf{f}}_{\mathrm{a}} + \frac{1}{\mu_3}\tilde{\mathbf{f}}_{\mathrm{a}}^{\mathrm{T}}\mathbf{Q}_3\tilde{\mathbf{f}}_{\mathrm{a}} + \mu_3\left(\mathbf{G}^{-1}\mathbf{TGf}_{\mathrm{a}}\right)^{\mathrm{T}}\mathbf{Q}_3^{-1}\left(\mathbf{G}^{-1}\mathbf{TGf}_{\mathrm{a}}\right) \tag{23}$$

To simplify the calculation, define the following transformation:

$$\upsilon = \left[\mathbf{e}^{\mathrm{T}} \quad \tilde{\mathbf{f}}_{\mathrm{a}}^{\mathrm{T}}\right]^{\mathrm{T}} \tag{24}$$

$$\theta = \frac{1}{\mu_1}\mathbf{T}_{\mathrm{d}}^{\mathrm{T}}\mathbf{Q}_1\mathbf{T}_{\mathrm{d}}^{\mathrm{T}} + \mu_2\left(\mathbf{G}^{-1}\dot{\mathbf{f}}_{\mathrm{a}}\right)^{\mathrm{T}}\mathbf{Q}_2^{-1}\left(\mathbf{G}^{-1}\dot{\mathbf{f}}_{\mathrm{a}}\right) + \mu_3\left(\mathbf{G}^{-1}\mathbf{TGf}_{\mathrm{a}}\right)^{\mathrm{T}}\mathbf{Q}_3^{-1}\left(\mathbf{G}^{-1}\mathbf{TGf}_{\mathrm{a}}\right) \tag{25}$$

$$\mathbf{E} = \begin{bmatrix}(-\mathbf{LC})^{\mathrm{T}}\mathbf{P} + \mathbf{P}(-\mathbf{LC}) + r^2\mathbf{PP} + \mathbf{I} + \mu_1\mathbf{PQ}_1^{-1}\mathbf{P} & \\ & \frac{1}{\mu_2}\mathbf{Q}_2 + \frac{1}{\mu_3}\mathbf{Q}_3 - 2\mathbf{G}^{-1}\mathbf{TG}\end{bmatrix} \tag{26}$$

Substituting (21)-(26) into (20), we can obtain that:

$$\dot{V} \le \upsilon^{\mathrm{T}}\mathbf{E}\upsilon + \theta \tag{27}$$

When $\mathbf{E} < 0$, we can obtain that $\dot{V} \le -\zeta\|\upsilon\|^2 + \theta$, where $\zeta = \lambda_{\min}(-\mathbf{E})$, It means that $\dot{V} \le 0$ as long as $\|\upsilon\|^2 \ge \frac{\theta}{\zeta}$. According to Lyapunov stability theory, $\left(\mathbf{e}, \tilde{\mathbf{f}}_{\mathrm{a}}\right)$ converges to a small set. That is to say, state estimation error and fault estimation error are uniformly bounded. This is the end of proof.

## 4. Simulation

In this section, numerical simulations are presented to illustrate the feasibility of the proposed fault identification scheme. Four reaction flywheels and magnetorquer are used to provide control moments. Without loss of generality, simulations just consider faults of the single flywheel installed on y-axis and magnetorquer. The satellite inertia matrix and flywheel inertia are as follows:

$$\mathbf{J}=\begin{bmatrix}0.5786 & 0.0003 & 0.0073\\ 0.0003 & 0.5810 & 0.0003\\ 0.0073 & 0.0003 & 0.2383\end{bmatrix} kg\cdot m^2 \quad J_w = 1.882e-6 kg\cdot m^2$$

The two observers design parameters of fault identification scheme are showed in Table 2. To illustrate the performance of the scheme, three fault scenarios are given in Table 3, where $f_y$ denotes the y-axis fault torque of magnetorquer.

**Table 2.** Observers design parameters.

| parameter | value |
|---|---|
| $L_1$ | 1.45 |
| $\tau_1$ | 0.4705 |
| $\tau_2$ | 3.764e-15 |
| $v_f$ | 1e-9 |
| $\mathbf{L}_2$ | diag([1.093,1.093,1.093]) |
| $\mathbf{G}$ | diag([0.08,0.11,0.01]) |
| $\mathbf{T}$ | diag([0.01,0.01,0.01]) |

**Table 3.** Specification of actuators failure scenarios.

| Scenario | Parameter | Time of failure(s) |
|---|---|---|
| Scenario 1(S1) | $k$=0.5,$v$=$10^{-9}$;$f_y$=$-10^{-6}$Nm | $t>20;t>20$ |
| Scenario 2(S2) | $k$=0,$v$=$5\times10^{-9}$;$f_y$=$e^{t/100}\times10^{-8}$Nm | $t>20;t>20$ |
| Scenario 3(S3) | $k$=0,$v$=$2\times10^{-10}$;$f_y$=$-10^{-6}$Nm | $t>20;40<t<60,100<t<120$ |

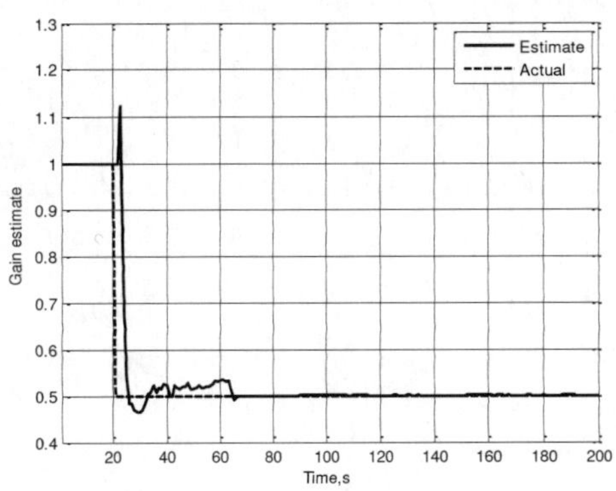


**Figure 2.** Gain estimate (S1).

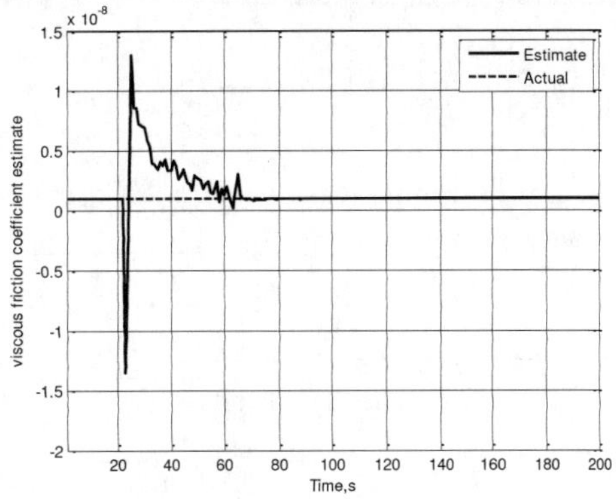


**Figure 3.** Viscous friction coefficient estimate (S1).

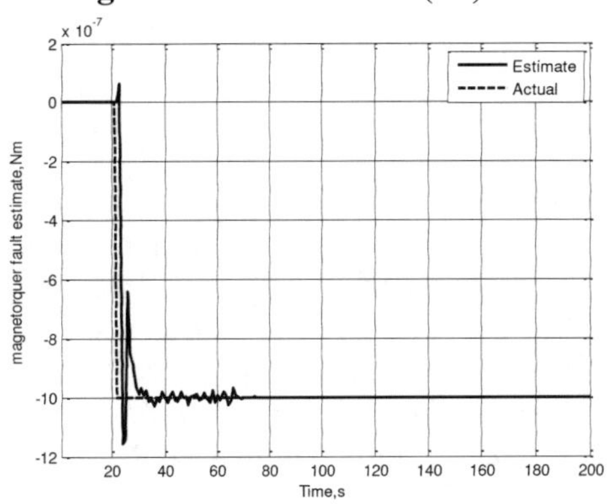


**Figure 4.** Magnetorquer fault estimate (S1).

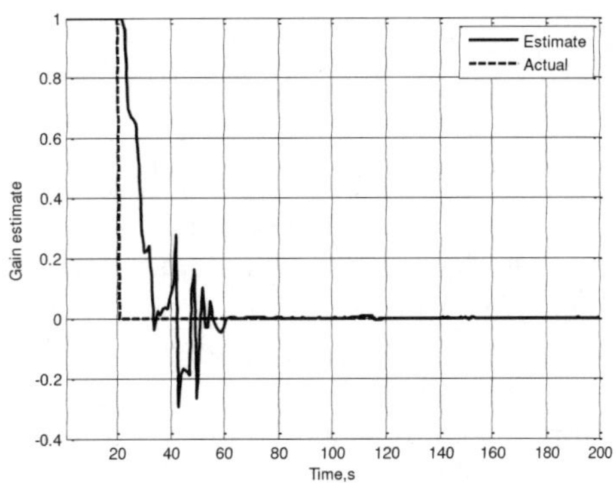


**Figure 5.** Gain estimate (S2).

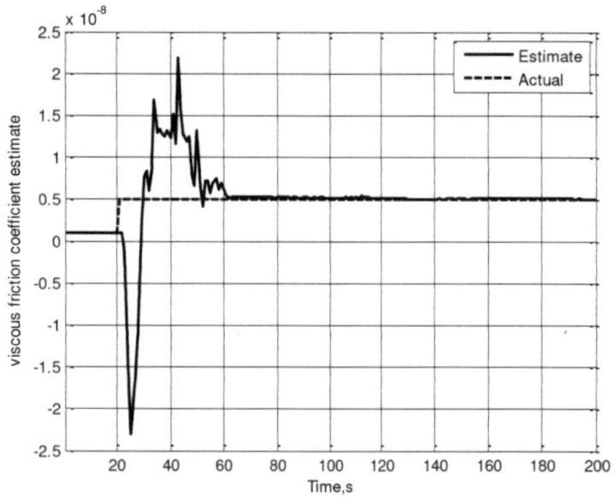


**Figure 6.** Viscous friction coefficient estimate (S2).

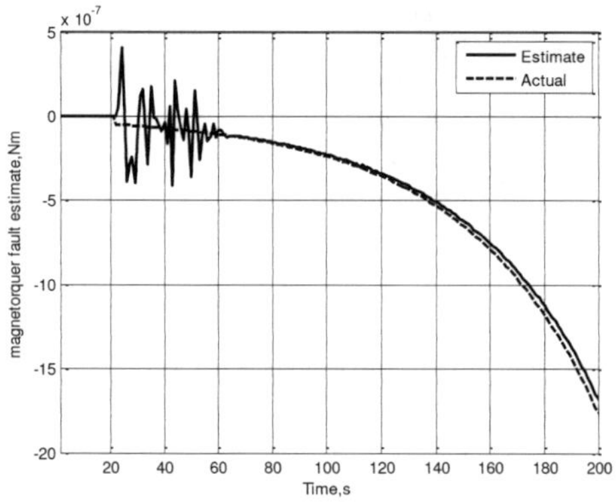


**Figure 7.** Magnetorquer fault estimate (S2).

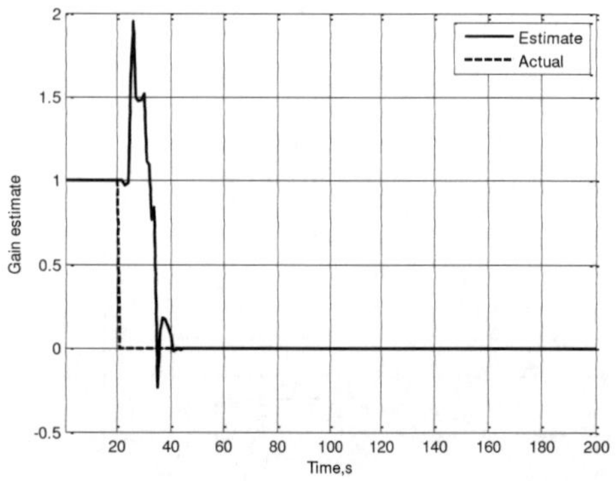


**Figure 8.** Gain estimate (S3).

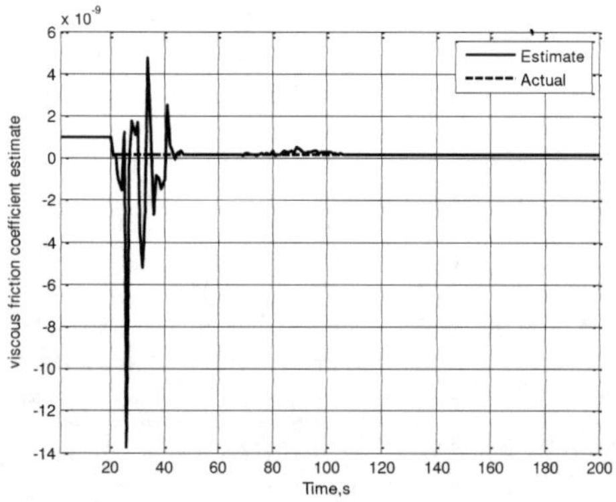


**Figure 9.** Viscous friction coefficient estimate (S3).

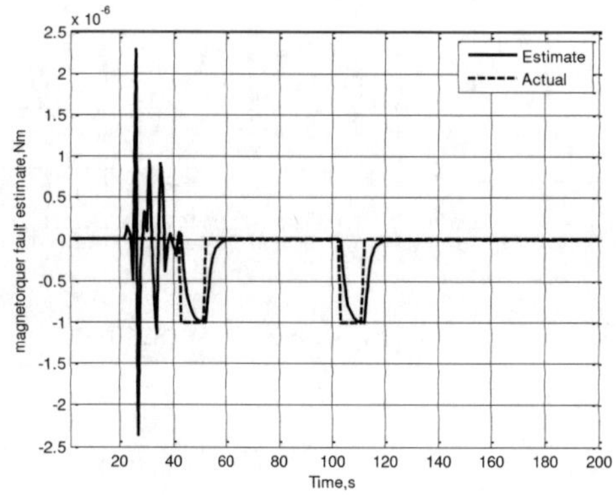


**Figure 10.** Magnetorquer fault estimate (S3).

In scenario 1, the gain decrease fault of the flywheel and the bias fault of the magnetorquer occur at 21s. The dashed line denotes the actual value and the solid line represents the estimated value. As shown in Fig.2 and Fig.3, the local observer can estimate output gain and viscous friction coefficient accurately within 40 seconds. It can be noted from Tab.1 that the flywheel occurred gain decrease fault, which is consistent with the actual fault. Given the global observer has removed flywheel fault torque, it can be seen from Fig.4 that magnetorquer fault is bias fault, and its magnitude is about $-10^{-6}$Nm.

In scenario 2, both flywheel and magnetorquer malfunction at 21s. It can draw from Fig.5 and Fig.6 that flywheel output gain is about 0 and viscous friction coefficient is about $5\times10^{-9}$. So the fault mode of flywheel is stalling fault referring to Tab.1. It can be seen from Fig.7 that the estimated fault torque gradually increase, in other words, magnetorquer occurs slow variation fault.

In scenario 3, the estimated output gain is about 0 and the estimated viscous friction coefficient is about $2\times10^{-10}$ in Fig.8 and Fig.9, which indicates that the flywheel occurs idling fault. It can be seen from Fig.10 that fault magnitude is up to $-10^{-6}$Nm during 40s~60s and 100s~120s, which shows the fault mode of magnetorquer is intermittent fault.

## 5. Conclusions

In this Note, an algorithm for adaptive identification of actuator faults in double observers is proposed. The parameterized description of flywheel faults is established. Using the flywheel speed information, the local adaptive observer can identify the flywheel parameters and realize fault identification. The global observer is designed based on the satellite dynamics model. The estimated output torque of flywheel is taken as the input information of global observer, which aims at eliminating the influence of the flywheel fault on the global observer, thereby fault identification of the magnetorquer can be realized. Simulation results demonstrate the effectiveness of the proposed scheme.

## References

[1] Wen X, Zhang X W, Qin Y Q 2014 Statistics and analysis of on-orbit failure modes of foreign spacecraft *Quality and reliability*. 13-18.
[2] Tan C L, Hu T B, Wang D P 2011 Analysis on foreign spacecraft in-orbit failures *Spacecraft Engineering*.
[3] Wen J J, Huang D 2007 Review on fault diagnosis *Advances in Computer Technology and Applications*.
[4] Xie M, Lou X 2016 Reviewed and developing trend of spacecraft fault diagnosis technology *Software*. **37(7)** 70-74.
[5] Su L, Shang C X, Liu W J 2010 Survey on the technology of fault diagnosis for spacecraft attitude control system *Journal of Changchun University of Science and Technology*, **33(4)**:23-27.
[6] Zhu S, Wang D, Shen Q 2017 Satellite attitude stabilization control with actuator faults *Journal of Guidance, Control, and Dynamics*. 1-11.
[7] Lu W G, Su Z H, Qi J 2018 Method of satellite fault diagnosis based on wavelet transform *Foreign Electronic Measurement Technology*.
[8] Gueddi I, Nasri O, Benothman K 2017 Fault detection and isolation of spacecraft thrusters using an extended principal component analysis to interval data *International Journal of Control Automation & Systems*. **15(2)** 1-14.
[9] Gueddi I, Nasri O, Benothman K 2015 VPCA-based fault diagnosis of spacecraft reaction wheels *2015 XXV International Conference on Information, Communication and Automation Technologies*.
[10] Su Z H, Lu W G, Qi J 2016 A method of satellite fault diagnosis based on BP neural network *Computer Measurement & Control*.
[11] Baldi P, Blanke M, Castaldi P 2015 Combined geometric and neural network approach to generic fault diagnosis in satellite reaction wheels *IFAC Papers On Line*. **48(21)** 194-199.
[12] Gao C Y 2013 Model-based actuator fault diagnosis and its application to satellite attitude control systems Harbin. 24-25.
[13] Jiang B, Marcel S 2002 Adaptive observer design for robust fault estimation *International Journal of Systems Science*. **33(9)** 9.
[14] Meng L Y, Jiang B 2008 Fault diagnosis based on the nonlinear adaptive observer *Systems Engineering and Electronics*. **30(7)** 1317-19.

# Local Information based Organization of Distributed Spacecraft Swarm Using Artificial Potential Field

**Wen Feng [1,2], Shuquan Wang [1,*] and Hongen Zhong [1]**

[1]Key laboratory of Space Utilization, Technology and Engineering Center for Space Utilization, Chinese Academy of Sciences, No. 9 Dengzhuang South Rd, Haidian Dist, Beijing, 100094, China
[2]University of Chinese Academy of Sciences, Beijing, 100094, China

[*]Correspondence author: Shuquan Wang, shuquan.wang@csu.ac.cn

**Abstract.** This paper investigates the problem of distributed organization of spacecraft swarm where each spacecraft is an independent agent that can interact with neighbouring members within a certain distance. According to the global objective of the organization, each agent makes decisions on its behaviour based on its own status and adjacent members within a certain distance. By a carefully designed strategy and under certain conditions, the swarm can accomplish the global objective. In the configuration reconstruction mission, only the expected global configuration is specified. The target position of each spacecraft is uncertain and not fixed. At the beginning of the reconstruction, every spacecraft can only determine the initial target point according to certain rules. It is possible to have conflicts of destination point selections since all spacecraft can only get local information. Each spacecraft constantly adjust their target point according to the local information in the process of accomplishing this task, until the group converges into the specified global configurations. In this paper, three different target allocation strategies are designed, so that each spacecraft can do continuous selection and optimization, and finally determine a non-conflict target matching to complete the formation reconstruction task. The artificial potential field method is utilized to control each spacecraft and complete the path planning, while ensuring that the spacecraft swarm to achieve collision avoidance and other requirements. In addition, the time and fuel consumption are compared. The numerical simulations show that these three strategies make obvious optimization effects.

## 1. Introduction

At present, some countries have proposed and gradually implemented various swarm spacecraft plans for astronomical observation, deep space exploration, earth exploration and space technology verification. Including The NASA Cyclone Global Navigation Satellite System (CYGNSS) Mission[1], "Tiantuo-3" micro-nano satellite cluster plan, Autonomous Nanotechnology Satellite[2]· DARPA F6 program[3], Edison Demonstration of SmallSat Networks[4] and others.

The distributed spacecraft swarm has the advantages of high functional density and technical performance, low investment and operation cost, strong flexibility, short development cycle and low risk. Multiple small spacecraft work together to complete complex space exploring missions has become a hotspot in the international space field. The application and development of spacecraft clusters will certainly become the strategic focus of future international space development.[5]

Intelligent Satellite swarm is essentially a problem of distributed organization where each spacecraft is an independent agent that can interact with neighbouring members. Due to the particularity of the information interaction among the spacecraft in the large-scale spacecraft formation, the architecture proposed for the small-scale formation composed of several spacecraft has some limitations when applied to the large-scale formation. Large-scale spacecraft formation control are limited by its architecture and inter-satellite communication performance, so it is difficult to obtain all state information of the system, which brings some difficulties to the collaborative control of the formation system and the strategy of spacecraft swarm formation based on local information needs to be considered.

Traditional control methods of spacecraft formation on relative orbit always base on circular orbit and ignores the influence of external perturbation, using Linear C-W Equation [6] to describe the relative motion model of spacecraft, and using consistency theory[7] and adaptive control[8] to complete formation reconstruction. In this paper, the thinking mode of swarm intelligence[9] is used as a reference. Each distributed agent in the swarm with simple strategy works together to achieve the target configuration based on local information.

## 2. Methodology

In this paper, only the expected global configuration is specified in the configuration reconstruction mission. The target position of each spacecraft is uncertain and not fixed. And three different target allocation strategies are designed so that each spacecraft can do continuous selection and optimization, and finally determine a non-conflict target matching to complete the formation reconstruction task.

This paper uses the artificial potential field method based on the information of the target configuration and the interaction between adjacent spacecraft within a certain range to control each spacecraft and complete the path planning, while ensuring that the spacecraft swarm to achieve collision avoidance and other requirements. At the beginning of the reconstruction, every spacecraft can only determine the initial target point according to certain rules. It is possible to exist conflicts of destination point selections since all spacecraft can only get local information. Each spacecraft constantly adjust their target point according to the local information in the process of accomplishing this task, until the group converges into the specified global configurations.

## 3. Artificial Potential Approach

In reconfiguration of distributed spacecraft swarm, each spacecraft can plan a feasible and safe path to reach the target point by using artificial potential field regulations after determining its target point. As an extended form of the Lyapunov function, the potential function can be used in the subsequent analyses. It is this function that may be defined analytically and used to drive the state vector of a dynamical system to the desired goal. **Frank McQuade** extended the artificial potential field method to on-orbit equipment.[10] That thesis, based on the Lyapunov's method[11] and extended to potential function theory, investigates an autonomous method of assembly large structures on Earth Orbit. **YAO Hong** extended the artificial potential field method to the cooperative control of spacecraft formations to solve the overall movement, global dispersion and configuration transformation.[12]

The collaborative control algorithm based on artificial potential field method has less computational complexity and is suitable for multi-agent real-time cooperative control. This paper extends the artificial potential field method to the cooperative control of spacecraft formations. In this scheme, multiple spacecraft in a swarm achieve formation control through attractive and repulsive forces among themselves. Based on the potential energy determined by the location and the Hill equation, the dynamic equations of the on-orbit spacecraft are established. With the orbital constraint, the artificial potential field method gets rid of the local minimum problem.

**Table 1.** Conditions of Completing Reconstruction

| Parameter | Limitation |
|---|---|
| the distance to the target point | 0 *m* |
| Speed when the spacecraft reaches the target | 0 *m/s* |

Because the potential energy function is determined in the zero point of the coordinate system, the potential energy of the object is a function related to the position coordinates and its speed. So for every spacecraft, its potential energy $P$ is as followed formula based on the conditions shown as Table 1 .

$$P=\frac{1}{2}k_1\vec{\rho}^T\cdot\vec{\rho}+\frac{1}{2}k_2\dot{\vec{\rho}}^T\cdot\dot{\vec{\rho}} \tag{1}$$

where $\vec{\rho}$ is the distance from initial position $\vec{r}$ to the target position $\vec{r}_{goal}$. $\dot{\vec{r}}$ and $\dot{\vec{r}}_{goal}$ are the speed of that spacecraft. $\ddot{\vec{r}}$ and $\ddot{\vec{r}}_{goal}$ are the accelerations. $k_1$ and $k_2$ are constant coefficients. The speed of the spacecraft is required to be $0m/s$ both at the starting point and at the target point. So that

$$\begin{gathered}\vec{\rho}=\vec{r}-\vec{r}_{goal}\\ \dot{\vec{r}}=0,\ \ddot{\vec{r}}=f(\vec{r},\dot{\vec{r}})+u\\ \dot{\vec{r}}_{goal}=0,\ \ddot{\vec{r}}_{goal}=f(\vec{r}_{goal},\dot{\vec{r}}_{goal})\end{gathered} \tag{2}$$

where $u$ is the control of spacecraft and $f$ is the orbital acceleration of gravity, related to the current position and velocity. And the derivative of equation (1) is:

$$\dot{P}=k_1\vec{\rho}^T\cdot\dot{\vec{\rho}}+k_2\dot{\vec{\rho}}^T\cdot\ddot{\vec{\rho}} \tag{3}$$

Equation (3) is always negative so that each spacecraft reaches the target point with zero speed. And the derivative of $P$ is given by

$$\dot{\vec{\rho}}^T\cdot\left(k_1\vec{\rho}+k_2\ddot{\vec{\rho}}\right)=-\dot{\vec{\rho}}^T\cdot\dot{\vec{\rho}} \tag{4}$$

From equation (2) and equation (4), the control u can be derived as:

$$\begin{gathered}k_2u=-\dot{\vec{\rho}}-k_1\vec{\rho}+k_2f(\vec{r}_{goal},\dot{\vec{r}}_{goal})-k_2f(\vec{r},\dot{\vec{r}})\\ u=-k_3\dot{\vec{\rho}}-k_4\vec{\rho}+f(\vec{r}_{goal},\dot{\vec{r}}_{goal})-f(\vec{r},\dot{\vec{r}})\end{gathered} \tag{5}$$

From equation (2) and equation (5), the acceleration of every spacecraft should be

$$\ddot{\vec{r}}=f(\vec{r},\dot{\vec{r}})+u=-k_3\dot{\vec{\rho}}-k_4\vec{\rho}+f(\vec{r}_{goal},\dot{\vec{r}}_{goal}) \tag{6}$$

However, for on-orbit spacecraft, its speed should not be determined only by the above variables, but also by the gravity. The well-known Hill Equation[13] can help a lot:

$$\begin{gathered}\ddot{x}=3w^2x+2w\dot{y}+u_x\\ \ddot{y}=-2w\dot{x}+u_y\\ \ddot{z}=-w^2z+u_z\end{gathered} \tag{7}$$

Where $\vec{r}=[x,y,z]=$, and one of the conditions for completion of the reconfiguration is that the speed of the spacecraft at the target point is $0m/s$, which is $\dot{\vec{r}}_{goal}=0$. So the orbital acceleration of gravity $f$ in equation (2) can be replaced by

$$f(\vec{r}_{goal},\dot{\vec{r}}_{goal})=\begin{bmatrix}3w^2 & 0 & 0\\ 0 & 0 & 0\\ 0 & 0 & -w^2\end{bmatrix}\vec{r}_{goal} \tag{8}$$

In Artificial Potential Fields, obstacle avoidance is important, which have different degrees of repulsive force for each spacecraft. At the same time, when the distance between spacecraft is less than a certain threshold, it will also exert some repulsive force on other spacecraft to avoid other spacecraft.

$$F_{rep}(q)=k_{rep}\left(\frac{1}{\|\vec{r}-\vec{r}_{obs}\|}-\frac{1}{\rho_0}\right)\frac{\vec{r}-\vec{r}_{obs}}{\|\vec{r}-\vec{r}_{obs}\|^3} \tag{9}$$

For low-altitude orbits (2000$km$), the gravity of each spacecraft at the target point in equation (7) is set as follows:

$$\ddot{\vec{r}}=f(\vec{r},\dot{\vec{r}})+u=-k_3\dot{\vec{\rho}}-k_4\vec{\rho}+\begin{bmatrix}3w^2 & 0 & 0\\ 0 & 0 & 0\\ 0 & 0 & -w^2\end{bmatrix}\vec{r}_{goal} \tag{10}$$

To sum up, in the artificial potential field, the target point produces "attraction" to the spacecraft and the obstacle produces "repulsive force", that is, acceleration controls the spacecraft's action to

IOP Conf. Series: Materials Science and Engineering **608** (2019) 012042 doi:10.1088/1757-899X/608/1/012042

make every spacecraft reach its determined target. And all the simulation results in this paper were obtained by setting $k_3 = 1s^{-1}$, $k_4 = 0.1s^{-2}, k_{rep} = 1 \cdot \frac{m^4}{s^2}$.

## 4. Target Assignment

In the configuration reconstruction mission of distributed spacecraft swarm, the target point of each spacecraft is uncertain and not fixed. Each spacecraft is an independent agent that can interact with surrounding members, and then according to the whole spacecraft group expectations, the status itself and adjacent spacecraft within a certain distance to make decisions and determine its own control behaviour, thus all the spacecraft can form the target configuration.

Task allocation problem is an important research content in multi-agent system, that is, how to allocate appropriate tasks to appropriate agents to achieve the overall optimal performance. This paper refers to some models of swarm intelligence to accomplish target point allocation, which is the collective behaviour of decentralized, self-organized systems, natural or artificial. The agents follow very simple rules, and although there is no centralized control structure dictating how individual agents should behave, local, and to a certain degree random, interactions between such agents lead to the emergence of "intelligent" global behaviour, unknown to the individual agents.

This paper assumes that the information of the spacecraft includes its position in the relative motion coordinate system and the selected target point can be shared to its neighbours within a certain distance.

### *4.1. 1st target matching strategy*

At the beginning of the reconfiguration, every spacecraft knows its own position and the location of all the target points in the relative coordinate system so that it can calculate a sequence of target points from the nearest to the farthest according the distance to each target position and choose the nearest one as the initial target.

In the process of the reconfiguration, it is possible to exist conflicts of destination point selections since all spacecraft can only get local information. Each spacecraft constantly adjust their target point according to the local information in the process of accomplishing this task, until the group converges into the specified global configurations.

Given that most of the target configurations are concentrated, when a spacecraft exchanges information with another spacecraft in the communication distance and finds that the target point of the two spacecraft is the same one, the spacecraft farther to this target point is still heading to the target point, and the other spacecraft choose sub-optimal target. Artificial potential field method can help spacecraft avoid collision with other spacecraft and reach the final target point.

**Table 2.** Fuel Consumption of 8-spacecraft

| | X Label (*m/s*) | Y Label (*m/s*) | Z Label (*m/s*) |
|---|---|---|---|
| 1 | 7.3392 | 6.5572 | 5.0379 |
| 2 | 9.3449 | 10.7258 | 7.2993 |
| 3 | 2.6644 | 11.3350 | 7.3319 |
| 4 | 1.8511 | 11.8738 | 6.7698 |
| 5 | 7.2117 | 7.2576 | 5.0244 |
| 6 | 3.9472 | 5.9216 | 6.7279 |
| 7 | 0.8330 | 4.4455 | 5.0126 |
| 8 | 4.1498 | 5.7681 | 5.0154 |

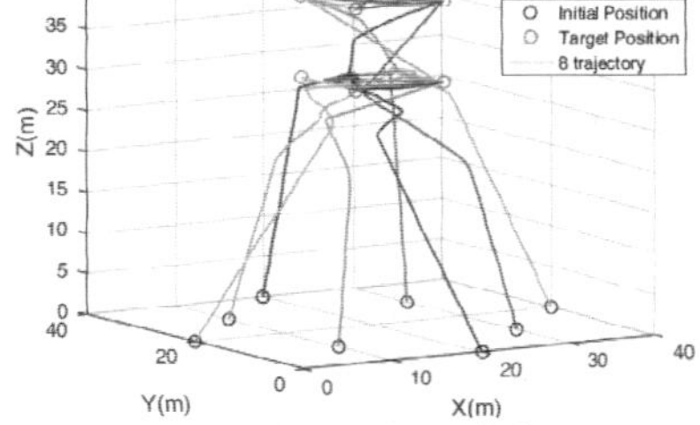

**Figure 1.** Trajectories of the swarm

This strategy was applied to the task of forming a cube from eight spacecraft in a plane on the orbit of 2000*km* and the simulation results is shown in Figure 1. It takes 184.78*s* for eight spacecraft to complete the reconfiguration, and the sum of fuel consumption (velocity increment) of all spacecraft is 178.011*m/s*. The fuel consumption of each spacecraft is shown in 0.

### *4.2. 2nd target matching strategy*

Data analysis found that many spacecraft have reached a certain target point, but then give this target to the later spacecraft for reducing additional fuel consumption.

In view of the above situation, at a certain moment, if the distance between a spacecraft to its target point less than a certain threshold at this moment, the target point is assigned to the spacecraft and is fixed. When other spacecraft that also choose this target communicates with this spacecraft and finds that the target point fixed, and then the others need to select the suboptimal target.

The other rules remain the same, every spacecraft calculate a sequence of target points from the nearest to the farthest according the distance to each target position and choose the nearest one as the initial target at the beginning. When a spacecraft exchanges information with another in the communication distance and finds that the target points are the same one, the spacecraft closer to this target point choose sub-optimal target, the other one is still heading to the target point.

This mode of assignment can be used in the above eight spacecraft mission and the simulation result is shown in Figure 2. The time required for the formation reconstruction of eight spacecraft is 108.42*s*, which is nearly half of the time using the first strategy. The sum of fuel consumption of the spacecraft is 120.8247*m/s*, which is reduced by nearly one third compared with the first strategy, and the fuel consumption of each is shown in Table 3.

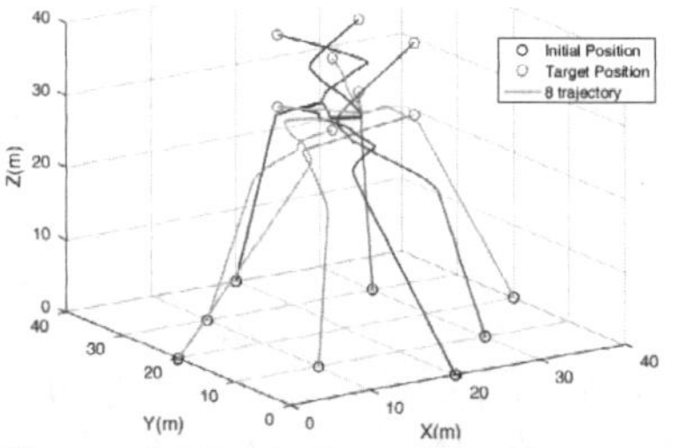

**Figure 2.**The trajectories of swarm

**Table 3.** Fuel Consumption of 8-spacecraft

| | X Label（*m/s*） | Y Label（*m/s*） | Z Label（*m/s*） |
|---|---|---|---|
| 1 | 7.3392 | 6.5572 | 5.0379 |
| 2 | 9.3449 | 10.7258 | 7.2993 |
| 3 | 2.6644 | 11.3350 | 7.3319 |
| 4 | 1.8511 | 11.8738 | 6.7698 |
| 5 | 7.2117 | 7.2576 | 5.0244 |
| 6 | 3.9472 | 5.9216 | 6.7279 |
| 7 | 0.8330 | 4.4455 | 5.0126 |
| 8 | 4.1498 | 5.7681 | 5.0154 |

### *4.3. 3^rd^ target matching strategy*

In the previous strategy, the priority order of the spacecraft to select the target point is determined by the distance between the initial position of the spacecraft and all the target points. With the spacecraft moving, the distance changes a lot. This paper proposed some improvements. Spacecraft choose sub-optimal target based on the distance between the current position and the target point other than the distance between initial position and the target point. And the target point that has been ceded to other spacecraft in the competition should be removed from the order of priority.

Such a method can be applied in the above task and the simulation result is shown in Figure 3. And it takes 98.6638*s* to complete the task, the fuel consumption of each spacecraft is shown in Table 4, and the sum of fuel consumption of the spacecraft is 112.4682*m/s*, which is slightly shortened.

**Table 4.** Fuel Consumption of 8-spacecraft

| | X Label（*m/s*） | Y Label（*m/s*） | Z Label（*m/s*） |
|---|---|---|---|
| 1 | 5.5099 | 4.5176 | 7.8811 |
| 2 | 3.8582 | 4.7856 | 5.8341 |
| 3 | 4.3568 | 7.6835 | 6.9687 |
| 4 | 4.8494 | 6.4218 | 7.0765 |
| 5 | 2.4779 | 2.5042 | 5.0065 |
| 6 | 3.5139 | 3.7904 | 5.8305 |
| 7 | 0.9437 | 3.6791 | 6.6368 |
| 8 | 2.5010 | 0.8347 | 5.0065 |

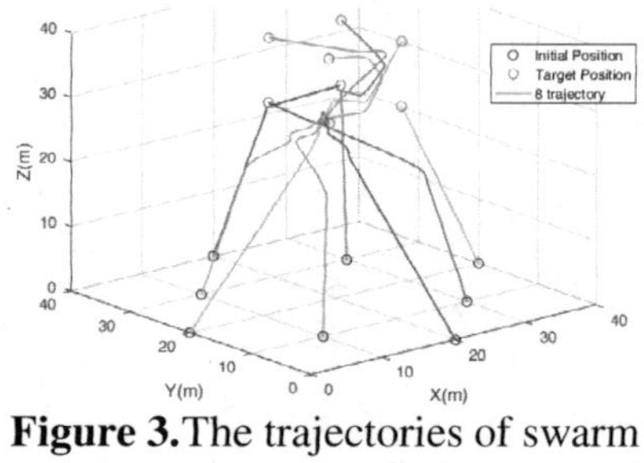

**Figure 3.**The trajectories of swarm

## 5. Comparison and conclusion

The above simulation results have proved the feasibility and effectiveness of the proposed method. By extending this method to the reconfiguration of more spacecraft, it is found that spacecraft can complete the selection of target points based on local information and get a safety trajectory to complete the mission. This section compared the time and fuel consumption in the reconfiguration of 42 spacecraft from a plane to a ball using above three strategies. And Table 5 is the result of the comparison.

On the basis achieving the target configuration, the three strategies showed significant improvement in the consumption of fuel and time.

Just because the target points are concentrated, this paper set that the spacecraft closer to this target point is still heading to the target point, and the other spacecraft choose sub-optimal target when the target points are consistent in most cases. These methods and strategies may not apply to all missions. Finding a general target allocation method is the next research focus. In addition, local information limits the optimization effect, and the algorithms of intelligent control and artificial intelligence need to be used to approach the global optimum.

**Table 5.** Fuel Consumption analysis of 42-spacecraft swarm organization

| | 1st strategy | 2nd strategy | 3rd strategy |
|---|---|---|---|
| Trajectories in LVLH frame | 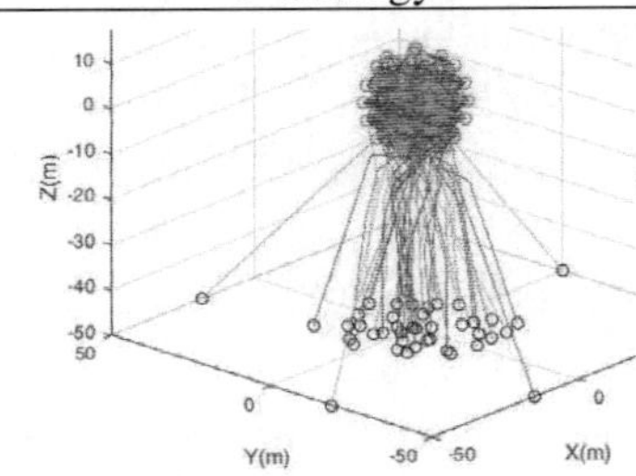 | 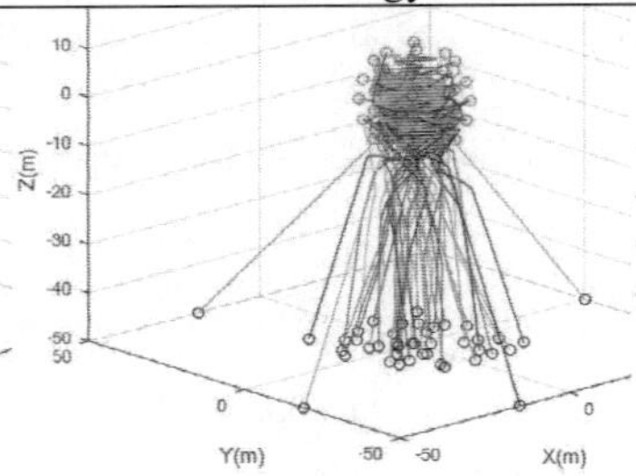 | 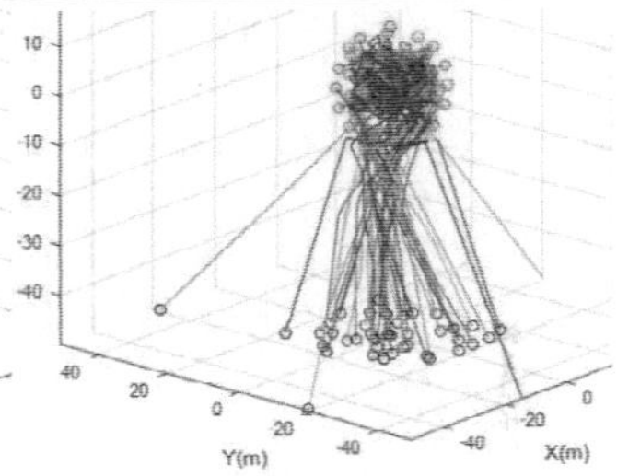 |
| Time(*s*) | 1239.2514 | 457.2380 | 397.7956 |
| Fuel(*m/s*) | 3876.656 | 2297.796 | 1656.546 |

## Acknowledgement

This research was supported by the National Natural Science Foundation under grant No. 11672294 and the Advanced Research Project of Manned Space Program of China.

## References

[1] Ruf C, Gleason S, Jelenak Z, et al. The NASA Cyclone Global Navigation Satellite System (CYGNSS) Mission[J]. 2013.

[2] Clark P E, Curtis S, Rilee M, et al. ANTS: Exploring the Solar System with an Autonomous Nanotechnology Swarm[C]// Lunar & Planetary Science Conference. 2002.

[3] Maciuca D, Chow J, Siddiqi A, et al. A Modular, High-Fidelity Tool to Model the Utility of Fractionated Space Systems[M]// Encyclopedia of Aerospace Engineering. 2010.

[4] Yost B. EDSN - Edison Demonstration for SmallSat Networks Overview[J]. 2013.

[5] Wen Xin. Space Exploration is Entering the Era of Spacecraft Cluster[J]. People's Forum, Academic Frontier, 2017(05):21-28.

[6] Clohessy W H. Terminal Guidance System for Satellite Rendezvous[J]. Aerospace Sci, 1960, 27(9).

[7] Ren W. Formation Keeping and Attitude Alignment for Multiple Spacecraft Through Local Interactions[J]. Journal of Guidance Control & Dynamics, 2012, 30(2):633-638.

[8] Ruiter A D. Adaptive spacecraft formation flying with actuator saturation[J]. Proceedings of the Institution of Mechanical Engineers Part I Jo urnal of Systems & Control Engineering, 2010, 224(4):373-385.

[9] Bonabeau E, Dorigo M, Theraulaz G. Swarm intelligence: from natural to artificial systems[M]. Oxford University Press, Inc. 1999.

[10] Mcquade, Frank. "Autonomous control for on-orbit assembly using artificial potential functions." University of Glasgow. 1998(1997).

[11] R. Kalman, and J. Bertram. "Control system analysis and design via the second method of lyapunov: (I) continuous-time systems (II) discrete time systems." Ire Transactions on Automatic Control. Vol. 4, No.3, 2003, pp. 112-112.

[12] Yao, Hong and Tang, Yafeng. "Coordinated control of spacecraft formation based on the artificial potential field "// China Intelligent Automation Conference. 2009, pp. 883-889.

[13] Clohessy, W. H. "Terminal Guidance System for Satellite Rendezvous." Aerospace Sci. Vol. 27, No. 9, 2012, pp. 653–658.

IOP Conf. Series: Materials Science and Engineering **608** (2019) 012043 doi:10.1088/1757-899X/608/1/012043

# Use of Tetrahedral Finite Element Method for Computing the Gravitation of Irregular-Shaped Asteroid

**Weidong Yin[1,2], Leizheng Shu[1,*], Yang Yu[3] and Yang Gao[1]**

[1] Key Laboratory of Space Utilization, Technology and Engineering Center for Space Utilization, Chinese Academy of Sciences, Beijing 100094, China
[2] University of Chinese Academy of Sciences, Beijing 100049, China
[3] School of Aeronautical Science and Engineering, Beihang University, Beijing 100091, China

[*] shuleizheng@csu.ac.cn

**Abstract.** Asteroids are one of the most important targets for deep space exploration. In previous asteroid missions, the accurate estimate of gravity has proved to have a strong influence on the design of the approach orbit and navigation strategy. The wield gravitational field of an asteroid is mainly caused by the irregular overall shape and possible heterogenous mass distribution the interior. We propose to use the finite element method to compute the gravity of irregularly shaped asteroids; this method combines the advantages of the conventional mascon method and the polyhedral method. The tetrahedral meshes can be generated following the conventional division technique. Taking asteroid 216 Kleopatra as an example, we calculate the exterior gravitational field using the above mentioned methodology. We then compare the results from the finite element method and the polyhedral method under a degenerated case, i.e., with constant density. Then, four different density distribution assumptions are given, and the gravitational fields are calculated respectively. The comparative study and the density distribution assumptions indicate that the proposed method is suitable for modeling an arbitrary asteroid with nonuniform mass distribution. This method is expected to provide reliable gravity data for the design of guidance, navigation, and control systems in future asteroid missions.

## 1. Introduction

The scientific exploration of asteroids has become a hot research topic in the field of deep space exploration. With the development of deep space exploration technology during the past 30 years, the technology of probing asteroids has been extended from Earth-based observation activities to launching unmanned probes to target asteroid sampling. The United States (NASA), Japan (JAXA), and the European Union have launched multiple probes for the probe of small objects since the 1990s. So far only three probes have successfully landed on asteroids, namely the NEAR Shoemaker [1], Muses-C [1] and Rosetta [1]. The Dawn probe [1] from NASA was launched in 2007 and is currently flying around Ceres. Japan launched Hayabusa 2 [2] in December 2014 and has reached the target asteroid 162173 Ryugu. Hayabusa 2 completed the first landing sampling at 23:06 (UTC) on February 21, 2019 and is expected to return to Earth in 2020. NASA launched OSIRIS-REx in 2016 and reached the target asteroid 101955 Bennu in 2018. OSIRIS-REx plans to land and sample in July 2020 and return to Earth in 2023 [3]. Moreover, the New Horizons New Field Vision was confirmed on

IOP Conf. Series: Materials Science and Engineering **608** (2019) 012043 doi:10.1088/1757-899X/608/1/012043

January 1, 2019 to have flown over the asteroid 486958 (Ultima Thule), which is the most distant celestial body ever probed by humans [4].

Compared with planets such as Earth and Jupiter, most asteroids are irregularly shaped; thus, their gravitational field is extremely complicated, which directly affects their orbital dynamics, and the navigation, guidance, and control of asteroid probe. The extended gravitational field of irregularly shaped asteroids is expressed and calculated through three main methods, namely, spherical harmonics, mascon method, and homogeneous polyhedron.

The spherical harmonics method is a classical method of expressing and calculating the gravitational field of any celestial body, and it has been widely used in the field of spacecraft orbital dynamics [5]. It uses infinite series to approximate the gravitational potential function; that is, a series of spherical harmonic functions is superimposed on the basis of the central gravitational term to describe the non-spherical perturbation of the gravitational field. The spherical harmonic function is widely used due to its rapid convergence and high computational efficiency. However, when it is used to describe the irregular gravitational field of asteroids, shortcomings are noticeable. First, the truncation error is enlarged as the distance to the asteroid shortens, and the convergence speed becomes slower and even becomes divergent within the Brillouin sphere. Second, the spherical harmonic method cannot determine whether the probe is inside or outside the asteroid and requires an extraordinary algorithm to determine it; then, its exact spherical harmonic coefficient must be determined by the actual data determined by the orbit. Although a series of improvements for the spherical harmonic function method, such as the ellipsoidal harmonic function method [6], the Brillouin interior/exterior spherical harmonic function [7], and the interior/exterior spherical Bessel function [8], has been proposed, it is the method itself that renders impossible the accurate description of the irregular gravitational field of an asteroid.

The mascon method [9] is an intuitive technique that describes the gravitational field of an asteroid; it uses the volume element (small sphere) to approximate the shape of the target asteroid, and each volume element is approximated as a mass point to calculate the gravitational field of the target asteroid. The mascon method is simple and easy to implement, and the precision of the gravitational field increases with the number of volume elements. The mass weight can be flexibly modified for asteroids with a nonuniform density distribution, and the method can simulate the structure of an asteroid realistically, but the drawback is also obvious. First, the convergence is slow. As the number of volume elements increases, the calculation increases sharply; thus, the error accumulation also becomes a problem. Second, as the calculation point approaches the surface of the asteroid, the error of the calculation increases [10], the spherical harmonic function rises; thus, collision detection is not possible.

The polyhedral method [11] uses a polyhedral model to approximate the shape of an irregular asteroid. With the assumption that the polyhedral density is uniform, the gravitational potential and gravitation are processed and obtained via integral transformation. Although assumed uniform density of the polyhedral method does not conform to the real situation, the gravitational force at the edge of the surface of the polyhedron will produce singularity, and the computational efficiency is lower than that of the spherical harmonic function, the expressions are all closed, thereby ensuring that the calculation error does not change with the relative distance of the asteroid. The error is related only to the model error (shape and density distribution errors), and collision detection can be performed to determine whether the calculation point enters the interior of the polyhedron. Therefore, the method has been used widely in gravitational field analysis, the approach orbit analysis and the orbital design of asteroid missions.

Numerous methods for calculating the gravitational field of irregular asteroids are available; however, when the density distribution is nonuniform, the above methods are not sufficient for accurately calculating the gravitational field. Therefore, by combining the advantages of the mascon and the homogeneous polyhedron methods, this study proposes a tetrahedral finite element method to estimate the gravitational field accurately under the condition of nonuniform density distribution; that is, the polyhedral model is filled with tetrahedrons based on the surface triangles. The gravitational potential and gravitation of each tetrahedral finite element are calculated by the homogeneous polyhedron method, and the total gravitational potential and gravitation are added by each value. Different density distribution model assumptions can be given according to the geological structure, or an accurate density distribution can be inverted by other data. Then, we can accurately obtain the gravitational field of an asteroid.

## 2. Methods

### *2.1. Mascon method*

The mascon method uses the volume element (small sphere) to approximate the shape of the target asteroid, and each volume element is approximated as a mass point to calculate the gravitational field of the target asteroid.

Equations (1) and (2) show the gravitational potential and the gravitational acceleration

$$U = \sum_{i=1}^{N} \frac{GM_i}{|r - d_i|} \tag{1}$$

$$a = \sum_{i=1}^{N} \frac{GM_i (r - d_i)}{|r - d_i|^3} \tag{2}$$

Each voxel mass $M_i$ satisfies

$$M_A = \sum_{i=1}^{N} M_i \tag{3}$$

where $N$ is the number of volume elements, $d_i$ is the position vector of the $i$th volume elements, $M_i$ is the mass of the $i$th volume element, $r$ is the position vector of the calculated point, $G$ is the gravitational constant and $M_A$ is the total mass of the asteroid.

### *2.2. Polyhedron*

The polyhedral method was first applied in the field of geophysical science. In the mid-1990s, Werner and Scheeres [11], assuming the polyhedral density to be uniform, used polyhedrons to approximate the shape of asteroids and derived the expressions of the gravitational potential and the gravitation near the polyhedron.

Equations (4) and (5) show the gravitational potential and the gravitational acceleration.

$$U = \frac{1}{2} G\sigma \sum_{e \in edges} \mathbf{r}_e \bullet \mathbf{E}_e \bullet \mathbf{r}_e \cdot L_e - \frac{1}{2} G\sigma \sum_{f \in faces} \mathbf{r}_f \bullet \mathbf{F}_f \bullet \mathbf{r}_f \cdot \omega_f \tag{4}$$

$$\nabla U = -G\sigma \sum_{e \in edges} \mathbf{E}_e \bullet \mathbf{r}_e \cdot L_e + G\sigma \sum_{f \in faces} \mathbf{F}_f \bullet \mathbf{r}_f \cdot \omega_f \tag{5}$$

where $G$ is the gravitational constant, $\sigma$ is the density of the asteroid, $e$ is the edge, and $f$ is the face of the polyhedron, $\hat{n}_f$ is the normal vector of the $f$th face, $F_f$ is the coefficient related to the face, $\hat{n}_e^f$ is the normal vector of the $e$th edge on the $f$th face, and $E_e$ is the coefficient related to the edge. $F_f = \hat{n}_f \hat{n}_f$, $E_e = \hat{n}_{f_1} \hat{n}_e^{f_1} + \hat{n}_{f_2} \hat{n}_e^{f_2}$

$$L_e = \int_e \frac{1}{r} ds = \int_{P_i}^{P_j} \frac{1}{r} ds = \ln \frac{r_i + r_j + e_{ij}}{r_i + r_j - e_{ij}} \tag{6}$$

$$\omega_f = \iint_{triangle} \frac{\Delta z}{r^3} dS = 2 \arctan \frac{\mathbf{r}_i \bullet \mathbf{r}_j \times \mathbf{r}_k}{r_i r_j r_k + r_i (\mathbf{r}_j \bullet \mathbf{r}_k) + r_j (\mathbf{r}_k \bullet \mathbf{r}_i) + r_k (\mathbf{r}_i \bullet \mathbf{r}_j)} \tag{7}$$

### *2.3. Tetrahedral finite element*

The tetrahedral finite element method combines the advantages of the mascon and the homogeneous polyhedron method, filling the polyhedral model with tetrahedrals based on the surface triangles. The gravitational potential and gravitation of each tetrahedral finite element are calculated by the homogeneous polyhedron method, and the total gravitational potential and gravitation is added by each value. Therefore, the expressions of the gravitational potential and the gravitational acceleration are

$$U = \frac{1}{2} G \sum_{l=1}^{n} \sigma_l \left( \sum_{e \in edges} \mathbf{r}_e^l \cdot \mathbf{E}_e^l \cdot \mathbf{r}_e^l \cdot L_e^l - \sum_{f \in faces} \mathbf{r}_f^l \cdot \mathbf{F}_f^l \cdot \mathbf{r}_f^l \cdot \omega_f^l \right) \tag{8}$$

$$\nabla U = -G \sum_{l=1}^{n} \sigma_l \left( \sum_{e \in edges} \mathbf{E}_e^l \cdot \mathbf{r}_e^l \cdot L_e^l - \sum_{f \in faces} \mathbf{F}_f^l \cdot \mathbf{r}_f^l \cdot \omega_f^l \right) \tag{9}$$

where the coefficients are similar to the coefficients of the polyhedral method.

### 2.4. The Polyhedron-FEM model

The modeling process of the tetrahedral finite element method requires the polyhedron to be filled up by the tetrahedral finite element. Based on a given polyhedral model, a tetrahedral unit of a certain size range is distributed to the interior until the internal space of the entire polyhedron is not empty, as shown in figure 1 (left).

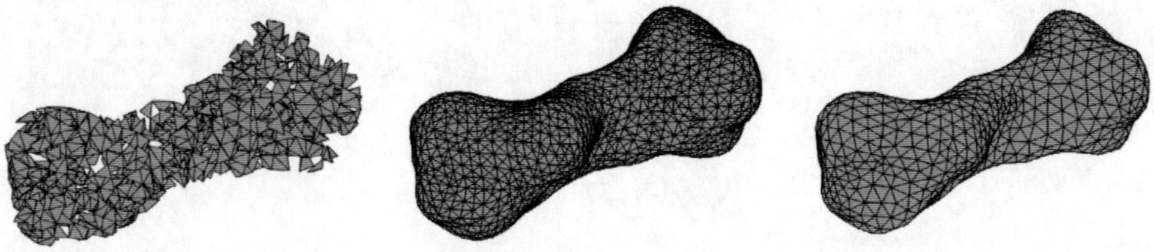

**Figure 1.** Model of tetrahedral finite element. (Left: tetrahedral distribution, middle: outer surface of original polyhedron model, right: outer surface of tetrahedral finite element model.)

However, the resulting triangular distribution of the outer surface of the polyhedron is not identical with the original polyhedral model, as shown in figure 1 (middle) and (right). Therefore, the data of the grid points and the triangle faces of the outer surface of the tetrahedral finite element model must be extracted to obtain the corresponding polyhedral model.

## 3. Discussion

In this section, the 216 Kleopatra asteroid is taken as an example for comparision with the polyhedron method to verify the effectiveness of the finite element tetrahedral method. As an example to make a reasonable density distribution hypothesis, the tetrahedral finite element can also calculate the gravitational field of asteroids accurately under the condition of nonuniform density.

The original polyhedral model consists of 2048 vertices, 4096 faces, and the dimensions of the three central inertia main axes are 217 km×94 km×81 km.

### 3.1. Case I: assumption of uniform density

First, the uniform density hypothesis is used to prove the validity of the tetrahedral finite element method. The average density is assumed to be 3.6 g/cm$^3$. The gravitational potential and gravitation are calculated at each field point from 150 km to the surface of the asteroid by the tetrahedral finite element and the polyhedron methods. As shown in figure 2, the finite element method divides three grids, with 6885, 37462, and 89029 tetrahedrons.

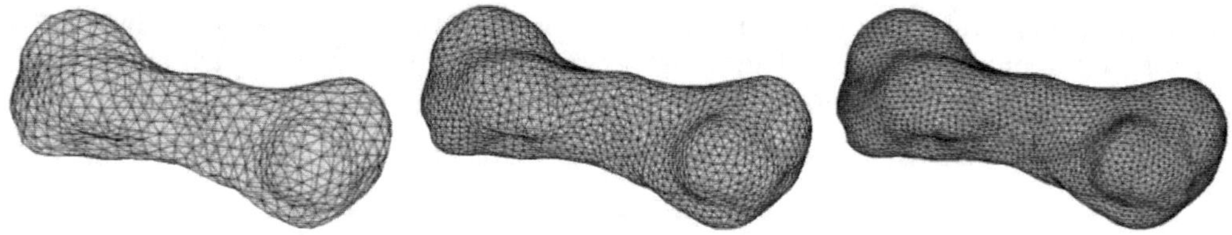

**Figure 2.** Three grids of finite element model. (Left: 6885 tetrahedrons, middle: 37462 tetrahedrons, right: 89029 tetrahedrons.)

Therefore, we need to know the accuracy of the calculation of the tetrahedral finite element.

We take the polyhedral method as the reference model. First, the accuracy between the tetrahedral finite element model under three kinds of meshes and the polyhedron model with the same outer surface (We call it finite element polyhedron below) are compared. Then the accuracy between the tetrahedral finite element model and the original polyhedron model under three kinds of grids are compared.

For the same field point, the error of the gravitational potential is defined as

$$\varepsilon = \frac{|U_t - U_p|}{|U_p|} \times 100\% \tag{10}$$

and the error of gravity is defined as

$$\nu = \frac{\|F_t - F_p\|}{\|F_p\|} \times 100\% \tag{11}$$

The direction error of the gravity is defined as

$$\delta = \langle F_p, F_t \rangle \tag{12}$$

where $U_t$ and $F_t$ are the gravitational potential and gravity calculated by the tetrahedral finite element method, respectively; $U_p$ and $F_p$ are the gravitational potential and gravity calculated by the polyhedral method, respectively; and $\langle *,* \rangle \in [0°,180°]$ is the angle between two gravitational vectors. Table 1 and table 2 present the error statistics for the two cases.

Figure 3 illustrates the contour of the gravitational potential of the x-y, x-z and y-z planes of the asteroid obtained by the finite element method of the 6885 tetrahedral elements.

**Table 1.** Gravity and gravitational potential error statistics of 216 Kleopatra calculated via the tetrahedral finite element and polyhedron models.

| | | Minimum | Maximum | Average | Standard deviation |
|---|---|---|---|---|---|
| **6885 tetrahedrons** | $\varepsilon$(%) | 0 | $2.24\times10^{-12}$ | $2.69\times10^{-13}$ | $2.31\times10^{-13}$ |
| | ν(%) | 0 | $4.90\times10^{-12}$ | $7.42\times10^{-13}$ | $5.17\times10^{-13}$ |
| | $\delta$(°) | 0 | $1.71\times10^{-6}$ | $2.05\times10^{-7}$ | $4.11\times10^{-7}$ |
| **37462 tetrahedrons** | $\varepsilon$(%) | 0 | $4.82\times10^{-12}$ | $5.27\times10^{-13}$ | $4.43\times10^{-13}$ |
| | ν(%) | 0 | $6.58\times10^{-12}$ | $1.12\times10^{-12}$ | $7.72\times10^{-13}$ |
| | $\delta$(°) | 0 | $1.71\times10^{-6}$ | $2.06\times10^{-7}$ | $4.12\times10^{-7}$ |
| **89029 tetrahedrons** | $\varepsilon$(%) | 0 | $5.85\times10^{-12}$ | $4.11\times10^{-13}$ | $4.58\times10^{-15}$ |
| | ν(%) | 0 | $1.13\times10^{-11}$ | $1.28\times10^{-12}$ | $1.03\times10^{-12}$ |
| | $\delta$(°) | 0 | $1.71\times10^{-6}$ | $2.04\times10^{-7}$ | $4.13\times10^{-7}$ |

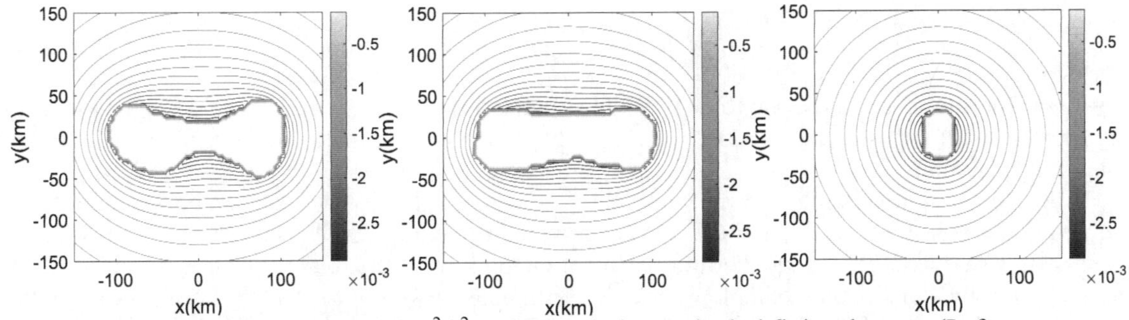


**Figure 3.** Gravitational potential ($km^2/s^2$) calculated via tetrahedral finite element. (Left: contour on x-y plane; middle: contour on y-z plane; right: contour on x-z plane)

Table 1 and figure 3 reveal that the results of the tetrahedral finite element method are nearly the same as those of the polyhedral method with the same polyhedron shape. The relative error values of gravitation and potential energy are within 10-12%, and the direction error in gravity is within 10-6° inches. We believe that the error comes from two sources. (1) Truncation error of used 64-bitcomputer: The gravitational value of each small tetrahedron is particularly small; hence, such an influence may occur. (2) Truncation error of the sum of the large and small numbers: The existing error is sufficient to prove the validity of the tetrahedral finite element method; thus, this error source is not further eliminated. In addition, the finer the mesh, the more the error accumulates; therefore, the mesh does not need to be further refined as long as the difference in gravity from the original polyhedron is appropriate.

Table 2 indicates that with the tetrahedral finite element mesh refinement, the average of the relative error of gravity and potential decreases from 1.1217% to 0.2474%, and the average of the direction error of the gravity decreases from 0.0477° to 0.0058°. This table shows that when the original polyhedral model is given, the gravitational field can be calculated by the tetrahedral finite element method. As the tetrahedral mesh becomes finer, the result becomes closer to the gravitational field of the original polyhedron.

**Table 2.** Gravity and gravitational potential error statistics of 216 Kleopatra calculated via tetrahedral finite element and original polyhedron models.

| | | Minimum | Maximum | Average | Standard deviation |
|---|---|---|---|---|---|
| **6885 tetrahedrons** | $\varepsilon$(%) | 0 | 1.5259 | 1.1217 | 0.1807 |
| | ν(%) | 0 | 3.7216 | 1.1082 | 0.2243 |
| | $\delta$(°) | 0 | 1.2591 | 0.0477 | 0.0544 |
| **37462 tetrahedrons** | $\varepsilon$(%) | 0 | 0.5482 | 0.4088 | 0.0648 |
| | ν(%) | 0 | 1.4306 | 0.4060 | 0.0776 |
| | $\delta$(°) | 0 | 0.3554 | 0.0141 | 0.0165 |
| **89029 tetrahedrons** | $\varepsilon$(%) | 0 | 0.2979 | 0.2489 | 0.0388 |
| | ν(%) | 0 | 0.6409 | 0.2474 | 0.04376 |
| | $\delta$(°) | 0 | 0.1805 | 0.0058 | 0.0088 |

### *3.2. Case II: assumption of density distribution model*

This study assumes four density distributions that may exist in the asteroids. Their gravity field is calculated by the tetrahedral finite element method to illustrate the influence of different density distributions on the external gravitational field.

The mass of the asteroid 216 Kleopatra is determined to be $4.64 \times 10^{18}$ kg, and the estimated average density given above is 3.6 $g/cm^3$. This condition is utilized as a constraint to generate a reasonable assumption of the density distribution.

The constraint of the total mass is

$$M = \rho_0 V_0 + \sum_{i=1}^{N} (\rho_i - \rho_0) V_i \tag{13}$$

where $M$ is the total mass of the asteroid, $V_0$ is the volume of the total asteroid shape, $\rho_0$ is the assumed nominal density of the asteroid, $V_i$ is the volume of the $i$th internal addition, and $\rho_i$ is the density of the $i$th internal addition.

216 Kleopatra is an M-type asteroid; thus, the part with the largest density can be assumed to be iron with a density of 7.5 g/cm$^3$, and the rest are assumed to be uniform density, which is constrained by mass (this study aims to verify the finite element method; thus, the total mass does not necessarily need to be strictly equal). And a simple way to determine the density of the tetrahedron is based on the centroid coordinates of the tetrahedron. The following four assumptions are made about the internal density distribution of 216 Kleopatra, as shown in figure 4.

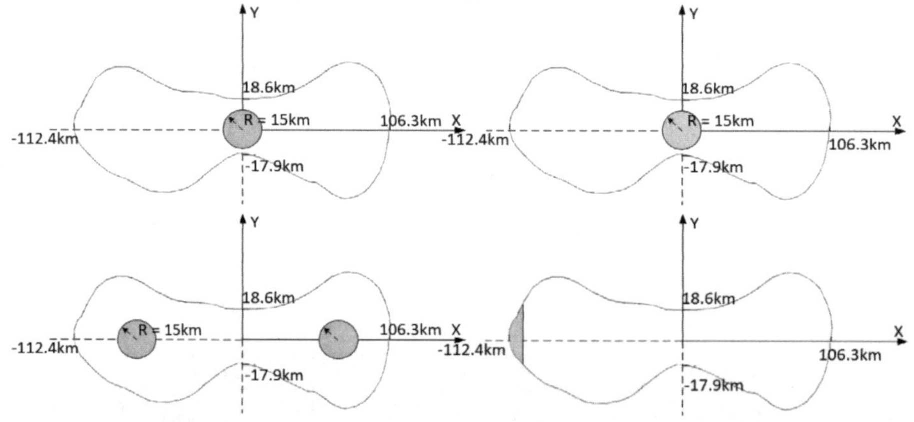


**Figure 4.** Single core model (top-left), air core model (top-right), binary core model (bottom-left), heavy mass model (bottom-right).

*3.2.1. Single core model.* Assume that the asteroid has a "nucleus" composed of iron with a density of 7.5 g/cm$^3$. As shown in the yellow part in figure 4 (top-left), the range of the core is assumed to be a sphere with a radius of 15 km, and the density of other portions is estimated to be 3.5 g/cm$^3$. Therefore, we can obtain the contour map of the gravitational potential on the equatorial plane in figure 5 and percentage contour map of the relative error compared with the normal density in figure 6. The figures show that the part with the largest relative error is approximately 4% near the core, whereas the error away from the core is within 0.5%, since the core density is significantly higher than the original density, which has a greater influence on the gravitational potential around it; however, the density concentration area in the core has a reduced influence on its surroundings.

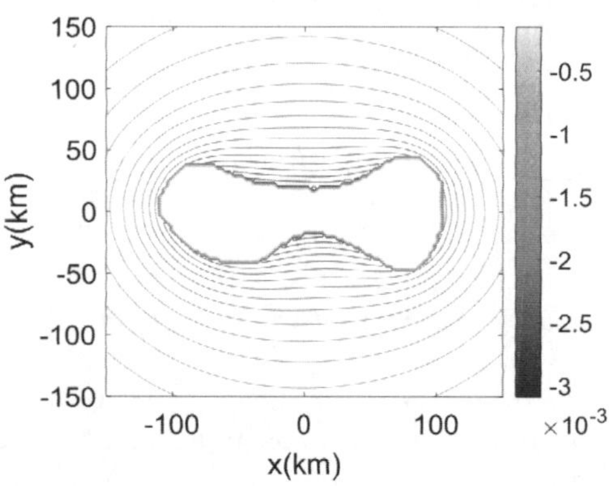


**Figure 5.** Potential contour map of single core model.

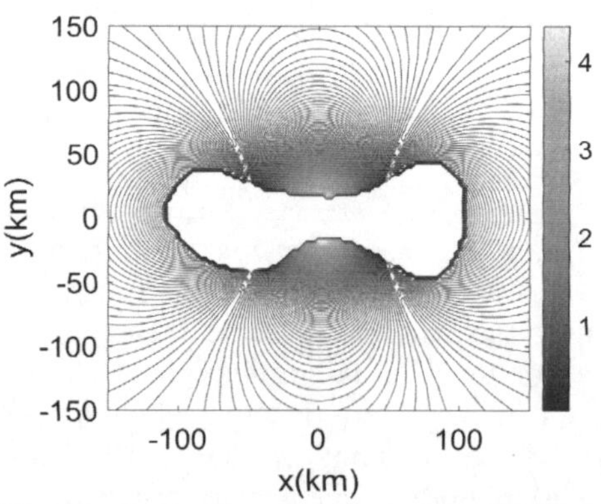


**Figure 6.** Error contour map of single core model.

*3.2.2. Air core model.* Assume that there is an air core exists inside the asteroid with a density of 0 g/cm$^3$, and the range of the air core is also assumed to be a sphere with a radius of 15 km. Then, the density of the other parts is estimated to be 3.7 g/cm$^3$, as shown in the gray part in figure 4 (top-right).Therefore, we can obtain the contour map of the gravitational potential on the equatorial plane in figure 7 and the percentage contour map of the relative error compared with the normal density in figure 8. It can be seen from the figures that the part with the largest relative error is near the air core approximately 3% and at both ends at about 2%, and the error away from the core is within 0.5%.It can be understood that the air core greatly influences the gravitational field near itself and has a certain influence overall. Unlike the mass under uniform density, the mass of the two ends accounts for a larger proportion of the total mass; thus, the difference in uniform density will also increase.

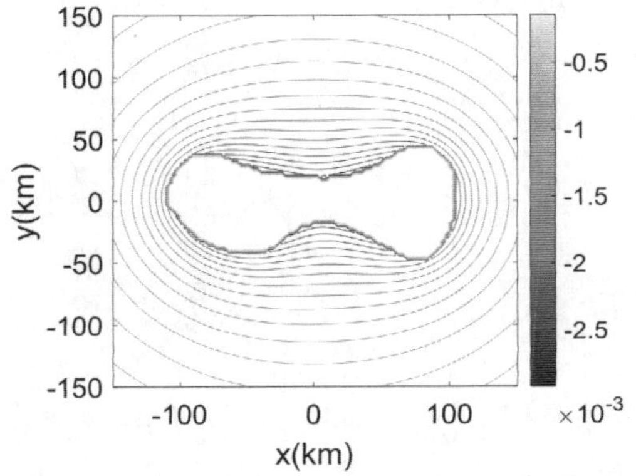


**Figure 7.** Potential contour map of air core model.

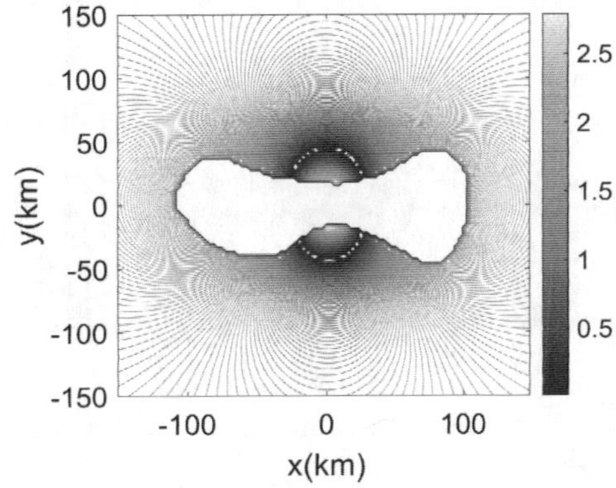


**Figure 8.** Error contour map of air core model.

*3.2.3. Binary core model.* Assume that the asteroid has two cores of iron on both sides, and its density is 7.5 g/cm$^3$. The range of each core is assumed to be a sphere with a radius of 15 km. As shown in the yellow part in figure 4 (bottom-left), the density of the other parts is estimated to be 3.5 g/cm$^3$.Therefore, we can obtain the contour map of the gravitational potential on the equatorial plane in figure 9 and the percentage contour map of the relative error compared with the normal density in figure 10 .The figures show that the part with the largest relative error is approximately 2.5% at both ends, and it can increase the error on the other parts. It can be understood that the "binary core" at both ends will not only have a great impact on the gravitational field near itself but will also have a certain impact overall. Nevertheless, the influence near the nucleus is relatively weaker than that of the single core, because part of the influence moves to other structures.

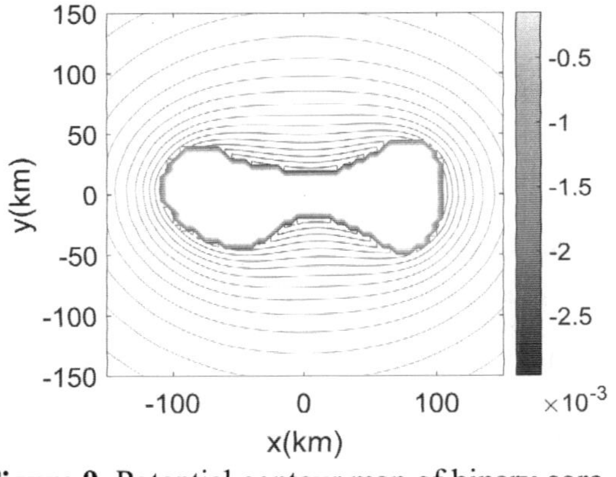


**Figure 9.** Potential contour map of binary core model.

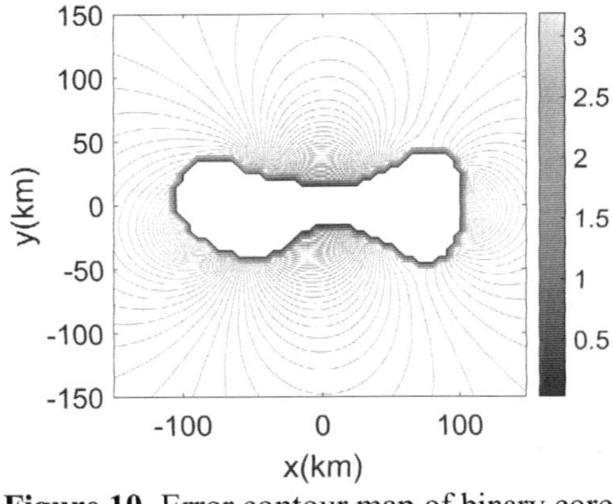


**Figure 10.** Error contour map of binary core model.

*3.2.4. Heavy mass model.* Assume that a heavy mass composed of iron exists on the surface of the asteroid, the density is 7.5 g/cm$^3$, and the heavy mass is located outside the −100km of the x-axis of the asteroid body coordinate system, as shown in the yellow part in figure 4 (bottom-right). The density of the other part is estimated to be 3.5 g/cm3. The contour map of the gravitational potential on its equatorial plane and the percentage contour map of the relative error compared with the normal density are obtained as shown in figures 11 and 12. The figures show that the part with the largest relative error of about 8% is near the heavy mass of the asteroid, whereas the relative error in other places is approximately 1%.

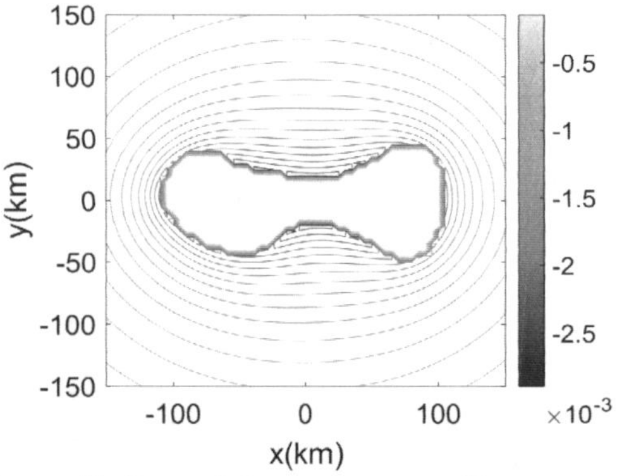


**Figure 11.** Potential contour map of heavy mass model.

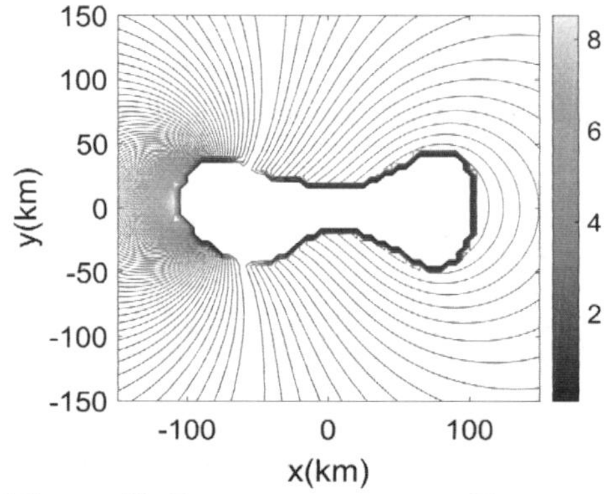


**Figure 12.** Error contour map of heavy mass model.

A large mass exists near the north pole of Ryugu, an asteroid being probed by JAXA [12]. It can be understood that when a heavy mass exists on the surface of the asteroid, the gravitational field near the mass will be highly different from the gravitational field with the uniform density. However, the mass only has a large effect on its vicinity. If the polyhedral method was used to calculate its gravitational field, it can only be assumed that its density is uniform. While the probe is approaching this asteroid, the true gravitational field will be different from the assumed gravitational field near this large mass. A mutation, which is highly dangerous for the probe, would occur in this field.

Therefore, if the density of the mass can be obtained, the tetrahedral finite element method can be used to obtain an accurate gravitational field of the Ryugu.

## 4. Conclusion

A tetrahedral finite element method is proposed to compute the gravity of irregularly shaped asteroids. Under the condition of constant density distribution, the tetrahedral finite element can be regarded as a refinement of the polyhedral method, and its essence is to calculate the gravitational field of the same polyhedral model. Therefore, in this case, the tetrahedral finite element method has the same advantages as the polyhedral method, but the calculation amount is slightly larger. The tetrahedral finite element model is unique advantageous when the density distribution is not uniform, because it can accurately calculate the external gravitational field; whereas the polyhedral model is not applicable, and the mascon and spherical harmonic methods have many limitations. The tetrahedral finite element model can also reflect the density distribution of the asteroid, which cannot be achieved by other methods.

When the density distribution inside the asteroid is not uniform, the part with the higher density will have a great influence on the gravitational field near itself, and it may also have a certain influence on the gravitational field of the whole asteroid. In particular, when the surface of an asteroid has a heavy mass, the gravitational field must only be accurately estimated by the tetrahedral finite element method.

**Acknowledgments**

This work was supported by National Natural Science Foundation of China through cooperative agreement No.11603057.

**References**

[1] Yu Y and BaoYin HX 2014 Review of Orbital Dynamics in the Vicinity of Solar System Small Celestial Bodies *J. Journal of Deep Space Exploration.* **1** 93-104

[2] JAXA (Japan Aerospace Exploration Agency). Issue 2019, available at http://www.hayabusa2.jaxa.jp/en/

[3] NASA (National Aeronautics and Space Administration). Issue 2018, available at https://www.asteroidmission.org/

[4] NASA (National Aeronautics and Space Administration). Issue 2019, available at http://pluto.jhuapl.edu/News-Center/News-Article.php?page=20190101

[5] Liu L 2000 *Orbit Throry of Spacecraft* (Beijing: National Defense Industry Press) pp 109-113

[6] Romain G and Jean-Pierre B 2001 Ellipsoidal harmonic expansions of the gravitational potential: Theory and application *J. Celestial Mechanics and Dynamical Astronomy.* **79** 235-275

[7] Takahashi Yu and Scheeres D J 2013 Surface gravity fields for asteroids and comets *J. Journal of guidance, control, and dynamics.* **36** 1825-44

[8] Takahashi Yu and Scheeres D J 2014 Small body surface gravity fields via spherical harmonic expansions *J. Celestial Mechanics and Dynamical Astronomy.* **119** 169-206

[9] Li J F, BaoYin H X and Jiang F H 2014 *Dynamic and Control of Interplanetary Flight* (Beijing: Tsinghua University Press) pp 308-310

[10] Xiao Y, Ruan X G and Wei R Y 2016 Research on the accuracy and operation time of polyhedron gravity model base on 433 Eros *J. Journal of Deep Space Exploration.* **3** 41–46

[11] Werner R A and Scheeres D J 1997 Exterior gravitation of a polyhedron derived and compared with harmonic and mascon gravitation representations of asteroid 4769 Castalia Celestial Mechanics and Dynamical Astronomy *J. Celestial Mechanics and Dynamical Astronomy.* **65** 313-344

[12] JAXA (Japan Aerospace Exploration Agency). Issue 2019, available at http://www.hayabusa2.jaxa.jp/topics/20180711je/index_e.html

# Research of High Resolution ISAR Algorithm Based on Coherent Registration and Fusion for Distributed Radar

**Xi Luo[1,2], Lixin Guo[2], Dawei Song[1], She Shang[1] and Xiaojun Li[1]**

[1]National Key Laboratory of Science and Technology on Space Microwave, Xi'an, 710010, China
[2]Xidian University, Xi'an, 710071, China

matrix02061016@126.com; lxguo@xidian.edu.cn; songdw0210@163.com; sir_shang@163.com; xjl@163.com

**Abstract.** According to the Inverse Synthetic Aperture Radar (ISAR), this thesis analyses the attributed to the physical restriction that the ISAR image resolution cannot be improved, therefore, a high resolution ISAR algorithm for distributed radar system is proposed. Depended on existing hardware development level, it combines multi-radar echo signals for one target and then utilize data fusion theory for multi-visual angle and multi-frequency band to obtain ultra-wide band and larger coherent accumulation angle, on this account, the much higher resolution 2D-image than the single radar is able to acquired. The main content illustrates the method for high resolution ISAR on target electromagnetic scattering two-exponential sum model, and the simulation is made to demonstrate the analysis is effective and validity.

## 1. Introduction

Since 2006, the Distributed Radar ISAR technology attracts dramatically attention, it is represented by More Input and More Output (MIMO) radar and Synthetic Impulse and Aperture Radar (SIAR)[1][2]. The principle of Distributed ISAR indicates that one target is emitted by a set of radar, where the observed information using width and angle relative to radar is utilized to realize associated imaging. In this way, the signal width and shifting angle is extended extremely to improving more two-dimension high resolution than single one[3][4]. In this paper, a two-dimension signal fusion ISAR on distributed radar method under the low SNR is proposed. What needs to be pointed out is that the method will organically combine distributed structure and data fusion theory in some way.

## 2. Geometric and Signal Model

Generally, there are three kinds of radar echo model , and we usually used including ideal scattering point model, index sum model and GTD model. Ideal scattering point model consider scattering intensity is fixed. But the GTD model take into account that the kind of scattering centre and position information[5]. In sum, the index sum model is applied to the followings, because the advantages about this method contain not only the simply and quickly mathematical operating, but also great matching when the observed width and angle is little needed.

### *2.1. One-Dimensional Model*

On radar observed frequency and filed, target can be decomposed into several discrete scattering centre group, so we regard radar back scattering as some kinds of these group's coherent composition. Thus, the total target frequency response consists of every scattering centre frequency[6][7].

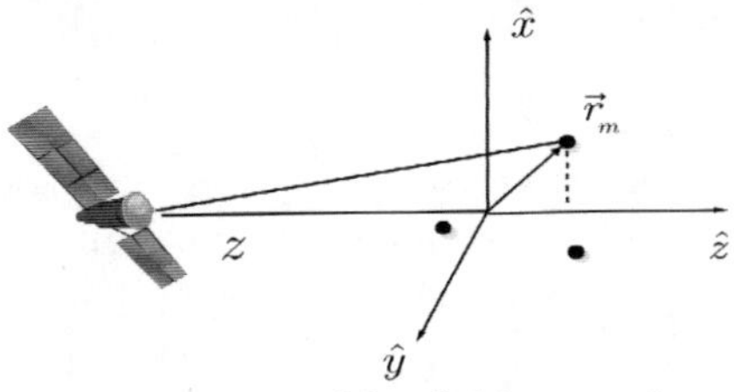


**Figure 1.** Geometry of far field scattering centre.

As shown in Figure 1, the proposed incident wave spreads along the z axis in positive direction, and the incident electrical field in $\vec{r}$ can be presented:

$$E(f)=\sum_{m=1}^{M}\phi_{m,f}e^{-j\frac{4\pi f}{c}\hat{z}\bullet\vec{r}_m} \tag{1}$$

The amplitude $\phi_{m,f}$ of scattering centre is affected by the change of frequency $(jf)^{\alpha}\leq$ from the GTD theory, and there $\alpha$ is model geometry type, hence, $\phi_{m,f}\partial$ can be represent as:

$$\phi_{m,f}=A_m(j\frac{f}{f_c})^{\alpha_m} \tag{2}$$

From equation (1) and equation (2), the following result is obtained:

$$E(f)=\sum_{m=1}^{M}A_m(j\frac{f}{f_c})^{\alpha_m}e^{-j\frac{4\pi f}{c}r_m} \tag{3}$$

Where $r_m=\hat{z}\cdot\vec{r}_m$, which is the target electromagnetic scattering model based on GTD theory. Model parameter $\{A_m,r_m,\alpha_m\}_{m=1}^{M}$ corresponds amplitude $A_m$, and the number of scattering center M.

The GTD frequency response model is established by the property of target electromagnetic scattering process in optics region, and it is difficult to solve the model parameters[8]. However, exponent sign $\beta_m^f$ can be substituted for exponential function $f^{\alpha_m}$ when the observed relative bandwidth is not wide enough. It is obtained that:

$$\begin{aligned}E(f)&\approx\sum_{m=1}^{M}A_m(j\frac{1}{f_c})^{\alpha_m}\beta_m^f e^{-j\frac{4\pi f}{c}r_m}\\&=\sum_{m=1}^{M}A_m^{'}\beta_m^f e^{-j\frac{4\pi f}{c}r_m}\end{aligned} \tag{4}$$

Where $A_m^{'}=A_m(j\frac{f}{f_c})^{\alpha_m}$.

For stepped frequency radar :

$$f=f_0+k\Delta f,\qquad k=0,\cdots,N-1 \tag{5}$$

Here $f_0$ the origins of stepped frequency, $\Delta f$ is stepped-frequency interval, N is step number, from equation (5) we can get:

$$\begin{aligned}E(f_0+k\Delta f)&=\sum_{m=1}^{M}A_m^{'}\beta_m^{f_0+k\Delta f}e^{-j\frac{4\pi(f_0+k\Delta f)}{c}r_m}\\&=\sum_{m=1}^{M}A_m^{'}\beta_m^{f_0}e^{-j\frac{4\pi f_0}{c}r_m}\beta_m^{k\Delta f}e^{-j\frac{4\pi k\Delta f}{c}r_m}\end{aligned} \tag{6}$$

If the target response $y(k)$ in frequency $f_0+k\Delta f$ is expressed by $E(f_0+k\Delta f)$, the equation (6) can be shown as:

$$y(k)=\sum_{m=1}^{M}d_m p_m^k,\qquad k=0,\cdots,N-1 \tag{7}$$

equation (7) is the one-dimension index sum model of target response. Model parameters $\{d_m, p_m\}_{m=1}^{M}$ correspond amplitude $d_m$, pole $p_m$, and model order M. Index sum model is applied to frequency response means make use of N frequency data to estimate parameter $\{d_m, p_m\}_{m=1}^{M}$.

### *2.2. Two-Dimension Index Sum Model*

the target's back scattering field can be seen as scattering of several individual centres. A single centre back scattering filed can be described as two-dimensional function containing frequency $f$ and angle $\theta$, thus the field will be presented by scattering synthesis of these points.

$$E(f,\varphi)=\sum_{m=1}^{M}\sigma_m \exp[-j\frac{4\pi f}{c}(x_m\cos\varphi+y_m\sin\varphi)] \tag{8}$$

Here, $c$ is light velocity, $x_m$ and $y_m$ is point two-dimensional coordinates. $\sigma_m$ is complex amplitude, the number of scattering centre is M. Only considering the frequency signal in a certain observed angle, the equation (8) is represented as:

$$E(f)=\sum_{m=1}^{M}\sigma_m \exp[-j\frac{4\pi f}{c}R_m] \tag{9}$$

Here, $R_m = x_m\cos\varphi + y_m\sin\varphi$, $c$ is light velocity.

From equation (8) it can indicate that under the certain observed angle it is correctly to get distance image with FFT transformation. But it is not correctly to get azimuth image, because phase is not a linear function for angle leading to image defocused with FFT transformation. Therefore, the phase decoupling is needed, and the polar coordinate transformation is:

$$\begin{aligned} u &= f\cos\varphi \\ v &= f\sin\varphi \end{aligned} \tag{10}$$

So equation (8) can be represented as:

$$E(u,v)=\sum_{m=1}^{M}\sigma_m \exp[-j\frac{4\pi}{c}(x_m u+y_m v)] \tag{11}$$

$$E(u,v)=\sum_{m=1}^{M}\sigma_m P_{xm}^{u} P_{ym}^{v} \tag{12}$$

In that, the estimation about $\sigma_m$, $P_{xm}$, $P_{ym}$ is applied to get echo data of any point in $(u,v)$ domain. Here $P_{xm}=\exp(-j\frac{4\pi x_m}{c})$, $P_{ym}=\exp(-j\frac{4\pi y_m}{c})$.

For sake of the radar data sampled uniform, the result transformed into $f$ and $\varphi$ domain is non-uniform sampling, so the bilinear interpolation method is adopted to imaging. Searching a nearby four points in $(f,\varphi)$ domain structure function:

$$\begin{aligned} a_1 f_1 + a_2\varphi_1 + a_3 f_1\varphi_1 + a_4 &= E(f_1,\varphi_1) \\ a_1 f_2 + a_2\varphi_2 + a_3 f_2\varphi_2 + a_4 &= E(f_2,\varphi_2) \\ a_1 f_3 + a_2\varphi_3 + a_3 f_3\varphi_3 + a_4 &= E(f_3,\varphi_3) \\ a_1 f_4 + a_2\varphi_4 + a_3 f_4\varphi_4 + a_4 &= E(f_4,\varphi_4) \end{aligned} \tag{13}$$

And the vector is:

$$\begin{bmatrix} f_1 & \varphi_1 & f_1\varphi_1 & 1 \\ f_2 & \varphi_2 & f_2\varphi_2 & 1 \\ f_3 & \varphi_3 & f_3\varphi_3 & 1 \\ f_4 & \varphi_4 & f_4\varphi_4 & 1 \end{bmatrix}\begin{bmatrix} a_1 \\ a_2 \\ a_3 \\ a_4 \end{bmatrix}=\begin{bmatrix} E(f_1,\varphi_1) \\ E(f_2,\varphi_2) \\ E(f_3,\varphi_3) \\ E(f_4,\varphi_4) \end{bmatrix} \tag{14}$$

It is easy to obtain $[a_1, a_2, a_3, a_4]$ from equation (14) and estimate the interpolation point value in order to extract echo data.

## 3. Multi-Radar and Multi-view ISAR High Resolution Imaging

Pointing at bistatic radar, on the one hand if the observed position is different as shown in Figure 2, there will be two radar scanning curved surface in the space, on the other hand, if two radar observe under small angle, the curved surface will be mostly flat. Generally, the scanning surface is not co-plane, namely different imaging surface.

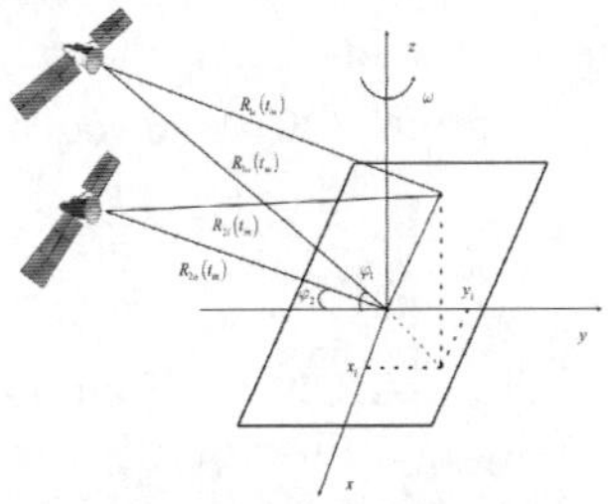

**Figure 2.** Geometry of two radar ISAR imaging.

Given objective real rotation axis and radar sight line remains unchanged during the whole imaging, thus the two radar scanning will form two conical surface around the axis of rotation. Usually obtaining the same under the view point of coherent accumulation time is more than the single imaging time, therefore, in such a short period of time actual rotation make approximation to the rotation is unreasonable. If the actual rotation axis is equal with two radar line of sight angle, so the radar scanning plane on the same cone surface as shown in Figure 3. Particularly, if the actual rotation axis is perpendicular to radar line of sight, both two radar scanning surface will be in $k_x - k_y$ surface which means coplanar imaging.

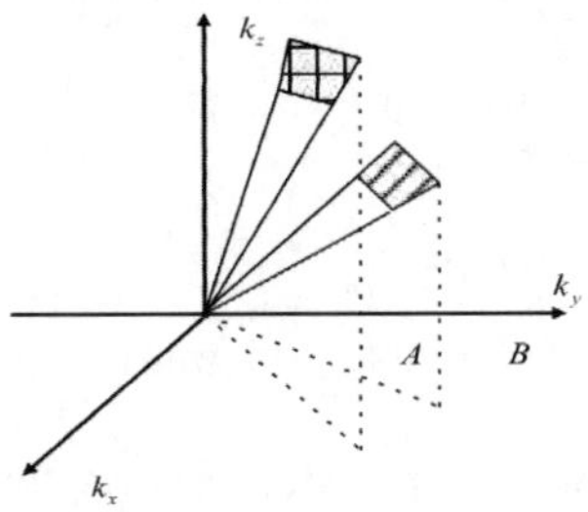


**Figure 3.** Bistatic radar observed scanning surface.

From the GTD scattering field theory, in the region of the high frequency radar, target backscatter can be treated as a scattering field of the coherent superposition of multiple independent scattering centre. When there is a relatively smaller bandwidth and observed angle, the scattering field can be expressed as:

$$E(f,\varphi) = \sum_{m=1}^{M} \sigma_m \exp[-j\frac{4\pi f}{c}(x_m \cos\varphi + y_m \sin\varphi)] \tag{15}$$

Here, $c$ is light velocity, $x_m$ and $y_m$ is the Two-Dimensional coordinates of the scattering points in the target coordinates. $\sigma_m$ is complex amplitude, and M is the number of scattering centre.

$$u = f\cos\varphi \\ v = f\sin\varphi \tag{16}$$

So, equation (15) can be written as:

$$E(u,v) = \sum_{m=1}^{M} \sigma_m \exp[-j\frac{4\pi}{c}(x_m u + y_m v)] \tag{17}$$

$$u = u_0 + \delta_u m, \quad m \in \{0,\dots,N-1\} \\ v = v_0 + \delta_v n, \quad n \in \{0,\dots,N-1\} \tag{18}$$

And then equation (18) is able to be simplified into Two-Dimensional all-pole model:

$$E(n_1,n_2) = \sum_{m=1}^{M} c_m \sigma_m \exp(-j\frac{4\pi}{c}(x_m \delta_u n_1 + y_m \delta_v n_2)) \\ = \sum_{m=1}^{M} a_m p_m^{n_1} q_m^{n_2} \tag{19}$$

$$p_m = \exp(-j\frac{4\pi}{c}x_m\delta_u) \tag{20}$$

$$q_m = \exp(-j\frac{4\pi}{c}y_m\delta_v) \tag{21}$$

Here , $p_m$ and $q_m$ in equation (19) is correspondingly indicate $u$ and $v$ directional pole, and $a_m$ means model amplitude coefficient. Equation (20) and equation (21) represent the $p_m$ and $q_m$ specifically.

## 4. Simulations

In order to validate the effectiveness of the ISAR high resolution in the multiple radar and multiple angle with Two-Dimensional fusion technology, First of all, some point targets are used to deal with simulation, and among them five scattering points is (-0.8,-0.8), (-0.8,0.8), (0.8,0.8), (0.8,-0.8), (1.6,0) and all of amplitude is 1, where the scattering points are shown in Figure 4. Working in different perspectives and different frequency bands of two radar parameters are shown in the following Table 1.

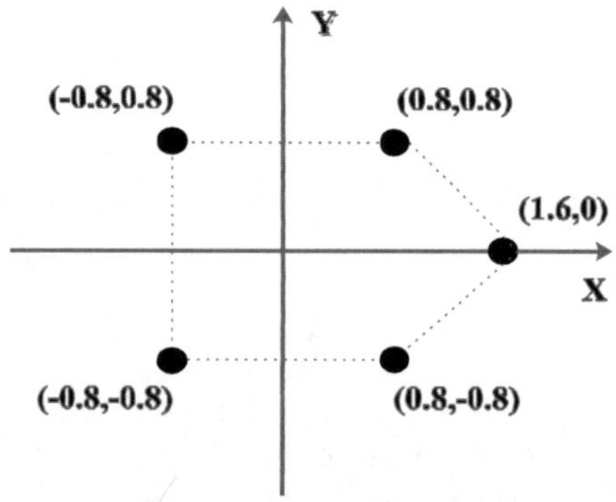


**Figure 4.** The schematic diagram of point target.

**Table 1.** Simulation Parameters.

| | Frequency Bandwidth | Frequency Interval | Angle | Angle Interval | SNR |
|---|---|---|---|---|---|
| **Radar1** | 5.9~6.0 GHz | 5MHz | -2°~0° | 0.1° | 13dB |
| **Radar2** | 6.0~6.1 GHz | 5MHz | 0°~2° | 0.1° | 13dB |
| **Radar3** | 5.9~6.1 GHz | 5MHz | -2°~2° | 0.1° | 13dB |

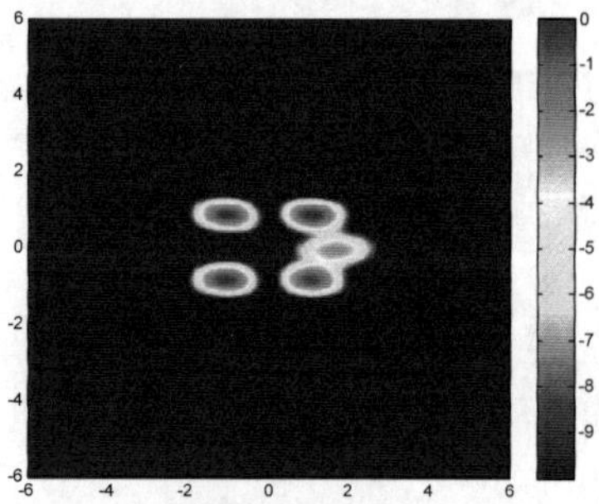


**Figure 5.** Radar 1 ISAR Image

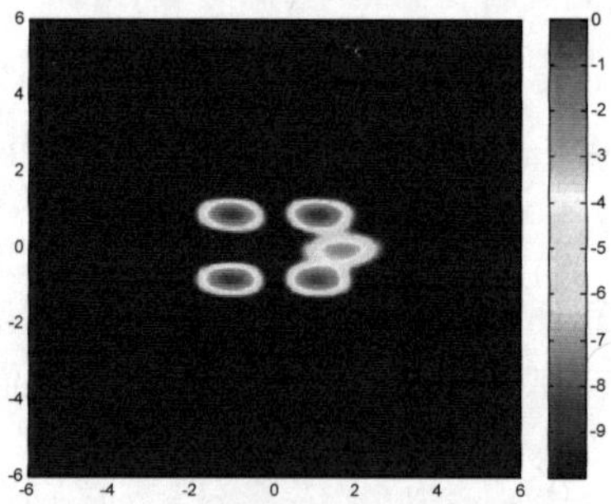


**Figure 6.** Radar 2 ISAR Image

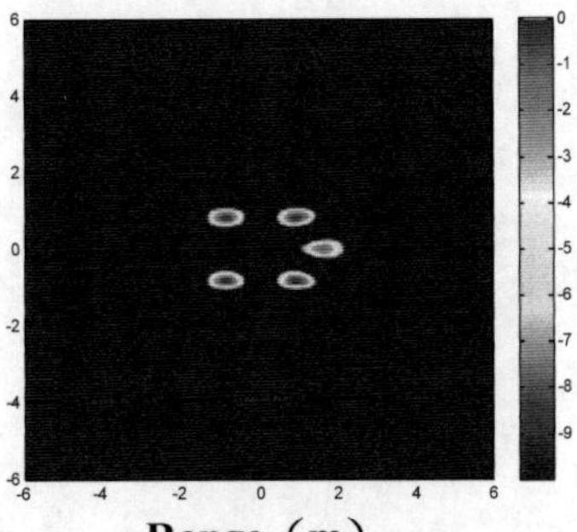


**Figure 7**. Full Spectrum Image.

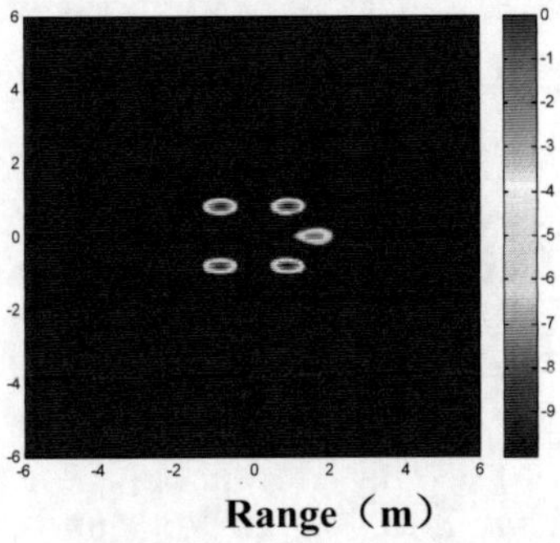


**Figure 8.** Fusion Image.

Figure 5~Figure 8 present the result in different method, whereby, Figure 7 shows the imaging result for the full spectrum domain signal in FFT, and then Figure 8 shows the result for Two-Dimensional fusion of the multiple radar and multiple angles. As a result, five scattering points are clearly visible, thus, it proves that both the imaging result in two method is roughly same. Figure 5 and Figure 6 are the imaging result for radar 1 and radar 2. Caused by a lack of resolution, there are three scattering points shown in the figure cannot be separated.

## 5. Conclusion

This paper devotes to studying the radar signal fusion of multiple bands and multiple sights high ISAR imaging technology. Firstly, the paper elaborates the electromagnetic scattering model, as well as the linear frequency modulation signal pre-processing and construct Two-Dimension index sum model. And then a correction method is proposed to extract model on the basis of the index sum model leading to improving the efficiency and stability pole estimation dramatically. In the end, the coefficient of model is achieved with least squares solution. The method in this paper is so useful to extrapolate or interpolation spectrum data for obtaining effectiveness of super resolution and high quality of ISAR image. The simulation demonstrates the method is effective.

## References

[1] Bao Zheng, Xing MengDao, Wang Tong. "Radar Imaging Technology", Electronic Industry Press, Beijing, China, (2008).

[2] Mancill C E. "Enhancement of radar imagery by maximum entropy processing", Proceedings of the 21st annual meeting of Human Factors Society, Santa, Monica, CA, pp. 241-243, (1977).

[3] Liu ChunBo, Chen BaiXiao, Zhang ShouHong. "Co-Channal Interference Suppression by Time and Range Adaptive Processing in Bistatic HF Surface Wave SIAR", IET Radar Sonar and Navigation, 3(6), pp. 638-645, (2015).

[4] Y. B. Hua. "High Resolution Imaging of Continuously Moving Target Using Steeped Frequency Radar", Signal Processing, 35(1), pp. 33-40, (2018)..

[5] Mayhan J, Burrows M L, Cuomo K M, et al. "High resolution 3D "Snapshot" ISAR imaging and Feature extraction", Aerospace and Electronic System, 37(2), pp. 630-641, (2017).

[6] Wang Jing, Zhou JianJiang, Wang Fei. "2D Scattering centers Extraction of Objects Based on GTD Model", Journal of Electronics & Information Technology, 31(4), pp. 958-962, (2016).

[7] Ye Fan, He Feng, Liang DianNong. "Multi-angular and Multi-band Fusion of Narrowband Signal Based on GTD Model", Signal Processing, 25(11), pp. 1675-1679, (2018).

[8] Chang P S and Willson A N. "Analysis of conjugate gradient algorithms for adaptive filtering", Signal Processing, 48(2), pp. 409-418, (2016).

# Deep Neural Networks Based Real-time Optimal Control for Lunar Landing

**Lingchao Zhu[1, 2], Jian Ma[1, 2] and Shuquan Wang[1,*]**

[1]Technology and Engineering Center for Space Utilization, Chinese Academy of Sciences, Beijing 100094, China
[2]University of Chinese Academy of Sciences, Beijing 100049, China

[*]Corresponding author: shuquan.wang@csu.ac.cn

**Abstract.** Recent research on deep learning control, a new control algorithm based on machine learning able to learn deep architectures, has shown excellent performance on robots and drones. With the development of intelligent control like deep learning and reinforcement learning, accuracy, real-time, adaptability, robustness and autonomy of control algorithm have been achieved by the intelligent controls. Traditional control methods have difficulties to achieve nice performance in complex situations. Deep learning offers powerful algorithms to real-time search near-optimal controllers of lunar landing spacecraft with nonlinear dynamics. In terms of lunar landing control system, deep architectures offer the possibility to get an approximate solution of co-state equation without time-consuming iterative process. Furthermore, real-time optimal thrust during lunar landing may be derived directly through deep neural networks. As a single infrastructure for machine learning in both production and research, TensorFlow is chosen for training the deep artificial neural networks in this paper. Numerical simulations demonstrate the effectiveness of deep neural networks. The results of deep neural networks based optimal control are contrasted with traditional optimal algorithm, whose main idea is to track the pre-designed optimal trajectory by ground station. This research provides an effective approach to cope with the lunar landing problem.

## 1. Introduction

The Moon is an astronomical body that orbits planet Earth and is Earth's only permanent natural satellite. Soft landing on large planetary bodies is a critical step in space exploration that enables future manned moon landing, lunar lander detection, and lunar base construction. The rise of strong willingness for sending humans back to the Moon within the next decade has also caused a lot of research boom. In particular, soft landing technology will continue to evolve to meet the demand for increasingly accurate and complex space exploration missions. Guidance and control are two independent systems for the mainstream landing system currently, wherein the guidance system designs an optimal trajectory by solving the Hamilton-Jacobi-Belmann (HJB) equations and the two point boundary value problem [1], and the goal of the control system is to control the spacecraft to track the optimal reference trajectory accurately. The improvement of direct and indirect methods [2] is the main research hotspot of optimal control problems (OCPs), the homotopy approach [3] could increase the probability of indirect methods in solving the optimal solutions. Additionally, a robust and precise controll techniques such as PID algorithm, sliding-mode control [4] and terminal sliding-mode control [5] are still needed to track the reference trajectory if the optimal trajectory is solved. Thanks to Moore 's Law that computing power has been boosted by several orders of magnitude and the cost has been greatly reduced over the years, it is possible to train a deep neural networks for real-time optimal control. Deep learning [6] has shown great excellent performance in feature extraction and model fitting in recent years. Its recognition rate in traditional recognition tasks has a significant increase [7].

For control systems with high dimensional data input, the introduction of deep neural networks (DNN) has a natural advantage. DNNs are artificial neural networks with an input layer, several hidden layers and an output layer, which can map the nonlinear relationship between input and output correctly [8]. Recently, some studies have focused on the application of deep learning in the field of aerospace control include two-dimensional moon optimal landings [9], two-dimensional quadcopter fuel optimal control, two-dimensional optimal orbit transfers, and asteroid irregular gravitational filed estimation using deep neural networks. A finite-time attitude tracking control scheme for spacecraft using Chebyshev neural network was proposed as the earliest research on introducing neural networks into spacecraft control [10], An RBF neural network was used to adaptively compensate for the plant uncertainties or estimate unknown nonlinearities in an adaptive output feedback control scheme [11,12], Stefan used neural reinforcement learning to control a spacecraft around a small celestial body with unknown gravity field [13], Dario proposed an early research of deep artificial neural networks to represent the optimal guidance profile of an interplanetary mission [14]. A deep Recurrent Neural Network and Long-Short Term Memory (RNN-LSTM) architecture was designed for predicting the fuel-optimal thrust from sequence of spacecraft's states during a powered planetary descent phase [15].

In this paper, the landing scenario is researched assuming overall information on the landing spacecraft state. Due to the assumptions considered in the models, a real-time fuel optimal control to the moon soft landing is proposed by utilizing the indirect methods with homotopy approach and training the feed-forward DNNs architectures directly in a supervised manner on a large amount of optimal trajectory data. The trained DNNs are suitable for the on-board generation of optimal control signal with no need solve OCPs on-board.

The paper is structured as follows: in Section II we introduce the dynamic model of spacecraft and mathematical form of the fuel OCP for lunar soft landing problems, then we give the definition of the fuel optimal control policy to be learned by the DNNs. In the following Section III, the two-point boundary value problem (TPBVP) is derived and solved by means of single shooting and continuation techniques. The training data was filtered before saved and normalized for DNNs. In the following section IV the network architectures and training procedures used to approximate the optimal solutions are trained with Tensorflow. A numerical simulation scenario based on MATLAB is proposed and the results are compared with the optimal trajectories. This paper is concluded in Section V.

## 2. Spacecraft dynamics and problem formulation

In this section, we consider a Lunar landing problem, in which the state comprises three components: the spacecraft dynamical state position and velocity, and mass. The control signal will have two component, one is the control thrust amplitude and one is the thrust direction . The equations of motion are the following:

$$\dot{\boldsymbol{r}} = \boldsymbol{v}, \quad \dot{\boldsymbol{v}} = c_1 \frac{u_1}{m} \hat{\boldsymbol{i}}_\theta - \mu_m \frac{\boldsymbol{r}}{r^3}, \quad \dot{m} = -c_2 u_1 \tag{1}$$

Here, $\boldsymbol{r}$ and $\boldsymbol{v}$ are position and velocity vector of spacecraft in Lunar inertial coordinate system. $m$ is the mass of the landing spacecraft, $\mu_m = 4.901793455 \times 10^{12} \, m^3/s^2$ is gravitational constant of Lunar. The constant $c_2 = c_1 / I_{sp} g_0$ represents the rocket engine efficiency in terms of its impulse $I_{sp} = 311s$ and $g_0 = 9.81 m/s^2$. The control $u_1 \in [0,1]$ models a thrust amplitude applied along the direction $\hat{\boldsymbol{i}}_\theta$. For fuel optimization problem, $u_1$ is thrust switch sequence that $u_1 = 0$ means engine off and $u_1 = 1$ means engine on with the maximum thrust magnitude $c_1$.

Where These equations have been integrated to minimize the following cost function:

$$J = (1-\alpha)\int_0^{t_f} \gamma_1 c_1 c_2 u_1^2 dt + \alpha \int_0^{t_f} c_2 u_1 dt \tag{2}$$

where $\gamma_1 = 1[1/N]$ is a trade-off coefficient for adjusting cost function between the two contributions. The parameter $\alpha \in [0,1]$, defines a continuation between a fuel optimal control problem (MOC) $\alpha = 1$ and a quadratic optimal control problem (QC) $\alpha = 0$. Consider the following Hamiltonian to solve the optimal control problem:

$$H = \boldsymbol{\lambda}_r \boldsymbol{v} + \boldsymbol{\lambda}_v \left( c_1 \frac{u_1}{m} \hat{\boldsymbol{i}}_\theta - \mu_m \frac{\boldsymbol{r}}{r^3} \right) + \lambda_m c_2 u_1 + (1-\alpha)\gamma_1 c_1 c_2 u_1^2 + \alpha c_2 u_1 \tag{3}$$

where we introduce several auxiliary co-state functions $\boldsymbol{\lambda}_r(t)$, $\boldsymbol{\lambda}_v(t)$ and $\lambda_m(t)$. From the maximum principle, the thrust direction must be in the opposite direction of $\boldsymbol{\lambda}_v$ for Hamiltonian function to be minimized, writing as follows:

$$\hat{\boldsymbol{i}}_\theta^* = -\boldsymbol{\lambda}_v / \lambda_v \tag{4}$$

where $\lambda_v = \|\boldsymbol{\lambda}_v\|_2$. If $\alpha \neq 1$, the optimal $u_1^*$ has the following finial expression:

$$u_1^* = \min\left( \max\left( \frac{c_1 \lambda_v / c_2 m + \lambda_m - \alpha}{2c_1(1-\alpha)}, 0 \right), 1 \right) \tag{5}$$

Note that in the corner case $\alpha = 1$, which correspond to a mass optimal control in this paper, the above expression results to be singular, it is necessary to rewrite $u_1^*$ by introducing the switching function as follows:

$$u_1^* = \begin{cases} 0 & S_1 < 0 \\ 1 & S_1 > 0 \end{cases} \tag{6}$$

where $S_1 = c_1 \lambda_v / c_2 m + \lambda_m - \alpha$ is the switching function for mass optimal control problem. The co-states must be a solution to the set of differential equations that are written using the Hamiltonian formalism as:

$$\begin{cases} \dot{\boldsymbol{\lambda}}_r = \mu_m \dfrac{\boldsymbol{\lambda}_v}{r^3} - 3\mu_m \dfrac{\boldsymbol{\lambda}_v \cdot \boldsymbol{r}}{r^5} \\ \dot{\boldsymbol{\lambda}}_v = -\boldsymbol{\lambda}_r \\ \dot{\lambda}_m = -\dfrac{c_1}{m^2} u_1 \lambda_v \end{cases} \tag{7}$$

Eventually, the following two points boundary value problem (TPBVP) is obtained by bring above equations into the dynamics model:

$$\dot{\boldsymbol{r}} = \boldsymbol{v}, \quad \dot{\boldsymbol{v}} = c_1 \frac{u_1^*}{m} \hat{\boldsymbol{i}}_\theta^* - \mu_m \frac{\boldsymbol{r}}{r^3}, \quad \dot{m} = -c_2 u_1^* \tag{8}$$

with boundary conditions $\boldsymbol{r}_0$, $\boldsymbol{v}_0$, $m_0$ at $t = 0$ and $\boldsymbol{r}_t = \boldsymbol{R}_{land}$, $\boldsymbol{v}_t = \boldsymbol{0}$ and $\lambda_{mt} = 0$ at $t = t_f$, where $\boldsymbol{R}_{land}$ is the position of the landing site.

## 3. Data generation

A fuel optimal trajectories data is obtained via continuation over parameter $\alpha$ through homotopy approach. Each optimal trajectory consists of a list of pairs $(\boldsymbol{x}^*, \boldsymbol{u}^*)$ where $\boldsymbol{x}^* = [\Delta\boldsymbol{r}, \Delta\boldsymbol{v}, m]$ is the position error, velocity error and the mass of spacecraft, $\boldsymbol{u}^*$ is the corresponding optimal thrust switch

IOP Conf. Series: Materials Science and Engineering **608** (2019) 012045 doi:10.1088/1757-899X/608/1/012045

sequence and its direction. For Lunar landing fuel-optimum control problem, to find the MOC solution, the corresponding quadratic control problem ($\alpha = 0$) is solved first and then a homotopy path is followed by continuously increasing $\alpha$ from 0 to 1, resulting in smooth changes of the control action. An initialization area $\Re$ is defined to draw the initial conditions from $x_0 \in \Re$. The initialization position area of $\Re$ is a cuboid whose three-axis length are [400,400,1000] meters, The initialization velocity area of $\Re$ is a cuboid whose three-axis length are [60,60,60] m/s. 80,000 optimal trajectories are generated, from each fuel-optimum trajectory 200 state-control pairs are uniformly selected along the trajectory and inserted in the training data.

It is worth noting that a good initial guess can greatly reduce the iterative solution time of the value of the co-state $\lambda_0$ and optimal control time $t_f$, thus it is worthy training an independent deep neural network with $\lambda_0$ and $t_f$ as the output of DNN and initial state $\boldsymbol{x}_0$ as input. An initial value $\lambda_0$ and $t_f$ close to the optimal solution can be given directly through a well-trained DNN, this DNN will greatly reduce the time for data generation. Limited by the length of the paper, this part will only shows a brief results. The DNN for co-state estimation is trained by a feedforward, fully-connected neural network with 5 hidden layers, 32 units/layer, Tanh activation function, linear output layer and MSE loss function, The results of the MSE after 10000 epoch is less than $1.6 \times 10^{-5}$. It could satisfactorily map the nonlinear relationship between initial spacecraft's state $\boldsymbol{x}_0$ and initial co-state $\lambda_0$.

After we generate the data containing 16,000,000 optimal state action pairs, the distribution of position error in the data are uneven as illustrated by **Figure 1**, The rest of the data is also unevenly distributed. This is because the data is too concentrated near the landing site. To accelerate the training process of deep neural networks, data pruning can smooth data distribution. It should be noted that the data input and output are simultaneously pruned, the principle of data pruning has two aspects. The first is to pruning the trajectory far away from most trajectories, that is, to remove the position and velocity error of the corresponding trajectory. The other is to perform random resections in places where a large number of data are concentrated at the end of the landing.

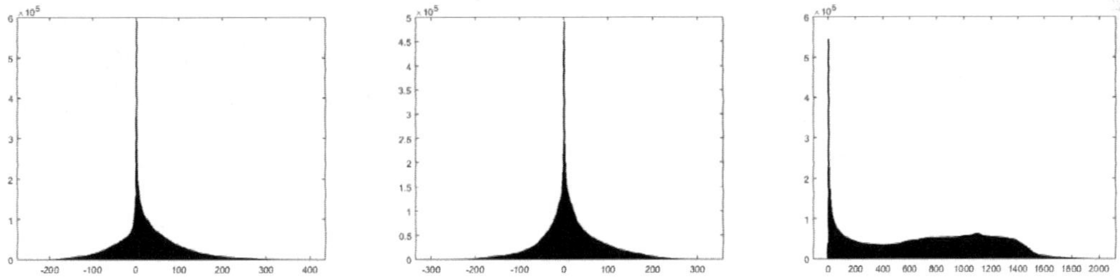

**Figure 1.** Position error distribution of raw data

The distribution of position error in the data after data pruning is illustrated in **Figure 2**. Normalize the data after pruning, then the finial state-action pairs are ready for DNNs to learn.

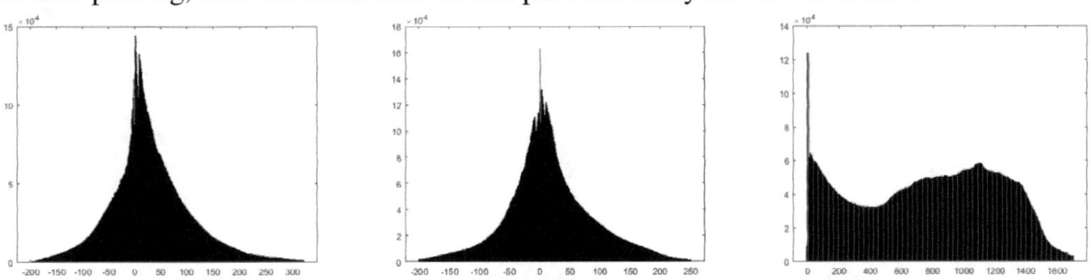

**Figure 2**. Position error distribution after data pruning

## 4. DNN training and simulation results

The Based on the fuel-optimum data generated by the indirect method with homotopy approach, DNN are used to learn the nonlinear relationship between the real-time state and optimal actions. Architectures with different numbers of layers and neurons per layer are considered. The control actions comprises two components, one is the magnitude of the thrust and the second one to the direction in which the thrust is applied. For the thrust magnitude, there are only two states (on for $u_1 = 1$ and off for $u_1 = 0$) as it has a bang-bang control action profile, Accordingly, the deep neural network architecture should be designed as a classification neural network. Moreover, the thrust direction is smooth and continuous. Consequently, the network shall be designed as a regression network to learn the optimal thrust direction.

The selection of activation function is one of the most important factors of DNNs, there are several choices such as ReLUs, Tanh, sigmoid, ELU, Leaky ReLU, PReLU etc., after repeated iteration simulations, we conclude that a proper architecture can give quite accurate results, the architectures of classification neural network and regression network are can be found in **Table 1**. The accuracy of classification deep neural network can reach 99.55%, test on data can at most reach a precision of 99.82%. The MSE of regression deep neural network can be as low as $6\times10^{-4}$ after 5000 epochs.

**Table 1** Selected hyper-parameters of classification and regression network

| | layers -units | activation function | Output layer | Loss function | Batch size | Learning rate | Optimizer |
|---|---|---|---|---|---|---|---|
| classification | 6-30 | ReLU | ReLU | Cross-entropy | 10000 | 0.001 | Adam |
| regression | 4-12 | tansig | Linear | MSE | 1000 | 0.001 | SGD |

The DNN-driven trajectories are evaluated as described in this section to demonstrate the performance. The landing site $\boldsymbol{R}_{land}$ is chosen as the North Pole of the moon, the expected velocity on the landing site is zero for a soft landing demand. The simulation parameters are given in section 2. Landing simulations are shown in **Figure 3**.

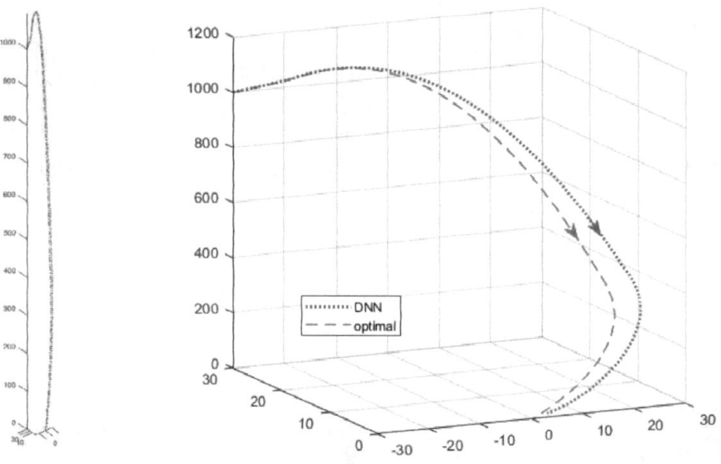


**Figure 3**. The optimal and DNN-driven trajectories

**Figure 3** shows the optimal and DNN-driven trajectories, the left subgraph displays the trajectories with equal ratio of axis, the right subgraph displays the magnified view. As we can see from the two subgraphs, the approximate optimal solutions obtained by the trained DNNs are very close to the optimal trajectory. We further quantitatively analyze the DNN-driven trajectory and find that the trajectory deviates from the optimal path with a maximum error of 5.963m. Therefore, a multi-scale DNN driven and nonlinear control cooperation strategy is proposed for lunar landing. **Figure 4** show the position and velocity error where the blue line is the DNN driven phase and the red line is the nonlinear control, the switching strategy is to switch to a non-linear control phase when the spacecraft reaches five meters above the landing site. **Figure 5** displays the switch sequence and its direction of trust, the statistical results of terminal fight errors is illustrated by **Table 2**.

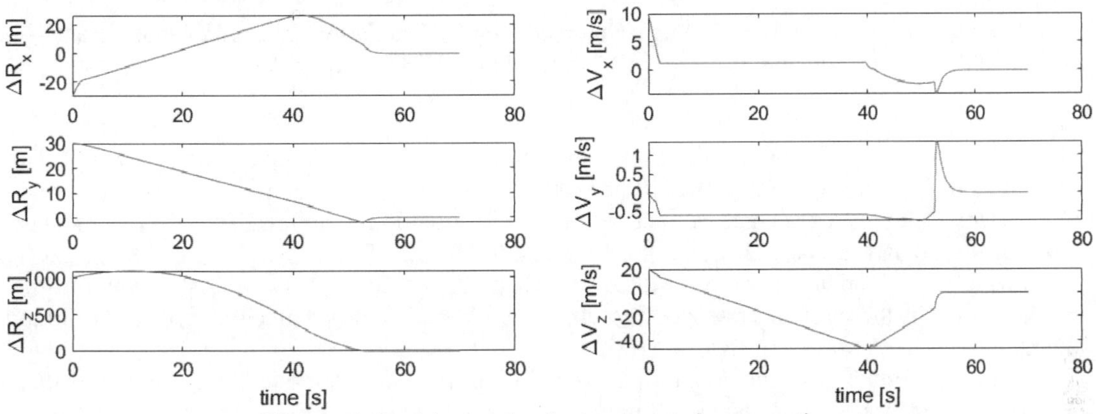


**Figure 4.** Position and velocity error of cooperation strategy

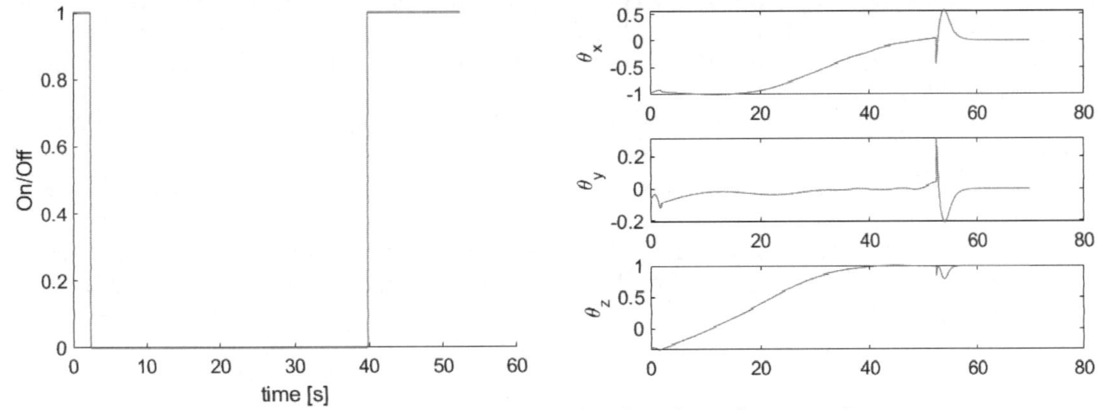


**Figure 5**. Switch sequence and its direction of cooperation strategy

**Table 2**. Position and velocity error of cooperation strategy

| | | | |
|---|---|---|---|
| Position error(m) | $1.732\times10^{-7}$ | $-6.303\times10^{-8}$ | $-4.657\times10^{-9}$ |
| Velocity error(m/s) | $-1.733\times10^{-7}$ | $6.38\times10^{-8}$ | $4.66\times10^{-9}$ |

As we can see from above simulation results that the DNN-based landing controller cooperated with nonlinear control is robust and achieves an almost fuel-optimum trajectory. The proposed DNN-driven controller can replace the time-consuming homogeneous polyhedral method and significantly improve the computational efficiency.

## 5. Conclusions

In this study, deep neural network technology is introduced to improve the computational efficiency for the fuel-optimum lunar landing problem. Generally, fuel-optimum problem can be found only for open-loop trajectories, closed-loop guidance algorithms are not available. By taking advantage of classification and regression neural networks with excellent generalization ability executed on Tensorflow, a real-time multi-scale DNN-driven and nonlinear control cooperation strategy is proposed for the landing problem. This DNN-driven optimal controller is trained off-line and outputs the real-time optimal control actions in the on-line application stage. Terminal landing accuracy is has been greatly improved by introducing a control switching strategy. Simulation results of fuel-optimum lunar landing problems are given to substantiate the effectiveness of real-time performance and illustrate the reliability of the multi-scale DNN driven and nonlinear control cooperation control strategy.

**Acknowledgement**

This research was supported by the National Natural Science Foundation of China (under grant No. 11672294) and the Advanced Research Project China's Manned Space Program.

**References**

[1] Todorov, E., "Optimality principles in sensorimotor control," Nature neuroscience, Vol. 7, No. 9, 2004, pp. 907–915, doi:DOI:10.1038/nn1309.
[2] J. T. Betts, "Survey of numerical methods for trajectory optimization," Journal of guidance, control, and dynamics, vol. 21, no. 2, pp. 193–207, 1998.
[3] C. Zhang, F. Topputo, F. Bernelli-Zazzera, and Y.-S. Zhao, "Low-thrust minimum-fuel optimization in the circular restricted three-body problem," Journal of Guidance, Control, and Dynamics, vol. 38, no. 8, pp. 1501–1510, 2015.
[4] Xiao, Bing, Qinglei Hu, and Youmin Zhang. "Adaptive sliding mode fault tolerant attitude tracking control for flexible spacecraft under actuator saturation." IEEE Transactions on Control Systems Technology 20.6 (2012): 1605-1612.
[5] Lee, Daero, and George Vukovich. "Adaptive finite-time control for spacecraft hovering over an asteroid." IEEE Transactions on Aerospace and Electronic Systems 52.3 (2016): 1183-1196.
[6] Schmidhuber, Jürgen. "Deep learning in neural networks: An overview." Neural networks 61 (2015): 85-117.
[7] Hunt, K. Jetal, et al. "Neural networks for control systems—a survey." Automatica 28.6 (1992): 1083-1112.
[8] Liu, Weibo, et al. "A survey of deep neural network architectures and their applications." Neurocomputing 234 (2017): 11-26.
[9] Sánchez-Sánchez, Carlos, and Dario Izzo. "Real-time optimal control via Deep Neural Networks: study on landing problems." Journal of Guidance, Control, and Dynamics 41.5 (2018): 1122-1135.
[10] Zou, An-Min, et al. "Finite-time attitude tracking control for spacecraft using terminal sliding mode and Chebyshev neural network." IEEE Transactions on Systems, Man, and Cybernetics, Part B (Cybernetics) 41.4 (2011): 950-963.
[11] Seshagiri, Sridhar, and Hassan K. Khalil. "Output feedback control of nonlinear systems using RBF neural networks." IEEE Transactions on Neural Networks 11.1 (2000): 69-79.
[12] Li, Yahui, et al. "Robust and adaptive backstepping control for nonlinear systems using RBF neural networks." IEEE Transactions on Neural Networks 15.3 (2004): 693-701.
[13] Willis, Stefan, Dario Izzo, and Daniel Hennes. "Reinforcement learning for spacecraft maneuvering near small bodies." AAS/AIAA Space Flight Mechanics Meeting. 2016.
[14] Izzo, Dario, Christopher Sprague, and Dharmesh Tailor. "Machine learning and evolutionary techniques in interplanetary trajectory design." arXiv preprint arXiv:1802.00180 (2018).
[15] Furfaro, Roberto, et al. "A Recurrent Deep Architecture for Quasi-Optimal Feedback Guidance in Planetary Landing." IAA SciTech Forum on Space Flight Mechanics and Space Structures and Materials. 2018.

IOP Conf. Series: Materials Science and Engineering **608** (2019) 012046 doi:10.1088/1757-899X/608/1/012046

# Numerical Simulation of Cooling System for Solar Simulator

**C HE[1], M ZHANG[2], Y ZHOU[1], G LI[1], W J DING[1], Z F GU[1], Y ZHANG[1] and A LI[1]**

[1] Beijing Institute of Spacecraft Environment Engineering, 100094 Beijing, China
[2] National Space Science Center, Chinese Academy of Sciences, 100190 Beijing, China

hechao85112@126.com

**Abstract.** The cooling system for the solar simulator was introduced, which contained wind-cooling system, water-cooling system, and refrigeration system. Then the numerical simulation of the wind-cooling system was numerical simulated. The solar simulator with 13 xenon lamp was modeled, and the flow and the heat transfer in the solar hood was simulated by the FLUENT. The flow field in the solar hood was studied, especially the flow near the xenon lamp and the collector. The result showed that the velocity of the most area in the solar hood was slow, but the area near the xenon lamp, because the flow area suddenly decreases, the velocity of the gas increased significantly, which is conducive to the heat exchange between the xenon lamp and the gas. The temperature of the xenon lamp was lower than 328 K, and the velocity on the surface of the xenon lamp was higher than 17 m/s, which met the operating requirement of the xenon lamp, and the temperature of the collector was lower than 308 K, and meet the operating requirement of the collector.

## 1. Introduction

Two methods has been applied to simulate the space heat flux during the thermal vacuum test: infrared radiation simulation method and solar simulation method [1-3]. In the solar simulation method, the energy radiated on the spacecraft from the sun is simulated by the solar simulator, which doesn't introduce any hypothesis, so the test results are reliable, and it is suitable for satellites with complex shape or large variations in surface optical properties.

Cooling system is a main subsystem of the solar simulator, to cooling the solar lamps and other optical modules, including the cathode and anode of the solar lamps, collectors, solar hood, water cooling plate, multiple lens units, etc.

## 2. System composition

The cooling system contains three subsystem: wind-cooling system, water-cooling system, and refrigeration system.

### *2.1. Wind-cooling system*

Wind-cooling system is applied to cool the modules in the solar hood and the multiple lens units, including solar lamps and collectors. The flowchart is shown in Figure 1, it contains fan, heat exchanger, filters, valves and pipes, etc. The GN2 (Nitrogen gas) exchange heat with the cool water in the heat exchanger, the cooled GN2 is send to the solar hood and multiple lens units by the fan to cool

IOP Conf. Series: Materials Science and Engineering **608** (2019) 012046 doi:10.1088/1757-899X/608/1/012046

the optical modules, the high temperature GN2 flow into the heat exchanger again to form the cycle. Two filters are applied to ensure the cleanliness of the GN2.

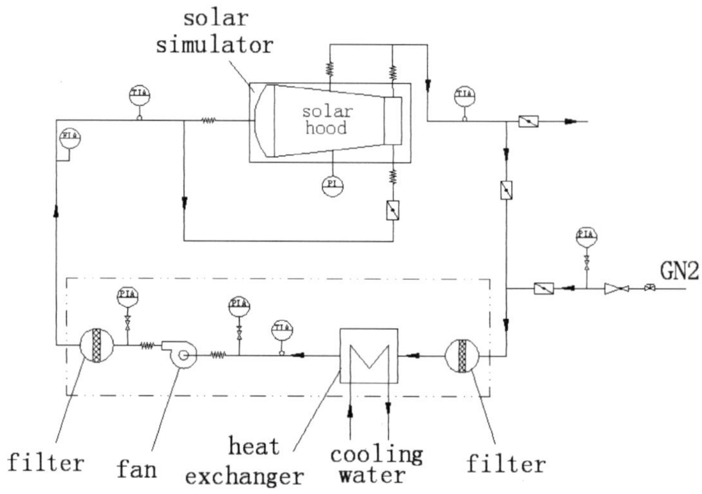


**Figure 1.** Flowchart of the wind-cooling system.

### *2.2. Water-cooling system*

Water-cooling system is applied to cool the solar hood, water cooling plate, and multiple lens units. As shown in Figure 2, the water cooling system is a closed cycle, and contains pump, heat exchanger, filter, tank, valves and pipes, etc. The pump can provide the power of the cycle, and the heat exchanger can cool the water in the system, and the tank can stabilify the pressure and supply water when turn on the pump.

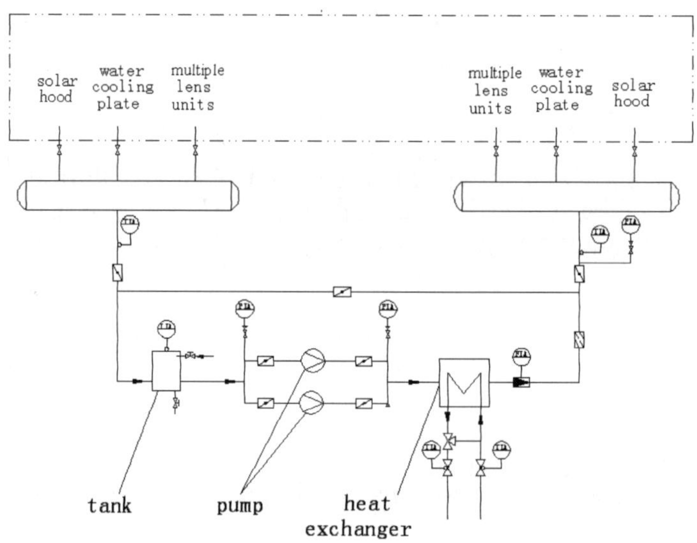


**Figure 2.** Flowchart of the water-cooling system.

### *2.3. Refrigeration system*

Refrigeration system contains pumps, refrigeration module, tanks, valves and pipes, as shown in the Figure 3, which can provide cold water to the heat exchangers of the wind-cooling system and water-cooling system to take away the heat from the solar simulator.

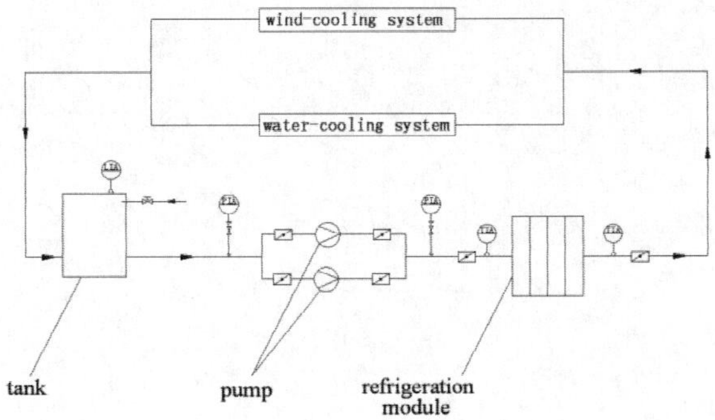


**Figure 3.** Flowchart of the refrigeration system

## 3. Modeling

The wind cooling system not only need take away all the heat of the xenon lamps, but also need ensure the cooling effect of every xenon lamp, especially the temperature and velocity around every xenon lamp. As the requirement of the solar simulator, the temperature of the xenon lamp should be lower than 503 K, and the velocity around xenon lamp should be faster than 17 m/s, and the temperature of the collector should be lower than 323 K. So the cooling effect should be numerical simulated, and the medium is GN2, the flow rate of the GN2 is 11000 $m^3/h$, the thermal load of the xenon lamps is 45 kw, and the thermal load of the collectors is 5 kw.

### *3.1. Mathematical modeling*

The GN2 of the wind cooling system is restricted by the laws of conservation of mass, momentum and energy [4, 5].

$$\frac{\partial \rho}{\partial t}+\frac{\partial}{\partial x_i}(\rho u_i)=0 \tag{1}$$

$$\frac{\partial}{\partial t}(\rho u_i)+\frac{\partial}{\partial x_j}(\rho u_i u_j)=\frac{\partial}{\partial x_j}\left(-\rho\overline{u_i' u_j'}\right)-\frac{\partial p}{\partial x_i}+\frac{\partial}{\partial x_j}\left[\mu\left(\frac{\partial u_i}{\partial x_j}+\frac{\partial u_j}{\partial x_i}-\frac{2}{3}\delta_{ij}\frac{\partial u_l}{\partial x_l}\right)\right] \tag{2}$$

$$\frac{\partial}{\partial t}(\rho E)+\frac{\partial}{\partial x_i}\left[u_i(\rho E+p)\right]=\frac{\partial}{\partial x_j}\left[\left(\lambda+\frac{c_p\mu}{\Pr}\right)\frac{\partial T}{\partial x_j}\right]+\frac{\partial}{\partial x_j}\left[u_i(\mu+\mu_t)\left(\frac{\partial u_i}{\partial x_j}+\frac{\partial u_j}{\partial x_i}-\frac{2}{3}\delta_{ij}\frac{\partial u_l}{\partial x_l}\right)\right] \tag{3}$$

### *3.2. Mesh of the solar simulator*

The solar simulator contains 13 xenon lamps is analysed as the model, and the mesh is shown in Figure 4 and Figure 5.

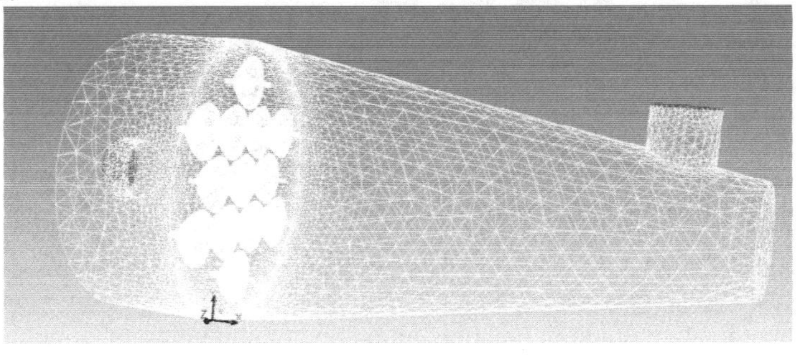

**Figure 4.** The mesh of the solar hood.

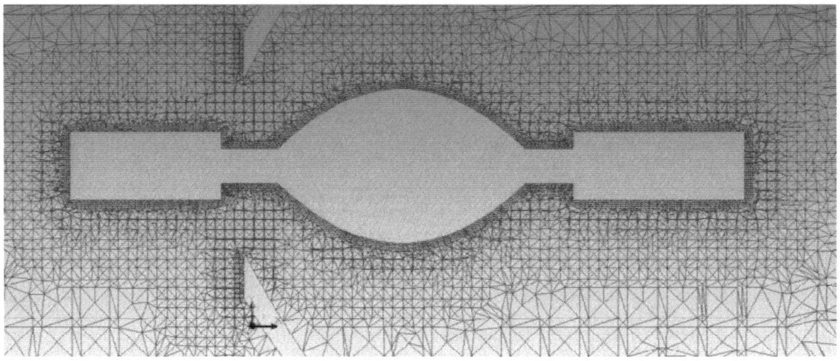

**Figure 5.** The mesh around the xenon lamp.

## 4. The result of the calculation

The flow in the solar hood is simulated by the FLUENT software [6, 7], the distribution of the velocity in the solar hood and around the xenon lamp are shown in Figure 6 and Figure 7. Most velocity in the solar hood is very slow, but because the area of the channel suddenly becomes smaller, the velocity near the xenon lamp increases significantly. The increase of the velocity can enhance the convective heat transfer on the surface of the xenon lamp, which is helpful to carry away the heat of the xenon lamp and the collector.

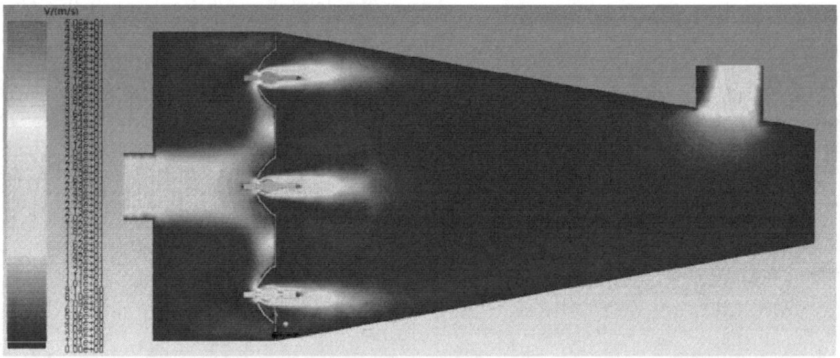

**Figure 6.** The distribution of velocity in the solar hood.

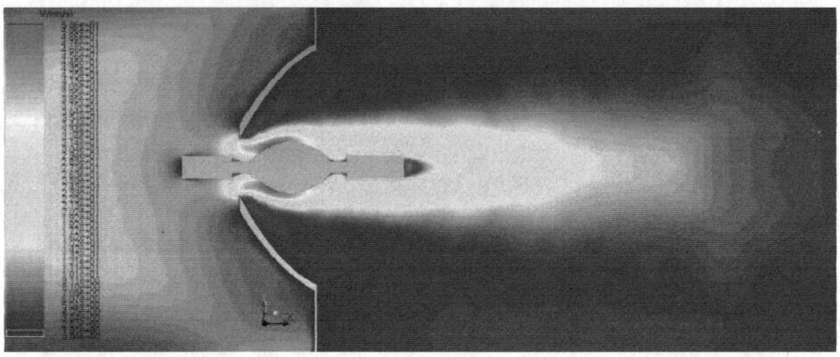

**Figure 7.** The distribution of velocity around the xenon lamp.

The velocity along the xenon lamp is shown in Figure 8. Because of the shape of the xenon lamp, the front velocity and the back velocity on the surface of the xenon lamp is slow, but most velocity on the surface of the xenon lamp is higher than 17 m/s.

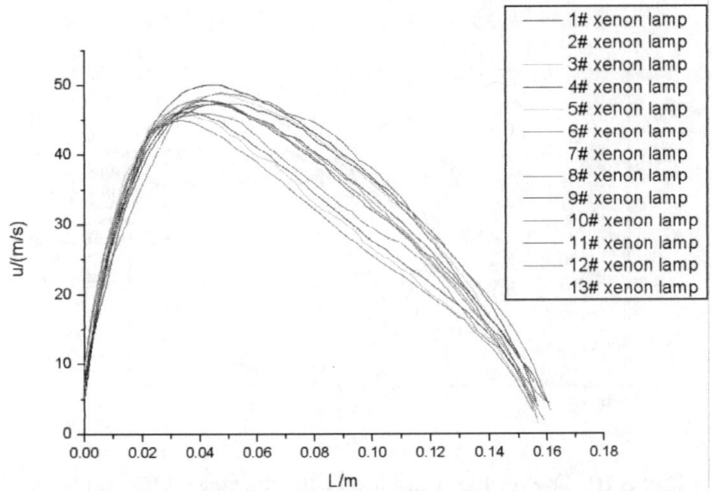


**Figure 8.** The distribution of the velocity on the surface of the xenon lamp

The distribution of the temperature on the surface of the xenon lamp and collector is shown in Figure 9. The heat flux density of the xenon is high. The velocity near the xenon lamp is fast, so the convective heat transfer coefficient on the surface of the xenon lamp is large, which is conducive to the heat exchange between the xenon lamp and the gas.

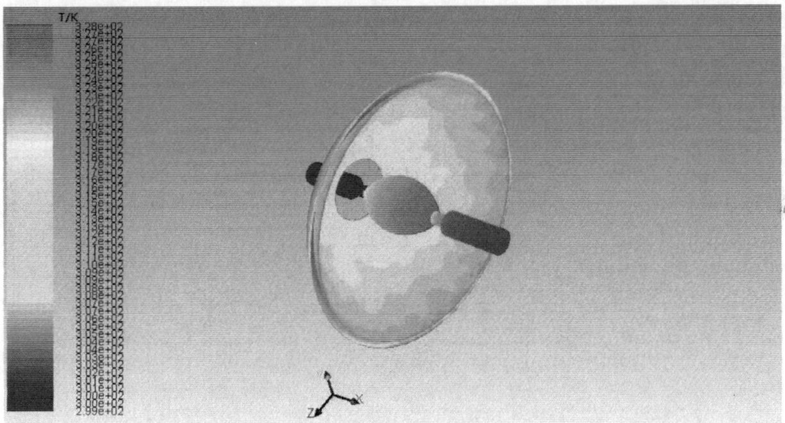


**Figure 9.** The distribution of the temperature of the xenon lamp and collector

The temperature along the surface of the xenon lamp is shown in Figure 10. Because the front velocity is faster, the convective heat transfer coefficient is larger, and the temperature of the gas is cooler, which make the front temperature of the xenon lamp cooler. The back velocity is slower which make the convective heat transfer coefficient smaller, and the gas temperature is higher after absorbing the heat of the xenon lamp and the collector, so the back temperature of the xenon is higher. As shown in the Figure 9 and Figure 10, the maximum temperature of the xenon lamp is 330 K, and the maximum temperature of the collector is 308 K, which meet the temperature requirement.

IOP Conf. Series: Materials Science and Engineering **608** (2019) 012046 doi:10.1088/1757-899X/608/1/012046

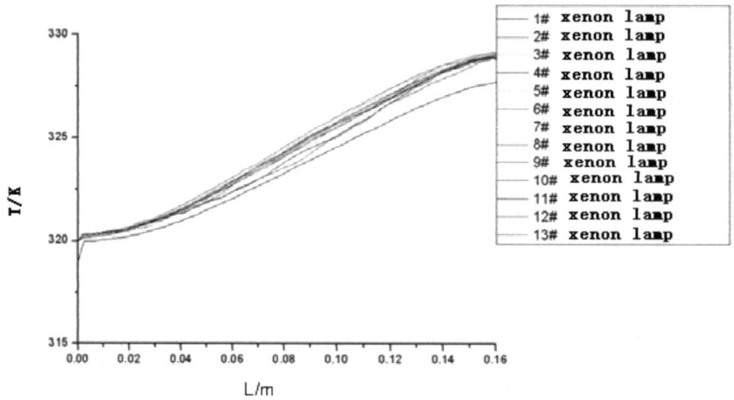


**Figure 10.** The temperature along the surface of the xenon lamp

## 5. Conclution

The cooling system for solar simulator was introduced, and the wind-cooling system is numerical simulated by the Fluent. The simulation result show that, when the flow rate of the GN2 was 11000 $m^3/h$, the thermal load of the xenon lamps was 45 kw, and the thermal load of the collectors was 5 kw, the velocity of the most area in the solar hood was slow, but the area near the xenon lamp, because the flow area suddenly decreases, the velocity of the gas increased significantly, which make the convective heat transfer coefficient on the surface of the xenon large. The xenon is the main equipment generating heat, and the heat flux density of the xenon lamp is high, but the convective heat transfer coefficient is high, which is conducive to the heat exchange between the xenon lamp and the gas, and the temperature of the xenon lamp is lower than 328 K, and the velocity on the surface of the xenon lamp is higher than 17 m/s, and meet the operating requirement of the xenon lamp. The velocity near the collector is low, which made the convective heat transfer coefficient low, but the heat flux density of the collector is low, the temperature of the collector is lower than 308 K, and meet the operating requirement of the collector.

## References

[1] Huang B C and Ma Y L 2002 *Technology of space environment test* (Beijing: National Defense Industry Press)

[2] Li G, Yu Q X and Yang L H 2006 *The cooling system of the KM6 solar simulator (Spacecraft Environment Engeering,* vol 4) p 293-296

[3] Yang L H and Li H S 2009 *The development of the foreign solar simulator (Spacecraft Environment Engeering* vol 2) p 162-167

[4] Yang S M and Tao W Q 1998 *Heat Transfer* (Beijing: Higher Education Press) p 130-202

[5] Tao Wen Quan 2001 *Numerical Heat Transfer* (Xian: Xian Jiaotong University Press)

[6] Zhou J J 2010 *Engineering technology and applying example of Fluent* (Beijing: China water conservancy and hydroelectric Press)

[7] Yu Y 2008 *The introduction and advanced course of Fluent* (Beijing: Beijing Institute of Technology press)

# Study on the Determination of Attitudes and Orbits of Micro-satellites only by 3-axis Magnetometer

**MyongChol Jang**

Automation Institute, the State Academy of Sciences, Unjong District, Pyongyang City, the Democratic People's Republic of Korea

songnammyong@gmail.com

**Abstract.** This paper explains a method to determine the attitude and orbit only using onboard magnetometer. This method, only using 3-axis magnetometer, makes it efficient and possible to get the attitude and orbit information, without high-priced and complicated sensors. This method comprises 2 parts- one is the attitude determination and the other is the orbit determination by the magnetometer. In the attitude determination, 2-stage filter (Linear Kalman filter and Unscented Kalman filter) is applied and in the orbit determination, the Extended Kalman filter. The attitude determination accuracy is below 1°, the attitude angular rate determination accuracy below 0.005°/s and the orbital position determination accuracy below 10 kms when the measurement error of magnetometer is about 200nT. Thus, this study contributes to the mission of the micro satellites.

## 1. Introduction

In order to decrease their total mass and development cost as much as possible, the general trend is not to use the high cost and complicated sensors and actuators but the high reliable and low cost ones. For this reason, nowadays, many countries are only using onboard 3-axis magnetometer to determine the attitudes and orbits of microsatellites.

## 2. The attitude determination of micro-satellite only using the magnetometer data

The attitude determination comprises 2-stage filter, firstly the LKF (Linear Kalman Filter) estimates the derivatives of the geomagnetic field. As rates of satellite are unknowns, Markov process is used to model the geomagnetic field.

$$\frac{d^3}{dt^3}\mathbf{B} = w, \tag{1}$$

where **B** is the geomagnetic field vector, w is the white Gaussian process noise. If the states are defined as

$$x = [B_x, B_y, B_z, \dot{B}_x, \dot{B}_y, \dot{B}_z, \ddot{B}_x, \ddot{B}_y, \ddot{B}_z]^T \tag{2}$$

the systematic equations can be written as follows.

$$\dot{x} = \frac{d}{dt}\left[B_x, B_y, B_z, \dot{B}_x, \dot{B}_y, \dot{B}_z, \ddot{B}_x, \ddot{B}_y, \ddot{B}_z\right]^T = \left[\dot{B}_x, \dot{B}_y, \dot{B}_z, \ddot{B}_x, \ddot{B}_y, \ddot{B}_z, 0,0,0\right]^T \tag{3}$$

$$y=\left[B_x, B_y, B_z\right]^T \tag{4}$$

Therefore, the state space model for estimating the derivatives of the geomagnetic field is as follows.

$$\dot{x}=Fx+w\,,\; y=Hx+v \tag{5}$$

where $w$ is the process noise, and v is the measurement noise.

Apply the LKF algorithm to the system above. After the determination of the magnetic field vector and its derivatives(pseudo-measurements), and the reference vector in inertial frame by the linear Kalman filter and the geomagnetic field model, only its attitudes and angular rates are remained unknown. If the state vector and measurement vector are selected as follows

$$x\equiv\left[q^i, \omega_{ib}^b\right]^T=[q_1, q_2, q_3, q_4, \omega_x, \omega_y, \omega_z]^T \tag{6}$$

$$y\equiv\left[B^b, \dot{B}^b\right]^T=[B_x, B_y, B_z, \dot{B}_x, \dot{B}_y, \dot{B}_z]^T \tag{7}$$

the system model for estimating attitude angles and rates is as follows.

$$\begin{aligned}\dot{x}&=f(x)+w\\ y&=h(x)+v\end{aligned} \tag{8}$$

$$f(x)=\left[\frac{1}{2}\Omega q^i, \left(I^b\right)^{-1}\left(T_c-\omega_{ib}^b\times\left(I^b\omega_{ib}^b\right)\right)\right]^T \tag{9}$$

$$h(x)=\left[R_i^b B^i, R_i^b\dot{B}^i-\omega_{ib}^b\times\left(R_i^b B^i\right)\right]^T \tag{10}$$

where

$$\Omega=\begin{bmatrix}0 & \omega_z & -\omega_y & \omega_1\\ -\omega_z & 0 & \omega_1 & \omega_y\\ \omega_y & -\omega_1 & 0 & \omega_z\\ -\omega_1 & \omega_y & -\omega_z & 0\end{bmatrix} \tag{11}$$

$$R_i^b=\begin{pmatrix}q_4^2+q_1^2-q_2^2-q_3^2 & 2(q_1q_2+q_3q_4) & 2(q_1q_3-q_2q_4)\\ 2(q_1q_2-q_3q_4) & q_4^2-q_1^2+q_2^2-q_3^2 & 2(q_2q_3+q_1q_4)\\ 2(q_1q_3+q_2q_4) & 2(q_2q_3-q_1q_4) & q_4^2-q_1^2-q_2^2+q_3^2\end{pmatrix} \tag{12}$$

$T_c$ : torque vector applied to the satellite

$I^b$ : inertia tensor of the satellite

$R_i^b$ : attitude transformation matrix from inertial frame to body frame

$B^i, \dot{B}^i$ : geomagnetic vector and its derivatives in inertial frame

The UKF (Unscented Kalman Filter) is applied to this non-linear system to estimate the states, the attitude quaternions and angular rates. [1,2]

## 3. The orbit determination only using the magnetometer data

For the orbit determination by the magnetometer data, the EKF (Extended Kalman Filter) is used. [3,4] The state variable X comprises 6 positions and velocities.

$$X=(\mathbf{r}, \mathbf{V})=(\mathbf{r}, \dot{\mathbf{r}}) \tag{13}$$

The state transition model is

IOP Conf. Series: Materials Science and Engineering **608** (2019) 012047 doi:10.1088/1757-899X/608/1/012047

$$\dot{X} = f(X(t), t) + w(t) \tag{14}$$

where the process noise w(t) can be related to the time, but the expectation value of its distribution is supposed zero at every moment.

$$E(w(t)) = 0 \tag{15}$$

In addition, the covariance of process noise can also be related to time, but it is supposed that the time correlation between observations does not exist.

$$E\left(w(t_j)w^T(t_k)\right) = Q(t_j)\delta_{jk} \tag{16}$$

where $Q(t_j)$ is the covariance of process noise at $t_j$ .

The state transition model for the satellite movement is the dynamics model, where the zonal harmonics is considered up to 4th order for its modelling. [5] Considering the atmospheric drag is very small at the height of 700km, this item is ignored. If the dynamics model is configured according to Newton's law, the state transition equation can be written as follows.

$$\dot{\mathbf{r}} = \mathbf{V}, \dot{\mathbf{V}} = \nabla U \tag{17}$$

where

$$U = \frac{GM_e}{r}\left[1 - J_2\left(\frac{a_e}{r}\right)^2\left(\frac{3}{2}\sin^2\varphi - \frac{1}{2}\right) - J_3\left(\frac{a_e}{r}\right)^3\left(\frac{5}{2}\sin^3\varphi - \frac{3}{2}\sin\varphi\right)\right.$$
$$\left. - J_4\left(\frac{a_e}{r}\right)^4\left(\frac{35}{8}\sin^4\varphi - \frac{15}{4}\sin^2\varphi + \frac{3}{8}\right)\right] \tag{18}$$

And φ is the latitude of space position, $M_e$ the mass of the earth and $r$ the distance from the center of the earth to a particular point of space. In standard earth model WGS84-EGM96, $GM_e$ = 3.986004418 $\times 10^5 km^3/s^2$, $a_e$ = 6378.137km. Using the relation $z = r\sin\varphi$ and expressing the earth gravitational potential by harmonic item,

$$U = U_0 + U_2 + U_3 + U_4 \tag{19}$$

$$U_0 = \frac{GM_e}{r} \tag{20}$$

$$U_2 = \frac{GM_e}{r}\left[-J_2\left(\frac{a_e}{r}\right)^2\left(\frac{3}{2}\left(\frac{z}{r}\right)^2 - \frac{1}{2}\right)\right] \tag{21}$$

$$U_3 = \frac{GM_e}{r}\left[-J_3\left(\frac{a_e}{r}\right)^3\left(\frac{5}{2}\left(\frac{z}{r}\right)^3 - \frac{3}{2}\frac{z}{r}\right)\right] \tag{22}$$

$$U_4 = \frac{GM_e}{r}\left[-J_4\left(\frac{a_e}{r}\right)^4\left(\frac{35}{8}\left(\frac{z}{r}\right)^4 - \frac{15}{4}\left(\frac{z}{r}\right)^2 + \frac{3}{8}\right)\right] \tag{23}$$

The calculation results of gradients of $U$ are given by items below.

$$\nabla U_0 = -GM_e\frac{1}{r^3}(x, y, z)^T \tag{24}$$

$$\nabla U_2 = GM_e A_2\left(\frac{15xz^2}{r^7} - \frac{3x}{r^5}, \frac{15yz^2}{r^7} - \frac{3y}{r^5}, \frac{15z^3}{r^7} - \frac{9z}{r^5}\right)^T \qquad A_2 = \frac{1}{2}J_2 a_e^2 \tag{25}$$

$$\nabla U_3 = GM_e A_3 \left( \frac{35xz^3}{r^9} - \frac{15xz}{r^7}, \ \frac{35yz^3}{r^9} - \frac{15yz}{r^7}, \ \frac{35z^4}{r^9} - \frac{30z^2}{r^7} + \frac{3}{r^5} \right)^T \quad A_3 = \frac{1}{2} J_3 a_e^3 \tag{26}$$

$$\nabla U_4 = GM_e A_4 \times \left( \frac{63xz^4}{r^{11}} - \frac{42xz^2}{r^9} + \frac{3x}{r^7}, \ \frac{63yz^4}{r^{11}} - \frac{42yz^2}{r^9} + \frac{3y}{r^7}, \ \frac{63z^5}{r^{11}} - \frac{70z^3}{r^9} + \frac{15z}{r^7} \right)^T$$
$$A_4 = \frac{5}{8} J_4 a_e^4 \tag{27}$$

The measurement model to decide the measurement Z from the state vector is

$$Z = h(X(t)) + v(t) \tag{28}$$

where $v(t)$ is zero-mean white noise.

$$E(v(t_j)) = 0, \qquad E(v(t_j)v^T(t_k)) = \mathbf{I}R^2\delta_{jk}$$

When using the magnetometer data, the magnitude(measurement) of the geomagnetic field is

$$B_{mes} = \left( B_{bx}^2 + B_{by}^2 + B_{bz}^2 \right)^{1/2} \tag{29}$$

where $B_{bx}$, $B_{by}$, $B_{bz}$ are the axial components of 3-axis magnetometer. Defining the magnitude of the geomagnetic field as measurement, the measurement model is as follows.

$$h(X(t)) = B(\mathbf{r}) = \left( B_r^2 + B_\theta^2 + B_\varphi^2 \right)^{1/2} \tag{30}$$

where $B_r$, $B_\theta$, $B_\varphi$ are the spherical coordinate components in ECEF frame.

$$B_r = \sum_{n=1}^{k} \left( \frac{a}{r} \right)^{n+2} (n+1) \sum_{m=0}^{n} \left( g^{n,m} \cos m\varphi + h^{n,m} \sin m\varphi \right) P^{n,m}(\theta) \tag{31}$$

$$B_\vartheta = -\sum_{n=1}^{k} \left( \frac{a}{r} \right)^{n+2} \sum_{m=0}^{n} \left( g^{n,m} \cos m\varphi + h^{n,m} \sin m\varphi \right) \frac{\partial P^{n,m}(\theta)}{\partial \theta} \tag{32}$$

$$B_\varphi = -\frac{1}{\sin\theta} \sum_{n=1}^{k} \left( \frac{a}{r} \right)^{n+2} \sum_{m=0}^{n} m \left( -g^{n,m} \cos m\varphi + h^{n,m} \sin m\varphi \right) P^{n,m}(\theta) \tag{33}$$

The meanings of symbols are as follows.

$(r, \theta, \varphi)$: earth centered spherical coordinates
$r$: geocentric distance
$\theta$: coangle of latitude
$\varphi$: longitude
$a$: standard radius(6371.2km)
$g^{n,m}$ : Gaussian coefficients
$P^{n,m}$ : Schmidt semi-orthogonalized Legendre polynomial

Gaussian coefficients in IGRF (International Geomagnetic Reference Field) are varied with time and it is supposed that they are varied at constant rate for 5 years. The state transition model and measurement model are as follows.

$$\dot{X} = f(X(t), t) + w(t) \tag{34}$$

$$Z = h(X(t)) + v(t) \tag{35}$$

IOP Conf. Series: Materials Science and Engineering **608** (2019) 012047 doi:10.1088/1757-899X/608/1/012047

where $w(t)$: zero-mean white process noise, $v(t)$ : zero-mean white measurement noise.

The EKF is applied to the state transition model and measurement model above.

## 4. Simulations

The micro-satellite circling the sun synchronous orbit at the height of 700km is supposed for simulation. The standard orbit data are as follows.

Epoch

| | |
|---|---|
| (UTC) | : 2017/10/10 02:33:33.6286 |
| (TLE) | : 17283.106639219914 |

Orbital elements(TLE)

| | |
|---|---|
| Semi-major radius | : 7064.80364867094 km |
| Eccentricity | : 1.04530705730871E-03 |
| Inclination | : 98.1581910505179° |
| Right ascension of ascending node | : 356.39561764568° |
| Argument of perigee | : 90° |
| Mean anomaly | : 50.4966213381126° |
| Mean motion | : 14.6110967250392 r/d |
| Ballistic drag coefficient | : 0 |

### *4.1 The simulation result of attitude determination*

The attitude determination algirithm of the micro-satellite using UKF is simulated for 20000s, and the effectiveness of this method is verified. Firstly, consider the estimation features of geomagnetic field vector and its derivatives.

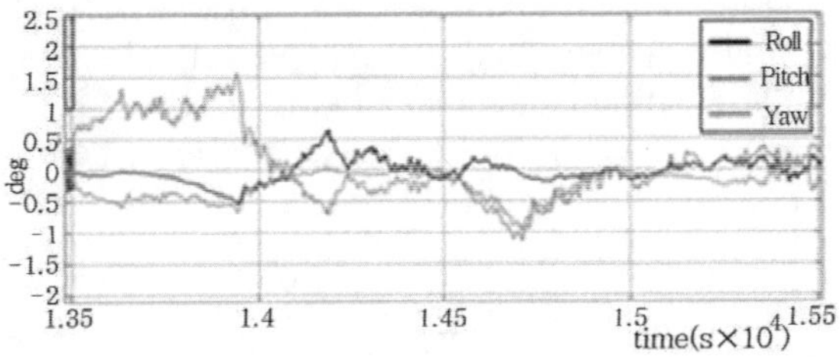


**Figure 1.** The micro-satellite attitude estimation error feature by UKF.

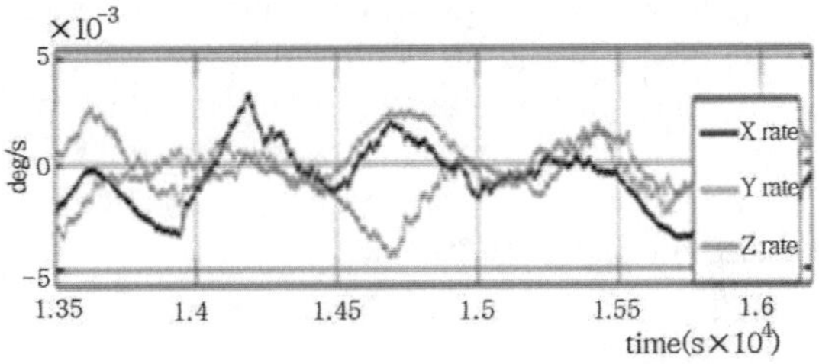


**Figure 2.** The micro-satellite rate estimation error feature by UKF.

Fig.1 shows the attitude estimation error of micro-satellite. The black, blue and red lines represent roll, pitch and yaw, respectively. Fig.2 shows the rate estimation error of micro-satellite by UKF. Black, blue and red lines represent rates of x, y and z axes respectively.

As shown above, one can know that the attitude determination error is below 1° and the rate determination error is below 0.005°/s at $1\sigma$, thus this method is very accurate.

### *4.2 Orbit determination simulation result*

The satellite position is obtained by propagating the satellite orbit at a second intervals on the standard orbit, and the geomagnetic field is calculated based on the IGRF model, and then the noise of 200nT is added for simulating the observation data. The covariance matrix of initial error is supposed for the position and velocity, respectively, as follows.

$$\sigma_r^2 = 100km^2$$

$$\sigma_{\dot{r}}^2 = 0.0001(km/s)^2$$

$$P_0 = \begin{pmatrix} \sigma_r^2 & 0 & 0 & 0 & 0 & 0 \\ 0 & \sigma_r^2 & 0 & 0 & 0 & 0 \\ 0 & 0 & \sigma_r^2 & 0 & 0 & 0 \\ 0 & 0 & 0 & \sigma_{\dot{r}}^2 & 0 & 0 \\ 0 & 0 & 0 & 0 & \sigma_{\dot{r}}^2 & 0 \\ 0 & 0 & 0 & 0 & 0 & \sigma_{\dot{r}}^2 \end{pmatrix}$$

The covariance matrix of process noise is set as $Q = 0.0001P_0$ at the beginning and after 2000s, it is forced to decrease by 1/10 with every 1000 seconds in the duration of 100 minutes. Different deviations are given to the standard orbit to simulate the initial position and velocity.

Fig.3 shows the positional deviation between the standard orbit and estimation orbit, when the initial position and velocity deviation compared with standard orbit is 10km and 10m/s. The observation time is about 114 minutes, when the positional deviation is less than 10km.

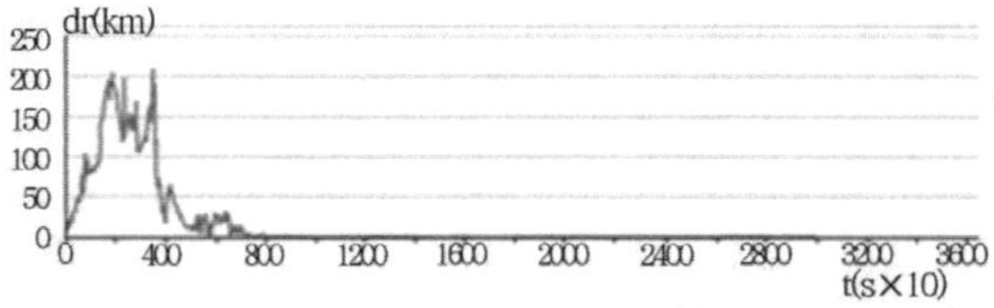


**Figure 3.** Positional deviation between standard orbit and estimation orbit (initial position and velocity deviation 10km, 10m/s).

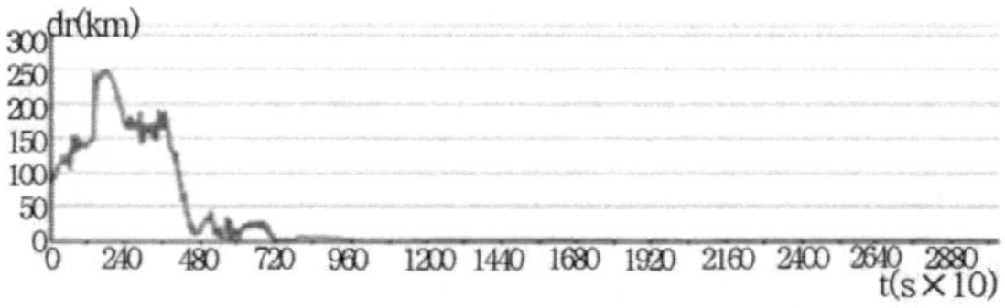


**Figure 4.** Positional deviation between standard orbit and estimation orbit (initial position and velocity deviation 100km, 10m/s).

Fig.4 shows the positional deviation between the standard orbit and estimation orbit when the initial position and velocity deviation with standard orbit is 100km and 10m/s, respectively. The observation time is about 116 minutes, when the positional deviation is less than 10km.

As shown in simulation, when the noise is less than 200nT, initial positional deviation is below 100km and velocity deviation is below 10m/s, the final positional deviation converges to below 10km after about 120 minutes.

## 5. Conclusion

In this paper, a method, only using the magnetometer data is presented to determine the attitudes and orbits. Different kinds of Kalman filters are used to solve these problems. Simulation results show that the attitudes and orbits determination accuracies are assured as expected for the micro-satellites.

## References

[1] Ma G and Jiang X 2005 *Unscented Kalman Filter for Spacecraft Attitude Estimation and Calibration Using Magnetometer Measurements*," Proc. of the Fourth Int. Conf. on Machine Learning and Cybernetics.

[2] Dyke V, M. C, Schwartz J and Hall L 2004 *Unscented Kalman Filtering for Spacecraft Attitude State and Parameter Estimation*, Proc. of the 14th AAS/AIAA Space Flight Mechanics Meeting, Maui, HI

[3] Zhou F, Zhao L, Zhou J, Chen Y 2001 *An Approach of Autonomous Navigation for Near-Earth-Orbit-Satellite Using Magnitude of Geomagnetic Field*, IFA-01-A.7.04,

[4] Mark L. P 2001 *Autonomous LEO Orbit Determination from Magnetometer and Sun Sensor Data*, AIAA-98-4308

[5] Joao Luis Marines, Xiaoping Yun, Eric R. Bachmann, Robert B. McGhee, and Michael J. Zyda 2001 *An Extended Kalman Filter for Quaternion-Based Orientation Estimation Using MARG Sensors* Proc. of the 2001 IEEE/RSJ, Int. Conf. on Intelligent Robots and Systems